AF874803

The Mycota

A Comprehensive Treatise on Fungi as Experimental Systems for Basic and Applied Research

Volume 14

Series Editors

Dee Carter, School of Life and Environmental Sciences, The University of Sydney, Sydney, NSW, Australia

Anuradha Chowdhary, Department of Medical Mycology, VP Chest Institute, University of Delhi, Delhi, India

Joseph Heitman, Duke University Medical Center, Durham, NC, USA

Ulrich Kück, General and Molecular Botany, Ruhr-University Bochum, Bochum, Germany

The fungi represent a heterogenous assemblage of eukaryotic microorganisms and have become favored organisms for research at the cellular and molecular level. Such research involvement has been stimulated by interest in the biotechnological application of fungi in processes related to industry, agriculture and ecology. Considering both yeasts and mycelial fungi, THE MYCOTA highlights developments in both basic and applied research and presents an overview of fungal systematics and cell structure. Foremost authorities in research on mycology have been assembled to edit and contribute to the volumes.

Stefanie Pöggeler • Timothy James
Editors

Evolution of Fungi and Fungal-Like Organisms

Second Edition

Springer

Editors
Stefanie Pöggeler
Institute for Microbiology and Genetics
Georg-August-University
Göttingen, Germany

Timothy James
Department of Ecology and Evolutionary Biology
University of Michigan
Ann Arbor, MI, USA

ISSN 2945-8048 ISSN 2945-8056 (electronic)
The Mycota
ISBN 978-3-031-29198-2 ISBN 978-3-031-29199-9 (eBook)
https://doi.org/10.1007/978-3-031-29199-9

© The Editor(s) (if applicable) and The Author(s), under exlusive licence to Springer Nature Switzerland AG 2011, 2023
This work is subject to copyright. All rights are solely and exclusively licensed by the Publisher, whether the whole or part of the material is concerned, specifically the rights of translation, reprinting, reuse of illustrations, recitation, broadcasting, reproduction on microfilms or in any other physical way, and transmission or information storage and retrieval, electronic adaptation, computer software, or by similar or dissimilar methodology now known or hereafter developed.
The use of general descriptive names, registered names, trademarks, service marks, etc. in this publication does not imply, even in the absence of a specific statement, that such names are exempt from the relevant protective laws and regulations and therefore free for general use.
The publisher, the authors, and the editors are safe to assume that the advice and information in this book are believed to be true and accurate at the date of publication. Neither the publisher nor the authors or the editors give a warranty, expressed or implied, with respect to the material contained herein or for any errors or omissions that may have been made. The publisher remains neutral with regard to jurisdictional claims in published maps and institutional affiliations.

This Springer imprint is published by the registered company Springer Nature Switzerland AG
The registered company address is: Gewerbestrasse 11, 6330 Cham, Switzerland

Paper in this product is recyclable.

Series Preface

The Mycota is an encyclopedic book series published by Springer as a comprehensive treatise spanning the Fungal Kingdom through a focus on fungi as experimental systems for basic and applied research. Articles in this series present introductory information in a comprehensive manner to broaden our general knowledge, along with state-of-the-art information on selected topics in fungal biology. Thus, these articles serve as a timeless reference source for historically relevant discoveries in fungal biology to drive further advances in the field.

The Mycota was founded in 1994 by Karl Esser and Paul Lemke. Since then, 15 volumes have been published, with several now in their third edition. The steadily growing interest in *The Mycota*, reflected in particular by the high number of e-book downloads, encouraged Springer to continue this publication with a new international Series Editorial board beginning in 2022.

Historically, the study of fungi originated as a sub-discipline of botany and was a largely descriptive discipline until the early 19th century. Subsequent experimental research has propelled the field of mycology, and achievements in the genetics and molecular biology of fungi have benefited studies in the related fields of fungal biochemistry and biophysics, plant pathology, medical mycology, and systematics. Ground-breaking research has been carried out using fungi as recognized by the receipt of nine Nobel Prizes, including the revolutionary 1945 Nobel Prize in Physiology and Medicine for the discovery of penicillin, and the foundational 2006 Nobel Prize in Chemistry for establishing the molecular basis of eukaryotic transcription. All such studies expand our knowledge of fungi and of fungal processes and improve our ability to understand, utilize, and control fungi for the benefit of humankind.

Fungi have invaded every conceivable ecological niche. Saprobic forms abound, especially in the decay of organic debris. Pathogenic forms exist with both plant and animal hosts. Fungi even grow on other fungi. They are found in aquatic as well as soil environments, and their spores have been found in the air of extreme environments. Fungi can be variously associated with plants as symbionts in the form of lichens, mycorrhizae, and endophytes and also occur as overt pathogens. Association with animal systems varies; examples include the predaceous fungi that trap nematodes, the micro-fungi that grow in the anaerobic environment of the rumen, and medically important pathogens afflicting humans. Currently, taxonomists have identified about 150,000 species, but it is thought that over 90% of fungal species remain undescribed, with conservative estimates ranging from 2 to 5 million fungal species on earth.

From this perspective, there are new topics in fungal biology to explore, which will expand the current volumes and take them in new directions. Further, there will be new volumes in areas (e.g., Cryptomycota) that were not covered before.

Our understanding of the evolution of fungi is still incomplete and mainly based on species that can be grown in culture. But recent environmental DNA analyses have revealed a highly diverse form of eukaryotic life that branches with the Fungi, and thus the resulting and highly diverse clades were named the cyptomycota. The discovery of novel intermediate forms will redefine the fungal tree of life. Other topics will consider the adaptation of fungi to climate changes, the occurrence of fungal pathogens in the environment, and the dispersal of fungi during global pandemics.

Fungal constituents of the microbiome have received much less attention thus far, yet recent findings from clinical and animal studies clearly establish fungi as a significant component of the oral, gastrointestinal, pulmonary, and skin microbiomes. Finally, the relevance of fungal biology to society is reflected by the increasing number of fungal-related human diseases.

In the history of pharmacology, fungi have always been sources of useful molecules for humans, but the diversity of molecules that can be obtained from fungi is still largely under explored. However, efforts in fungal genomics provide resources for data-driven genome mining and large-scale comparisons to explore the molecular repertoires produced by fungi. The result will be new compounds with applications in the pharmaceutical and agricultural science industries.

Fungi can also help the entire planet and may, for example, be relevant in specific sectors, such as that linked to pollution from plastic. Fungi produce a wide range of enzymes that have the potential to break down the chemical bonds of plastic polymers, and in this context the potential role of marine fungi in plastic degradation may be of major relevance. Finally, new biomaterials from fungal species may open the door to alternatives to fossil-based materials, and thus reduce environmental pollution.

For consistency throughout this series of volumes, the names adopted for major groups of fungi should be followed according to the following paper, which gives an overview of all of the orders in the fungal kingdom: https://pubmed.ncbi.nlm.nih.gov/32660385/

We are grateful to Springer for continuing *The Mycota* series and are especially thankful to all of the Volume Editors in selecting topics and assembling experts from diverse fields of fungal biology.

Sydney, Australia — Dee Carter
Delhi, India — Anuradha Chowdhary
Durham, USA — Joseph Heitman
Bochum, Germany — Ulrich Kück
July 15, 2022

Volume Preface to the Second Edition

Since the first edition of Volume XIV of *The Mycota*, new developments have affected the scale of and the types of questions that can be asked in fungal evolution. Various genome sequences and metagenomic sequences allow questions to be addressed concerning the origin of the fungal kingdom and fungal evolution at a level of analytical refinement that have never been possible before. Most of the 13 articles are new contributions to this volume. The remaining two articles were rewritten and new information have been introduced to update the overviews.

The second edition of *The Mycota* XIV provides a selection of state-of-the-art reviews dealing with major aspects of fungal evolution. The four parts comprise Evolutionary Roots of Fungi (Chaps. 1–3), Evolution of Pathogenetic Strategies (Chaps. 4–6), Evolution of Mutualistic Interactions (Chaps. 7–10), and Evolution of Metabolism and Development in Fungi (Chaps. 11–13).

Fungi are among the oldest clades of eukaryotic organisms in the living world and thus it is important to understand their history and diversification on the way toward the origin of their more recent sister group, the Metazoa. Chapter 1 gives an overview about the protistan relatives of animals and fungi. Dictyostelids were once grouped with fungi because their fruiting structures are superficially similar. However, these cellular slime molds evolved quite distinct strategies to reach this mode of species propagation. Chapter 2 describes the cell communication systems that cause Dictyostelid amoebas to aggregate and that regulate cell-type specialization and cell movement during fruiting body formation. This part is completed by Chap. 3 about the evolution of mitochondrial genomes. Fungal mitochondrial genomes are highly variable while they contain a certain set of conserved genes. When sampling the full diversity of the fungal tree of life, it is clear that the mitochondrial genome is constantly rearranging a set of highly conserved genes through the action of recombination mediated by introns.

Part II deals with the evolution of pathogenic strategies. Chapter 4 reviews the evolution of yeast and hyphal dimorphism in fungi, which is often associated with virulence and pathogenicity. Dimorphic transitions occur not only in Ascomycota and Basidiomycota but also in Mucoromycotina in species of the genus *Mucor*. Chapter 5 provides an overview of recent research on genome evolution in fungal pathogens. It presents an overview of drivers of diversification including transposable elements and epigenetic modifications regulating pathogenicity factors and genomic defense mechanisms. Population genomics data revealing short-term dynamics in plant pathogens are reviewed as well. In plant pathogenic fungi, the interaction with the host plant is the major ecological determinant of fitness and host switches are a major driver in the diversification of pathogens. Chapter 6 summarizes the current state of knowledge on range expansions and host jumps in plant pathogens, focusing on (hemi)biotrophic fungi and oomycetes.

The third part focuses on the evolution of mutualistic interactions. Viruses have a huge impact on their host species, changing host population sizes, behaviors, and forcing the evolution of complex defensive mechanisms. Although under study for almost 60 years, the impact of mycoviruses as hyperparasites of ecologically and agriculturally important pathogens, symbionts and, potentially, tools for combating diseases and genetic engineering is just being recognized. Chapter 7 reviews the current knowledge on the diversity of mycoviruses, their phenotypic impacts on hosts, their mode of transmission, and impacts on host evolution. Like viruses, bacterial endosymbionts can influence fungal growth, transcription, protein abundance, metabolism, and sporulation. The possibility for them to act as mutualists to their hosts needs to be explored. Chapter 8 gives an overview on bacterial endosymbionts of Mucoromycota fungi and summarizes the diversity and function of bacterial and fungal interactions. The endosymbionts fall into two major groups of bacteria related to *Burkholderia* and *Mycoplasma*. Both types influence the host fungal reproduction and growth.

Filamentous fungi and yeasts are exposed to micropredators in their natural environment. This predator–prey relationship can impose selection pressure on both partners. Chapter 9 focuses on micropredators, bacteria, nematodes, and especially the ubiquitous amoebae that have adapted highly specific tools to find, engulf, or open fungal cells. It describes laboratory-controlled bipartite interaction studies aiming to reveal the mechanistic cellular, biochemical, and molecular details and the coevolution between fungi and micropredators that have triggered fungal defense mechanisms.

The diversity and distribution of fungal communities play critical roles in many ecosystem processes. Understanding the extent of global fungal biodiversity and the factors that determine it can provide crucial information about the stability and functioning of ecosystems. Chapter 10 addresses how the ability to recognize and classify fungi based on appropriate molecular markers obtained by high-throughput DNA sequencing has opened up the opportunity to explore the composition and diversity of fungal communities and describes a unique resource the GlobalFungi database that aggregates numerous studies of soil fungal communities gathered from high-throughput DNA sequencing.

The last part sets its sight on the evolution of fungal metabolism and development. Metabolism is one of the key characteristics of life, implicating that evolution always had an impact on metabolic processes. Studies on fungal metabolism have traditionally concentrated on metabolites of specific interest, namely mycotoxins, pathogenicity factors, antibiotics, and other compounds displaying interspecific activity. Analyses of fungal genomes revealed that fungi have the potential to produce far more secondary metabolites than have been identified yet. Chapter 11 summarizes the recent research on the activation of silent metabolic gene clusters by mimicking naturally inducing conditions as well as the elucidation of the ecological roles of secondary metabolites.

Fungi are able to emit light emission by a luciferin/luciferase chemical reaction. Chapter 12 reviews the current knowledge on how bioluminescent fungi have evolved, retained, or lost their luciferase cluster. Genomics has revealed a single origin of the genes involved in bioluminescence in gilled mushrooms followed by multiple losses, which was a surprising finding given the patchy distribution in the mycenoid and marasmioid fungi.

Fruiting body evolution follows a complex pattern, which we are only beginning to understand. Chapter 13 summarizes evolutionary theories on the evolution of secotioid and gasteroid basidiomes in the Basidiomycota which have evolved from gilled forms through changes in the development of the basidiome such that the hymenium is never exposed. The evolution of this development is considered to have evolved by paedomorphosis or the retention of juvenile traits or early stages in the development of agaric mushrooms.

We are very grateful to the authors for their contribution to the second edition of The Mycota XIV, all are extremely busy; and we appreciate their willingness to commit valuable time to this project. We hope that the readers enjoy reading this volume of The Mycota.

Göttingen, Germany — Stefanie Pöggeler
Ann Arbor, MI, USA — Timothy James
January 2023

Contents

Editors and Contributors

About the Series Editors

Dee Carter is Professor of Microbiology at the University of Sydney. She has a degree in microbiology and biochemistry from Otago University, New Zealand, and a PhD from Imperial College, London, UK. After postdoctoral fellowships in Montpellier, France, and UC Berkeley, California, USA, she moved to Australia in 1995 to take up a lectureship at the University of Sydney. Dee's research has encompassed (1) the population genetics and ecology of medically important fungi, including yeast and mold pathogens; (2) responses of fungi to host and antifungal stress, including the production of variant morphological forms; and (3) the transcriptome and proteome response to antifungal therapy, particularly during synergistic interactions between antifungals and natural products. As well as research, Dee teaches microbiology including mycology, epidemiology, and molecular biology at Sydney University.

Anuradha Chowdhary is a Professor of Medical Mycology at the Vallabhbhai Patel Chest Institute, University of Delhi, India. She received her medical degree in 1992 and MD (Microbiology) degree in 1996 from the University of Delhi, India and PhD, from the Faculty of Medical Sciences, Radboud University Medical Center, Netherlands. Dr Chowdhary's research interests include molecular ecology and population genetics of pathogenic fungi, antifungal drug resistance mechanisms, especially Aspergillus and Candida spp., and the epidemiology of systemic mycoses. She is currently working on molecular epidemiology of Candida auris, terbinafine-resistant dermatophytes, and azole-resistant Aspergillus fumigatus.

Joseph Heitman is James B. Duke Professor and Chair, Department of Molecular Genetics and Microbiology, Duke University, Durham, NC, USA. His research studies model and pathogenic fungi addressing fundamental questions of scientific and medical importance. Pioneering studies with Baker's yeast revealed how immunosuppressive natural products interdict signaling cascades via FKBP12-drug complexes and discovered TOR as a globally conserved nutrient sensor targeted by the immunosuppressive, antiproliferative drug rapamycin. His research discovered unisexual reproduction of pathogenic microbes, with implications for pathogen emergence, how sex generates diversity, and how sex evolved. Dr. Heitman's lab has further developed genetic and genomic approaches elucidating molecular principles of fungal virulence, identifying therapeutic targets, and illustrating convergent evolution of fungal mating-type loci with mammalian, insect, and plant sex chromosomes, defined the calcium-activated protein phosphatase calcineurin as a globally conserved fungal virulence factor, and elucidated functions of RNAi in microbial pathogen genome integrity, hypervirulent outbreak lineages, and drug resistance via epimutation.

Ulrich Kück is a Professor of General and Molecular Botany at the Ruhr-University in Bochum, Germany. He is being graduated in Biology and Chemistry and has a long-standing experience in the molecular biology of fungi, algae, and plants. His research with fungi has focused on two general aspects: 1st, Genetic engineering of the secondary metabolism of biotechnically relevant filamentous fungi, including functional genomics and proteomics. 2nd, Molecular genetic analysis of cellular growth in filamentous fungus, with the focus on sexual development. In particular, he is interested in the function of mating-type loci and the involvement of conserved signaling complexes, such as the striatin-interacting phosphatases and kinases (STRIPAK) complex, in the control of cellular and developmental processes.

About the Volume Editors

Stefanie Pöggeler (born 1963) studied Biology at the Ruhr-Universität in Bochum (Germany). In 1993, she graduated with a thesis on intron-encoded polypeptides in plastids and mitochondria under the supervision of Prof. Ulrich Kück. She later on completed her "Habilitation" at the Ruhr-Universität Bochum in 2000 and was awarded the Venia Legendi in Botany. Between 2001 and 2003, she had a stand-in assistant professorship in Botany at the Wilhelms Universität in Münster (Germany). In 2006, she was appointed as professor for Genetics of Eukaryotic Microorganisms at the Georg-August-Universität Göttingen (Germany).
Her work is focused on the analysis of mating-type genes and autophagy in sexual development of filamentous ascomycetes. In a second line of research, she is interested in the evolution of fungal inteins.

Timothy Y. James (born 1973) studied Biology at Duke University in Durham, North Carolina (United States of America) in the lab of Prof. Rytas Vilgalys. He graduated in 2003 with a thesis on the evolution of mating-type genes in mushroom forming fungi. He completed postdoctoral fellowships at Duke University, Uppsala University (Sweden), and McMaster University (Canada) before starting as an assistant professor in Ecology and Evolutionary Biology at the University of Michigan (U.S.A.). He is the Curator of Fungi at the University of Michigan Fungarium (MICH) and is the Lewis E. Wehmeyer and Elaine Prince Wehmeyer Chair in Fungal Taxonomy. His work is focused on resolving the fungal tree of life, in particular discovering the hidden phylogenetic diversity of the tree by analysis of zoosporic and unculturable lineages of fungi. He is also interested in the evolution of ploidy and mitotic recombination.

Contributors

Leen Nanchira Abraham Laboratory of Evolutionary Genetics, Institute of Biology, University of Neuchâtel, Neuchâtel, Switzerland

M. Catherine Aime Department of Botany and Plant Pathology, Purdue University, West Lafayette, IN, USA

Kevin R. Amses Department of Botany and Plant Pathology, Oregon State University, Corvallis, OR, USA

Petr Baldrian Laboratory of Environmental Microbiology, Institute of Microbiology of the Czech Academy of Sciences, Praha 4, Czech Republic

Paola Bonfante Department of Life Sciences and Systems Biology, University of Torino, Torino, Italy

Gregory Bonito Department of Plant Soil and Microbial Sciences, Michigan State University, East Lansing, MI, USA

Axel A. Brakhage Department of Molecular and Applied Microbiology, Leibniz Institute for Natural Product Research and Infection Biology (Leibniz-HKI), Jena, Germany
Institute for Microbiology, Friedrich Schiller University Jena, Jena, Germany

Martin Carr Department of Biological & Geographical Sciences, University of Huddersfield, Queensgate, Huddersfield, West Yorkshire, UK

Chiung-Chih Chang Biodiversity Research Center, Academia Sinica, Taipei, Taiwan
Biodiversity Program, Taiwan International Graduate Program, Academia Sinica and National Taiwan Normal University, Taipei, Taiwan
Department of Life Science, National Taiwan Normal University, Taipei, Taiwan

Daniel Croll Laboratory of Evolutionary Genetics, Institute of Biology, University of Neuchâtel, Neuchâtel, Switzerland

Franz Essl Division of Bioinvasions, Global Change, Macroecology, Department of Botany and Biodiversity Research, University of Vienna, Vienna, Austria

Alice Feurtey Laboratory of Evolutionary Genetics, Institute of Biology, University of Neuchâtel, Neuchâtel, Switzerland
Plant Pathology, D-USYS, ETH Zurich, Zurich, Switzerland

Alexis Garcia Department of Molecular Microbiology and Immunology, South Texas Center for Emerging Infectious Diseases (STCEID), The University of Texas at San Antonio, San Antonio, TX, USA

Michael L. Ginger Department of Biological & Geographical Sciences, University of Huddersfield, Queensgate, Huddersfield, West Yorkshire, UK

Hanna Maren Glad Laboratory of Evolutionary Genetics, Institute of Biology, University of Neuchâtel, Neuchâtel, Switzerland

Emile Gluck-Thaler Laboratory of Evolutionary Genetics, Institute of Biology, University of Neuchâtel, Neuchâtel, Switzerland

Falk Hillmann Junior Research Group Evolution of Microbial Interactions, Leibniz Institute for Natural Product Research and Infection Biology – Hans Knöll Institute (HKI), Jena, Germany
Biochemistry/Biotechnology, Faculty of Engineering, Wismar University of Applied Sciences Technology, Business and Design, Wismar, Germany

Kayleigh Hopkins Department of Biological & Geographical Sciences, University of Huddersfield, Queensgate, Huddersfield, West Yorkshire, UK

Chen Hsiao Biodiversity Research Center, Academia Sinica, Taipei, Taiwan

Timothy Y. James Department of Ecology and Evolutionary Biology, University of Michigan, Ann Arbor, MI, USA

Huei-Mien Ke Department of Microbiology, Soochow University, Taipei, Taiwan

Levente Kiss Centre for Crop Health, University of Southern Queensland, Toowoomba, QLD, Australia

Petr Kohout Laboratory of Environmental Microbiology, Institute of Microbiology of the Czech Academy of Sciences, Praha 4, Czech Republic

Alexandra M. Kortsinoglou Section Genetics and Biotechnology, Department of Biology, National and Kapodistrian University of Athens, Athens, Greece

Vassili N. Kouvelis Section Genetics and Biotechnology, Department of Biology, National and Kapodistrian University of Athens, Athens, Greece

Mario K. C. Krespach Department of Molecular and Applied Microbiology, Leibniz Institute for Natural Product Research and Infection Biology (Leibniz-HKI), Jena, Germany
Institute for Microbiology, Friedrich Schiller University Jena, Jena, Germany

Francisco Kuhar Instituto Multidisciplinario de Biología Vegetal (CONICET), FCEFyN, Universidad Nacional de Córdoba, Córdoba, Argentina

Soo Chan Lee Department of Molecular Microbiology and Immunology, South Texas Center for Emerging Infectious Diseases (STCEID), The University of Texas at San Antonio, San Antonio, TX, USA

Min R. Lu Biodiversity Research Center, Academia Sinica, Taipei, Taiwan

Jillian M. Myers Department of Ecology and Evolutionary Biology, University of Michigan, Ann Arbor, MI, USA

Eduardo Nouhra Instituto Multidisciplinario de Biología Vegetal (CONICET), FCEFyN, Universidad Nacional de Córdoba, Córdoba, Argentina

Ursula Oggenfuss Laboratory of Evolutionary Genetics, Institute of Biology, University of Neuchâtel, Neuchâtel, Switzerland
Department of Microbiology and Immunology, University of Minnesota, Minneapolis, MN, USA

Jie-Hao Ou Department of Plant Pathology, National Chung Hsing University, Taichung, Taiwan

Laila P. Partida-Martínez Department of Genetic Engineering, Centro de Investigación y de Estudios Avanzados del Instituto Politécnico Nacional (Cinvestav) - Unidad Irapuato, Irapuato, México

Donald H. Pfister Department of Organismic and Evolutionary Biology, Harvard University, Cambridge, MA, USA

Guido Puccetti Laboratory of Evolutionary Genetics, Institute of Biology, University of Neuchâtel, Neuchâtel, Switzerland
Syngenta Crop Protection AG, Stein, Switzerland

Silvia Radosa Junior Research Group Evolution of Microbial Interactions, Leibniz Institute for Natural Product Research and Infection Biology – Hans Knöll Institute (HKI), Jena, Germany

Claudia Sarai Reyes-Avila Laboratory of Evolutionary Genetics, Institute of Biology, University of Neuchâtel, Neuchâtel, Switzerland

Maira Rosin Department of Molecular and Applied Microbiology, Leibniz Institute for Natural Product Research and Infection Biology (Leibniz-HKI), Jena, Germany
Institute for Microbiology, Friedrich Schiller University Jena, Jena, Germany

Nauman Saeed Junior Research Group Evolution of Microbial Interactions, Leibniz Institute for Natural Product Research and Infection Biology – Hans Knöll Institute (HKI), Jena, Germany

Alessandra Salvioli Department of Life Sciences and Systems Biology, University of Torino, Torino, Italy

Sandra Milena González Sáyer Laboratory of Evolutionary Genetics, Institute of Biology, University of Neuchâtel, Neuchâtel, Switzerland
Laboratory of Molecular Characterization, Biotechnology Institute - Universidad Nacional de Colombia, Bogotá, DC, Colombia

Anna Schertler Division of Bioinvasions, Global Change, Macroecology, Department of Botany and Biodiversity Research, University of Vienna, Vienna, Austria

Christina Schilde School of Life Sciences, D'Arcy Thompson Unit, Carnelley Building, University of Dundee, Dundee, UK

Volker Schroeckh Department of Molecular and Applied Microbiology, Leibniz Institute for Natural Product Research and Infection Biology (Leibniz-HKI), Jena, Germany

Hanno Seebens Senckenberg Biodiversity and Climate Research Centre, Frankfurt, Germany

Roger G. Shivas Centre for Crop Health, University of Southern Queensland, Toowoomba, QLD, Australia

Courtney P. Smith Department of Molecular Microbiology and Immunology, South Texas Center for Emerging Infectious Diseases (STCEID), The University of Texas at San Antonio, San Antonio, TX, USA

Matthew E. Smith Department of Plant Pathology, University of Florida, Gainesville, FL, USA

Luzia Stalder Laboratory of Evolutionary Genetics, Institute of Biology, University of Neuchâtel, Neuchâtel, Switzerland

Maria C. Stroe Department of Molecular and Applied Microbiology, Leibniz Institute for Natural Product Research and Infection Biology (Leibniz-HKI), Jena, Germany
Department of Microbiology, Institute of Applied Biosciences, Karlsruhe Institute of Technology (KIT) – South Campus, Karlsruhe, Germany

Yuichi Taneyama Nagano, Japan

Marco Thines Goethe Universität Frankfurt am Main, Frankfurt, Germany
Senckenberg Biodiversity and Climate Research Centre, Frankfurt, Germany

Sabina Moser Tralamazza Laboratory of Evolutionary Genetics, Institute of Biology, University of Neuchâtel, Neuchâtel, Switzerland

Isheng Jason Tsai Biodiversity Research Center, Academia Sinica, Taipei, Taiwan

Jessie K. Uehling Department of Botany and Plant Pathology, Oregon State University, Corvallis, OR, USA

Tomáš Větrovský Laboratory of Environmental Microbiology, Institute of Microbiology of the Czech Academy of Sciences, Praha 4, Czech Republic

Lukas M. Zehner Department of Molecular and Applied Microbiology, Leibniz Institute for Natural Product Research and Infection Biology (Leibniz-HKI), Jena, Germany
Institute for Microbiology, Friedrich Schiller University Jena, Jena, Germany

Part I

Evolutionary Roots of Fungi

The Protistan Origins of Animals and Fungi

1

Martin Carr, Kayleigh Hopkins, and Michael L. Ginger

Abstract

Fungi and Metazoa (animals) are two major multicellular kingdoms of life and both are positioned in the eukaryotic Opisthokonta. Within the supergroup Fungi and Metazoa fall into either side of the opisthokont root, in the major sub-groups Holomycota and Holozoa. In this chapter, we cover recent advances in the understanding of opisthokont biology, in particular looking at their diversity and where opisthokonts fall in the eukaryotic tree. Although much uncertainty remains over how different eukaryotic supergroups are related to each other, the closest relatives of Opisthokonta are now widely recognised.

The composition of Opisthokonta has been revised due to the discovery of new species, as well as the reassignment of taxa on the basis of phylogenetic analyses. We consider common traits and characteristics found in opisthokonts. The explosion of genomic and transcriptomic sequencing since the turn of the century has allowed the identification of genes involved in multicellularity in both Metazoa and Fungi; molecular phylogenies show multicellularity has independently evolved in multiple lineages across the opisthokonts. Annotated gene complements from species spanning the group highlight that gene loss and gain is a dynamic process in the opisthokonts.

Keywords

Holomycota · Holozoa · Multicellularity · Opisthokonta · Taxonomy

1.1 Introduction

Fungi and Metazoa constitute two of the major multicellular eukaryotic lineages and a large body of robust data confirms that they are close relatives (Brown et al. 2018; Derelle et al. 2015; Wainright et al. 1993). Together, along with a number of clades comprising unicellular taxa, Fungi and Metazoa make up the eukaryotic supergroup Opisthokonta (Adl et al. 2019). This chapter sets out to describe our current knowledge on the species and biology of the opisthokonts, with a particular emphasis on the unicellular representatives.

The composition of Opisthokonta has been disputed since the existence of the group was first proposed; however, a more settled view has emerged within recent years, with between seven to ten major lineages recognised. In addition to the multicellular Fungi and Metazoa, Opisthokonta also contains the unicellular

M. Carr (✉) · K. Hopkins · M. L. Ginger
Department of Biological & Geographical Sciences, University of Huddersfield, Queensgate, Huddersfield, West Yorkshire, UK
e-mail: m.carr@hud.ac.uk; kayleigh.hopkins@hud.ac.uk; m.ginger@hud.ac.uk

© The Author(s), under exclusive license to Springer Nature Switzerland AG 2023
S. Pöggeler, T. James (eds.), *Evolution of Fungi and Fungal-Like Organisms*, The Mycota 14,
https://doi.org/10.1007/978-3-031-29199-9_1

lineages Choanoflagellates, Filasterea, Ichthyosporea, and Opisthosporidia, as well as the nucleariid amoebae (Adl et al. 2019). The monophyly of Opisthosporidia has been questioned (Karpov et al. 2014a, b), with Aphelida potentially forming a clade independent of the other opisthosporidians (Torruella et al. 2018). A putative ninth taxon, Pluriformea, has also been proposed (Hehenberger et al. 2017), but, at present, the relationships between the two known pluriform species and other opisthokonts have not been resolved and recognition of the group is not universal (Torruella et al. 2015). The most recently discovered independent opisthokont group is the genus *Tunicaraptor*, currently only known by the type species *T. unikontum* (Tikhonenkov et al. 2020).

The relationships amongst the opisthokont groups, and of those of the opisthokonts with other eukaryotic supergroups, are slowly becoming clearer. We discuss here the major taxonomic groups within Opisthokonta and their relationships with each other. It is now clear that the deepest bifurcation within the opisthokonts resulted in two major lineages, referred to as Holozoa and Holomycota (the latter has also been labelled by authors as the holofungi (Lara et al. 2009) and Nucletmycea (Brown et al. 2009)). Holomycota is composed of the Fungi, Opisthosporidia, and nucleariid amoebae (Lara et al. 2009). Metazoa, choanoflagellates, filastereans, pluriformeans, *Tunicaraptor*, and the ichthyosporeans are collectively known as Holozoa (Lang et al. 2002; Shalchian-Tabrizi et al. 2008).

The lack of any universal diagnostic characteristics means that membership of Opisthokonta is often based upon phylogenetic trees. Early molecular phylogenetic studies of eukaryotes were prone to generating erroneous topologies, which lead to conflicting theories on how groups were related to each other. In particular, in many studies there was a paucity of species for which sequence data were available. Limited taxa sampling can lead to species being present on isolated long branches; this, in turn, may lead to problems when reconstructing phylogenies due to the phenomenon of long-branch attraction (Felsenstein 1978; Hendy and Penny 1989). When distantly related sequences share a relatively high number of homoplasies (shared characters, present due to convergence rather than common ancestry) the true phylogenetic signal may be overwhelmed and long-branched sequences incorrectly clustered together. Long-branch effects can also be produced by unequal rates of evolution; therefore, when possible, it is advisable to screen taxa and select those most suitable for phylogenetic reconstruction. This is not always possible, particularly in the case of less well-studied eukaryotes where some lineages are only represented by a single species. A further problem was that early phylogenetic studies mainly relied upon single gene phylogenies. This was often the small subunit ribosomal (SSU) RNA gene, which had the advantages of being ubiquitous across all domains of life, present in multiple copies per genome and amplifiable with universal PCR primers (Medlin et al. 1988). Limitations associated with single gene phylogenies include insufficient phylogenetically informative sites and lineage-specific rate changes which have the potential to produce long-branch artefacts.

Studies have subsequently shown that increasing the length of alignments, through concatenating multiple gene sequences, can reduce long-branch issues (reviewed in Bergsten 2005), as the effects of gene-specific homoplasies are diluted within the greater volume of data. Furthermore, increasing the number of taxa in a phylogenetic analysis reduces the average branch length across a tree, lessening the impact of long-branch attraction by dispersing homoplasies which would otherwise be concentrated on long internal branches (DeBry 2005; Zwickl and Hillis 2002).

With the advent of high-throughput sequencing it is now comparatively inexpensive to sequence an organism's genome or transcriptome, allowing the generation of phylogenetic datasets made up from hundreds of protein sequences (e.g. see Brown et al. 2018; Gawryluk et al. 2019; Derelle et al. 2016). This field of phylogenomics has resolved many previously unknown relationships within the eukaryotic tree; however, artefacts remain a major issue (Betancur-R et al. 2014; Philippe et al. 2011) and conflicting topologies are still regularly published in academic papers. The use of sequences from multigene families can be problematic unless the chosen genes are carefully screened. Members of gene families that diverged when species diverged, termed orthologues, are suitable for species phylogenetic reconstruction. In contrast, paralogues, which are the product of gene duplication events, have different evolutionary histories from the species that encode them. The inclusion of paralogues in datasets used to create species trees can therefore result in misleading topologies. Paralogy may be difficult to identify when gene duplication events are ancient, or in cases where a gene family is evolving under weak selective constraint. Further complications in phylogenetic analyses may arise due to the choice of the amino acid substitution model used, as well as the partitioning of alignments, with incorrect models resulting in both erroneous relationships and inaccurate node support values (Philippe et al. 2011; Young and Gillung 2020).

A well-resolved phylogeny is essential in order to determine how traits and characteristics have evolved within groups of species. Due to the presence of both Fungi and Metazoa, the opisthokonts have been intensively studied with regard to the origins of multicellularity. Two forms of multicellularity are known to have evolved within eukaryotes. In clonal multicellularity a single initial cell undergoes rounds of cell division, with the daughter cells remaining adhered to each other, resulting in a multicellular organism made up from genetically identical cells. In aggregate multicellularity, exemplified by *Dictyostelium discoideum* (Schilde and Schaap 2013), genetically unrelated unicellular individuals of the same species assemble together to form a multicellular "superorganism". Although both metazoans and fungi exhibit clonal multicellularity, they have fundamentally different developmental pathways and early opisthokont phylogenies confirmed that multicellularity evolved independently in the two groups, as both groups are more closely related to unicellular relatives than each other (Ruiz-Trillo et al. 2004). As we will set out in this chapter, it is clear that multicellularity, both aggregate and clonal, has evolved independently in a diverse range of opisthokont lineages.

1.2 Opisthokonta

The original Opisthokont group, proposed by Vischer (1945), unified chytrid fungi with the choanoflagellates on the basis of both groups possessing a single posterior flagellum that pushes swimming cells through water. Gams (1947) subsequently expanded the group with the inclusion of Metazoa, under the name Opisthokonten. In contrast to modern views on opisthokonts, both Vischer and Gams placed uniflagellate algae with their groupings. Perhaps surprisingly, the next amendment of the group, proposed by Copeland (1956) under the phylum Opisthokonta, was not a further expansion but to restrict membership to only chytrid fungi. Copeland dismissed an evolutionary link between the chytrids with the choanoflagellates and metazoans as "far-fetched" and there was limited acceptance of the group, with debate continuing over the relationships between the multicellular kingdoms of animals, fungi, and plants for the next four decades. The taxon was revived, again on the basis of morphological characteristics, as the informal group Opisthokonta by Cavalier-Smith (1987) to encompass fungi, metazoans, and the choanoflagellates.

Molecular studies in the 1980s produced varied and equivocal results on the placement of both Fungi and Metazoa within the eukaryotic tree (Gouy and Li 1989; Sogin et al. 1986). Through a combination of a greater number of taxa and increasingly sophisticated phylogenetic analyses, robust molecular support for the opisthokont group emerged through a trio of papers in the early 1990s (Baldauf and Palmer 1993; Hasegawa et al. 1993; Wainright et al. 1993). The Wainright et al. (1993) study was of particular importance, as, in addition to highlighting the close relationship between fungi and metazoans, it confirmed, through the phylogenetic position of choanoflagellates, the existence of unicellular opisthokonts. Whilst the Wainright phylogeny recovered a strongly supported Opisthokonta, the relationships between the three represented lineages remained unresolved.

As the volume of molecular data increased, it became clear that the opisthokonts harboured much greater diversity, with multiple unicellular lineages being closely related to the metazoans (Cavalier-Smith and Allsopp 1996; Herr et al. 1999; Kerk et al. 1995). The first indications of unicellular holomycotans were discovered at a similar time (Edlind et al. 1996; Keeling and Doolittle 1996); however, clear evidence for unicellular relatives of Fungi was later produced through the phylogenetic placement of species that were previously believed to belong to Amoebozoa and Holozoa (Amaral-Zettler et al. 2001; Brown et al. 2009; Ruiz-Trillo et al. 2004).

Early evidence for the existence of the opisthokont grouping was based upon morphological traits. However, as the depth of diversity in opisthokont lineages has been uncovered, it has become apparent that there are no recognised universal morphological characters unique to this group. The posterior flagellum is not present in all opisthokont groups; loss of the flagellum must have occurred on multiple occasions within both Holomycota and Holozoa (Adl et al. 2019; Galindo et al. 2021; James et al. 2006). Throughout the opisthokonts, the morphology of mitochondrial cristae is predominantly flat, but this appears to be a plastic trait, with tubular and discoidal cristae also present (Adl et al. 2019; Amaral-Zettler et al. 2001; Ragan et al. 1996; Wylezich et al. 2012). Nonetheless, the posterior flagellum and flat cristae are widespread across the opisthokonts and point to the ancestral states of the group. The abilities to produce amoeboid cells and also to engulf particles by phagocytosis are present in most of the major lineages. Moreover, Metazoa is the only major opisthokont lineage that does not contain species with cell walls, and it has been suggested that the last common ancestor of the opisthokonts also possessed the potential to produce a cell wall (Mendoza et al. 2002).

1.3 The Placement of Opisthokonta in the Eukaryotic Tree of Life

The improvements in phylogenetic analyses set out in Sect. 1.2 have also led to a much greater understanding of the overall eukaryotic tree. For much of the last two decades, most eukaryotes were believed to fall into one of seven supergroups, namely the Alveolata, Amoebozoa, Archaeplastida, Excavata, Opisthokonta, Rhizaria, and the stramenopiles. In addition to the supergroups, a number of minor, or orphan, groups of uncertain phylogenetic position were recognised (reviewed in Adl et al. 2005, 2012, 2019). Robust relationships between the supergroups have been less certain, but the stramenopiles, alveolates, and rhizarians were frequently recovered as clade known as the SAR group (Burki et al. 2007). Furthermore, a close relationship between the amoebozoans and opisthokonts was also reported (Stechmann and Cavalier-Smith 2003a, b). As both groups contain uniflagellate species, the combined Amoebozoa +Opisthokonta taxon was defined as unikont, a term originally coined by Cavalier-Smith (2002) to describe species which possess a single flagellum and single centriole. The remaining eukaryotic supergroups, which predominantly contain biflagellate taxa, were labelled as bikonts. It is now clear that these terms are no longer appropriate, since nested within the unikont grouping are a small number of biflagellate species such as *Apusomonas proboscidea* (Kim et al. 2006; Vickerman et al. 1974).

Recent findings have cast doubt on the simplicity of the supergroup system (Burki et al. 2020), with neither Archaeplastida nor Excavata being recovered as monophyletic in large-scale phylogenomic studies (Cavalier-Smith et al. 2014; Gawryluk et al. 2019; Heiss et al. 2018). Advances in phylogenetics have also resulted in the recovery of further novel lineages, such as Ancyromonadida, Hemimastigophora, and Malawimonadida, which appear to fall outside of the previously recognised supergroups (Fig. 1.1). The greatest issue currently hindering

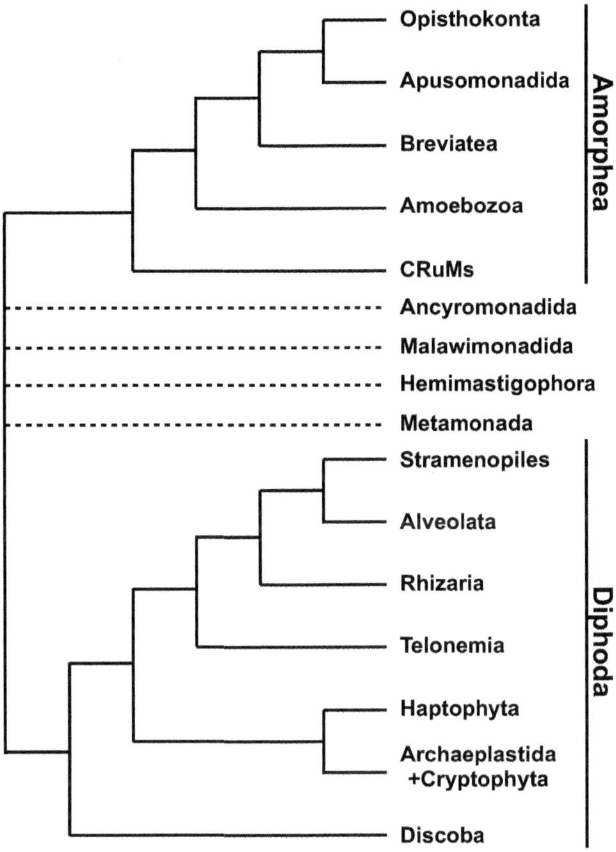

Fig. 1.1 Representative cladogram of Eukaryota. The Amorphea and Diphoda are written in black font. Due to the lack of a resolved root for the eukaryotic tree, four lineages, shown by dotted grey branches, are of uncertain position. The topology is based upon phylogenies presented in Brown et al. (2018), Heiss et al. (2018), Lax et al. (2018), and Strassert et al. (2019)

eukaryotic deep phylogenetics is the unknown position of the root, or earliest branching point, of the eukaryotes. Until the position of the root is established, determining the order of divergence events and establishing sister groups cannot be confidently achieved.

Eukaryotes appear to have evolved from a symbiosis between an alphaproteobacterium and an Asgard archaeon (Gray et al. 1999; Zaremba-Niedzwiedzka et al. 2017), with the eukaryotic crown group estimated to have arisen between 1.6 and 2.5 billion years ago (Parfrey et al. 2011). This great antiquity leads to long branches positioned between eukaryotes and archaea in phylogenetic trees, with long-branched artefacts pulling rapidly evolving species to the base of the eukaryotic group (Brinkmann et al. 2005; Williams and Embley 2014). Eukaryotic phylogenies created with bacterial genes tend to recover accepted eukaryotic supergroups but fail to find a consistent root (Derelle et al. 2015; He et al. 2014). A particularly contentious issue with deep eukaryotic phylogenetics, highlighted in the He and Derelle studies, is the position of Excavata. The supergroup is proposed to consist of three lineages, in Discoba, Malawimonadida, and Metamonada, all of which contain multiflagellate species that possess a ventral feeding groove (Adl et al. 2012, 2019). Multiple studies have both recovered (Hampl et al. 2009; He et al. 2014) and rejected (Brown et al. 2018; Heiss et al. 2018) the monophyly of the excavates. Trees which fail to recover excavate monophyly often place excavate lineages on either side of the eukaryotic root (Derelle et al. 2015; Heiss et al. 2018). This raises the possibility of the excavates being a paraphyletic grouping, with two flagella and a feeding groove being ancestral traits for all eukaryotes.

Whilst the position of the eukaryotic root, as well as the monophyly of the excavate lineages, remains uncertain, two major domains are now recognised. The Diaphoretickes consists of the SAR group, now unified with Telonemia to form the TSAR clade (Strassert et al. 2019), Archaeplastida, Cryptista, and Haptista (Adl et al. 2012, 2019). The relationships between latter three lineages have yet to be resolved; however, a number of recent phylogenies suggest that Cryptista may in fact be a clade within the archaeplastids (Gawryluk et al. 2019; Strassert et al. 2019). Derelle et al. (2015) further unified Diaphoretickes with the excavate Discoba lineage under the name Diphoda (Fig. 1.1).

The opisthokonts fall into the second major eukaryotic domain, Amorphea (Adl et al. 2012; Fig. 1.1). A clade of gliding, biflagellate heterotrophs termed Apusomonadida (Karpov and Mylnikov 1989) are robustly recovered as the sister group to the opisthokonts, highlighting that the flagellum loss to the uniflagellate state must have occurred in the opisthokont stem group. Breviatea (Cavalier-Smith et al. 2004) is a clade of gliding flagellated amoebae, which may possess either one or two flagella. The phylogenetic position of this group has had something of a troubled history, being placed initially with excavates (Cavalier-Smith et al. 2004) and then amoebozoans (Minge et al. 2009) before being finally recognised as an independent lineage related to both the opisthokonts and apusomonads in the taxon Obazoa (Brown et al. 2013). The earliest branching Amorphea lineage, and sister taxon to the Obazoa, is the Amoebozoa (Adl et al. 2019).

The recently proposed CRuMs supergroup, comprising free-swimming Collodictyonidae, amoeboid Rigifilida, and the gliding *Mantamonas* has been recovered as the putative closest relative to Amorphea (Brown et al. 2018; Lax et al. 2018); however, this proposition is controversial due to the use of unrooted phylogenetic trees and the uncertainty over the position of the eukaryotic root.

1.4 Holomycota

Molecular phylogenies in the early 1990s confirmed that Fungi is a member of Opisthokonta; however, it was later in the decade that unicellular close relatives of Fungi were identified. Due to weak phylogenetic support, early phylogenies could not differentiate between unicellular holomycotans being close relatives of, or being nested within, Fungi (Edlind et al. 1996; Fast et al. 1999; Gill and Fast 2006; Keeling and Doolittle 1996). The expansion of known unicellular holomycotan diversity has been mainly driven by the phylogenetic analysis, and subsequent taxonomic reassignment, of previously known taxa, such as microsporidians, nucleariids, and aphelids. Whilst not universally agreed upon (Richards et al. 2017), there has also been a movement of groups from Fungi, into sister lineages of the group. Two groups of early-branching holomycotans are recognised, in the nucleariid amoeba and the opisthosporidians; however, the latter may be paraphyletic and not a true clade (Fig. 1.2). At present, there is also no consensus upon whether opisthosporidians should be considered members of an enlarged Fungi kingdom, or whether they should be classified as the sister group to "true" or "classical" Fungi. Within this chapter, we refer to Opisthosporidia as the closest protistan relatives to Fungi and not as members of the group.

1.4.1 Nucleariid Amoebae

The nucleariid amoebae is the informal name assigned, initially, to two distinct genera, *Fonticula* and *Nuclearia*, of holomycotans that were unified through multigene amino acid phylogenies (Brown et al. 2009). The group is also known under the formalised synonym Rotosphaerida, which was originally proposed as a taxon within Heliozoa (Rainer 1968). As is the case with numerous unicellular opisthokont groups, both genera were previously erroneously

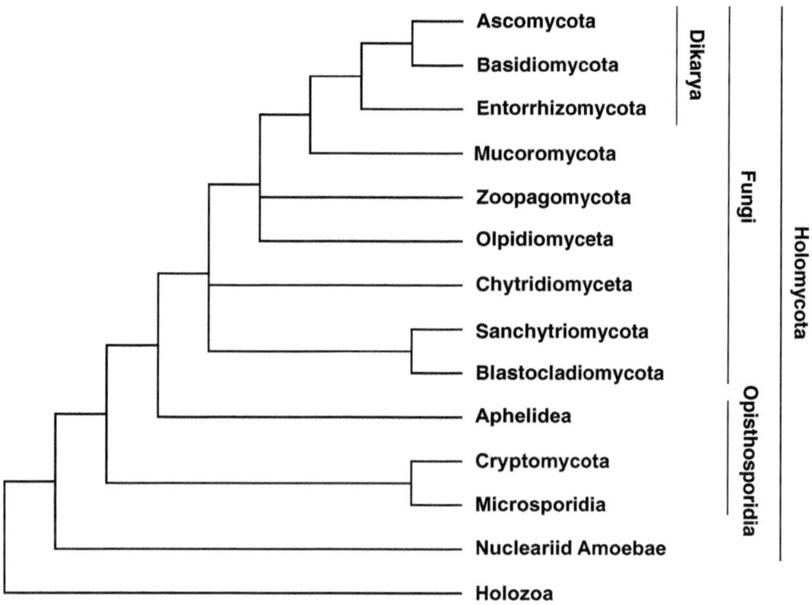

Fig. 1.2 Representative cladogram of Holomycota. The holomycotans are rooted with Holozoa. Polytomies highlight areas of uncertainty in fungal phylogenetic studies and Opisthosporidia is shown as paraphyletic. The topology is based upon phylogenies presented in Bauer et al. (2015), James et al. (2020), Tikhonenkov et al. (2020), and Torruella et al. (2018)

assigned to other taxonomic groups before molecular studies placed them within Holomycota. Nucleariids are slow grazers that can feed on a variety of organisms depending on the species; these include unicellular bacteria and algae. It is suggested that their growth rate may be directly proportional to the availability of their prey, with growth most observable after algal blooms (Dirren et al. 2017).

Discovered over 150 years ago (Cienkowski 1865), to date fewer than 10 nucleariid species have been described (López-Escardó et al. 2017). Thriving in both eutrophic and polluted environments, *Nuclearia* species can feed on toxic cyanobacteria owing to their bacterial endosymbionts which are believed to provide toxicity protection (Dirren and Posch 2016; Dirren et al. 2017). All known species have been isolated from freshwater habitats; individuals floating within the water column possess a protoplast (cell body) that is spherical and non-flagellated, whilst cells in contact with a substratum become amoeboid. Both forms of protoplast exhibit thin filopodia, which are used to contact bacterial and algal prey prior to phagocytosis (Yoshida et al. 2009). Mitochondrial cristae are either flattened or discoidal (Amaral-Zettler et al. 2001; Dyková et al. 2003), whilst species may be uninucleate, multinucleate, or capable of switching between the two states (Dirren and Posch 2016).

Page (1987) and Cavalier-Smith (1993) both placed *Nuclearia* into the order Cristidiscoidida, as members of the amoebozoan phylum Rhizopoda due to morphological characteristics. This placement was questioned by Patterson (1999), on the basis that branching filopodia are likely to have evolved on multiple occasions within eukaryotes as well as the unexpected presence of three mitochondrial cristae morphologies in the group. Through the use of rRNA phylogenetic trees, Amaral-Zettler et al. (2001) placed *Nuclearia* within Opisthokonta; however, their phylogenies could not resolve whether the group was more closely related to either Metazoa or Fungi. Ruiz-Trillo et al. (2004) finally robustly placed *Nuclearia* as members of Holomycota, as the then recognised sister group to Fungi.

Fonticula alba was isolated from dog faeces in 1960, but was not named for almost two decades, until Worley et al. (1979) described the species as a bacterivorous acrasiomycete slime mold. Individual cells are small (7–13 μm), enclosed in a mucosal glycocalyx, possess mitochondria with discoidal cristae, and exhibit pseudopodia that may project several body lengths from the protoplast (Brown et al. 2009). Cells may produce either filose pseudopodia, when feeding, or a single lobose pseudopodium when migrating (Toret et al. 2022). Individual cells predominantly possess a single nucleus, but larger bi- and trinucleate cells are infrequently observed in laboratory cultures (Worley et al. 1979). In a form of multicellularity superficially similar to that observed in dictyostelid amoebozoans, trophic cells may form aggregates when local prey becomes depleted (Worley et al. 1979). Aggregates become enveloped in a slime-based extracellular matrix and form into a mound; the height of the mound increases and subsequently develops into a globular sorocarp (fruiting body) on volcano-like stalk. The majority of cells in the aggregate migrate to the sorocarp and encyst to become spores; however, a small number of cells remain at the base of the stalk. After spores have germinated the stalk structure subsides, allowing juvenile amoeboid cells to disperse (Worley et al. 1979).

It has recently been discovered that multicellular aggregates also develop when *F. alba* is presented with a new prey source (Toret et al. 2022). When encountering a bacterial biofilm a leader cell directs an elongated collective of connected cells into the prey. Cell-to-cell contacts are well defined with regions enriched with actin; however, individual cells may leave or join the predatory collective during its migration.

The original description in Acrasiomycetes was problematic, as *F. alba* did not appear to be a member of any known acrasiomycete group (Worley et al. 1979). Cavalier-Smith (1993) placed *F. alba* alongside *Nuclearia* in Cristidiscoidia within Filosea; however, Brown et al. (2009) showed that *Fonticula* clusters with *Nuclearia* as the sister group to Fungi. Morphological studies have associated a number of other filose amoebae with the nucleariids (Mikrjukov 1999; Patterson 1985; Patterson et al. 1987). The monotypic genus *Vampyrellidum* is similar to *Nuclearia* in that cells are surrounded by a mucosal glycocalyx. In contrast, species in the genera *Lithocolla* and *Pompholyxophrys* possess assemblages of scales surrounding the protoplast (Galindo et al. 2019); *Pompholyxophrys* taxa produce their own siliceous scales, whilst in *Lithocolla* cells acquire sand particles or diatom frustules (Gabaldón et al. 2022). Molecular phylogenies have confirmed *Lithocolla* and *Pompholyxophrys*, as well as *Parvularia* (originally deposited in the American Type Culture Collection as a *Nuclearia* taxon), as members of the nucleariid amoebae (Galindo et al. 2019). At present no sequence data are available for *Vampyrellidum perforans* so the taxonomic affinity of this species remains unclear.

Galindo et al. (2019) reconstructed states for ancestral nucleariid amoebae on the basis of a multigene phylogeny. They proposed that the ancestral opisthokont single flagellum was lost in the stem lineage of the nucleariids and that last common ancestor was a freshwater protist that possessed filose pseudopodia and a mucosal glycocalyx. Aggregate multicellularity evolved in the *Fonticula* lineage, whilst scale bearing evolved an ancestor of *Lithocolla* and *Pompholyxophrys*.

1.4.2 Opisthosporidia

The superphylum Opisthosporidia, which comprises three endoparasitic lineages in the Aphelida, Microsporidia, and Cryptomycota (the latter also described as Rozellida and the Rozellomycota), was only proposed in 2014 on the basis of molecular phylogenies (Karpov et al. 2014a). The taxon is a controversial one, as its placement in Fungi or alternatively the sister group to Fungi is disputed. Furthermore, the validity of Opisthosporidia is unclear, as it has been recovered as paraphyletic in some molecular phylogenies; whilst Cryptomycota and Microsporidia are recognised as sister groups, Aphelida has been recovered in a variety of positions in Holomycota (Corsaro et al. 2014; Galindo et al. 2019, 2021; James et al. 2006; Karpov et al. 2014a; Torruella et al. 2018).

Initially described in the 1880s by Balbiani (1882) microsporidians are a group of predominantly obligate intracellular parasites. Their non-flagellated cells are distinguished by the presence of a unique structure, the polar tube, which is present within their spores. Upon infecting a new host the polar tube breaks through the chitin-based spore wall and penetrates the plasma membrane of a host cell, allowing the unwalled sporoplasm to pass down the polar tube and enter the cytoplasm of the infected cell. The sporoplasm, which may be uninucleate or binucleate, then proliferates within the new host cells, resulting in a new generation of infective spores (Franzen 2005; Wadi and Reinke 2020).

The first described microsporidians were parasites of metazoans, notable for their highly reduced nuclear genomes, small ribosomes, and lack of mitochondria (Fast et al. 1999). Cavalier-Smith (1983) proposed that they were amongst the earliest branching eukaryotes and placed them in the kingdom Archezoa along with other amitochondriate eukaryotes. Rapidly evolving gene sequences appeared to confirm their antiquity (Vossbrinck et al. 1987); however, the use of more functionally conserved genes showed that the phylogenetic recovery of Archezoa was an artefact due to long-branch attraction, with the microsporidia being either fungi or closely related to fungi (Edlind et al. 1996; Hirt et al. 1999).

Microsporidians are now known to associate with non-metazoan hosts, acting as endosymbionts and hyperparasites for multiple species in the SAR eukaryotic supergroup (Bass et al. 2018). Phylogenetic studies including environmental DNA (eDNA) have uncovered a far greater diversity of microsporidians than was previously known (Bass et al. 2018; Bojko et al. 2022). It is now clear that many of the traits present in the "canonical" or long-branch microsporidians, which infect metazoans, are absent in other lineages. In particular, mitochondria are widespread outside of the canonical microsporidians. The derived *Paramicrosporidium saccamoebae* possesses a mitochondrion which has a genome similar to those of typical fungi (Quandt et al. 2017). In contrast, the early-branching *Mitosporidium daphniae* has lost the genes for the mitochondrial respiratory chain complex I, highlighting independent mitochondrial reduction occurring across Microsporidia (Haag et al. 2014).

Despite being recognised in 2011, the Cryptomycota remain an enigmatic group within Holomycota with the composition of the group in a state of flux (Bass et al. 2018; Corsaro et al. 2014; Jones et al. 2011; Letcher et al. 2013). *Rozella* is the most extensively studied genus within Cryptomycota, with 27 species recognised by Letcher and Powell (2018). Species possess uniflagellated, unwalled zoospores and were for a long time considered to be chytrid fungi (Adl et al. 2005; James et al. 2006); however, rozellids differ from chytrids in that they employ phagocytosis to consume the cytoplasmic contents of their hosts, rather than osmotrophic absorption of nutrition (Powell 1984). Host species include early-branching fungi, as well as green algae and oomycete stramenopiles (Letcher and Powell 2018). Upon contact with a new host cell the *Rozella* zoospore withdraws its flagellum and forms a cyst which possesses a chitin-based wall. The infective cell then enters the host cell via a penetration tube (James and Berbee 2012). Sequencing of the *R. allomycis* mitochondrial genome revealed a 12 kb circular chromosome that has undergone extensive gene loss and a loss of functional constraint on the remaining genes (James et al. 2013), mirroring the reduction of mitochondrial genomes observed in the closely related microsporidians.

The initial proposal of Cryptomycota by Jones et al. (2011) was based in part upon a phylogenetic analysis which contained a large number of eDNA sequences. This tree, as well as a number of others published in the 2010s, indicated that Cryptomycota possessed a high level of species diversity (Corsaro et al. 2014; Karpov et al. 2014a, b, 2018). The inclusion of early-branching microsporidian sequences in phylogenetic trees, however, revealed that much of this diversity was not actually cryptomycotan but microsporidian (Bass et al. 2018; Bojko et al. 2022), suggesting that the diversity of the group may not be much greater than the *Rozella* taxa which have currently been described.

The aphelids are parasites of archaeplastid and stramenopile algae and show a number of similarities with rozellids with regard to their infection of host cells. Infective propagules make contact with host cells and their pseudopodia search for a break in the algal cell wall. The propagule subsequently forms a cyst with a chitin-based wall and develops a penetration tube at the site of the hole in the host cell wall. The alga is breached by the tube, allowing the amoeboid trophic stage to pass from the cyst into the host cytoplasm. The aphelid phagocytoses the alga cytoplasm, ultimately killing the host cell, and undergoes nuclear division to become a plasmodium. Multiple rounds of cell division subsequently occur to produce the next generation of uninucleate zoospores which then emerge through the host cell wall via the hole produced by the penetration tube (reviewed in Karpov et al. 2014a; Letcher and Powell 2019; Torruella et al. 2018).

The type species *Aphelidium deformans* was described by Zopf (1885) and subsequently 14 further species have been described (Letcher and Powell 2019). Aphelid taxa have been assigned to four ecologically distinct genera: *Amoebaphelidium* and *Aphelidium* are freshwater genera and *Paraphelidium* and *Pseudaphelidium* form marine genera (Letcher and Powell 2019). Whilst all described rozellids possess flagellated zoospores, aphelid infective cells may be flagellated, amoeboflagellated, or amoeboid cells with an immobile flagellum (Letcher and Powell 2019). Studies suggest that mitochondrial cristae morphologies may vary across species and also between the different stages of lifecycles, with flat, tubular, and lamellar cristae reported (Karpov et al. 2014a; Letcher et al. 2013).

The taxonomic placement of aphelids has proved to be controversial, with varying authors considering them to be members of Rhizopoda, Phycomycetes, Mesomyceteozoea, Rozellidea, and Cryptomycota (Gromov 2000; Karpov et al. 2014a; Letcher et al. 2013; Letcher and Powell 2019). Karpov et al. (2013) recovered *Amoeboaphelidium protococcarum* in a clade with the cryptomycotan *Rozella allomycis* and nine microsporidians, with the taxon Opisthosporidia erected the following year (Karpov et al. 2014a). More recent phylogenomic studies have placed Aphelida as the sister group to Fungi and on this basis Galindo et al. (2022) proposed two novel holomycotan taxa. Phytophagea unifies Fungi with Aphelida, whilst Cryptomycota and Microsporidia make up Opisthophagea.

1.4.3 Fungi

Fungi are a highly diverse group of heterotrophs, which may be saprotrophs, commensal symbionts, or parasites; species employ osmotrophy, gaining nutrition through absorption from their environment. Fungi often produce multinucleate hyphae and possess cell walls that comprise both β-glucan and chitin (Adl et al. 2019; Cavalier-Smith 1998a). In a process that appears to be mirrored in Metazoa and Microsporidia the mitochondrial genomes of fungi are reduced in comparison to the ancestral state of opisthokonts, typically encoding 30–40 genes. This reduction in mitochondrial genome size and gene content has been shown to have begun in the fungal stem group, but is a continuing process in the crown group (Bullerwell and Lang 2005). Within the early-branching Neocallimastigomycota the capacity for oxidative phosphorylation has been completely lost and the mitochondrion has evolved into a hydrogenosome; this organelle produces ATP anaerobically and has convergently evolved in eukaryotes on multiple occasions (Embley et al. 2003). Some members of Chytridiomyceta have also undergone tRNA gene loss in their mitochondria and rely upon extensive tRNA editing or the import of tRNA molecules into mitochondria in order to facilitate localised protein translation (Bullerwell and Lang 2005).

The taxonomy of Fungi has undergone substantial revisions since the turn of the century, due to molecular phylogenetic studies. In particular, the number of phyla has increased considerably from the four generally recognised in the 1990s, in the Ascomycota, Basidiomycota, Chytridiomycota, and Zygomycota, to between 8 and 18 phyla (Galindo et al. 2021; James et al. 2020; Tedersoo et al. 2018). The disagreements in the number of phyla are predominantly due to the taxonomic rank assigned, either phylum or sub-phylum, to fungal groups, rather than the validity of the groups themselves. However, disagreements over phylum number are also due to classification systems, such as that proposed by Tedersoo et al. (2018), which include opisthosporidian groups as fungal phyla.

The Ascomycota and Basidiomycota remain recognised as valid taxa, but the early-branching Chytridiomycota has now been split into Cryptomycota, Blastocladiomycota, Monoblepharidomycota, Neocallimastigomycota and Olpidiomyceta, as well as a reduced Chytridiomycota (James et al. 2020; Spatafora et al. 2016). The Chytridiomycota, Monoblepharidomycota, and Neocallimastigomycota have been shown to form a monophyletic grouping and make up the subkingdom Chytridiomyceta. The former Zygomycota has now been divided into the Mucoromycota and Zoopagomycota (Fig. 1.2; James et al. 2020; Spatafora et al. 2016). The branching order of the non-Dikarya fungi has yet to be resolved, with Blastocladiomycota, Chytridiomyceta, or a unified Blastocladiomycota+Chytridiomyceta clade being recovered as the earliest branching fungal lineage in different studies (Chang et al. 2015; Galindo et al. 2021; Tedersoo et al. 2018).

The subkingdom Dikarya, also described as Neomycota (Cavalier-Smith 1998b), is defined by species that possess cells with unfused haploid nuclei called dikaryons. Dikarya contains the two major fungal phyla Ascomycota and Basidiomycota and encompasses ~98% of known fungal diversity (James et al. 2006). A recent phylogenetic study erected a third dikaryon phylum in Entorrhizomycota, a group which had previously been considered a member of the Basidiomycota (Bauer et al. 2015). The Mucoromycota are now recognised as the sister clade to Dikarya (James et al. 2020).

A robust phylogeny of Fungi is vital in order to understand how the group evolved. Phagocytosis was ancestral to opisthokonts, and the trait is present in early-branching holomycotans, including the opisthosporidians, but absent from all fungi, indicating it was lost in the stem lineage after the divergence of Fungi from Aphelida. The ancestral single flagellum is present in early-branching fungi; however, it has been lost on multiple occasions across the kingdom. Within Chytridiomyceta, *Hyaloraphidium curvatum* appears to be aflagellate (Ustinova et al. 2000), highlighting a loss within early-branching taxa. Sanchytriomycota is a novel phylum erected to accommodate two recently discovered flagellated early-branching fungi (Karpov et al. 2018, 2019b). Unusually, their flagella are non-motile, with evidence indicating that the structure has evolved to become a light sensing organelle (Galindo et al. 2021). Species within Dikarya, Mucoromycota, and Zoopagomycota all appear to lack a flagellum (James et al. 2006); however, their relationships with the flagellated Olpidiomyceta remain uncertain. As a result, it is not clear if the ancestral flagellum was lost within these species on one or two occasions.

Whilst a comprehensive phylogeny of the deepest branches of Fungi has yet to be agreed upon, initiatives such as the Joint Genomes Initiative's MycoCosm program are generating the data required for an accurate reconstruction of ancestral fungal and holomycotan metabolism as well as a resolved phylogeny of the group.

1.5 Holozoa

The second major lineage of Opisthokonta, Holozoa, was proposed in 2002, on the basis of the phylogenetic unification of Metazoa with two unicellular groups, in the choanoflagellates and ichthyosporeans, to the exclusion of Fungi (Lang et al. 2002). Since this initial proposal the known diversity of the clade has expanded, with the identification of the filasterean group (Shalchian-Tabrizi et al. 2008). Two additional putative lineages have recently been described in Pluriformea (Hehenberger et al. 2017) and *Tunicaraptor* (Tikhonenkov et al. 2020); however, neither of these groups have robustly resolved phylogenetic positions and their validity has not been confirmed.

1.5.1 Ichthyosporea

The ichthyosporeans are an ecologically and morphologically diverse group of predominantly parasitic, unicellular holozoans. The first reports of this clade came from Ragan et al. (1996) and Spanggaard et al. (1996), with the former naming the group the DRIP clade as an acronym based on the first known members—*Dermocystidium*, rosette agent, *Ichthyophonus,* and *Psorospermium.* Cavalier-Smith (1998a) described these species as Ichthyosporea, due to the four known species all infecting fish hosts. Herr et al. (1999) noted, however, that non-fish hosts had subsequently been discovered and so recommended the name Mesomycetozoa to reflect the phylogenetic position of the group, that is holozoans located between Fungi and Metazoa. Adl et al. (2005) expanded Mesomycetozoa to include unicellular holozoans and holomycotans; however, this status, as a paraphyletic dustbin taxon, was not widely accepted and the term Mesomycetozoa has generally reverted to being a synonym of Ichthyosporea.

A lack of unifying morphological characters means that membership of Ichthyosporea is mainly based upon molecular phylogenies. Prior to the emergence of molecular phylogenetics, a number of ichthyosporean taxa, such as species in the Amoebidiidae and Eccrinidae, were considered to be trichomycete fungi; however, these earlier placements were considered somewhat controversial at the time due to the presence of amoeboid stages in their lifecycles and a lack of a chitinous cell wall (Lichtwardt 1986; Trotter and Whisler 1965).

All known ichthyosporeans are symbionts of metazoans, with relationships varying between commensalism, mutualism, and parasitism; however, there has also been speculation that some species have saprotrophic stages within their lifecycles (Glockling et al. 2013; Mendoza et al. 2002). Species are unicellular and frequently multinucleate, exhibiting a great variety of morphologies. Across the group mitochondrial cristae are flat, with a single known exception in *Ichthyophonus hoferi* which possesses tubular cristae (Ragan et al. 1996). Phylogenetic studies have consistently recovered the ichthyosporeans as comprising two robustly supported clades, labelled Ichthyophonida and Dermocystida (Cafaro 2005; Grau-Bové et al. 2017; Lohr et al. 2010; Pereira et al. 2005) that are both morphologically and ecologically distinct.

Erected by Cavalier-Smith (1998a), the order Dermocystida has also been described as the family Rhinosporidiaceae (Mendoza et al. 2001). Most described species are parasites of vertebrates; however, the host species and site of infection vary across dermocystid species. *Dermocystidium* and *Sphaerothecum* species infect fish, with the latter apparently restricted to infecting internal organs whilst the former may additionally be found on external structures, such as gills and fins (Glockling et al. 2013; Ramaiah 2006). *Rhinosporidium* species cause rhinosporidiosis (Thompson 2016), a disease resulting in granulated polyps in the sinonasal tract, conjunctiva, and urethra in mammals and birds (Kennedy et al. 1995; Seeber 1900). Six genera of dermocystids are known to infect amphibians, predominantly infecting the skin but also being associated with the heart and liver (reviewed in Borteiro et al. 2018).

The infectious agents of dermocystids are endospores. Fish parasites tend to have uniflagellate zoospores, whilst flagellum loss appears to have occurred in species with amphibian hosts (Glockling et al. 2013). Upon infecting a new host the endospore encysts and produces a walled sporangium (cyst) in the host. The sporangia then increase in size to 200–400 μm in diameter, whilst cells undergo division to produce thousands of zoospores. Endospores range across 7–15 μm diameter (Herr et al. 1999) and are either released directly into the tissue of the original host or into the environment to infect a new host (for a review of the life cycle, see Mendoza et al. 2002). Whilst Dermocystida is consistently recovered as monophyletic, the internal relationships within the clade are poorly resolved (Fig. 1.3; González-Hernández et al. 2010). As a result of this lack of phylogenetic

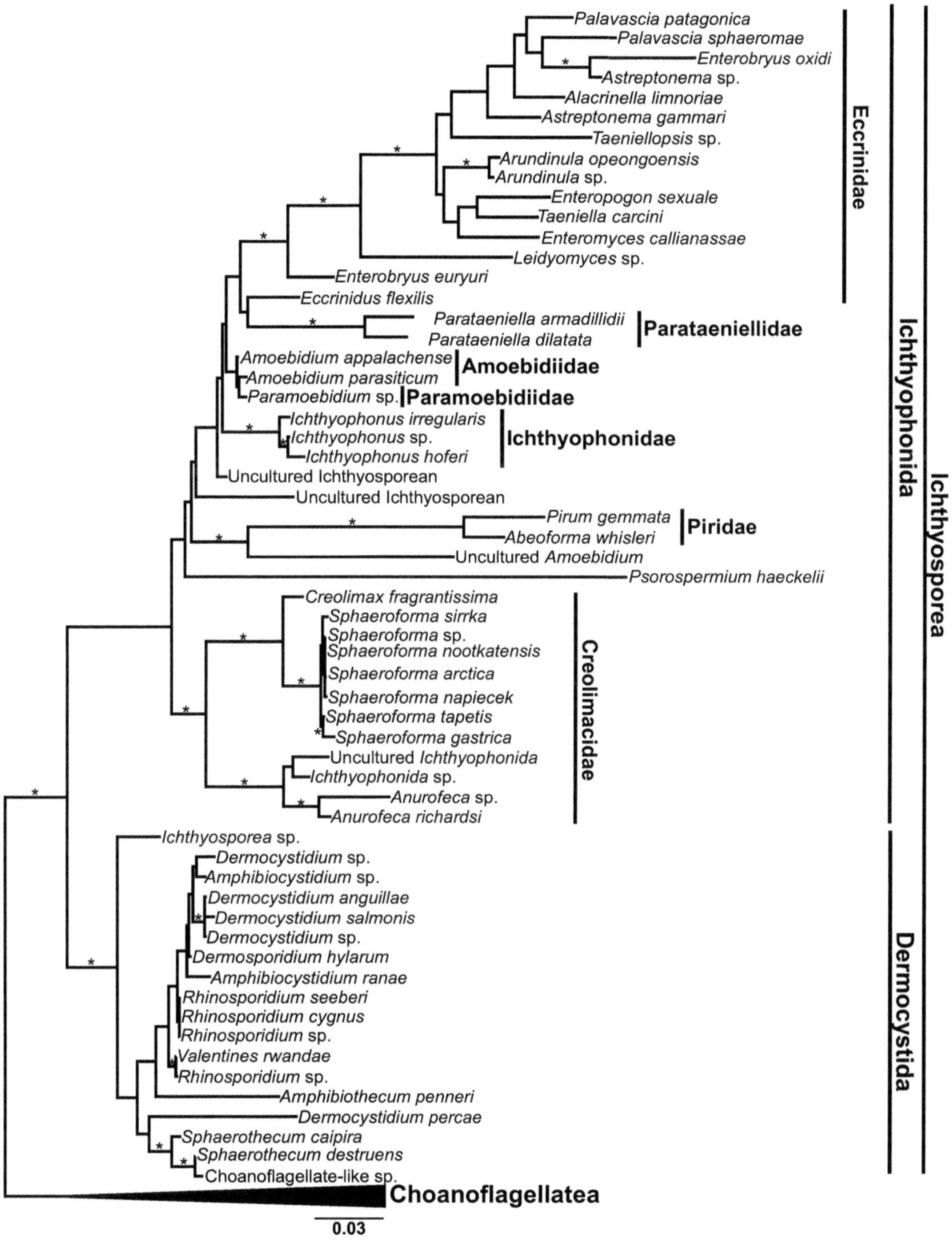

Fig. 1.3 Maximum likelihood phylogeny of Ichthyosporea. The phylogeny was created with RAxMLGUI 2.0 (Edler et al. 2020) from a SSU RNA alignment of 1538 sites sequences generated in MAFFT 7.309 (Katoh and Standley 2013). The phylogenetic analysis used the TIM2 model (Posada 2003) with a four-category gamma distribution and a proportion of invariant sites. Support values were calculated as bootstrap percentages from 1000 replicates. A Bayesian inference phylogeny was created using the same alignment with MrBayes 3.2.6 (Ronquist et al. 2012) The ichthyosporeans are rooted with SSU sequences from four choanoflagellates (*Diaphanoeca grandis*, *Monosiga brevicollis*, *Savillea parva*, and *Stephanoeca diplocostata*). Asterisks highlight nodes strongly supported with both methodologies (≥70% maximum likelihood bootstrap percentage, ≥0.97 Bayesian inference posterior probability). The scale bare represents the number of substitutions per site. Family names are taken from Reynolds et al. (2017)

definition, to date it has been difficult to reconstruct the evolution of observed traits within the order.

In contrast to Dermocystida, phylogenetic relationships within Ichthyophonida are well resolved and the order is made up from multiple recognised families (Fig. 1.3; Glockling et al. 2013). Whilst vertebrate parasitism is prevalent in the dermocystids, only two genera of ichthyophonids are known to have vertebrate hosts. *Anurofeca richardsi* is a gut pathogen of frogs and toads and infection results in inhibited larval growth (Baker et al. 1999; Beebee and Wong 1992); *Ichthyophonus* species are also known to parasitise a range of amphibians, as well as both freshwater and marine fish hosts (Herman 1984; Raffel et al. 2006; Rowley et al. 2013). The majority of the investigated ichthyophonids associate with arthropods and molluscs, although species also infect echinoderms, peanut worms, and tunicates (Lu et al. 2020; Marshall and Berbee 2011, 2013; Reynolds et al. 2017). Not all ichthyophonids are believed to be parasitic; despite its name, the arthropodophilous *Amoebidium parasiticum* is a non-pathogenic symbiont which attaches itself to the external exoskeleton of its insect hosts (Benny and O'Donnell 2000). A further contrast to dermocystids is found in the morphology of the motile dispersal stage. Flagellated cells have yet to be reported, with all motile dispersal cells showing an amoeboid morphology (Mendoza et al. 2002). However, amoeboid dispersal cells are not universal across the group and appear to have been lost on multiple occasions (Lord et al. 2012; Reynolds et al. 2017).

The well-resolved phylogeny of Ichthyophonida allows the evolutionary reconstruction of traits in the group. In particular, it can be seen that *Anurofeca* and *Ichthyophonus*, the two genera known to parasitise vertebrate hosts, are distantly related (Fig. 1.3), showing that ichthyophonids have undergone at least two transitions from non-vertebrate to vertebrate hosts. Based upon their phylogenetic trees, Reynolds et al. (2017) speculated that the ancestors of *Ichthyophonus* were externally attached commensals of non-vertebrate hosts that subsequently evolved into internal parasites of vertebrates.

Whilst the monophyly of Ichthyosporea is widely accepted, the relationships of the group with other holozoans are far from resolved. Since the recognition of the group in 1996 many studies have recovered the ichthyosporeans as both an independent lineage and the earliest branching group in Holozoa (Hehenberger et al. 2017; Paps et al. 2013; Ragan et al. 1996; Torruella et al. 2012). However, discoveries of novel holozoans in recent years have clouded this view and the identities of the most basal holozoan group and the closest relatives of the ichthyosporeans are not presently clear (Grau-Bové et al. 2017; Tikhonenkov et al. 2020; Torruella et al. 2015).

1.5.2 Pluriformea

One of two enigmatic holozoan taxonomic groups Pluriformea was only erected in 2017 and, at present, contains two highly disparate species (Hehenberger et al. 2017). The first species identified in the group, *Corallochytrium limacisporum*, is a unicellular saprotroph isolated from coral lagoons in the Arabian Sea (Raghukumar 1987). To date flagellated cells have not been observed; however, transcriptome analyses have shown presence of the genes required for flagellum assembly (Torruella et al. 2015), suggesting that the *C. limacisporum* may possess more morphological diversity than that witnessed in laboratory cultures. Like most opisthokonts the mitochondria of *C. limacisporum* possess flat cristae (Mendoza et al. 2002). Mature cells are spherical, ranging from 4.5 to 20.0 μm in diameter, and possess a filamentous cell wall that contains multiple pores (Cavalier-Smith and Allsopp 1996; Mendoza et al. 2002). Cell division occurs within the cell wall, resulting in the release of up to 32 elongated, amoeboid daughter cells (Mendoza et al. 2002). The juvenile cells show motility via a slow rocking movement (Cavalier-Smith and Allsopp 1996). Unlike many eukaryotes, cell division is decoupled from nuclear division and *C. limacisporum* is binucleate for most of the lifecycle as observed in laboratory cultures. The majority of binucleate cells undergo cell division to produce uninucleate daughter cells; however, approximately 1% of observed binucleate cells continue to undergo nuclear division to produce multinucleate coenocytes. Coenocytes eventually bud off daughter cells to return to a uninucleate state (Kożyczkowska et al. 2021).

Upon its discovery *C. limacisporum* was described by Raghu-kumar (1987) as a thraustochytrid. Cavalier-Smith and Allsopp (1996) disputed this classification, on the basis that *C. limacisporum* lacks any of the diagnostic morphological characters of either thraustochytrids or fungi. With a SSU rRNA phylogeny, they recovered *C. limacisporum* as a holozoan and a close relative of the choanoflagellates. Later multigene phylogenies, based upon three or four genes, provided conflicting positions with *C. limacisporum* recovered as a close relative of either choanoflagellates or ichthyosporeans (Carr et al. 2008; Ruiz-Trillo et al. 2006; Steenkamp et al. 2006). Torruella et al. (2015) found a sister relationship between two strains of *C. limacisporum* and the ichthyosporeans, as the earliest branching group of Holozoa, and suggested the name Teretosporea (meaning "rounded spore") for this basal clade.

The discovery of *Syssomonas multiformis*, as a close relative of *C. limacisporum* in 2017, led to the proposal of the Pluriformea clade (Hehenberger et al. 2017). *S. multiformis*, a freshwater unicellular holozoan isolated from a pond in Vietnam, shares very little morphological or ecological similarities with *C. limacisporum*. *S. multiformis* can switch between amoeboid and amoeboflagellate cells, as well as form cysts and multicellular clusters. The species is a predator, which ingests the cytoplasmic content of eukaryotic prey cells (Hehenberger et al. 2017). A 225-gene phylogenetic analyses placed Ichthyosporea as an independent basal holozoan lineage, with Pluriformea being recovered in a more derived position; however, the authors acknowledged their datasets could not exclude the possibility of an enlarged Teretosporea clade made up from Ichthyosporea+Pluriformea (Hehenberger et al. 2017).

1.5.3 Tunicaraptor

The second, and most recently described, of the enigmatic holozoan lineages, *Tunicaraptor*, was only discovered in 2020 and is represented by a single species in *T. unikontum* (Tikhonenkov et al. 2020). Isolated from marine waters from the coast of Chile, *T. unikontum* is a small (3.5–5.1 μm in length) predator of unicellular eukaryotes. As with the pluriform *S. multiformis*, *T. unikontum* exhibits a broad range of morphologies. Observed cells are predominantly uniflagellate, with most of the cell enclosed in a rigid theca that possesses long hair-like structures. The flagellum protrudes from the posterior end of the theca, whilst an anterior aperture of the theca exposes a mouth-like structure. To date no other unicellular opisthokonts are known to possess a feeding mouth, although superficially similar structures are widespread amongst biflagellate eukaryotes (Steinert and Novikoff 1960; Verni and Gualtieri 1997). Short filopodia may exude from the protoplast and aggregations of 3–6 cells have been observed, particularly during feeding.

Despite the availability of a full transcriptome, the phylogenetic position of *T. unikontum* has not been established. Indeed, the inclusion of *T. unikontum* genes into phylogenomic datasets leads to instability in phylogenetic reconstructions of Holozoa. Tikhonenkov et al. (2020) presented a consensus Bayesian inference tree placing the *Tunicaraptor* genus as a sister group to a clade comprising Metazoa, Choanoflagellatea, and Filasterea. However, the authors acknowledged that this position was not well supported and their own maximum likelihood phylogeny using the same data united *Tunicaraptor* in a weakly supported clade with Filasterea, Ichthyosporea, and Pluriformea. Additional analyses by Tikhonenkov et al. (2020), varying both the number of genes and taxa, resulted in poorly resolved phylogenies with no consistent placement of *T. unikontum*.

1.5.4 Filasterea

As with most of the taxonomic groups within Holozoa, Filasterea is a clade based upon phylogenetic relationships, with member species harbouring a high level of morphological and ecological diversity. Erected in 2008 by Cavalier-Smith (Shalchian-Tabrizi et al. 2008) through the phylogenetic clustering of two *incertae sedis* holozoan genera, the group has expanded in recent years to its current membership of six taxa.

Capsaspora owczarzaki was the first species discovered, as an endosymbiont of the pulmonate snail *Biomphalaria glabrata* (Owczarzak et al. 1980; Stibbs et al. 1979); however, the species was not named and described for a further 23 years, when it was recognised as a member of Holozoa (Hertel et al. 2002). *C. owczarzaki* protoplasts are typically spherical and amoeboid (3–5 μm in diameter), possessing numerous thin, unbranching filopodia (Hertel et al. 2002; Stibbs et al. 1979). Cells encyst due to overcrowding (Hertel et al. 2002) and mature cells may form aggregates (Sebé-Pedrós et al. 2013). Flagellated cells have never been observed and the sequenced genome revealed the loss of over 80 genes necessary for flagellum formation (Suga et al. 2013). As with all known filastereans, mitochondria possess flattened cristae (Amaral-Zettler et al. 2001; Urrutia et al. 2022).

Within the *B. glabrata* host *C. owczarzaki* predates the sporocysts of the parasitic trematode flatworm *Schistosoma mansoni*, which uses *B. glabrata* as a vector in its lifecycle (Stibbs et al. 1979). A specialised filopodium extends from the protoplast and penetrates the sporocyst when *C. owczarzaki* comes into contact with *S. mansoni* (Stibbs et al. 1979). It is unknown if *C. owczarzaki* provides a pathogenic burden to its snail host, or if the symbiosis is one of mutualism; however, *C. owczarzaki* can be cultured in axenic media, highlighting the existence of nutritional mechanisms other than *S. mansoni* predation (Stibbs et al. 1979).

Two further filastereans were discovered in the mid-1990s, prior to the recognition of the group. *Ministeria marisola* (Patterson et al. 1993) and *M. vibrans* (Tong 1997) are both cosmopolitan marine bacteriovores, which capture prey through phagocytosis. Both species are characterised by a small spherical protoplast (1.5–3.6 μm in diameter) that possesses an array of straight, rigid microvilli (approximately 9 μm in length) that radiate symmetrically around the cell body (Mylnikov et al. 2019; Patterson et al. 1993). *M. vibrans* may attach to surfaces via a peduncle, previously believed to be a derived flagellum (Cavalier-Smith and Chao 2003); however, this appears to be a plastic trait with many *M. vibrans* cultures lacking the peduncle (Mylnikov et al. 2019). The presence of unequivocally flagellated protoplasts has recently been confirmed in *M. vibrans*; however, less than 1% of observed cells in culture were observed to possess a flagellum (Mylnikov et al. 2019).

Three recently discovered species have further expanded the known morphological and ecological diversity of filastereans. Hehenberger et al. (2017) isolated two freshwater unicellular, flagellated eukaryovores from Chile and Vietnam, which they assigned to the novel genus *Pigoraptor*. Whilst being distinct from *C. owczarzaki* as a flagellated species, both *Pigoraptor* species show clear similarities to *C. owczarzaki*, as all three species feed through the capture of the cytoplasmic contents of their eukaryotic prey and have the capacity to encyst as well as form multicellular aggregates (Hehenberger et al. 2017). A second symbiotic, and potentially parasitic, filasterean was recently identified by Urrutia et al. (2022). In contrast to *C. owczarzaki*, which habituates the haemolymph of its snail host, *Txikispora philomaios* is an intracellular symbiont of multiple genera of amphipod crustaceans. Individual infected host cells may harbour up to 10 *T. philomaios* cells. Host cells become necrotic as a result of infection, with the host individuals becoming lethargic and unresponsive to stimuli (Urrutia et al. 2022). *T. philomaios* cells are typically spherical (2–4 μm in diameter) and frequently are enclosed within a cell wall. Multicellular groupings of 3–4 cells are also observed within host cells; however, it is unclear if these are aggregates of unrelated cells or the result of clonal cell division. Naked protoplasts, lacking a cell wall, may produce microvilli and a flagellum (Urrutia et al. 2022).

Phylogenetic analyses have consistently recovered Filasterea as a robust clade. The amoeboid symbiont *C. owczarzaki* has been shown to cluster with the flagellated *Pigoraptor* predators, whilst the parasitic *T. philomaios* appears to be the closest known relative of the free-living *Ministeria* species (Hehenberger et al. 2017; Urrutia et al. 2022).

1.5.5 Choanoflagellatea

With approximately 300 species described in over 50 recognised genera, the choanoflagellates are acknowledged as the most speciose group of unicellular holozoans. The first records of choanoflagellates came in the mid-nineteenth century (Ehrenberg 1831, 1838; von Fresenius 1858); however, due to the limitations of microscopy at the time, the microanatomy of the protoplast was not clearly visible. Choanoflagellate cells are characterised by a distinctive collar of 30–40 actin-based microvilli that surrounds a single apical flagellum (see Carr et al. 2008 for a review of choanoflagellate morphology). These features were not observed until the 1860s when James-Clark described three species, whilst also noting the morphological similarity between choanoflagellate cells and the choanocyte feeding cells of poriferans (James-Clark 1866, 1867). Independently Bütschli (1878), Kent (1878, 1880–1882) and von Stein (1878) all recognised that the various species of recently described collared flagellates could be assigned to a single taxonomic group, which they respectively named as Cylicomastiges, Choanoflagellata, and Craspedomonadina. The latter two names have both persisted in the literature, and Kent's Choanoflagellata has been adapted to provide the group's common name. In the most recent major taxonomic revision, Nitsche et al. (2011) raised the taxonomic rank of the group to the class Choanoflagellatea.

The choanoflagellates are aquatic protists, although some species may also be present in heavily hydrated soils, where they play an important role in microbial communities and food webs as filter feeders (Leadbeater 2015). The beating of their flagellum creates a water current that sweeps bacteria and eukaryotic picoplankton onto the collar of microvilli, which acts as a filter by trapping food particles on the outside of collar (Pettitt 2001). Pseudopodia, which emerge from the base of the collar, engulf trapped food particles; however, captured cells may also be translocated down the microvilli to be phagocytosed closer to the cell body. Food vacuoles formed within the pseudopodia move to the base of the cell where digestion occurs (Fig. 1.4; Leadbeater 1983).

Despite the choanoflagellates showing a similar phylogenetic diversity to Metazoa (Richter et al. 2018), the morphology of their protoplasts shows very little variation. Cells tend to be spherical or ovoid with a collar surrounding a posterior flagellum. All species have a periplast (extracellular coat) that encloses at least some part of the protoplast. One function of the periplast in many choanoflagellate species is to secure the cell to a surface. In motile cells the beating of the flagellum drives locomotion and reduces the volume of water flow over the collar (Lighthill 1976). As a result, swimming cells are less efficient feeders in comparison to sedentary cells, which can filter greater water volumes through each beat of the flagellum (Leadbeater 2008a).

Kent's description (1880–1882) of Choanoflagellata contained three families, namely the Codonosigidae, Phalansteriidae, and Salpingoecidae; however, of these, only the Salpingoecidae remains accepted as a valid choanoflagellate taxon (Nitsche et al. 2011). In addition to Salpingoecidae, the Acanthoecidae and Stephanoecidae are morphologically distinct families (Nitsche et al. 2011).

Three different forms of periplast have been described, which were traditionally the basis underpinning choanoflagellate taxonomy. Two forms of periplast are purely organic based in construction, in the glycocalyx and the theca (reviewed in Leadbeater 2015). The glycocalyx is a mucilaginous and flexible investment, which may be made from either one or two layers of fine

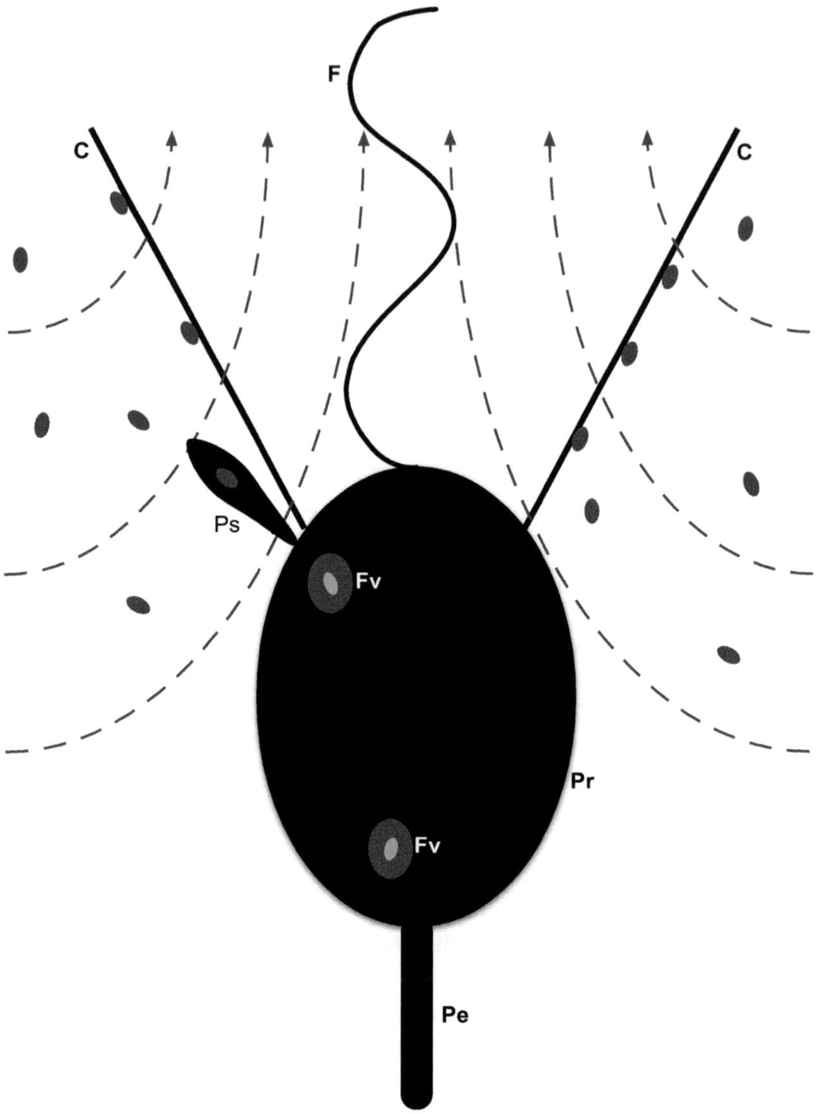

Fig. 1.4 Feeding mechanism of a sedentary choanoflagellate cell. The protoplast (Pr) is secured to a surface by a peduncle (Pe) which extends from an external cell coat. The beating of the flagellum (F) creates a water current (grey dotted arrows) through the microvilli of the collar (C). Food particles (grey ovals) are trapped on the outside of the collar and are phagocytosed by pseudopodia (Ps) which extend from the protoplast at the base of the collar. Food vacuoles (Fv) are transported from the apical pole of the cell to the base of the cell for digestion. Figure based upon Lapage (1925) and Pettitt et al. (2002)

fibrils (Carr et al. 2017). Species across the known diversity of choanoflagellates possess a glycocalyx and the structure appears to be both a universal and ancestral morphological trait of choanoflagellates (Leadbeater 2008a). In many species the glycocalyx extends into a peduncle which secures the protoplast to a surface (Fig. 1.4). Species which exhibit a single cell per peduncle are referred to as *monosigid*, whereas in *codosigid* species multiple clonal cells share the same peduncle. Phylogenetic studies have shown that both the monosigid and codosigid morphologies are polyphyletic, with convergent evolution occurring across the choanoflagellate tree (Carr et al. 2017; Nitsche et al. 2011; Fig. 1.5).

The theca is a more substantial and robust periplast than the glycocalyx, being composed of carbohydrate-based microfibrils (Leadbeater 2008a). The theca exhibits sufficient morphological variation across species for it to have been used as a taxonomic character, with cup, flask, ovoid, and tube shaped thecae being possessed by choanoflagellate species. However, species possessing particular thecae forms do not form monophyletic groups, indicating that theca

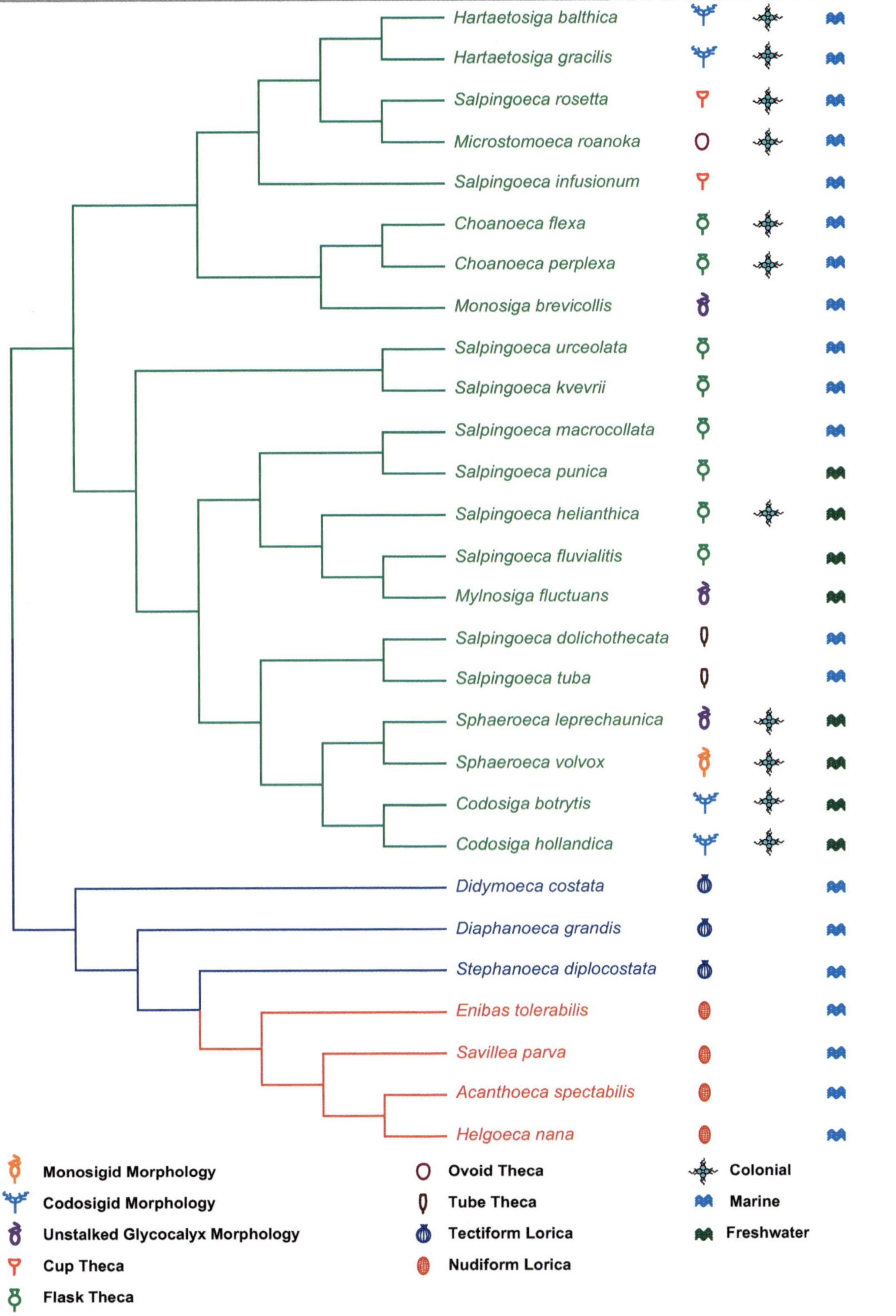

Fig. 1.5 Maximum likelihood 14-gene cladogram of 28 choanoflagellate species. The phylogeny was created with RAxMLGUI 2.0 using a mixed alignment of 4242 nucleotide positions, from partial sequences of SSU and LSU, and 5941 amino acid positions from 40S Ribosomal Protein S8, 60S Ribosomal Protein L10-B, Ribosomal

morphology is a plastic trait with multiple examples of convergent evolution and loss (Carr et al. 2017; Fig. 1.5). Whilst cell division may occur within the confines of the flexible glycocalyx, the rigid nature of the theca means that cells must become amoeboid and emerge before undergoing cell division outside of the thecae (Carr et al. 2008). After cell division, daughter cells from both glycocalyx and theca-bearing species may undergo a motile stage, with collared, flagellated swimming cells (Carr et al. 2008).

Species presenting only a glycocalyx were traditionally assigned to the family Codonosigidae, with theca-bearing species placed into the Salpingoecidae. Molecular phylogenies indicated neither family was monophyletic (Carr et al. 2008; Medina et al. 2003; Steenkamp et al. 2006), and Nitsche et al. (2011) subsumed the polyphyletic Codonosigidae into the paraphyletic Salpingoecidae. The Salpingoecidae is the lone recognised family within the choanoflagellate order Craspedida (originally described by Cavalier-Smith 1997), which contains all species that possess a purely organic-based periplast.

Codonosigidae and Salpingoecidae were both described in Kent's pioneering taxonomic work of the 1880s; however, the family Acanthoecidae was not described until 1965, under botanical nomenclature, with the name Acanthoecaceae (Norris 1965). Due to the diversity of species recognised within the group, the taxon was raised from a family to the order Acanthoecida by Nitsche et al. (2011). The order is characterised by the lorica, a distinctive periplast comprising two layers of silica costal strips which form a basket-like structure around the cell (Fig. 1.6a; Leadbeater et al. 2009). Organic microfibrils, similar to those present in the theca, are frequently arranged between the costal strips; these facilitate the adhesion of the protoplast to the lorica and, in some species, act to funnel water over the collar (Leadbeater 2008a; Leadbeater et al. 2009).

In contrast to the glycocalyx and theca, loricae show considerable morphological variation between species and are considered reliable diagnostic taxonomic characters (Nitsche et al. 2017; Schiwitza and Nitsche 2021). Loricate taxa can be divided into two groups based upon their mechanisms of cell division, as well as the structure of their loricae. *Nudiform* taxa are similar to craspedid species in that after cell division one daughter cell remains with the original periplast, whilst the second daughter cell will undergo a "naked" motile dispersal stage (Manton et al. 1981). The motile stage is relatively brief in nudiform species, compared to that observed in craspedids, and swimming cells rapidly settle onto a surface and construct a new lorica from costal strips that have accumulated in membrane-bound vesicles within the protoplast (Leadbeater 2008b). The loricae of nudiform species possess both longitudinal and helical costal strips but lack transverse rings of strips (Leadbeater et al. 2008). In contrast to the nudiforms, *tectiform* species do not have a naked dispersal stage. Prior to cell division the parental cell exocytoses all of the costal strips required to produce a new, second lorica and stores the strips at the top of its collar (Manton et al. 1981). Cell division occurs within the parental lorica and immediately after cytokinesis one daughter cell exits the lorica, taking the previously deposited costal strips and assembles

Fig. 1.5 (continued) Protein S5, Rps15A, EFL, EF1A, Hsp70, Hsp90, RNA Pol, RNA Pol II, TubA, and TubB. Each gene was aligned individually in MAFFT 7.309 and then concatenated. For nucleotide sites, the TN93 model (Tamura and Nei 1993) was employed with maximum likelihood-derived base frequencies and proportion of invariant sites, as well as a gamma distribution of rate variation. For amino acid sites, the LG model (Le and Gascuel 2008) was employed with maximum likelihood-derived base frequencies, as well as a gamma distribution of rate variation. Parameters were calculated through optimization in the RAxMLGUI. Craspedid species are shown in green and nudiform acanthoecid species are shown in red and tectiform acanthoecids are in blue. The key defines morphological and ecological traits

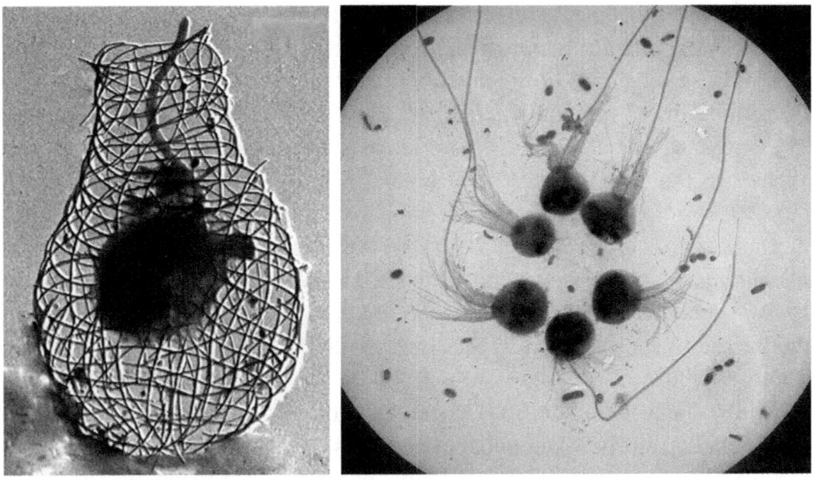

Fig. 1.6 Morphological variation in choanoflagellates. (**a**) The acanthoecid choanoflagellate *Savillea parva* surrounded by a lorica of silica strips. (**b**) Clonal colony of the choanoflagellate *Salpingoeca punica*. The flagella and collars of all cells face outward to maximise prey capture. Pictures presented with kind permission from Barry Leadbeater

its own new lorica (Leadbeater 2010; Manton et al. 1981). Tectiform loricae are often more elaborate than nudiform loricae in their construction, with costal strips being organised in longitudinal, helical, and transverse ring arrangements. The presence of transverse rings provides greater structural stability than helical strips, which has resulted in many tectiform species evolving lightweight, open baskets, allowing species to become planktonic and free-floating within the water column (Leadbeater 2010).

An excess of 100 tectiform species have been described; however, fewer than 10 nudiform species are recognised (Nitsche et al. 2017; Schiwitza and Nitsche 2021). Within Acanthoecida, the nudiforms and tectiforms have both been assigned family status, with the former described as Acanthoecidae and the latter as Stephanoecidae (Nitsche et al. 2011). Whilst both taxa are recognised as distinct families, the phylogenetic relationship between the two is far from clear. The use of nucleotide or amino acid datasets, as well as different phylogenetic methodologies, may result in either two monophyletic sister families or a monophyletic Acanthoecidae being nested within a paraphyletic Stephanoecidae (Carr et al. 2008, 2017). Carr and Leadbeater (2022) proposed that the monophyly of Stephanoecidae is an artefact produced due to convergent phylogenetic signals present in synonymous third codon positions and through a 14-gene phylogeny showed that the nudiform species evolved within a lineage of tectiform choanoflagellates.

Over 70 species of choanoflagellate have publicly available gene sequences; however, the majority of described species lack sequence data and cannot be placed into molecular phylogenies. Environmental DNA studies have indicated that a number of diverse choanoflagellate clades exist that currently have no known representative species (del Campo and Massana 2011; del Campo and Ruiz-Trillo 2013). However, as eDNA surveys tend to be based upon a single gene, the phylogenetic support for the novel groups is frequently weak and fewer novel lineages are recovered when more stringent analyses are undertaken (Carr et al. 2017).

A large number of choanoflagellate species have been observed to switch from a unicellular state to colonial forms (Fig. 1.6b). The number of species known to have the capacity to form colonies is probably an underestimate of the true number, as environmental conditions appear to play a major role in colony formation and laboratory culture conditions may not always provide the necessary cues for coloniality (Carr et al. 2017; Ireland et al. 2020). The morphologies of colonies vary between choanoflagellate taxa and may even vary within the same species (Dayel et al. 2011). Sedentary species possessing either a glycocalyx or theca may generate colonies on a single peduncle, whilst free-swimming species may produce spherical globular colonies or raftlike chains of cells (Dayel et al. 2011). Such colonies are clonal, rather than aggregate, in nature (Dayel et al. 2011) and also show evidence for cellular differentiation (Laundon et al. 2019). The presence of a spacious, restrictive lorica appears to limit the capacity of acanthoecid species to form connections between individual protoplasts (Carr et al. 2017). However, *Diaphanoeca sphaerica* and *Parvicorbicula socialis* produce colonies through the linkage of the loricae of individual cells (Leadbeater 2015; Thomsen 1982).

Whilst James-Clark (1867) speculated on a close relationship between choanoflagellates and metazoans, the exact relationship between choanoflagellates and other eukaryotic groups was a long source of considerable debate. Since their discovery, it has been speculated that the choanoflagellates may be highly simplified metazoans (Maldonado 2004), a paraphyletic grouping ancestral to both Metazoa and Fungi (Cavalier-Smith 1987), the sister group to Ichthyosporea or *Corallochytrium* (Cavalier-Smith and Chao 2003; Medina et al. 2003; Mendoza et al. 2002), or even a form of algae (Bourrelly 1968; Chadefaud 1960). Early molecular phylogenies lacked phylogenetic support or omitted important holozoan groups (Baker et al. 1999; Jiménez-Guri et al. 2007; Medina et al. 2003; Steenkamp et al. 2006; Wainright et al. 1993). The first phylogenetic analysis to include the major recognised lineages within the Holozoa recovered the choanoflagellates as the sister group of Metazoa (Carr et al. 2008), and this relationship has been consistently recovered in subsequent phylogenomic studies (Hehenberger et al. 2017; Tikhonenkov et al. 2020). Molecular clock analyses indicate that the last common ancestor of choanoflagellates and metazoans existed approximately one billion years ago (Berbee and Taylor 2010; Parfrey et al. 2011).

Based upon traits shared between both Choanoflagellatea and Metazoa, a number of putatively ancestral characteristics have been recovered. Carr et al. (2008) proposed the last common ancestor of both groups was a marine organism that possessed a microvilli collar and apical flagellum employed for filter-feeding. Due to the widespread nature of coloniality in choanoflagellates, it is also possible that the ancestor of both metazoans and choanoflagellates may also have been able to transition between unicellular and colonial states (Carr et al. 2008). The cytoplasmic bridges formed between cells in choanoflagellate colonies resemble those present between metazoan cells and similar bridges may have been present in a putative colonial ancestor of both groups (Dayel et al. 2011).

Based upon 16 species, Carr et al. (2008) highlighted a marine origin of the choanoflagellates, with freshwater species falling into a single phylogenetic group. Subsequent phylogenetic studies have confirmed one major freshwater invasion occurred early in the evolutionary history of choanoflagellates. There have been a limited number of more recent, minor freshwater incursions which appear to involve individual species (Carr et al. 2017; Nitsche et al. 2011; Paul 2012). Marine-freshwater transitions in the group appear to be rare, which is not uncommon in unicellular eukaryotes (Logares et al. 2009), with only a single species, *Salpingoeca macrocollata*, known to have reverted to a marine habitat from freshwater ancestors (Carr et al. 2017).

1.5.6 Metazoa

Uniquely amongst the opisthokonts, Metazoa only contains species that have multicellular stages within their lifecycles. Species exhibit multiple epithelial layers which are linked by connective tissue that generally contains collagen fibres (Cavalier-Smith 1998b). Mitochondrial cristae are flat and metazoans typically possess a 13–19 kb circular mitochondrial chromosome which encodes a suite of 37 highly conserved protein-coding and non-coding RNA genes (Burger et al. 2003). The reduction in mitochondrial genome size must have occurred in the metazoan stem group and contrasts with the large genomes present in Choanoflagellatea, Filasterea, and Ichthyosporea (Burger et al. 2003; Suga et al. 2013). As noted in Sect. 1.4, a similar reduction in the mitochondrial genome has also occurred within fungi; however, fungal mitochondrial genomes are far more variable in length and gene content in comparison to metazoan genomes (Bullerwell and Lang 2005).

Despite being highly conserved across most metazoans, the ancestral state of a reduced circular mitochondrial chromosome has been independently lost in a number of cnidarian lineages. Linear mitochondrial chromosomes, with telomeres, have evolved in both medusozoan and anthozoan cnidarians (Smith et al. 2011; Stampar et al. 2019). Furthermore, the anthozoan *Protanthea simplex* has its mitochondrial genome divided into two circular mito-chromosomes (Dubin et al. 2019), whilst the myxozoan *Henneguya salminicola* possesses a mitochondria-related organelle but lacks both a mitochondrial genome and many of the genes required for aerobic respiration (Yahalomi et al. 2020).

The number of recognised phyla within Metazoa is disputed, with between 30 and 35 being recognised (Erwin and Valentine, 2013). A clear division in phyla is observed in the number of cell layers present within organisms. In four phyla, the Porifera (sponges), Ctenophora (comb jellies), Placozoa (a phylum consisting of three described marine species possessing highly simplified bodyplans), and Cnidaria (corals, jellyfish, sea anemones, and sea pens), species possess two layers of cells, generally referred to as the endoderm (internal layer) and ectoderm (outer layer), giving rise to the informal name *diploblasts* (Kobayashi et al. 1996).

All other metazoan phyla have body plans that contain a third cell layer, the mesoderm, located between the endoderm and ectoderm, resulting in them being referred to as *triploblasts* (Christen et al. 1991). In at least part of their lifecycles triploblasts show left:right (bilateral) symmetry across the sagittal plane of their body plans, resulting in an alternative, formalised, name of Bilateria (Hatschek 1888). Bilateral symmetry is also present in a number of anthozoan and hydrozoan cnidarians, whilst radial symmetry is observed in ctenophores, as well as some cnidarians and poriferans (Hyman 1940; Malakhov 2016).

Within diploblasts, the internal and external cells layers are separated by a gelatinous matrix termed mesophyll in poriferans (Bonasoro et al. 2001) and mesoglea in both cnidarians and ctenophores (Hyman 1940). The mesoglea of cnidarians and ctenophores contains both muscle cells and a simple neural network (Hyman 1940), but neither of these tissue types are present in either placozoans or poriferans (Burkhardt and Sprecher 2017). A further similarity between cnidarians and ctenophores is the presence of a blind gut, whereas both the placozoans and poriferans lack any form of gut (Hyman 1940). The *Hox* gene family plays a major role in body patterning during bilaterian development, establishing cell identity along the anterior-to-posterior axis (Ferrier and Holland 2001). Cnidarians and placozoans also possess a limited repertoire of *Hox* or *Hox*-like genes, but orthologues have not been identified to date in any poriferan or ctenophore (Ramos et al. 2012; Moroz et al. 2014; Pastrana et al. 2019; Srivastava et al. 2010).

Molecular phylogenetics has yet to produce an unequivocal tree of metazoan phyla; however, it is clear the bilaterians form a robustly supported clade. The Porifera, Ctenophora, Placozoa, and Cnidaria are recovered at the base of the metazoan tree and, as such, can be described as early-branching metazoans. The four early-branching phyla do not form a single clade and their branching order remains unresolved. Cnidaria is now recognised as the sister taxon to Bilateria (Philippe et al. 2019; Rouse et al. 2016; Cannon et al. 2016). Similarities in embryonic gene expression patterns between the groups, as well as the presence of bilateral symmetry in a number of cnidarian species, have led to speculation that the common ancestor of Bilateria and Cnidaria exhibited bilateral symmetry and that the condition has been lost on multiple occasions within cnidarians (Matus et al. 2006). Most phylogenies that contain Placozoa only include one of the three described species, in *Trichoplax adhaerens*; however, the phylum is consistently recovered in phylogenomic studies as the sister group to the clade of Bilateria+Placozoa (Philippe et al. 2019; Pisani et al. 2015).

Molecular phylogenies of the earliest branching metazoan group have proved to be extremely controversial (Telford 2016; Moroz and Halanych 2016). Recent large phylogenomic studies have recovered either the poriferans (Kapli and Telford 2020; Pick et al. 2010; Pisani et al. 2015; Simion et al. 2017) or the ctenophores (Dunn et al. 2008; Moroz et al. 2014; Ryan et al. 2013; Whelan et al. 2015) as the first branching metazoan lineage. The internal branch leading to Ctenophora in phylogenetic trees is considerably longer than the branches leading to other metazoan phyla, giving rise to concerns that the recovery of ctenophores at the base of the metazoans may be due to long-branch attraction (Kapli and Telford 2020; Simion et al. 2017; Telford 2016). The effects of long-branch artefacts can be weakened through the use of more appropriate substitution models in phylogenetic reconstruction. In general, site-homogenous amino acid substitution models tend to place ctenophores as the basal lineage, whilst more sophisticated site-heterogenous models recover the poriferans as the first branching group. At present, however, it is not clear whether either the homogenous or heterogenous models more accurately estimate amino acid changes within deep metazoan evolution.

This controversy is not simply one of taxonomy, as the correct identification of the first branching metazoan group has a considerable impact on the reconstruction of ancestral traits. The ctenophore-first scenario indicates either that neuronal cells, musculature, and possibly a gut were present in the metazoan last common ancestor (LCA) or that these characters have evolved independently in different metazoan phyla. The distinctive nature of neurons and striated muscle in ctenophores may indicate convergent evolution within metazoans (Moroz and Halanych 2016); however, such convergence could have occurred under both the ctenophore-first and poriferan-first scenarios. The poriferan-first model would suggest that the metazoan LCA was a relatively simple organism which existed prior to the evolution of true tissue layers. The similar morphologies of choanoflagellate protoplasts and choanocytes have been argued to be homologous and therefore present in the metazoan LCA. However, a detailed study of cell ultrastructure and flagella beating, albeit based only upon two species, has raised the possibility that the collared flagellated cells may not be homologous but could have independently evolved in the ancestors of both extant choanoflagellates and poriferans (Mah et al. 2014). Furthermore, gene expression profiles across different poriferan cell types and unicellular holozoans also failed to recover evidence for homology between choanoflagellates and choanocytes (Sogabe et al. 2019). However, as remarked upon by Laundon et al. (2019), the two cell types have been evolving under different evolutionary pressures since the choanoflagellate and metazoan lineages diverged; therefore, differences between choanoflagellates and choanocytes should not be unexpected even if they are homologous.

The majority of metazoan diversity and phyla are assigned to Bilateria. Based upon embryonic development patterns, Grobben (1908) divided Bilateria in protostomes and deuterostomes. During gastrulation, in the former group the primary opening (blastopore) typically develops into the mouth, whilst in the latter group the blastopore develops into the anus. Multigene phylogenies have generally recovered both Deuterostomia and Protostomia as monophyletic (Bourlat et al. 2008; Dunn et al. 2008; Laumer et al. 2019); however, a number of recent phylogenomic studies indicate that the deuterostomes may be paraphyletic (Kapli and Telford 2020; Kapli et al. 2021; Philippe et al. 2019). In these studies, the Chordata (metazoans that possess a notochord) can either cluster with Protostomia, resulting in deuterostome paraphyly, or within a monophyletic Deuterostomia that is separated from Protostomia by a very short internal branch in trees. If the latter topology is correct, it indicates that the deuterostome radiation commenced only a very short time after the origin of Bilateria. In contrast, deuterostome paraphyly would point to the ancestral bilaterian possessing deuterostome traits, such as blastopore fate and the presence of pharyngeal slits, which would have been subsequently lost in the derived protostomes.

1.6 Genomic Evolution

The advent of high-throughput technologies has made sequencing whole genomes relatively rapid and low cost. RNA-Seq, the sequencing of transcriptomes, is an alternative approach that may be employed when obtaining pure genomic DNA is technically challenging or when sequencing intergenic DNA is not required. As a result, a large body of data now exists on the genomes of the unicellular relatives of both fungi and metazoans. These data are providing new insights into the origins of multicellularity.

Fungi are characterised by the presence of both chitin and 1,3-β-glucan in their cell walls, and the gene complements of unicellular holomycotans have allowed the evolution of these cell wall components to be reconstructed. Chitin synthases are present in fungi and all opisthosporid lineages, indicating chitin walls are an ancestral character (James and Berbee 2012; Torruella et al. 2018). The early-branching opisthosporids, Microsporidia and Cryptomycota, possess chitin, but only in the walls of their infective or resting cyst stages (Torruella et al. 2018). The infective cysts of aphelids also contain chitin and their genomes encode a broad range of genes involved in chitin synthesis, modification, and degradation, some of which may have been acquired through horizontal gene transfer (Torruella et al. 2018). Microsporidians and cryptomycotans lack 1,3-β-glucan, as do chytrid fungi. However, the presence of 1,3-β-glucan synthase genes in the transcriptome of the aphelid *Paraphelidium tribonemae* (Torruella et al. 2018) indicates that 1,3-β-glucan was present in the cell walls of the ancestor of fungi and aphelids, with its subsequent loss in the chytrids. Chang et al. (2015) highlighted a major expansion in pectinase gene diversity within fungal lineages, which they proposed occurred after the divergence of the Cryptomycota lineage in Holomycota. Their explanation for this expansion was that ancient fungi evolved to digest the cell walls of algae. The putative sister group to Fungi is Aphelida, a group made up of algal parasites, and the findings from the Chang et al. (2015) study would indicate that the last common ancestor of Aphelida and Fungi was a flagellated, freshwater parasite of archaeplastid algae. The aphelids have retained this ancestral nutritional mode, as have some extant fungal lineages, whilst other fungi have diversified to parasitise other host groups or become saprotrophs and symbionts.

The parasitic lifecycles of microsporidians and cryptomycotans appear to have resulted in genome reduction; this is seen in terms of genome size and gene content, with both groups showing a loss of genes involved in metabolism (Bass et al. 2018; James et al. 2013). Gene loss has occurred across the microsporidian crown group, with different lineages having lost different ancestral genes; genes involved in processes such as lipid metabolism, glycolysis, and mRNA splicing have been independently lost on multiple occasions across the group (Wadi and Reinke 2020). Microsporidian genomes have also undergone considerable expansions of copies in some gene families, through both individual gene and whole-genome duplication events, resulting in species within the same genus having large numbers of different genes (Reinke et al. 2017). Despite being closely related parasites, the aphelids do not appear to have undergone a similar loss to those observed in cryptomycotans and microsporidians, with *P. tribonemae* showing a similar diversity of metabolic genes to fungi (Torruella et al. 2018); however, further genome sequences are required to determine if *P. tribonemae* is typical of aphelids. In general, holomycotan genomes are characterised by gene reduction, with notable losses of genes involved in signal transduction. The repertoire of metabolism genes began to expand in the common ancestors of Fungi and Aphelida, a process which continued within the fungal stem lineage (Ocaña-Pallarès et al. 2022).

Gene complements from all major holozoan lineages have been accrued over the last decade (Denbo et al. 2018; Fairclough et al. 2013; Hehenberger et al. 2017; López-Escardó et al. 2019; Richter et al. 2018; Suga et al. 2013). One striking finding from these studies is that many gene families previously believed to be metazoan-specific are now known to have much greater antiquity with origins deeper in Holozoa. Both the metazoan and choanoflagellate stem groups underwent considerable expansions of gene number, as both lineages gained approximately 2000 novel families (Richter et al. 2018), with a burst of innovation in genes involved in both gene transcription and signal transduction occurring in the metazoan stem group (Ocaña-Pallarès et al. 2022). The expansion of the metazoan gene complement appears to have been through both gene duplication and the rearrangement of existing domains, rather than a result of evolving novel protein domains (López-Escardó et al. 2019; Richter et al. 2018). Holozoan genome evolution is a dynamic process, as, in addition to the large-scale gene gain that occurred, over 1500 gene families, including many involved in energy production, as well as the metabolism of amino acids, carbohydrates, and lipids, were lost in premetazoan genomes (Ocaña-Pallarès et al. 2022). Gene loss has continued in metazoans as, in a screen of crown-group taxa, fewer than 40 of the novel genes that evolved in the stem group were found to be universally retained (Richter et al. 2018). As a result of these changes, extant metazoans, unlike fungi, have gene complements which are distinct from their protistan relatives (Ocaña-Pallarès et al. 2022). As the number of genomes from early-branching holomycotan and holozoan genomes increases, ancestral reconstructions of gene loss and gain will become more accurate and fine-tuned in comparison to current studies.

The increased volume of genomic data now available has highlighted the role that horizontal gene transfer has played in the holozoan evolution. Hundreds of genes identified in choanoflagellates, filastereans, ichthyosporeans, and pluriformeans have been shown to be acquired from donor species outside of Opisthokonta (Betat et al. 2015; Carr et al. 2010; Matriano et al. 2021; Southworth et al. 2019; Yue et al. 2013). The identified donor species consist of bacteria and algal alveolates, archaeplastids, haptophytes, rhizarians, and stramenopiles, all of which are prey items of predatory unicellular holozoans. Horizontal transfer may occur if partially degraded chromosomal DNA escapes from the food vacuoles of holozoan predators, passes into the nucleus, and then integrates into chromosomes. Transferred genes have been shown to provide new functions to the recipient cell and, on some occasions, replace ancestral holozoan homologues (Carr and Leadbeater 2022; Yue et al. 2013). Highlighting the dynamic nature of genome evolution, horizontally acquired genes may subsequently be lost in some descendent lineages whilst retained in others (Carr et al. 2010; Suga et al. 2013).

1.7 Conclusions

1.7.1 Molecular Phylogenetic Analyses of Opisthokonta and the Eukaryotic Tree of Life

Advances in both DNA sequencing technology and phylogenetic methodologies have allowed many of the deeper branches in the eukaryotic tree to be resolved. This greater resolution has been somewhat offset by the discovery that a number of orphan lineages, such as Hemimastigophora and Malawimonadida, appear to fall outside of the generally recognised supergroups (Brown et al. 2018; Lax et al. 2018). The ongoing failure to locate the root of eukaryotic tree means that the composition of the Amorphea clade, to which the opisthokonts belong, remains unclear. Despite this uncertainty, the placements of the biflagellate Apusomonadida and uniflagellate Breviatea as the closest relatives of Opisthokonta are now broadly recognised.

Within Opisthokonta, the view that the root lies between Holomycota and Holozoa remains unchallenged, despite numerous novel lineages being identified since Holomycota and Holozoa were originally proposed. Areas of uncertainty remain in Holomycota, with the monophyly or paraphyly of Opisthosporidia still to be robustly resolved (Karpov et al. 2014a, b; Torruella et al. 2018). As a result, the sister group to Fungi is not universally agreed upon. Despite major revisions to the groups positioned in Holozoa, the choanoflagellates appear to be the closest relatives of Metazoa. The recent inclusion of the predatory *Tunicaraptor* in holozoan phylogenies disrupts the support for the deeper branches in Holozoa (Tikhonenkov et al. 2020); however, it is at present unclear if *Tunicaraptor* sequences are introducing phylogenetic artefacts or allowing a weak, but genuine, phylogenetic signal to be observed within trees. Furthermore, the earliest branching lineages in Metazoa and "classical" Fungi have not been identified, preventing reliable ancestral state reconstruction in both of these major multicellular kingdoms.

1.7.2 Determining Opisthokont Diversity

Like most, if not all, eukaryotic supergroups, the phylogenetic diversity of the opisthokonts remains uncertain. Since the turn of the century, novel lineages have been identified in both Holomycota and Holozoa, with the discovery of small predatory holozoans and slowly evolving microsporidians (Bass et al. 2018; Hehenberger et al. 2017; Tikhonenkov et al. 2020). Furthermore, eDNA studies have highlighted putative enigmatic clades of opisthokonts that have no recognised representatives. Our increasing knowledge of opisthokont diversity is well exemplified by the class Filasterea. Since its description in 2008, the number of known species has doubled; however, phylogenies which include eDNA sequences have identified clades containing unknown aquatic and terrestrial filastereans (Urrutia et al. 2022), highlighting the presence of taxa that have yet to be isolated and described.

The phylogenetic support of eDNA trees is however often weak; therefore, whilst such studies are useful for highlighting novel diversity, their potential for producing accurate phylogenetic placements is more limited. Novel approaches, such as the use of eDNA sequences as probes to identify cells of unknown species (Jones et al. 2011) and single-cell genomics (López-Escardó et al. 2019), may assist in expanding knowledge of opisthokont diversity.

1.7.3 Reconstructing Opisthokont Evolution and the Multiple Origins of Multicellularity

The last decade has seen an explosion in the sequencing of opisthokont genomes and transcriptomes. Whilst the emphasis has been on the multicellular Fungi and Metazoa, the increasing data available from their unicellular relatives have begun to shine a light on the origins of both kingdoms. Candidates for the closest relatives of Fungi and Metazoa have now appear to have been identified, in the Aphelida and Choanoflagellatea, respectively (Carr et al. 2008; Galindo et al. 2021). With two whole genomes and 20 transcriptomes, a greater volume of gene data is currently available for the choanoflagellates (Richter et al. 2018), whilst at the time of writing only a single publicly available transcriptome, from *Paraphelidium tribonemae*, has been generated for the aphelids (Torruella et al. 2018). Comparative genomic and morphological studies have highlighted previously unidentified ancestral traits (Booth et al. 2018; Karpov et al. 2019a; Southworth et al. 2018; Torruella et al. 2018); however, many uncertainties remain. For example, whilst coloniality is widespread across craspedid choanoflagellates it is not clear if this trait was ancestral to choanoflagellates. Morphological similarities between their cytoplasmic bridges, as well as choanoflagellate rosette colonies and metazoan larval blastulae, raise the possibility that the common ancestor of metazoans and choanoflagellates may have exhibited facultative coloniality (Carr et al. 2008; Brunet and King 2017). Intriguingly, bacterial-mediated coloniality is observed in both Holomycota and Holozoa, whilst bacteria have been shown to be essential to the normal development of some metazoans (Fraune and Bosch 2010). Future work may uncover whether this was an ancestral character, or if similar mechanisms have evolved convergently in the two major branches of Opisthokonta.

Functional studies are required to show if there are any common genetic pathways in the development of metazoan bodies and choanoflagellate colonies. Furthermore, such studies may also identify shared developmental pathways across multicellularity in Choanoflagellatea, Filasterea, Fungi, Ichthyosporea, Metazoa, and nucleariid amoebae which may have evolved in the opisthokont stem. The advent of transfection techniques in unicellular opisthokonts (Booth et al. 2018; Kożyczkowska et al. 2021; Parra-Acero et al. 2018) promises to reveal the function of fungal and metazoan multicellularity genes in their protistan relatives.

The ongoing drive to sequence genomes across the eukaryotic tree of life will in the next few years provide accurate gene complements for the LCAs of Holomycota, Holozoa, and Opisthokonta. In combination with studies of gene expression and function, these gene complements can be expected to identify the origins of multicellularity in Fungi and Metazoa, as well as colony formation in opisthokont protists.

References

Adl SM, Simpson AGB, Farmer MA et al (2005) The new higher level of classification of eukaryotes with emphasis on the taxonomy of protists. J Eukaryot Microbiol 52:399–451

Adl SM, Simpson AGB, Lane CE, Lukeš J, Bass D et al (2012) The revised classification of eukaryotes. J Eukaryot Microbiol 59:429–493

Adl SM, Bass D, Lane CE et al (2019) Revisions to the classification, nomenclature, and diversity of eukaryotes. J Eukaryot Microbiol 66:4–119

Amaral-Zettler LA, Nerad TA, O'Kelly J, Sogin ML (2001) The nucleariid amoebae: more protists at the animal–fungal boundary. J Eukaryot Microbiol 48: 293–297

Baker GC, Beebee TJ, Ragan MA (1999) *Prototheca richardsi*, a pathogen of anuran larvae, is related to a clade of protistan parasites near the animal–fungal divergence. Microbiology 145:1777–1784

Balbiani G (1882) Sur les microsporidies ou psorospermies des Articulés. C R Acad Sci 95:1168–1171

Baldauf SL, Palmer JD (1993) Animals and fungi are each others closest relatives: congruent evidence from multiple proteins. Proc Natl Acad Sci U S A 90:11558–11562

Bass D, Czech L, Williams BAP et al (2018) Clarifying the relationships between Microsporidia and Cryptomycota. J Eukaryot Microbiol 65:773–782

Bauer R, Garnica S, Oberwinkler F et al (2015) Entorrhizomycota: A new fungal phylum reveals new perspectives on the evolution of Fungi. PLoS One 10: e0128183

Beebee TJC, Wong AL-C (1992) *Prototheca*-mediated interference competition between anuran larvae operates by resource diversion. Physiol Zool 65:815–831

Benny GL, O'Donnell K (2000) *Amoebidium parasiticum* is a protozoan, not a Trichomycete. Mycologica 92: 1133–1137

Berbee ML, Taylor JW (2010) Dating the molecular clock in fungi – how close are we? Fungal Biol Rev 24:1–16

Bergsten J (2005) A review of long-branch attraction. Cladistics 21:163–193

Betancur-R R, Naylor GJP, Ortí G (2014) Conserved genes, sampling error, and phylogenomic inference. Syst Biol 63:257–262

Betat H, Mede T, Tretbar S et al (2015) The ancestor of modern Holozoa acquired the CCA-adding enzyme from Alphaproteobacteria by horizontal gene transfer. Nucleic Acids Res 43:6739–6746

Bojko J, Reinke AW, Stentiford GD et al (2022) Microsporidia: a new taxonomic, evolutionary, and ecological synthesis. Trends Parasitol 38:642–659

Bonasoro F, Wilkie IC, Bavestrello G et al (2001) Dynamic structure of the mesohyl in the sponge *Chondrosia reniformis* (Porifera, Demospongiae). Zoomorphology 121:109–121

Booth DS, Szmidt-Middleton H, King N (2018) Transfection of choanoflagellates illuminates their cell biology and the ancestry of animal septins. Mol Biol Cell 29: 3026–3038

Borteiro C, Baldo D, Maronna MM et al (2018) Amphibian parasite of the Order Dermocystida (Ichthyosporea): current knowledge, taxonomic review and new records from Brazil. Zootaxa 4461:499–518

Bourlat SJ, Nielsen C, Economou AD, Telford MJ (2008) Testing the new animal phylogeny: a phylum level molecular analysis of the animal kingdom. Mol Phylogenet Evol 49:23–31

Bourrelly P (1968) Les algues d'eau douce. Tome II: Les algues jaunes et brunes. N. Boubée et Cie, Paris

Brinkmann H, van der Giezen M, Zhou Y et al (2005) An empirical assessment of long-branch attraction artefacts in deep euakaryotic phylogenomics. Syst Biol 54:743–757

Brown MW, Spiegel FW, Silberman JD (2009) Phylogeny of the "forgotten" cellular slime mold, *Fonticula alba*, reveals a key evolutionary branch within Opisthokonta. Mol Biol Evol 26:2699–2709

Brown MW, Sharpe SC, Silberman JD et al (2013) Phylogenomics demonstrates that breviate flagellates are related to opisthokonts and apusomonads. Proc R Soc B 280:20131755

Brown MW, Heiss AA, Kamikawa R et al (2018) Phylogenomics places orphan protistan lineages in a novel eukaryotic super-group. Genome Biol Evol 10: 427–433

Brunet T, King N (2017) The origin of animal multicellularity and cell differentiation. Dev Cell 43:124–140

Bullerwell CE, Lang BF (2005) Fungal evolution: the case of the vanishing mitochondrion. Curr Opin Microbiol 8:362–369

Burger G, Forget L, Zhu Y et al (2003) Unique mitochondrial genome architecture in unicellular relatives of animals. Proc Natl Acad Sci U S A 100:892–897

Burkhardt P, Sprecher SG (2017) Evolutionary origin of synapses and neurons – bridging the gap. BioEssays 39:1700024

Burki F, Shalchian-Tabrizi K, Minge M et al (2007) Phylogenomics reshuffles the eukaryotic supergroups. PLoS One 8:e790

Burki F, Roger AJ, Brown MW, Simpson AGB (2020) The new tree of eukaryotes. Trends Ecol Evol 35:43–55

Bütschli O (1878) Beiträge zur Kenntnis der Flagellaten und einiger verwandter Organismen. Z Wiss Zool Abt A 30:219–281

Cafaro MJ (2005) Eccrinales (Trichomycetes) are not fungi, but a clade of protists at the early divergence of animals and fungi. Mol Phylogenet Evol 35:21–34

Cannon JT, Cossermelli Vellutini B, Smith J III et al (2016) Xenacoelomorpha is the sister group to Nephrozoa. Nature 530:89–93

Carr M, Leadbeater BSC (2022) Re-evaluating loricate choanoflagellate phylogenetics: molecular evidence points to the paraphyly of tectiform species. Protist 173:125924. https://doi.org/10.1016/j.protis.2022.125924

Carr M, Leadbeater BSC, Hassan R et al (2008) Molecular phylogeny of choanoflagellates, the sister group to Metazoa. Proc Natl Acad Sci U S A 105:16641–16646

Carr M, Leadbeater BSC, Baldauf SL (2010) Conserved meiotic genes point to sex in the choanoflagellates. J Eukaryot Microbiol 57:56–62

Carr M, Richter D, Fozouni P et al (2017) A six-gene phylogeny provides new insights into choanoflagellate evolution. Mol Phylogenet Evol 107:166–178

Cavalier-Smith T (1983) A 6-kingdom classification and a unified phylogeny. In: Schenk HEA, Schwemmler WS (eds) Endocytobiology. II. Intracellular space as oligogenetic. Walter de Gruyter, Berlin, pp 1027–1034

Cavalier-Smith T (1987) The origin of fungi and pseudofungi. In: Rayner ADM, Brasierand M, Moore D (eds) Evolutionary biology of fungi. Cambridge University Press, Cambridge, pp 339–353

Cavalier-Smith T (1993) Kingdom protozoa and its 18 phyla. Microbiol Rev 57:953–994

Cavalier-Smith T (1997) Amoeboflagellates and mitochondrial cristae in eukaryote evolution: megasystematics of the new protozoan subkingdoms Eozoa and Neozoa. Arch Protistenkd 147:237–258

Cavalier-Smith T (1998a) Neomonada and the origin of animals and fungi. In: Coombs GH, Vickerman K, Sleigh MA, Warren A (eds) Evolutionary relationships among Protozoa. Chapman & Hall, London, pp 375–407

Cavalier-Smith T (1998b) A revised six kingdom system of life. Biol Rev 73:203–266

Cavalier-Smith T (2002) The phagotrophic origin of eukaryotes and phylogenetic classification of protozoa. Int J Syst Evol Biol 52:297–354

Cavalier-Smith T, Allsopp MTEP (1996) *Corallochytrium*, an enigmatic non-flagellate protozoan related to choanoflagellates. Eur J Protistol 32: 306–310

Cavalier-Smith T, Chao EE-Y (2003) Phylogeny of Choanozoa, Apusozoa, and other protozoa and early eukaryote megaevolution. J Mol Evol 56:540–563

Cavalier-Smith T, Chao EE-Y, Oates B (2004) Molecular phylogeny of Amoebozoa and the evolutionary significance of the unikont *Phalansterium*. Eur J Protistol 40:21–48

Cavalier-Smith T, Chao EE, Snell EA et al (2014) Multigene eukaryote phylogeny reveals the likely protozoan ancestors of opisthokonta (animals, fungi, choanozoans) and Amoebozoa. Mol Phylogenet Evol 81:71–85

Chadefaud M (1960) Les végétaux non vasulaires (Cryptogamie). In: Chadeauf M, Emberger I (eds) Traité de botanique systématique, vol 3. Masson et Cie, Paris, pp 1–1018

Chang Y, Wang S, Sekimoto S et al (2015) Phylogenomic analyses indicate that early fungi evolved digesting cells walls of algal ancestors of land plants. Genome Biol Evol 7:1590–1601

Christen R, Ratto A, Baroin A et al (1991) An analysis of the origin of metazoans, using comparisons of partial sequences of the 28S RNA, reveals an early emergence of triploblasts. EMBO J 10:499–503

Cienkowski L (1865) Beiträge zur Kenntniss der Monaden. Arch mikr Anat 1:203–232

Copeland HF (1956) The classification of lower organisms. Pacific Books, Palo Alto, CA

Corsaro D, Walochnik J, Venditti D et al (2014) Rediscovery of *Nucleophaga amoebae*, a novel member of the Rozellomycota. Parasitol Res 113:4491–4498

Dayel MJ, Alegado RA, Fairclough SR et al (2011) Cell differentiation and morphogenesis in the colony-forming choanoflagellate *Salpingoeca rosetta*. Dev Biol 357:73–82

DeBry RW (2005) The systematic component of phylogenetic error as a function of taxonomic sampling under parsimony. Syst Biol 54:432–440

del Campo J, Massana R (2011) Emerging diversity within chrysophytes, choanoflagellates and bicosoecids based upon molecular surveys. Protist 162:435–448

del Campo J, Ruiz-Trillo I (2013) Environmental survey meta-analysis reveals hidden diversity among unicellular opisthokonts. Mol Biol Evol 30:802–805

Denbo S, Aono K, Kai T et al (2018) Revision of the *Capsaspora* genome using read mating information adjusts the view on premetazoan genome. Dev Growth Differ 61:34–42

Derelle R, Torruella G, Klimeš V et al (2015) Bacterial proteins pinpoint a single eukaryotic root. Proc Natl Acad Sci U S A 112:E693–E699

Derelle R, López-Garcia P, Timpano H, Moreira D (2016) A phylogenomic framework to study the diversity and evolution of stramenopiles. Mol Biol Evol 33:2890–2898

Dirren S, Posch T (2016) Promiscuous and specific bacterial symbiont acquisition in the amoeboid genus *Nuclearia* (Opisthokonta). FEMS Microbiol Ecol 92: fiw105

Dirren S, Pitsch G, Silva MOD, Posch T (2017) Grazing of *Nuclearia thermophila* and *Nuclearia delicatula* (Nucleariidae, Opisthokonta) on the toxic cyanobacterium *Planktothrix rubescens*. Eur J Protistol 60:87–101

Dubin A, Chi SI, Emblem Å et al (2019) Deep-water sea anemone with a two-chromosome mitochondrial genome. Gene 692:195–200

Dunn CW, Hejnol A, Matus DQ et al (2008) Broad phylogenomic sampling improves resolution of the animal tree of life. Nature 452:745–749

Dyková I, Veverkov M, Fiala I et al (2003) *Nuclearia pattersoni* sp. n. (Filosea), a new species of amphizoic amoeba isolated from gills of roach (Rutilus rutilus), and its rickettsial endosymbiont. Folia Parasitol 50: 161–170

Edler D, Klein J, Antonelli A, Silvestro D (2020) raxmlGUI 2.0: a graphical interface and toolkit for phylogenetic analyses using RAxML. Methods Ecol Evol 12:373–377

Edlind TD, Li J, Visvesvara GS et al (1996) Phylogenetic analysis of β-tubulin sequences from amitochondrial protozoa. Mol Phylo Evol 5:359–367

Ehrenberg CG (1831) Über die Entwickelung und Lebensdauer der Infusionthiere: nebst ferneren Beiträgen zu einer Vergleichung ihrer organischen Système, vol 1831. Abh Königl Akad Wiss Berlin, pp 1–154

Ehrenberg CG (1838) Die Infusionsthierchen als vollkommene Organismen. Voss, Leipzig

Embley TM, van der Giezen M, Horner DS et al (2003) Hydrogenosomes, mitochondria and early eukaryotic evolution. IUBMB Life 55:387–395

Erwin DH, Valentine JW (2013) The Cambrian explosion: the construction of animal biodiversity. Roberts & Company, Calgary, AB

Fairclough SR, Chen Z, Kramer E et al (2013) Premetazoan genome evolution and the regulation of cell differentiation in the choanoflagellate *Salpingoeca rosetta*. Genome Biol 14:R15

Fast NM, Logsdon JM Jr, Doolittle WF (1999) Phylogenetic analysis of the TATA box binding protein (TBP) gene from *Nosema locustae*: evidence for a microsporidia-fungi relationship and spliceosomal intron loss. Mol Biol Evol 16:1415–1419

Felsenstein J (1978) Cases in which parsimony or compatibility methods will be positively misleading. Syst Biol 27:401–410

Ferrier DEK, Holland PWH (2001) Ancient origin of the *Hox* gene cluster. Nat Rev Genet 2:33–38

Franzen C (2005) How do microsporidia invade cells. Folia Parasitol 52:36–40

Fraune S, Bosch TCG (2010) Why bacteria matter in animal development and evolution. BioEssays 32: 571–580

Gabaldón T, Völcker E, Torruella G (2022) On the biology, diversity and evolution of nucleariid amoebae (Amorphea, Obazoa, Opisthokonta). Protist 173: 125895

Galindo LJ, Torruella G, Moreira D et al (2019) Combined cultivation and single-cell approaches to the phylogenomics of nucleariid amoebae, close relatives of fungi. Philos Trans R Soc B 374:20190094

Galindo LJ, López-García P, Torruella G et al (2021) Phylogenomics of a new fungal phylum reveals multiple waves of reductive evolution across Holomycota. Nat Commun 12:4973

Galindo LJ, Torruella G, López-García P et al (2022) Phylogenomics supports the monophyly of aphelids and fungi and identifies new molecular synapomorphies. Syst Biol. https://doi.org/10.1093/sysbio/syac054

Gams H (1947) Die Protochlorinae als autotrophe Vorfahren von Pilzen und Tieren? Mikroskopie 2: 383–387

Gawryluk RMR, Tikhonenkov DV, Hehenberger E et al (2019) Non-photosynthetic predators are sister to red algae. Nature 572:241–243

Gill EE, Fast NM (2006) Assessing the microsporidia-fungi relationship: combined phylogenetic analysis of eight genes. Gene 375:103–109

Glockling SL, Marshall WL, Gleason FH (2013) Phylogenetic interpretations and ecological potentials of the Mesomycetozoae (Ichthyosporea). Fungal Ecol 6: 237–247

González-Hernández M, Denoël M, Duffus AJL et al (2010) Dermocystid infection and associated skin lesions in free-living palmate newts (*Lissotriton helveticus*) from Southern France. Parasitol Int 59: 344–350

Gouy M, Li W-H (1989) Molecular phylogeny of the kingdoms Animalia, Plantae and Fungi. Mol Biol Evol 6:109–122

Grau-Bové X, Torruella G, Donachie S et al (2017) Dynamics of genomic innovation in the unicellular ancestry of animals. elife 6:e26036

Gray MW, Burger G, Lang BF (1999) Mitochondrial evolution. Science 283:1476–1481

Grobben K (1908) Die systematische Einteilung des Tierreiches, vol 58. Verhandlungen der kaiserlich-königlichen zoologisch-botanischen Gesellschaft in Wien, Vienna, pp 491–511

Gromov BV (2000) Algal parasites of the genera *Aphelidium*, *Amoeboaphelidium* and *Pseudoaphelidium* from the Cienkovski's "Monadinea" group as representatives of new class. Zool Zhurn 79:517–525

Haag KL, James TY, Pombert J-F, Larsson R, Schaer TMM, Refardt D, Ebert D (2014) Evolution of a morphological novelty occurred before genome compaction in a lineage of extreme parasites. Proc Natl Acad Sci U S A 111:15480–15485

Hampl V, Hug L, Leigh JW et al (2009) Phylogenomic analyses support the monophyly of Excavata and resolve relationships among eukaryotic "supergroups". Proc Natl Acad Sci U S A 106:3859–3864

Hasegawa M, Hashimoto T, Adachi J et al (1993) Early branchings in the evolution of eukaryotes: ancient divergence of Entamoeba that lacks mitochondria revealed by protein sequence data. J Mol Evol 36: 380–388

Hatschek B (1888) Lehrbuch der Zoologie: eine morphologische Übersicht des Thierreiches zur Einführung in das Studium dieser Wissenschaft. Georg Fischer, Jena

He D, Fiz-Palacios O, Fu C-J et al (2014) An alternative root for the eukaryotic tree of life. Curr Biol 24:456–470

Hehenberger E, Tikhonenkov DV, Kolisko M et al (2017) Novel predators reshape Holozoan phylogeny and reveal the presence of a two-component signaling system in the ancestor of animals. Curr Biol 27:2043–2050

Heiss AA, Kolisko M, Ekelund F et al (2018) Combined morphological and phylogenomic re-examination of malawimonads, a critical taxon for inferring the evolutionary history of eukaryotes. R Soc Open Sci 5: 171707

Hendy MD, Penny D (1989) A framework for the quantitative study of evolutionary trees. Syst Zool 38:297–309

Herman RL (1984) Ichthyophonus-like infection in newts (*Notophthalmus viridescens* Rafinesque). J Wildl Dis 20:55–56

Herr RA, Ajello L, Taylor JW et al (1999) Phylogenetic analysis of *Rhinosporidium seeberi*'s 18S small-subunit ribosomal DNA groups this pathogen among members of the protoctistan Mesomycetozoa clade. J Clin Microbiol 37:2750–2754

Hertel LA, Bayne CJ, Loker ES (2002) The symbiont *Capsaspora owczarzaki*, nov. gen. nov. sp., isolated from three strains of the pulmonate snail *Biomphalaria glabrata* is related to members of the Mesomycetozoa. Int J Parasitol 32:1183–1191

Hirt RP, Logsdon JM Jr, Healy B et al (1999) Microsporidia are related to fungi: evidence from the largest subunit of RNA polymerase II and other proteins. Proc Natl Acad Sci U S A 96:580–585

Hyman LH (1940) The invertebrates: protozoa through ctenophora. McGraw Hill, New York

Ireland EV, Woznica A, King N (2020) Synergistic cues from diverse bacteria enhance multicellular development in a choanoflagellate. Appl Environ Microbiol 86:e02920

James TY, Berbee ML (2012) No jacket required – new fungal lineage defies dress code. BioEssays 34:94–102

James TY, Kauff F, Schoch L et al (2006) Reconstructing the early evolution of fungi using a six-gene phylogeny. Nature 443:818–822

James TY, Pelin A, Bonen L et al (2013) Shared signatures of parasitism and phylogenomics unite Cryptomycota and Microsporidia. Curr Biol 23:1548–1553

James TY, Stajich JE, Hittinger CT, Rokas A (2020) Toward a fully resolved fungal tree of life. Annu Rev Microbiol 74:291–313

James-Clark H (1866) Note on the Infusoria Flagellata and the Spongiae Ciliatae. Am Sci 1:113–114

James-Clark H (1867) On the Spongiae Ciliatae as Infusoria Flagellata; or observations on the structure, animality and relationship of *Leucosolenia botryoides*, Bowerbank. Ann Mag Nat Hist 1:133–142

Jiménez-Guri E, Philippe H, Okamura B, Holland PWH (2007) *Buddenbrockia* is a cnidarian worm. Science 317:116–118

Jones MDM, Forn I, Gadelha C et al (2011) Discovery of novel intermediate forms redefines the fungal tree of life. Nature 474:200–203

Kapli P, Telford MJ (2020) Topology-dependent asymmetry in systematic errors affects phylogenetic placement of Ctenophora and Xenacoelomorpha. Sci Adv 6: eabc5162

Kapli P, Natsidis P, Leite DJ et al (2021) Lack of support for Deuterostomia prompts reinterpretation of the first Bilateria. Sci Adv 7:eabe2741

Karpov SA, Mylnikov AP (1989) Biology and ultrastructure of colorless flagellates Apusomonadida ord. n. Zool Zhurn 68:5–17

Karpov SA, Mikhailov KV, Mirzaeva GS et al (2013) Obligately phagotrophic aphelids turned out to branch with the earliest-diverging Fungi. Protist 164:195–205

Karpov SA, Mamkaeva MA, Aleoshin VV et al (2014a) Morphology, phylogeny, and ecology of the aphelids (Aphelidae, Opisthokonta) and proposal for the new superphylum Opisthosporidia. Front Microbiol 5:112

Karpov SA, Mamkaeva MA, Benzerara K et al (2014b) Molecular phylogeny and ultrastructure of *Aphelidium aff. Melosirae* (Aphelida, Opisthosporidia). Protist 165:512–526

Karpov SA, López-Garcia P, Mamkaeva MA et al (2018) The chytrid-like parasites of algae *Amoeboradix gromovi* gen. et sp. nov. and *Sanchytrium tribonematis* belong to a new fungal lineage. Protist 169:122–140

Karpov SA, Cvetkova VS, Annenkova NV, Vishnyakov AE (2019a) Kinetid structure of *Aphelidium* and *Paraphelidium* (Aphelida) suggests the features of the common ancestor of Fungi and Opisthosporidia. J Eukaryot Microbiol 55:911–924

Karpov SA, Vishnyakov AE, Moreira D, López-García P (2019b) The ultrastructure of *Sanchytrium tribonematis* (Sanchytriaceae, Fungi *incertae sedis*) confirms its close relationship to *Amoeboradix*. J Eukaryot Microbiol 66:892–898

Katoh K, Standley DM (2013) MAFFT multiple sequence alignment software version 7: improvements in performance and usability. Mol Biol Evol 30:772–780

Keeling PJ, Doolittle WF (1996) Alpha-tubulin from early-diverging eukaryotic lineages and the evolution of the tubulin family. Mol Biol Evol 13:1297–1305

Kennedy FA, Buggage RR, Ajello L (1995) Rhinosporidiosis: a description of an unprecedented outbreak in captive swans (*Cygnus* spp.) and a proposal for revision of the ontogenic nomenclature of *Rhinosporidium seeberi*. J Med Vet Mycol 33:157–165

Kent WS (1878) A new field for the microscopist. Pop Sci Rev 2:113–132

Kent WS (1880–1882) A manual of the Infusoria, vol 1–3. D Bogue, London

Kerk D, Gee A, Standish M, Wainwright PO et al (1995) The rosette agent of chinook salmon (*Oncorhynchus tshawytscha*) is closely related to choanoflagellates, as determined by the phylogenetic analyses of its small ribosomal subunit RNA. Mar Biol 122:187–192

Kim E, Simpson AGB, Graham LE (2006) Evolutionary relationships of apusomonads inferred from taxon-rich analyses of 6 nuclear encoded genes. Mol Biol Evol 23:2455–2466

Kobayashi M, Wada H, Satoh N (1996) Early evolution of the Metazoa and phylogenetic status of diploblasts as inferred from amino acid sequence of Elongation Factor-1. Mol Phylo Evol 5:414–422

Kożyczkowska A, Najle S, Ocaña-Pallarès E et al (2021) Stable transfection in protist *Corallochytrium limacisporum* identifies novel cellular features among unicellular animals relatives. Curr Biol 31:1–7

Lang BF, O'Kelly C, Nerad T et al (2002) The closest unicellular relatives of Animals. Curr Biol 12:1773–1778

Lapage G (1925) Notes on the choanoflagellate, *Codosiga botrytis*, Ehrbg. Q J Microsc Sci 69:471–508

Lara E, Moreira D, López-Garcia P (2009) The environmental clade LKM11 and *Rozella* form the deepest branching clade of Fungi. Protist 161:116–121

Laumer CE, Fernández R, Lemer S et al (2019) Revisiting metazoan phylogeny with genomic sampling of all phyla. Proc R Soc B 286:20190831

Laundon D, Larson BT, McDonald K et al (2019) The architecture of cell differentiation in choanoflagellates and sponge choanocytes. PLoS Biol 17:e3000226

Lax G, Eglit Y, Eme L et al (2018) Hemimastigophora is a novel supra-kingdom-level lineage of eukaryotes. Nature 564:410–414

Le SQ, Gascuel O (2008) An improved general amino acid replacement matrix. Mol Biol Evol 25:1307–1320

Leadbeater BSC (1983) Distribution and chemistry of microfilaments in choanoflagellates, with special reference to the collar and other tentacle systems. Protistologica 19:157–166

Leadbeater BSC (2008a) Choanoflagellate evolution: the morphological perspective. Protistology 5:256–267

Leadbeater BSC (2008b) Choanoflagellate lorica construction and assembly: The nudiform condition. I. *Savillea* species. Protist 159:259–268

Leadbeater BSC (2010) Choanoflagellate lorica construction and assembly: The tectiform condition. *Volkanus costatus* (=*Diplotheca costata*). Protist 161:160–176

Leadbeater BSC (2015) The Choanoflagellates: evolution, biology and ecology. Cambridge University Press, Cambridge, UK

Leadbeater BSC, Hassan R, Nelson M et al (2008) A new genus, *Helgoeca* gen. nov., for a nudiform choanoflagellate. Eur J Protistol 44:227–237

Leadbeater BSC, Yu QB, Kent J, Stekel DJ (2009) Three-dimensional images of choanoflagellate loricae. Proc R Soc B 276:3–11

Letcher PM, Powell MJ (2018) A taxonomic summary and revision of *Rozella* (*Cryptomycota*). IMA Fungus 9: 383–399

Letcher PM, Powell MJ (2019) A taxonomic summary of *Aphelidiaceae*. IMA Fungus 10:4

Letcher PM, Lopez S, Schmieder R et al (2013) Characterization of *Amoeboaphelidium protococcarum*, an algal parasite new to the Cryptomycota isolated from an outdoor algal pond used for the production of biofuel. PLoS One 8:e56232

Lichtwardt RW (1986) The Trichomycetes: Fungal associates of arthropods. Springer, New York

Lighthill J (1976) Flagellar hydrodynamics. SIAM Rev 18:161–230

Logares R, Bråte J, Bertilsson S et al (2009) Infrequent marine-freshwater transitions in the microbial world. Trends Microbiol 17:414–422

Lohr JN, Laforsch C, Koerner H, Wolinsk J (2010) A *Daphnia* parasite (*Caullerya mesnili*) constitutes a new member of the Ichthyosporea, a group of protists near the animal-fungi divergence. J Eukaryot Microbiol 57:328–336

López-Escardó D, López-García P, Moreira D et al (2017) *Parvularia atlantis* gen. et sp. nov., a nucleariid filose amoeba (Holomycota, Opisthokonta). J Eukaryot Microbiol 65:170–179

López-Escardó D, Grau-Bové X, Guillaumet-Adkins A et al (2019) Reconstruction of protein domain evolution using single-cell amplified genomes of uncultured choanoflagellates sheds light on the origin of animals. Philos Trans R Soc B 374:20190088

Lord JC, Hartzer KL, Kambhampati S (2012) A nuptially transmitted ichthyosporean symbiont of *Tenebrio molitor* (Coleoptera: Tenebrionidae). J Eukaryot Microbiol 59:246–250

Lu Y, Ocaña-Pallarès E, López-Escardó D et al. (2020) Revisiting the phylogenetics position of *Caullerya mesnili* (Ichthyosporea), a common *Daphnia* parasite, based on 22 protein-coding genes. Mol Phylogenet Evol 151:106891

Mah JL, Christensen-Dalsgaard C, Leys SP (2014) Choanoflagellate and choanocyte collar-flagellar systems and the assumption of homology. Evol Dev 16:25–37

Malakhov VV (2016) Symmetry and the tentacular apparatus in Cnidaria. Russ J Mar Biol 42:287–298

Maldonado M (2004) Choanoflagellates, choanocytes, and animal multicellularity. Invert Biol 123:1–22

Manton I, Bremer G, Oates K (1981) Problems of structure and biology in a large collared flagellate (*Diaphanoeca grandis* Ellis) from Arctic Seas. Proc R Soc B 213:15–26

Marshall WL, Berbee ML (2011) Facing unknowns: living cultures (*Pirum gemmate* gen. nov., sp. nov., and *Abeoforma whisleri*, gen. nov., sp. nov.) from invertebrate digestive tracts represent an undescribed clade within the unicellular opisthokont lineage Ichthyosporea (Mesomycetozoea). Protist 162:33–57

Marshall WL, Berbee ML (2013) Comparative morphology and genealogical delimitation of cryptic species of sympatric isolates of *Sphaeroforma* (Ichthyosporea, Opisthokonta). Protist 164:287–311

Matriano DM, Alegado RA, Conaco C (2021) Detection of horizontal gene transfer in the genome of the choanoflagellate *Salpingoeca rosetta*. Sci Rep UK 11: 5993

Matus DQ, Pang K, Marlow H et al (2006) Molecular evidence for deep evolutionary roots of bilaterality in animal development. Proc Natl Acad Sci U S A 103: 11195–11200

Medina M, Collins AG, Taylor JW et al (2003) Phylogeny of Opisthokonta and the evolution of multicellularity and complexity in Fungi and Metazoa. Int J Astrobiol 2:203–211

Medlin L, Elwood HJ, Stickel S et al (1988) The characterization of enzymatically amplified eukaryotic 16S-like rRNA-coding regions. Gene 71:491–499

Mendoza L, Ajello L, Taylor JW (2001) The taxonomic status of *Lacazia loboi* and *Rhinosporidium seeberi* has been finally resolved with the use of molecular tools. Rev Iberoam Micol 18:95–98

Mendoza L, Taylor JW, Ajello L (2002) The Class Mesomycetozoa: a heterogeneous group of microorganisms at the animal–fungal boundary. Annu Rev Microbiol 56:315–344

Mikrjukov KA (1999) Taxonomic revision of scale- bearing Heliozoon-like amoebae (Pompholyxophryidae, Rotosphaerida). Acta Potozool 38:119–131

Minge MA, Silberman JD, Orr RJS et al (2009) Evolutionary position of breviate amoebae and the primary eukaryote divergence. Proc R Soc B 276:597–604

Moroz LL, Halanych KM (2016) A sisterly dispute. Nature 529:286–287

Moroz LL, Kocot KM, Citarella MR et al (2014) The ctenophore genome and the evolutionary origins of neural systems. Nature 510:109–114

Mylnikov AP, Tikhonenkov DV, Karpov SA, Wylezich C (2019) Microscopical studies on *Ministeria vibrans* Tong, 1997 (Filasterea) highlight the cytoskeletal structure of the common ancestor Filasterea, Metazoa and Choanoflagellata. Protist 170:385–396

Nitsche F, Carr M, Arndt H, Leadbeater BSC (2011) Higher level taxonomy and molecular phylogenetics of the Choanoflagellatea. J Eukaryot Microbiol 58: 452–462

Nitsche F, Thomsen HA, Richter DJ (2017) Bridging the gap between morphological species and molecular barcodes – exemplified by loricate choanoflagellates. Eur J Protistol 57:26–37

Norris RE (1965) Neustonic marine Craspedomonadales (Choanoflagellata) from Washington and California. J Protozool 12:589–612

Ocaña-Pallarès E, Williams TA, López-Escardó D et al (2022) Divergent genomic trajectories predate the origin of animals and fungi. Nature 609:747–753

Owczarzak A, Stibbs HH, Bayne CJ (1980) The destruction of *Schistosoma mansoni* mother sporocysts in vitro by amoebae isolated from *Biomphalaria glabrata*: an ultrastructural study. J Invertebr Pathol 35:26–33

Page FC (1987) The classification of 'naked' amoebae (phylum rhizopoda). Arch Protistenkd 133:199–217

Paps J, Medina-Chacón LA, Marshall W, Suga H, Ruiz-Trillo I (2013) Molecular phylogeny of unikonts: new insights into the position of apusomonads and ancyromoands and the internal relationships of opisthokonts. Protist 164:2–12

Parfrey LW, Lahr DJG, Knoll AH, Katz LA (2011) Estimating the timing of early eukaryotic diversification with multigene molecular clocks. Proc Natl Acad Sci U S A 108:13624–13629

Parra-Acero H, Ros-Rocher N, Perez-Posada A et al (2018) Transfection of *Capsaspora owczarzaki*, a close unicellular relative of animals. Development 145:dev162107

Pastrana CC, DeBiasse MB, Ryan JF (2019) Sponges lack ParaHox genes. Genome Bio Evol 11:1250–1257

Patterson DJ (1985) On the organization and affinities of the amoeba, *Pompholyxophrys punicea* Archer, based on ultrastructural examination of individual cells from wild material. J Eukaryot Microbiol 32:241–246

Patterson DJ (1999) The diversity of eukaryotes. Am Nat 154(Suppl):S86–S124

Patterson DJ, Surek B, Melkonian M (1987) The ultrastructure of *Vampyrellidium perforans* Surek & Melkonian and its taxonomic position among the naked filose Amoebae. J Protozool 34:63–67

Patterson DJ, Nygaard K, Steinberg G, Turley CM (1993) Heterotrophic flagellates and other protists associated with oceanic detritus throughout the water column in the mid North Atlantic. J Mar Biol Assoc UK 73:67–95

Paul M (2012) *Acanthocorbis mongolica* nov. spec. – description of the first freshwater loricate choanoflagellate (Acanthoecida) from a Mongolian lake. Eur J Protistol 48:1–8

Pereira CN, Di Rosa I, Fagotti A et al (2005) The pathogen of frogs *Amphibiocystidium ranae* is a member of the order Dermocystida in the class Mesomycetozoea. J Clin Microbiol 43:192–198

Pettitt ME (2001) Prey capture and ingestion in choanoflagellates. PhD thesis, University of Birmingham

Pettitt ME, Orme BAA, Blake JR, Leadbeater BSC (2002) The hydrodynamics of filter feeding in choanoflagellates. Eur J Protistol 38:313–332

Philippe H, Brinkmann H, Lavrov DV et al (2011) Resolving difficult phylogenetic questions: why more sequences are not enough. PLoS Biol 9:e1000602

Philippe H, Poustka AJ, Chiodin M et al (2019) Mitigating anticipated effects of systematic errors supports sister-group relationship between Xenacoelomorpha and Amubulacraria. Curr Biol 29:1818–1826

Pick KS, Philippe H, Schreiber F et al (2010) Improved phylogenomic taxon sampling noticeably affects nonbilaterian relationships. Mol Biol Evol 27:1983–1987

Pisani D, Pett W, Dohrmann M et al (2015) Genomic data do not support comb jellies as the sister group to all other animals. Proc Natl Acad Sci U S A 112:15402–15407

Posada D (2003) Using Modeltest and PAUP* to select a model of nucleotide substitution, current protocols in bioinformatics. Wiley, New York

Powell MJ (1984) Fine structure of the unwalled thallus of *Rozella polyphagia* in its host *Polyphagus euglenae*. Mycologia 76:1039–1048

Quandt CA, Beaudet D, Corsaro D et al (2017) The genome of an intracellular parasite, *Paramicrosporidium saccamoebae*, reveals alternative adaptations to obligate intracellular parasitism. eLife 6: e29594

Raffel TR, Dillard JR, Hudson PJ (2006) Field evidence for leech-borne transmission of amphibian *Ichthyophonus* sp. J Parasitol 92:1256–1264

Ragan MA, Goggins CL, Cawthorn RJ et al (1996) A novel clade of protistan parasites near the animal–fungal divergence. Proc Natl Acad Sci U S A 93:11907–11912

Raghu-kumar S (1987) Occurrence of the thraustochytrid, *Corallochytrium limacisporum* gen. et sp. nov. in the coral reef lagoons of the Lakshadweep Islands in the Arabian Sea. Bot Mar 30:83–89

Rainer H (1968) Urtiere, Protozoa, Wurzelfu �ßler, Rhizopoda, Sonnentierchen, Heliozoa. Systematik und Taxonomie, Biologie, Verbreitung und Ökologie der Arten der Erde. In: Dahl M, Peus F (eds) Die Tierwelt Deutschlands und der angrenzenden Meeresteile, vol 56. VEB Gustav Fischer Verlag, Jena

Ramaiah N (2006) A review on fungal diseases of algae, marine fishes, shrimps and corals. Indian J Mar Sci 35: 380–387

Ramos OM, Barker D, Ferrier DEK (2012) Ghost loci imply *Hox* and *ParaHox* existence in the last common ancestor of animals. Curr Biol 22:1951–1956

Reinke AW, Balla KM, Bennett EJ et al (2017) Identification of microsporidia host-exposed proteins reveals a repertoire of rapidly evolving proteins. Nat Commun 8: 14023

Reynolds NK, Smith ME, Tretter ED et al (2017) Resolving relationships at the animal-fungal divergence: a molecular phylogenetic study of the protist trichomycetes (Ichthyosporea, Eccrinida). Mol Phylogenet Evol 109:447–464

Richards TA, Leonard G, Wideman JG (2017) What defines the "kingdom" Fungi. Microbiol Spectr 5. https://doi.org/10.1128/microbiolspec.FUNK-0044-2017

Richter DJ, Fozouni P, Eisen MB, King N (2018) Gene family innovation, conversation and loss on the animal stem lineage. eLife 7:e34226

Ronquist F, Teslenko M, van der Mark P et al (2012) MrBayes 3.2: efficient Bayesian phylogenetic inference and model choice across a large model space. Syst Biol 61:539–542

Rouse GW, Wilson NG, Carvajal JI et al (2016) New deep-sea species of *Xenoturbella* and the position of Xenacoelomorpha. Nature 530:94–97

Rowley JJL, Gleason FH, Andreou D et al (2013) Impacts of mesomycetozoean parasites on amphibian and freshwater fish populations. Fungal Biol Rev 27:100–111

Ruiz-Trillo I, Inagaki Y, Davis LA et al (2004) *Capsaspora owczarzaki* is an independent opisthokont lineage. Curr Biol 14:R946–R947

Ruiz-Trillo I, Lane CE, Archibald JM et al (2006) Insights into the evolutionary origin and genome architecture of the unicellular opisthokonts *Capsaspora owczarzaki* and *Sphaeroforma arctica*. J Eukaryot Microbiol 53: 379–384

Ryan JF, Pang K, Schnitzler CE et al (2013) The genome of the ctenophore *Mnemiopsis leidyi* and its implications for cell type evolution. Science 342: 1242592

Schilde C, Schaap P (2013) The Amoebozoa. Methods Mol Biol 983:1–15

Schiwitza S, Nitsche F (2021) A needle in the haystack – mapping sequences to morphology exemplified by the loricate choanoflagellate *Enibas thessalia* sp. nov. (Acanthoecida, Acanthoecidae). Protist 172:125782

Sebé-Pedrós A, Irimia M, del Campo J et al (2013) Regulated aggregative multicellularity in a close unicellular relative of metazoa. eLife 2:e01287

Seeber GR (1900) Un nuevo esporozoario parásito del hombre. Dos casos encontrados en pólipos nasales. Graduation thesis, Facultad de Ciencias Médicas, Universidad Nacional de Buenos Aires, Buenos Aires

Shalchian-Tabrizi K, Minge MA, Espelund M et al (2008) Multigene phylogeny of Choanozoa and the origin of animals. PLoS One 3:e2098

Simion P, Philipe H, Baurain D et al (2017) A large and consistent phylogenomic dataset supports sponges as the sister group to all other animals. Curr Biol 27:958–967

Smith DR, Kayal E, Yanagihara AA et al (2011) First complete mitochondrial genome sequence from a box jellyfish reveals a highly fragmented linear architecture and insights into telomere evolution. Genome Biol Evol 4:52–58

Sogabe S, Hatleberg WL, Kocot KM et al (2019) Pluripotency and the origin of animal multicellularity. Nature 570:519–522

Sogin ML, Elwood HJ, Gunderson JH (1986) Evolutionary diversity of eukaryotic small-subunit rRNA genes. Proc Natl Acad Sci U S A 83:1383–1387

Southworth JS, Armitage P, Fallon B et al (2018) Patterns of ancestral animal codon usage bias revealed through holozoan protists. Mol Biol Evol 35:2499–2511

Southworth JS, Grace CA, Marron AO et al (2019) A genomic survey of transposable elements in the choanoflagellate *Salpingoeca rosetta* reveals selection on codon usage. Mob DNA 10:44

Spanggaard B, Skouboe P, Rossen L et al (1996) Phylogenetic relationships of the intercellular fish pathogen *Ichthyophonus hoferi*, and fungi, choanoflagellates and the rosette agent. Mar Biol 126:109–115

Spatafora JW, Chang Y, Benny GL et al (2016) A phylum-level phylogenetic classification of zygomycete fungi based on genome-scale data. Mycologia 108:1028–1047

Srivastava M, Simakov O, Chapman J et al (2010) The *Amphimedon queenslandica* genome and the evolution of animal complexity. Nature 466:720–726

von Fresenius G (1858) Beiträge zur Kenntnis mikroskopischer Organismen. Abh Senckenb Naturforsch Ges 2:211–242

Stechmann A, Cavalier-Smith T (2003a) The root of the eukaryote tree pinpointed. Curr Biol 13:R665–R666

Stechmann A, Cavalier-Smith T (2003b) Phylogenetic analysis of eukaryotes using heat-shock protein Hsp90. J Mol Evol 57:408–419

Stampar SN, Broe MB, Macrander J et al (2019) Linear mitochondrial genoe in Anthozoa (Cnidaria): a case study in Ceriantharia. Sci Rep UK 9:6094

Strassert JFH, Jamy M, Mylnikov AP et al (2019) New phylogenomic analysis of the enigmatic phylum Telonemia further resolves the eukaryote tree of life. Mol Biol Evol 36:757–765

Steenkamp ET, Wright J, Baldauf SL (2006) The protistan origins of animals and fungi. Mol Biol Evol 23:93–106

von Stein FR (1878) Der Organismus der Infusionthiere. W. Engelmann, Leipzig

Steinert M, Novikoff AB (1960) The existence of a cytostome and the occurrence of pinocytosis in the trypanosome, *Trypanosoma mega*. J Biophys Biochem Cytol 8:563–569

Stibbs HH, Owczarzak A, Bayne CJ, DeWan P (1979) Schistosome sporocyst-killing amoebae isolated from *Biomphalaria glabrata*. J Invertebr Pathol 33:159–170

Suga H, Chen Z, de Mendoza A et al (2013) The *Capsaspora* genome reveals a complex unicellular prehistory of animals. Nat Commun 4:2325

Tamura K, Nei M (1993) Estimation of the number of nucleotide substitutions in the control region of mitochondrial DNA in humans and chimpanzees. Mol Biol Evol 10:512–526

Tedersoo L, Sánchez-Ramírez S, Kõljalg U et al (2018) High-level classification of the Fungi and a tool for evolutionary ecological analyses. Fungal Divers 90: 135–159

Telford MJ (2016) Fighting over a comb. Nature 529:286

Thompson LD (2016) Rhinosporidiosis. Ear Nose Throat J 95:101

Thomsen HA (1982) Planktonic choanoflagellates from Disco Bugt, West Greenland, with a survey of the marine nanoplankton of the area. Meddelelser om Gronland Bioscience 8:3–35

Tikhonenkov DV, Mikhailov KV, Hehenberger E et al (2020) New lineage of microbial predators adds complexity to reconstructing the evolutionary origin of animals. Curr Biol 30:4500–4509

Tong SM (1997) Heterotrophic flagellates and other protists from Southampton Water, U.K. Ophelia 47: 71–131

Toret C, Picco A, Boiero-Sanders M et al (2022) The cellular slime mold *Fonticula alba* forms a dynamic, multicellular collective while feeding on bacteria. Curr Biol 32:1961–1973

Torruella G, Derelle R, Paps J et al (2012) Phylogenetic relationships within the Opisthokonta based on phylogenomic analyses of conserved single-copy protein domains. Mol Biol Evol 29:531–544

Torruella G, Grau-Bové X, Moreira D et al (2018) Global transcriptome analysis of the aphelid *Paraphelidium tribonemae* supports the phagotrophic origin of fungi. Commun Biol 1:231

Torruella G, Mendoza A, Grau-Bové X et al (2015) Phylogenomics reveals convergent evolution of lifestyles in close relatives of animals and fungi. Curr Biol 25:2404–2410

Trotter MJ, Whisler HC (1965) Chemical composition of the cell wall of *Amoebidium parasiticum*. Can J Bot 43: 869e876

Urrutia A, Mitsi K, Foster R et al (2022) *Txikispora philomaios* n. sp., n. g., a micro-eukaryotic pathogen of amphipods, reveals parasitism and hidden diversity in Class Filasterea. J Eukaryot Microbiol 69:e12875

Ustinova I, Krienitz L, Huss VA (2000) *Hyaloraphidium curvatum* is not a green alga, but a lower fungus; *Amoebidium parasiticum* is not a fungus, but a member of the DRIPs. Protist 151:253–262

Verni F, Gualtieri P (1997) Feeding behaviour in ciliated protists. Micron 28:487–504

Vickerman K, Darbyshire JF, Ogden CG (1974) *Apusomonas proboscidae* Aléxéieff 1924, an unusual phagotrophic flagellate from soil. Arch Protistenkd 116:254–269

Vischer W (1945) Über einen pilzähnlichen, autotrophen Mikroorganismus, *Chlorochytridion*, einige neue Protococcales und die systematische Bedeutun der Chloroplasten. Verhandlungen der Naturforschenden Gesellschaft in Basel 6:41–49

Vossbrinck CR, Maddox JV, Friedman S et al (1987) Ribosomal RNA sequence suggests microsporidia are extremely ancient eukaryotes. Nature 326:411–414

Wadi L, Reinke AW (2020) Evolution of microsporidia: An extremely successful group of eukaryotic intracellular parasites. PLoS Pathog 16:e1008276

Wainright PO, Hinkle G, Sogin ML, Stickel SK (1993) Monophyletic origins of the Metazoa: an evolutionary link with fungi. Science 260:340–342

Whelan NV, Kocot KM, Moroz LL, Halanych KM (2015) Error, signal, and the placement of Ctenophora sister to all other animals. Proc Natl Acad Sci U S A 112:5773–5578

Williams TA, Embley TM (2014) Archaeal "Dark Matter" and the origin of eukaryotes. Genome Biol Evol 6: 474–481

Worley AC, Raper KB, Hohl M (1979) *Fonticula alba*: a new cellular slime mold (Acrasiomycetes). Mycologia 71:746–760

Wylezich C, Karpov SA, Mylnikov AP et al (2012) Ecologically relevant choanoflagellates collected from hypoxic water masses of the Baltic Sea have untypical mitochondrial cristae. BMC Microbiol 12:271

Yahalomi D, Atkinson SD, Neuhof M et al (2020) A cnidarian parasite of salmon (Myxozoa: *Henneguya*) lacks a mitochondrial genome. Proc Natl Acad Sci U S A 117:5358–5363

Yoshida M, Nakayama T, Inouye I (2009) *Nuclearia thermophila* sp. nov. (Nucleariidae), a new nucleariid species isolated from Yunoko Lake in Nikko (Japan). Eur J Protistol 45:147–155

Young AD, Gillung JP (2020) Phylogenomics – principles, opportunities and pitfalls of big-data phylogenetics. Syst Entomol 45:225–247

Yue J, Sun G, Hu X, Huang J (2013) The scale and evolutionary significance of horizontal gene transfer in the choanoflagellate *Monosiga brevicollis*. BMC Genomics 14:729

Zaremba-Niedzwiedzka K, Caceres EF, Saw JH et al (2017) Asgard archaea illuminate the origin of eukaryotic cellular complexity. Nature 541:353–358

Zopf W (1885) Zur Morphologie und Biologie der niederen Pilztiere (Monadinen). University of Strasbourg, Leipzig

Zwickl DJ, Hillis DM (2002) Increased taxon sampling greatly reduces phylogenetic error. Syst Biol 51:588–598

Evolution of Signalling and Morphogenesis in the Dictyostelids

2

Christina Schilde

Abstract

Dictyostelids were once grouped with fungi because their fruiting structures are superficially similar. However, either group evolved quite distinct strategies to reach this mode of species propagation. This chapter describes the cell communication systems that cause dictyostelid amoebas to aggregate and regulate cell-type specialisation and cell movement during fruiting body formation. It also highlights how increasingly complex morphologies appeared during dictyostelid evolution and how some of these morphological innovations were caused by modification of specific cell–cell signalling systems.

Keywords

Amoebozoa · Camp · Cyst · Dictyostelia · Fruiting body · Evolution · Multicellularity · Signalling · Social amoebas · Spore

2.1 Introduction

Like fungi, dictyostelids build fruiting bodies for the dispersal of spores. Because of the sessile nature of their fruiting bodies and their cellulose-walled vacuolised stalk cells, they were initially considered to belong to the plant kingdom and hence follow plant taxonomic nomenclature rules. However, they are only distant relatives to fungi and not related to plants at all. The dictyostelids, or social amoebas, are members of the Amoebozoa, a sister group to Opisthokonts, the group that comprises the animals and fungi (Baldauf et al. 2000). Within the Amoebozoa, Dictyostelia form the class Mycetozoa together with Myxogastria and Protosporangiida. Dictyostelids themselves are subdivided into the two orders: Acytosteliales (branch I) and Dictyosteliales (branch II) (Sheikh et al. 2018) (Fig. 2.1). Both branches consist of two major groups and some minor clades. Acytosteliales comprise the Cavenderiaceae (group 1), Acytosteliaceae (group 2), and the "polycarpum" complex. Dictyosteliales encompass the "polycephalum" complex, the Raperosteliaceae (group 3), the "polysphondylium" complex, and the Dictyosteliaceae (group 4).

Unlike fungi, which mostly feed by absorption of low-molecular-weight compounds through cell walls and membranes, dictyostelids are active predators of bacteria and other microbes. Other differences are the absence of a cell wall during most of the dictyostelid life cycle, and the fact that multicellularity is achieved by cell aggregation and not but by cell division. In fact, multicellularity and morphogenesis in dictyostelids are highly dependent on the fact that their unwalled

C. Schilde (✉)
School of Life Sciences, D'Arcy Thompson Unit, Carnelley Building, University of Dundee, Dundee, UK
e-mail: c.schilde@dundee.ac.uk

© The Author(s), under exclusive license to Springer Nature Switzerland AG 2023

S. Pöggeler, T. James (eds.), *Evolution of Fungi and Fungal-Like Organisms*, The Mycota 14,
https://doi.org/10.1007/978-3-031-29199-9_2

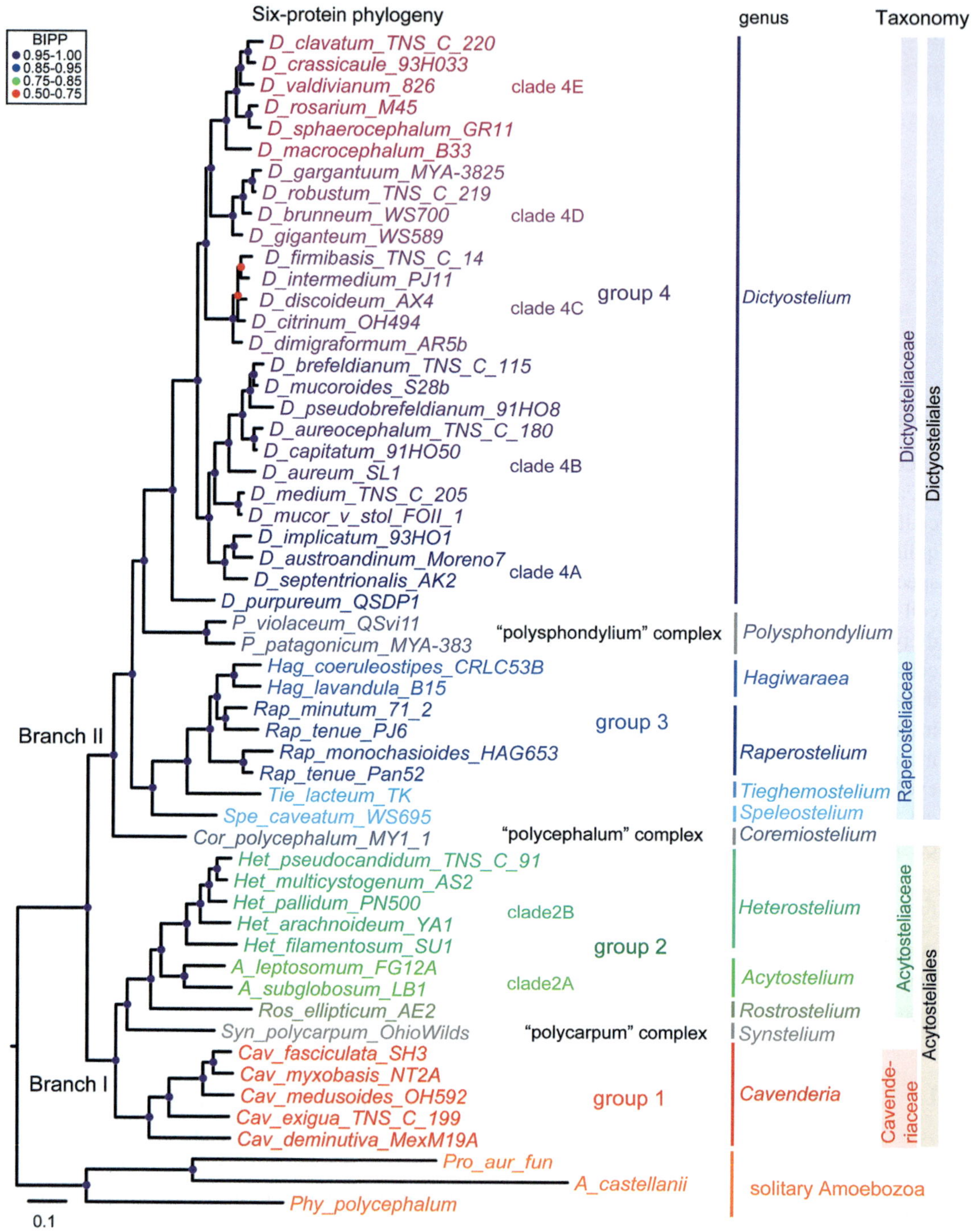

Fig. 2.1 Overview of the phylogenetic relationships within the Dictyostelia (after Schilde et al. 2019). The phylogenetic tree of the Dictyostelia based on a Bayesian inference analysis of six concatenated conserved proteins (Agl, AmdA, PurD, PurL, RpaA, SmdA) is shown on the left with Bayesian inference posterior probabilities indicated as (coloured) dots on the nodes. Three species of solitary Amoebozoa (*Acanthamoeba castellanii*, *Physarum polycephalum*, and *Protostelium aurantium* var. *fungivorum*) have been included as outgroups (orange). The phylogeny shows subdivision of all species into four major groups and some minor group-intermediate complexes and the within-group relationships and groups. Species names and strain designations (are colour-coded to) reflect their affiliation with groups or clades and use the nomenclature proposed by Sheikh

amoeboid cells remain motile until the fruiting body is completed. As single-celled amoebae they feed on bacteria that live on decomposing leaves or animal waste. They have an efficient machinery for engulfing and digesting bacteria, at a rate of about 300 per hour per cell (Gerisch 1960; Peracino et al. 1998). Amoebas feed on many different bacterial species but can also be infected by and serve as host for bacterial pathogens (Steinert 2011).

Dictyostelids are found in all types of soils including those in arctic regions and above the tree line. However, species diversity is higher in forest soils and particularly those at tropical latitudes (Swanson et al. 1999). Some species show a very widespread distribution, like *Dictyostelium mucoroides,* whereas others are more restricted, like the cannibalistic species *Speleostelium caveatum* which has only ever been found on bat guano of caves in Arkansas (Waddell 1982; Baldauf et al. 2018). Around 150 dictyostelid species have been isolated that grow under laboratory conditions, whereas the actual diversity of dictyostelids estimated from sequencing environmental DNA samples is much higher (Baldauf et al. 2018). Some species, such as *Heterostelium oculare* (Cavender et al. 2005) and *Tieghemostelium menorah* (Vadell and Cavender 2007), only form fruiting bodies efficiently when co-cultured with other microorganisms, indicating that symbiotic relationships may exist. Other species may not be growing under laboratory conditions at all.

In addition to the vegetative reproduction cycle, where haploid amoebas divide by binary fission, dictyostelids have three responses to environmental stress, such as starvation. Similar to a range of other free-living amoebas, many dictyostelid species can form cysts (microcysts), where individual amoebae encapsulate without aggregation (Raper 1984). The microcyst, which has a two-layered cellulose wall, can endure relatively short periods of nutrient depletion and hatches a single haploid amoeba when food becomes available again. Conversely, the macrocyst is a sexual structure and can survive for long periods. Dictyostelid strains can be homothallic (self-fertile) or heterothallic (self-sterile) and can have multiple mating types (Bloomfield 2019). Macrocyst formation starts with the fusion of two (or often more) haploid amoebas of opposing mating types to form a diploid zygote, which then attracts other amoebas by chemotaxis (Bloomfield 2019). Most of these cells are cannibalised by the zygote, and their resources are used to build a three-layered thick wall around the >100 μm large structure. The zygote goes through sexual recombination and rounds of meiotic and mitotic divisions and eventually numerous haploid amoebae emerge (Raper 1984). Although the macrocyst stage should allow application of traditional genetic approaches to *Dictyostelium* research, its usefulness is hampered by the absence of protocols to induce efficient germination of the macrocyst.

Multicellularity is the third response to nutrient depletion, which triggers regulated secretion of a chemoattractant by the amoebas (Fig. 2.2). Cells aggregate to form a mound, which can consist of up to a million amoebas. Sophisticated cell–cell signalling mechanisms between the amoebas orchestrate the differentiation of up to five different cell types and coordinate an intricate progression of cell movements. In combination with the synthesis of a flexible skin-like matrix, cell differentiation and cell movement first generate the formation of a motile structure, called the "slug". The slug is attracted by light and warmth, which cause it to move to the soil's top layer. Here, the slug cell mass projects upwards and

Fig. 2.1 (continued) et al. (2018). Genus, family, and order designations are shown on the right-hand side and genera are indicated by vertical lines. There are four families (Cavenderiaceae, Acytosteliaceae, Raperosteliaceae, Dictyosteliaceae) each comprised of one to four genera. Group 4 so far is the species-richest group containing only a single genus *Dictyostelium*

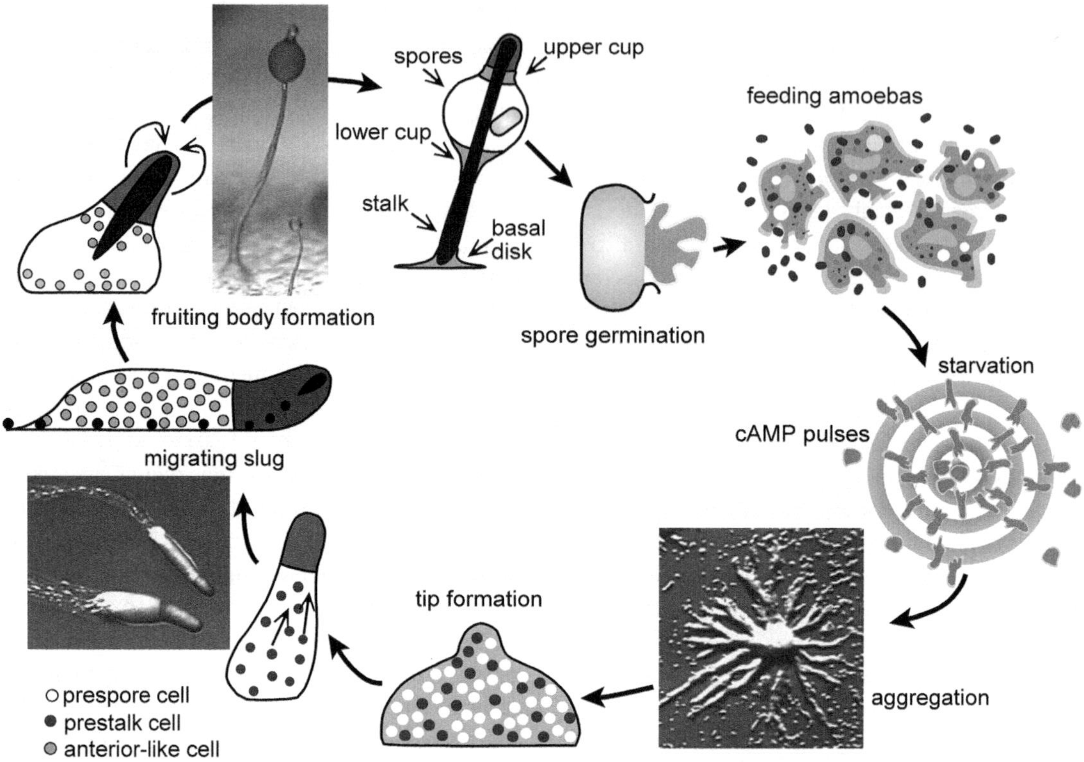

Fig. 2.2 Life cycle of *Dictyostelium discoideum*. *D. discoideum* amoebas emerge from spores and start feeding on bacteria. When food runs out, amoebas secrete cAMP pulses, which trigger chemotactic movement and aggregation of cells. Once aggregated, the amoebas differentiate into prestalk and prespore cells in a regulated ratio. The organising tip continues to emit cAMP pulses, which shape the cell mass by coordinating cell movement. The cAMP pulses also cause the prestalk cells, which are chemotactically most responsive to cAMP, to move towards the front. At the posterior, prespore cells dedifferentiate into anterior-like cells, which replenish the prestalk cells when needed. At the onset of culmination, the cells synthesise a cellulose tube, the apical prestalk cells move into the tube and mature into stalk cells, and the remaining prestalk cells form support structures, such as the upper and lower cup and the basal disc. The prespore cells move up the stalk and mature into spores

forms the fruiting body, which consists of a column of stalk cells topped by a globular mass of spores. Depending on the species, the stalk can show different patterns of side branches and/or be decorated with disc-, root-, or cup-shaped support structures (Raper 1984).

The developmental process of dictyostelids with its strong dependence on coordinated cell migration is more reminiscent of animal development than fungal or plant development, which depends mostly on coordinated outgrowth. The ease of culture and genetic manipulation of *Dictyostelium discoideum*, combined with its excellent accessibility to biochemical and cell biological approaches, has therefore encouraged many workers to use this organism as a model for studies of chemotaxis, cell migration, developmental cell signalling and cell-type specialisation, and cellular processes such as vesicle trafficking, phagocytosis, and cytokinesis.

More recently, dictyostelids have also become popular to resolve evolutionary questions. Most multicellular organisms start life as a zygote, where after cell division cells remain together and, through coordinated cell division, cell differentiation, and cell migration, generate structures, organs, and tissues that consist of millions of cells. Although these cells may vary greatly in form and function, they are all genetically identical. Dictyostelia become multicellular by aggregation and the contributing cells are not necessarily genetically identical. Since only part of the cells propagate as spores, the genes of the non-propagating stalk cells are lost. This generates genetic conflict in the population with cells avoiding stalk fate (cheating), leading ultimately to both physical and metaphorical collapse of the multicellular assembly. This raises the question how cheating is suppressed by counteracting mechanisms (Strassmann et al. 2000; Castillo et al. 2005; Forget et al. 2021).

The dictyostelids with their facultative multicellular lifestyle are very suited to answer another fundamental question. What caused multicellularity to evolve in the first place? Multicellular organisms are typically dependent on intercellular communication, whereas unicellular organisms respond mainly to environmental signals. A major aspect of the evolution of multicellularity is therefore how the sensory and secretory systems of unicellular organisms were adapted to allow for signalling between cells.

In this chapter, I firstly present an overview of the signalling mechanisms that control development of the model organism *D. discoideum.* Secondly, I will summarise how phenotypic complexity evolved in the dictyostelids and highlight changes in developmental signalling processes that may have caused novel forms and behaviours to appear.

2.2 Development of *Dictyostelium discoideum*

2.2.1 Cell-Type Specialisation and Morphogenesis

2.2.1.1 Aggregation

Upon nutrient depletion, many dictyostelid species have a choice of three alternative pathways. They can aggregate to form fruiting bodies, encyst individually, or fuse with a mate and then attract other cells to form a macrocyst. Apart from starvation there are apparently other cues to guide their choice. In general terms, high local cell densities, light, and low humidity favour fruiting body formation, whilst macrocyst formation occurs under dark and humid conditions when compatible mating types are present, using ethylene as a specific trigger (Chang et al. 1983; Amagai 1984). Microcyst formation is promoted by high osmolyte or high ammonia levels and also occurs at lower cell densities (Toama and Raper 1967; Lonski 1976).

Like all group 4 members, *D. discoideum* has lost the microcyst pathway, but nevertheless depends on other signals besides starvation to initiate development. Two of these signals are the glycoproteins PSF (prestarvation factor) and CMF (conditioned medium factor). They act as quorum sensors; cells sense that a cell density appropriate for aggregation has been reached, once PSF and CMF concentrations reach a specific threshold. The response to PSF is inhibited by bacteria and allows cells to monitor their own density relative to food. CMF acts later and is secreted by starving cells (Clarke and Gomer 1995). PSF stimulates basal expression of genes that encode proteins required for aggregation, such as the G-protein coupled cAMP receptor, cAR1, which detects the chemoattractant 3′,5′--cyclic adenosine monophosphate (cAMP), and the extracellular cAMP phosphodiesterase PdsA (Rathi et al. 1991). CMF contributes to this response, but also directly facilitates cAMP signalling (Brazill et al. 1997). In other species, the chemoattractant used for aggregation can be the dipeptide glorin, folate, pterins, or so far uncharacterised diffusible molecules (Shimomura et al. 1982; De Wit and Konijn 1983a, b).

Once basal levels of cAMP signalling genes are expressed, some cells start to spontaneously secrete pulses of cAMP. In surrounding cells, this evokes two responses: (1) extension of a pseudopod and movement towards the cAMP source and (2) synthesis and secretion of a cAMP pulse. The latter response causes the original pulse to travel outward through the population and the former causes the cells to move inwards and collect at the oscillating centre (Fig. 2.2).

In addition to mediating aggregation, the cAMP pulses also serve another function. They trigger rapid upregulation of all genes that are involved in the aggregation process and induce competence for the subsequent stages of development (Gerisch et al. 1975; Schaap et al. 1986). Continued pulsatile cAMP signalling from the tip organiser also holds the cell mass together (Dormann et al. 1998).

2.2.1.2 Formation and Migration of Slugs

The formation of well-proportioned fruiting bodies from loose aggregations of cells depends on an interplay between cell differentiation and cell movement. Cell movement remains under control of cAMP signalling until the end of development (Singer et al. 2019). Even though other dictyostelid species use different chemoattractants for aggregation, oscillatory waves of cAMP may organise their morphogenesis, too, and cAMP receptors have been found in all dictyostelid species (Alvarez-Curto et al. 2005). The tip of the aggregate takes over the function of the aggregation centre and emits waves of cAMP (Dormann et al. 1998). Cells become more adhesive and secrete matrix proteins and cellulose, which form an elastic sheath around the aggregate (Wilkins and Williams 1995). The cells, which are now confined, can only move upwards in response to cAMP waves and form the sausage-shaped slug. Gravity causes the slugs to topple over and they start migrating towards light and warmth (Bonner 1950). In nature, this serves to reach the top layer of the soil, the most propitious spot for fruiting body formation and spore dispersal (Bonner and Lamont 2005). Migration of slugs coincides in most species with stalk formation and such slugs leave a stalk behind them on the substratum ("stalked migration"). Only *D. discoideum* and a few close relatives have evolved to stalk-free migration (Romeralo et al. 2013; Schilde et al. 2019).

Meanwhile, cells are differentiating into spore and stalk precursors: prespore and prestalk cells. In the mound, these cell types appear at first intermixed (Fig. 2.2). Intrinsic factors, such as nutritional state and cell cycle phase at the onset of starvation, influence the initial choice in such a manner that the prespore cells differentiate from the largest, best-fed cells and the prestalk cells from the more starved cells (Leach et al. 1973; Weijer et al. 1984). In a starving population the most starved cells are also the first to initiate aggregation and to express aggregation-associated genes (Wang et al. 1988a). Additionally, cells which have suffered DNA damage are shed from the rear of the slug (Miermont et al. 2019). Cell–cell adhesion in the aggregate is mediated by cognate immunoglobulin-like cell surface receptors TgrB1 and TgrC1 that act in specific trans pairs (Chen et al. 2013; Hirose et al. 2011). Even though prespore cells chemotax faster towards cAMP, they are highly polarised by existing TgrB1/C1 cell–cell contacts (Fujimori et al. 2019). Prestalk cells are less polarised and form more pseudopods and therefore move more chemotactically towards the oscillating cAMP at the tip whilst the prespore cells show greater adhesion between them, thus creating an anterior–posterior prestalk–prespore pattern (Fujimori et al. 2019; Sternfeld and David 1981). Initially, approximately 80% of cells differentiate into prespore cells, and this results in a relatively stable 70–80% prespore and 20–30% prestalk population patterning. Additionally, there are cells with prestalk-like characters found scattered within the prespore region, called anterior-like cells (ALCs) that will replenish lost prestalk cells and contribute to several auxiliary structures of the fruiting body (Sternfeld and David 1981).

The differentiation of the prespore cells involves the expression of many spore-coat genes (Fosnaugh and Loomis 1991); the cells fill up with Golgi-derived vesicles that bear a thick internal coating of spore-coat glycoproteins and contain precursors and enzymes for spore coat synthesis (West 2003). The prestalk cells express genes encoding sheath and stalk matrix proteins (McRobbie et al. 1988) and are characterised by autophagic vacuoles (Schaap 1983; Yamamoto and Takeuchi 1983). These vacuoles will later fuse to form the large central vacuole of the stalk cells similar to the tonoplast of plant cells. This and the synthesis of the cellulose stalk cell wall require extensive energy and material resources, making autophagy an essential process for development (Otto et al. 2003; Mesquita et al. 2017).

In most species, transdifferentiation of prespore cells at the posterior boundary of the prestalk zone replenishes the prestalk cells which are constantly differentiated into stalk cells (Schilde et al. 2014). Rapid transdifferentiation of prespore cells is also responsible for regeneration of a lost tip after injury of the slug (Mohri et al. 2020). In *D. discoideum*, the prestalk region is replenished by the forward movement of ALCs that are generated by dedifferentiation of the most starved prespore cells (Sternfeld and David 1982; Brown and Firtel 1999; Thompson and Kay 2000a). Until the point of terminal differentiation is reached, dedifferentiation into vegetative cells can also occur when cells are dissociated from the aggregate and presented with a fresh food source. For about 60% of genes dedifferentiation is a complete reversal of the differentiation pathway albeit with differing kinetics (Nichols et al. 2020). Dedifferentiation may also occur in cells that are continuously shed from the rear of the slug if they encounter a fresh food source. This dual strategy of fruiting body construction, as well as leaving some cells able to rapidly respond to a new nutrient source, may be a strategy of bet-hedging in changeable environments.

2.2.1.3 Fruiting Body Construction

Fruiting bodies reach heights of 70 μm (*A. minutissimum*) to 1 cm (*D. giganteum*) with the typical height being 1–2 mm for *D. discoideum*. Incident light and low local ammonia levels trigger fruiting body formation (Schindler and Sussman 1977). The slug contracts and points its tip upwards (Fig. 2.2). Prestalk cells start to synthesise a central cellulose tube (Raper and Fennell 1952). They continue their movement towards the cAMP waves emitted by the tip, but now move into the upper open end of the stalk tube and differentiate into stalk cells. This involves synthesis of a cellulose cell wall, uptake of water, formation of a large central vacuole, and eventually cell death. The prespore cells climb up the newly formed stalk and mature into spores, starting from the prestalk boundary downwards. ALCs sort both downwards and upwards from the prespore region (Fig. 2.2). The downward moving cells form a basal disc surrounding the base of the stalk. A portion of the ALCs moves behind the spore mass and together with cells from the prestalk region forms a ‘lower cup’ that supports the spore head, whilst the upward moving cells from the posterior prestalk boundary cap the spore head from above as the ‘upper cup’ (Sternfeld 1998; Bichler and Weijer 1994). Cup cells, which are only found in a subset of group 4, are evolutionarily derived from stalk cells (Kin et al. 2018). Cup cells remain motile until late in development and may have functions in spore survival as well.

A remarkable exception to the construction of the stalk of dead cells is the genus *Acytostelium*. As the name suggests, *Acytostelium* species only secrete an acellular stalk tube and do not differentiate into stalk cells. As a consequence, their relatively weak stalks can only support fruiting bodies less than 2 mm tall (Cavender and Vadell 2000). The cells aggregate and secrete a thin cellulose stalk through temporal division of labour and all cells eventually ascend the stalk and differentiate into spores (Mohri et al. 2013; Urushihara et al. 2015).

Spore maturation is a rapid and tightly regulated process that involves fusion of prespore vesicles with the plasma membrane. Those Golgi-derived prespore vesicles only form in prespore cells and contain cell-wall precursors and enzymes. Their fusion with the plasma membrane is triggered only in the physical presence of prestalk cells. This brings the first layer of the spore coat in situ at the surface, whilst the remaining two layers are rapidly being produced by secreted enzymes and prefabricated wall materials (West 2003). Spores are kept dormant by high osmolarity in the spore head. Additionally, cryoprotective and antimicrobial compounds in the spore head may protect spores from infection and freezing conditions.

2.2.2 Signals that Control *D. discoideum* Development

2.2.2.1 Signalling Before and During Aggregation

The most remarkable aspect of *D. discoideum* development is the very dominant role of cAMP. As a secreted signal, it controls cell movement and differentiation throughout the developmental programme, but in its more common role as intracellular messenger, it mediates the effect of many other developmental signals. The main intracellular target for cAMP is the cAMP-dependent protein kinase A (PKA), which, like the fungal PKA, consists of a single catalytic (PKA-C) and a single regulatory subunit (PKA-R) (Mutzel et al. 1987).

During most of the life cycle, PKA is activated in a manner similar to other eukaryotes. cAMP binds to PKA-R, which causes PKA-R to dissociate from PKA-C, leaving PKA-C in its active form. However, in early development, PKA-C is regulated at the translational level. In feeding cells, PKA-C translation is inhibited by binding of the translational repressor PufA to the 3′--region of the *pkaC* mRNA (Souza et al. 1999). Upon starvation, this repression is relieved by YakA, a member of a deeply conserved protein kinase family that also regulates the decision between growth and differentiation in animals and fungi (Hartley et al. 1994; Souza et al. 1998; Mercer and Friedman 2006). Active PKA is essential for expression of early genes such as the tetrameric surface lectin *discoidin* I, and for basal expression of the aggregation genes, the cAMP receptor *cAR1*, the secreted cAMP phosphodiesterase *pdsA*, and adenylate cyclase A (*acaA*) (Schulkes and Schaap 1995). The second quorum signal CMF affects early gene expression through another route, which involves the receptor CMFR1 (conditioned media factor receptor 1) (Deery et al. 2002).

The **biochemical network that enables pulsatile cAMP signalling** involves the following sequence of events (Fig. 2.3):

1. The adenylate cyclase A (ACA) produces a low basal level of cAMP, which is rapidly secreted.
2. Secreted cAMP binds to the cAMP receptor cAR1 and upregulates ACA activity. This involves activation of the heterotrimeric G-protein, G2, and a series of intermediate steps. Binding of cAMP to cAR1 also leads to activation of the extracellular mitogen-activated kinase ErkB, which in turn inhibits the intracellular cAMP-degrading phosphodiesterase RegA resulting in an increase in intracellular cAMP concentration (Adhikari et al. 2020). This, in turn, activates cAMP-dependent protein kinase A (PKA) by dissociation of its regulatory subunit, but activated PKA also relieves the inhibition of RegA by ErkB via a negative feedback loop contributing to the cAMP oscillations (Maeda et al. 2004).
3. The cAMP pulse lasts for 3 minutes and reaches its peak when persistent stimulation with cAMP brings cAR1 in an inactive desensitised state, blocking further activation of ACA.
4. Hydrolysis of extracellular cAMP by the cAMP phosphodiesterase PdsA frees up cAR1 and together with relief of ErkB inhibition on RegA resets the system for the next pulse (Devreotes 1994; Kriebel and Parent 2004; Saran and Schaap 2004).

The combination of positive and negative feedback signalling circuits in a field of cells creates an excitable medium. Such media can generate spontaneous oscillations and waves of excitation. In a field of starving cells, randomly pulsing cells appear first, followed by circular waves that break up to form single- or multi-armed spirals. The oscillatory cAMP waves direct cell movements through regulation of cytoskeletal components (Pal et al. 2019; Schaap 2021). Spiral cAMP waves persist in mounds once aggregation is completed. They evolve into scroll waves in front of the slug and standing waves in

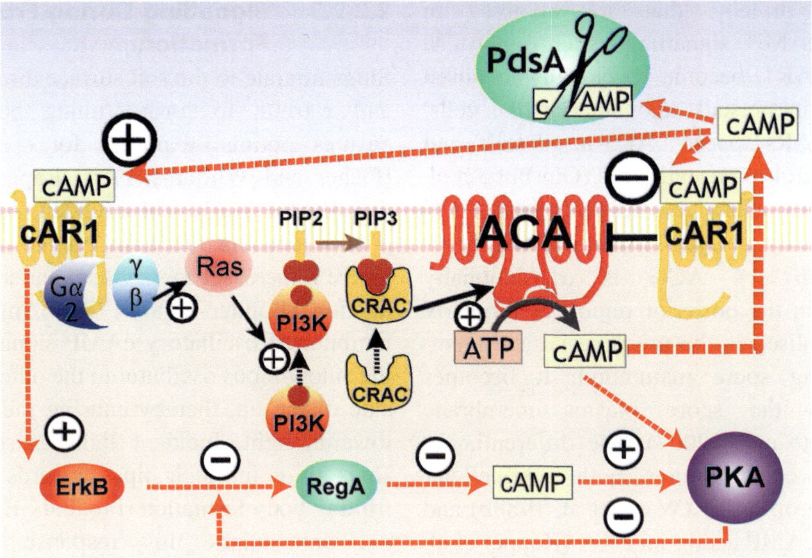

Fig. 2.3 Production of cAMP pulses. Pulsatile cAMP signalling results from positive and negative feedback loops acting on cAMP production. Cells secrete a low basal level of cAMP, which binds to the cAMP receptor cAR1 to activate the heterotrimeric G-protein G2, which dissociates into its α and βγ subunits. The latter activates PI3-kinase, which phosphorylates the membrane phospholipid PIP2 to form PIP3. PIP3 brings the Cytosolic Regulator of Adenylate Cyclase (CRAC) to the membrane, where it activates the adenylate cyclase ACA. This positive feedback loop terminates because cAMP also causes desensitisation of cAR1, which blocks further cAMP accumulation. Removal of extracellular cAMP by the phosphodiesterase PdsA resets the system for the next pulse. Binding of cAMP to cAR1 also activates extracellular signal-regulated kinase B (ErkB) which inhibits the intracellular phosphodiesterase RegA by phosphorylation. This allows for an increase in intracellular cAMP and activation of the cAMP-dependent protein kinase A (PKA) which initiates changes in gene transcription. Activated PKA also inhibits ErkB inhibition of RegA in a negative feedback loop contributing to oscillatory pulses of cAMP

the rear, the three-dimensional equivalents of spiral and concentric waves, respectively (Dormann et al. 1998; Dormann and Weijer 2001). cAMP waves originating from the tip shape the organism during slug and fruiting body formation by coordinating the movement of its component cells. Other species create similar oscillatory aggregation patterns using chemoattractants other than cAMP. Extracellular cAMP also gains other roles in cell type specification, and new layers of complexity in cAMP signalling become apparent after aggregation.

In a system of aggregative multicellularity where the cells are non-clonal, genetic conflicts can arise and kin-recognition systems are in place to guide aggregation of only genetically compatible strains. The *tgrB1/C1* genes are expressed from early aggregation to mound stage and form cognate pairs of cell surface receptors which have highly polymorphic loci (Hirose et al. 2011). Allorecognition through activation of the intracellular part of the receptors is necessary for subsequent gene activation and cells without compatible cell adhesion *tgrB1/C1* genes are shed from the aggregate (Ho and Shaulsky 2015). This ensures that only closely related strains will co-aggregate and protects the aggregate from potential cheaters that do not contribute to the formation of the stalk (Ho et al. 2013). The size of aggregates is regulated by a secreted high-molecular-weight complex called counting factor (CF) which modulates cell–cell adhesion properties (Roisin-Bouffay et al. 2000; Gomer et al. 2011).

2.2.2.2 Signalling in Mounds and Slugs

Once collected in mounds, phenotypic differences between cell types become apparent.

Firstly, the proteins that are involved in oscillatory cAMP signalling, such as ACA, PdsA, and cAR1, become specifically localised or enriched in prestalk and anterior-like cells. Three new cARs appear: cAR3 in all cells, and cAR2 and cAR4 in prestalk cells (Ginsburg et al. 1995). Two new adenylate cyclases appear, ACG and AcrA (adenylate cyclase with response regulator domain A). ACG is translationally upregulated in the posterior region of the early slug and localised in the prespore vesicle membrane. During spore maturation, it becomes localised in the spore plasma membrane (Alvarez-Curto et al. 2007). The differentiation of prespore cells requires both extracellular cAMP acting on cAR1 (Wang et al. 1988b) and intracellular cAMP acting on PKA (Hopper et al. 1993). ACG produces cAMP for both purposes. AcrA is preferentially expressed in prestalk cells; its activity increases at early fruiting body formation, where it is required for both spore and stalk cell maturation (Alvarez-Curto et al. 2007; Meima and Schaap 1999; Soderbom et al. 1999).

Dictyostelids possess large families of polyketide synthases (PKS) that show extensive clade-specific gene gain and loss. These enzymes produce polyketide compounds, some of which have signalling functions, such as the differentiation-inducing factors, DIF-1-3 and 4-methyl-5-pentylbenzene-1,3-diol (MPBD) in *D. discoideum* (Zucko et al. 2007; Ghosh et al. 2008). DIF-1, produced by the prespore cells, triggers differentiation of a subpopulation of ALCs that will sort downward during fruiting body formation to form the basal disc and lower cup (Saito et al. 2008). DIF-1, DIF-2, and MPBD also act to modulate the cAMP relay (Kuwayama and Kubohara 2009; Narita et al. 2014; Wang et al. 1986). DIF-1 has different effects in *Polysphondylium violaceum* where a mutant that cannot produce DIF-1 overproduces misshapen stalk cells (Narita et al. 2020). Species of groups 1–3 may utilise chemically distinct polyketide compounds.

2.2.2.3 Signalling During Fruiting Body Formation

Slugs migrate to the soil surface directed by light and warmth to form fruiting bodies, which ensures optimal exposure for spore dispersal (Fisher and Williams 1981; Raper 1940; Poff and Skokut 1977). The rounded slug tip acts as a lens focusing the light on the opposite side where it increases metabolism and ammonia production (Bonner et al. 1988). Ammonia is an inhibitor of oscillatory cAMP signalling driving the autonomous oscillator to the other light facing side of the tip, thereby causing the slug to turn towards light. Incident light causes migrating slugs to point their tips upwards and initiate fruiting body formation (Fig. 2.4). Excess ammonia antagonises this response and causes prolonged migration (Schindler and Sussman 1977). This is at least partly due to inhibition of terminal stalk and spore differentiation by ammonia (Hopper et al. 1993). Ammonia is sensed through the sensor histidine kinase DhkC (Fig. 2.5) and the ammonia transporters AmtC and AmtA (Kirsten et al. 2005; Singleton et al. 1998, 2006).

Sensor-linked histidine kinases are common sensors for external stimuli in bacteria, fungi, and plants but not in animals (Shaulsky et al. 1998; Thomason et al. 1998, 1999). The histidine kinase moiety can also act as a histidine phosphatase. Ligand binding to the sensor domain can trigger either the kinase or phosphatase activity. Active kinase phosphorylates itself on a histidine residue, which starts a series of His-Asp-His transfers. In *Dictyostelium*, the phosphoryl group is deposited via the phosphorelay intermediate RdeA onto the response regulator of the intracellular cAMP phosphodiesterase RegA, thereby activating cAMP hydrolysis (Fig. 2.5). When ligand binding activates the histidine phosphatase, phosphorelay runs in reverse and inactivates RegA. Ammonia activates the histidine kinase DhkC and consequently RegA, resulting in degradation of intracellular cAMP

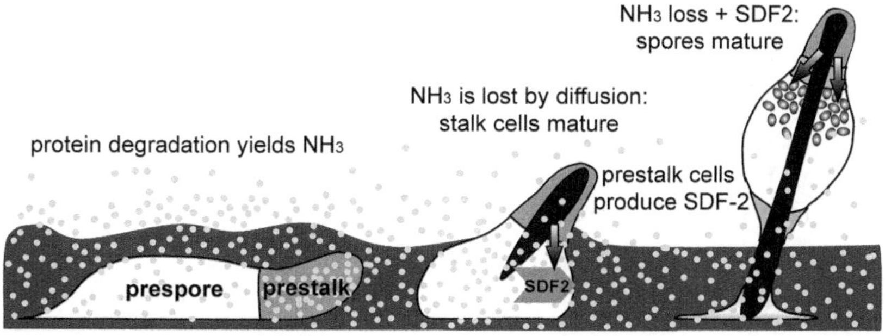

Fig. 2.4 Induction of terminal differentiation. Protein degradation in starving cells yields large amounts of ammonia (NH_3), which accumulates in soil moisture. Once slugs reach the soil surface light from above redirects the tip upwards across the air–water interface. Ammonia is now lost by gaseous diffusion and ammonia depletion allows stalk cells to mature. Maturing prespore cells secrete a protein AcbA that is processed by prestalk cells to produce the spore differentiation factor SDF2, which together with ammonia depletion triggers spore encapsulation

which prevents activation of PKA (Singleton et al. 1998). PKA activation is essential for spore and stalk cell maturation, and ammonia activation of RegA therefore effectively blocks both processes (Harwood et al. 1992; Hopper et al. 1993; Mann et al. 1994).

Culmination is induced by a sudden drop in ammonia concentration as the slug tip penetrates the surface water film (Fig. 2.4). This allows the accumulation of intracellular cAMP to activate PKA and activation of the transcription factor STATa (signal transducer and activator of transcription A) (Mohanty et al. 1999). The drop in ammonia concentration is sensed through the ammonia transporter AmtC as no nuclear accumulation of STATa in prestalk cells is observed in the *amtC*-null mutant (Kirsten et al. 2005). STATa induces expression of another transcription factor, CudA (Culmination Deficient A), which is localised in the nucleus of a cone of prestalk cells within the tip and is considered to induce stalk formation (Fukuzawa et al. 1997) (Fig. 2.5).

Another stalk cell inducing signal has been identified in *D. discoideum* as cyclic di-(3′,5′)-guanosine monophosphate (cyclic di-GMP) which is produced by diguanylate cyclase (DGC) (Chen and Schaap 2012). A role for cyclic di-GMP had so far only been established as a prokaryotic second messenger mainly involved in the transition from free-living bacteria to sessile or biofilm forming species, and no DCG proteins have ever been identified in other eukaryotes (Römling et al. 2013). Whereas DIF-1 is mainly responsible for induction of basal disc and lower cup cells, cyclic di-GMP acts to induce stalk cells of the main stalk by locally upregulating cAMP production through ACA specifically in the tip cells (Chen et al. 2017). Mutants of DGC in *D. discoideum* fail to enter culmination and are arrested at the slug stage. DGC homologues have been identified in all groups of dictyostelids (Chen and Schaap 2012), but it remains to be seen if functions of cyclic di-GMP are conserved as well.

Prespore cells require the presence of prestalk cells to proceed with encapsulation. Prespore cells secrete the protein AcbA, which is processed by the protease TagC localised on prestalk cells to yield the peptide SDF-2 that induces sporulation (Fig. 2.5). SDF-2 in turn binds to the sensor histidine kinase DhkA on prespore cells and activates its phosphatase activity, resulting in reverse phosphorelay and inactivation of RegA. cAMP can then accumulate and activate PKA, which triggers spore encapsulation (Anjard et al. 1998; Wang et al. 1999; Anjard and Loomis 2005).

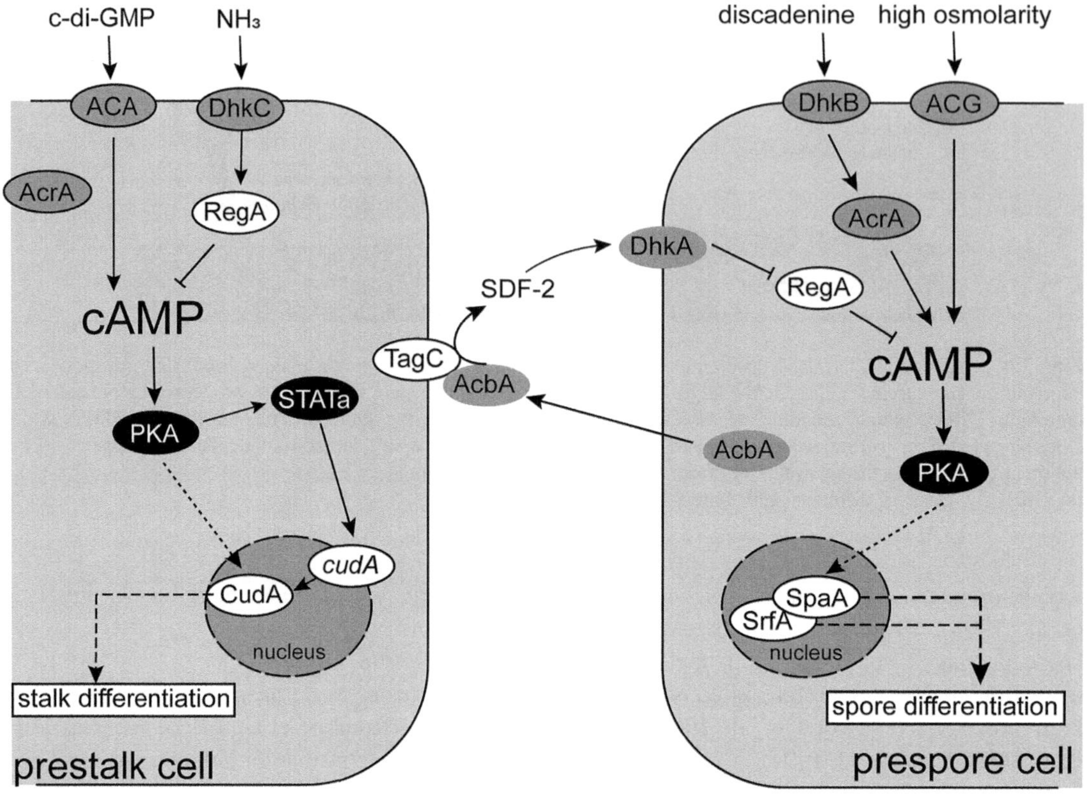

Fig. 2.5 Regulation of terminal differentiation of prestalk and prespore cells. Terminal spore and stalk maturation requires high levels of PKA activation by cAMP. In prestalk cells at the tip, cAMP is mostly produced by the adenylate cyclase ACA which is hyperactivated by the stalk inducing signal cyclic di-GMP (c-di-GMP). In prespore cells, cAMP is produced by the adenylate cyclases ACG, activated by high osmolarity, and AcrA, activated by the cytokine discadenine. The cAMP phosphodiesterase RegA requires to be phosphorylated to be active. The phosphorylation state is controlled by a sensor histidine kinase/phosphatase that either adds or removes the phosphate when a ligand binds to their sensor domain (DhkA and DhkC). Ammonia activates the histidine kinase activity of DhkC, thereby causing activation of RegA and preventing cAMP accumulation. A drop in NH_3 at the tip allows cAMP to accumulate and activate PKA. This causes activation of the transcription factor STATa leading to expression of the transcription factor CudA, which is then the most likely indirect target for activation by PKA resulting in the expression of stalk genes. Prespore cells secrete the precursor of SDF-2, AcbA. The protease TagC, which is expressed on prestalk cells, cleaves AcbA to yield SDF-2 which activates the phosphatase activity of the histidine kinase DhkA. This results in dephosphorylation of RegA and consequent cAMP accumulation and PKA activation in prespore cells. Solid lines indicate direct activations and inhibitions, dotted arrows potential protein phosphorylation cascades, and dashed angled arrows gene transcription cascades leading to cellular responses (boxed)

2.2.2.4 Signalling in the Spore Head

Dictyostelid survival critically depends on spore dispersal to fresh feeding grounds. The dispersal mechanism is thought to be surface water, but accidental or targeted dispersal by insects has also been discussed (Smith et al. 2014) and different fruiting body morphologies including pigmented spore heads may be an adaptation to animal dispersal. Spore germination is therefore tightly regulated. High intracellular cAMP levels and strong PKA activation maintain spores in the dormant state (Van Es et al. 1996; Virdy et al. 1999). High osmolarity of the spore head fluid, caused by high concentrations of ammonium phosphate, blocks spore germination (Cotter 1977; Cavallo et al. 1999). High osmolarity has two targets: it acts on the intrinsic osmosensor of ACG to activate cAMP synthesis (Saran and Schaap 2004; Van Es et al. 1996), and it inhibits cAMP degradation by RegA by triggering reverse phosphorelay through the sensor histidine kinase DokA (*Dictyostelium* osmosensing kinase) (Oehme and Schuster 2001; Schuster et al. 1996).

The cytokine discadenine is another strong inhibitor of spore germination in *D. discoideum* (Abe et al. 1976; Obata et al. 1973) and also promotes spore maturation (Anjard and Loomis 2008). Discadenine acts via activation of the sensor histidine kinase DhkB (Anjard and Loomis 2008; Zinda and Singleton 1998). Activation of DhkB results in an increase in intracellular cAMP concentration, thus activating PKA to rapidly cause encapsulation of spores and keep them dormant (Anjard and Loomis 2008). However, unlike DhkA, the DhkB pathway is dependent on the adenylate cyclase AcrA (which also has histidine kinase function) and is independent of the phosphorylation intermediate RdeA (Anjard and Loomis 2008). Activation of DhkB leads to heterodimerisation with AcrA and activation of adenylate cyclase activity. This produces a rise in cAMP and activation of PKA induces the MADS-box transcription factor SrfA, which is essential for spore differentiation (Escalante and Sastre 2002), and also BzpF, essential for both correct spore differentiation and dormancy (Huang et al. 2011). The transcription factor Spores Absent A (SpaA) is acting as PKA-dependent regulator for both spore differentiation and spore dormancy and acts upstream of SrfA (Yamada et al. 2018). Thus, there are several pathways acting in parallel to ensure spore differentiation and spore dormancy in the spore head (Fig. 2.5).

In conclusion, the transduction mechanisms of known signals that control the final stages of development all converge at regulating intracellular cAMP levels and the activation of PKA. cAMP is the most deeply conserved signalling intermediate in both pro- and eukaryotes, with a broad repertoire of functions in environmental sensing, hormone action, neurotransmission, and developmental control. Whilst intracellular use of cAMP is common to most organisms, extracellular cAMP signalling is thus far unique to social amoebas. This suggests that the extracellular use of cAMP is derived from its intracellular use. PKA is not essential for growth in social amoebas but is thereafter required for all stages of multicellular development. It is therefore likely that cAMP signalling in social amoebas has evolved from a stress response in their unicellular ancestors.

2.3 Evolution of Morphogenesis and Developmental Signalling

2.3.1 The Evolution of Morphology

The evolution of novel forms in multicellular organisms results from alterations in the developmental programmes that generate these forms. For most groups of multicellular organisms, more is known of developmental signalling in the extant, complex members of the group than in the often extinct unicellular ancestors. The evolutionary origins of these processes can therefore only be retraced by a top-down approach. To assess what are ancestral and derived traits, it is essential to know the phylogenetic relationships of the group of organisms under study. Apart from minor inconsistencies, the most recent phylogenies of dictyostelids based on either rDNA or multiple protein sets have reached agreement and recognise 4 major groups and some minor group intermediates (Schilde et al. 2019; Sheikh et al. 2015; Singh et al. 2016) (Fig.2.1). By plotting morphological characters of individual species to their family tree, the order in which these characters evolved can be assessed. Computational methods can be used to reconstruct trait probabilities in the last common ancestor (Romeralo et al. 2013; Schilde et al. 2019).

The four groups do not coincide with the earlier subdivision into genera that was based on fruiting body morphology (Raper 1984). The genus *Polysphondylium,* characterised by similar fruiting bodies bearing regular whorls of spore heads and confusingly using the same chemoattractant (glorin), appeared at two different locations: once in group 2 and once between groups 3 and 4. This and other inconsistencies have been addressed in the revised taxonomy of dictyostelids and the genus *Polysphondylium* of group 2 was renamed *Heterostelium* (Sheikh et al. 2018).

Social amoeba species differ greatly in the shape and size of the fruiting bodies, which can be solitary or clustered, unbranched or carrying branches with specific topologies. Cellular structures such as basal discs and triangular supporters may support the stalk, whilst upper and lower cups may support the spore head. The stalk may be constructed of dead cells or secreted material. A migrating slug stage can be present, but more often a stalk is laid down during migration. Alternative life cycles, such as the microcyst or macrocyst, can be present. Spores vary in size and shape and may carry conspicuous granules at their poles. The chemoattractant used for aggregation may be cAMP, folic acid, glorin, or a pterin but is unknown for most species. Twenty consistently described characters were plotted to the family tree (Romeralo et al. 2013). The characters that showed clear evolutionary trends are summarised in Fig. 2.6.

Sexual macrocysts are found in all groups, but microcysts appear to be lost from the most derived group 4. Spores are smallest in group 1 and largest in group 4, where the polar granules are smaller and more dispersed compared to their usually clustered position at the spore poles (Romeralo et al. 2013; Lawal et al. 2020). Clusters of fruiting bodies usually arise from a single aggregate in groups 1–3, and these fruiting bodies are also often branched. Aggregates usually yield solitary unbranched fruiting bodies in group 4 species; however, more recently discovered *Dictyostelium* species have clustered or coremiform (brush-like) fruiting bodies with polar spore granules indicating a greater diversity in fruiting body morphologies within this group than previously assumed (Romeralo et al. 2011). In general, group 4 fruiting bodies are significantly larger than those of other groups with species like the aptly named *Dictyostelium gargantuum* reaching heights of up to 8 mm when erect and up to 30 mm when prone (Romeralo et al. 2011). The larger fruiting bodies of group 4 species are supported by additional cell types such as upper and lower cup cells and basal discs or triangular supporters. Outside group 4 there is only a single species with a basal disc in group 1 (*Dictyostelium mexicanum*) (Cavender et al. 1981), and a cluster of species in group 3 with a crampon-shaped support (Romeralo et al. 2011). Remarkably, for all tested species in group 4, the chemoattractant is cAMP, whilst the closest related clade "polysphondylium" complex uses glorin (Shimomura et al. 1982). All tested species in the other groups use glorin or other chemoattractants but not cAMP. There is apparently a correlation between the use of cAMP during aggregation with additional cell-type specialisation, large unbranched fruiting body typus, large spore size, and loss of microcysts (Romeralo et al. 2013). This suggests that elaboration of cAMP signalling in group 4 species may have contributed to these phenotypic changes.

In contrast, in members of group 2A, the genus *Acytostelium* seems to have undergone miniaturisation and all cells of the small aggregates develop into spores which are crawling up a collectively secreted stalk tube. The largest species of *Acytostelium* reach only 2 mm in length and spore number in the spore head is amongst the lowest of all dictyostelids, because the secreted stalk can only support fewer spores. Since the group 1 species as well as the other genus of group 2B, *Heterostelium*, display a cellular stalk, the most parsimonious explanation is that the cellular stalk was present in the last common ancestor of group 2 (Traub et al. 1981). The loss of a cellular stalk in *Acytostelium* might have been the consequence of a catastrophic cheating event where the only stable solution to the dilemma of a cheater finding itself without any partner to cheat on was the cooperation of all cells to secrete a stalk.

Cell-type determination in group 1–3 species is predominantly achieved by transdifferentiation of prespore cells to replenish a continuously diminishing prestalk cell population during stalked sorogen formation (Schilde et al. 2014). The early determination of intermixed prespore

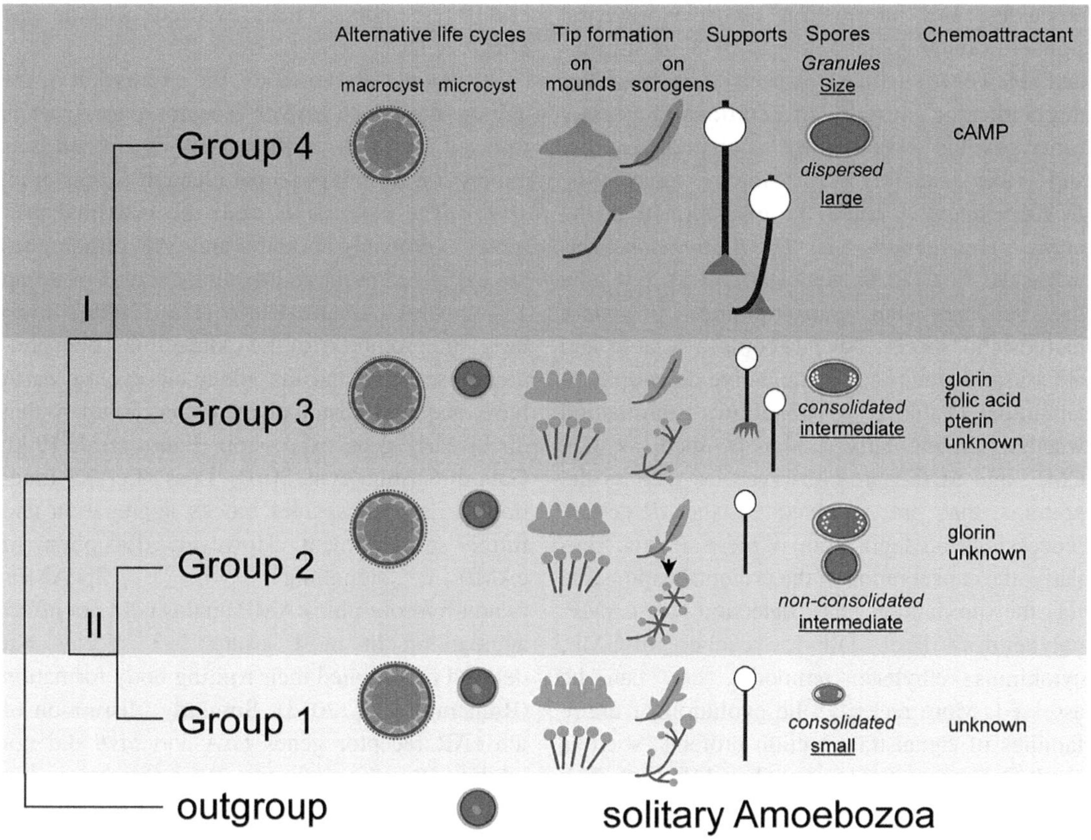

Fig. 2.6 A total of 20 described morphological characters were mapped to the phylogenetic tree. Many of those are more or less randomly scattered over the four taxon groups. An example of such a trait is sexual macrocyst formation, which occurs in multiple species in each group. However, encystation of solitary amoebas to form microcysts is lost in the most derived group 4. Species in groups 1–3 generally form multiple tips on aggregates and sorogens, which give rise to clusters of small, branched fruiting bodies. Group 4 species usually form only one or a few tips per aggregate and fruiting bodies are consequently solitary, unbranched, and large. Many group 4 species form cellular supporters or basal discs to hold up the stalk. A cluster of four species in group 3 forms unique crampon-like supports. Spore size is smallest in group 1 and largest in group 4, where spore granules are not as conspicuous as in spores of groups 1–3. None of the species in groups 1–3 use cAMP as chemoattractant, but all investigated species in group 4 do

and prestalk cells in the mound in appropriate proportions according to their later fate (80% spore and 20% stalk cells) is an evolutionary innovation of group 4 dictyostelids. We find additional triangular supporter structures at the base of the stalk in many members of group 4 and a few of group 3. A basal disc that is induced by DIF-1 is only observed in a very restricted subgroup of group 4 (clade 4C) (Romeralo et al. 2013). Upper and lower cup cells are derived from stalk cells but share transcriptional signatures with both prespore and prestalk cells (Kin et al. 2018). They also remain motile and have their own expression signature indicating additional functions at late stages of fruiting body formation.

2.3.2 The Evolution of Developmental Signalling

The model system *D. discoideum* has been almost exclusively used to study developmental processes, and information on developmental signals in other social amoebas is therefore limited. However, in many aspects, like free slug migration, the presence of additional supporter cell types, the use of cAMP as chemoattractant, and the inability to form microcysts, *D. discoideum* is rather the exception than the norm. The genome is *D. discoideum* was published first (Eichinger et al. 2005), but now the genomes and transcriptomes of more dictyostelid species have become available and are a useful source for comparative developmental studies (Table 2.1). This allows examination whether protein-derived signals such as CF, CMF, and SDF-2 are conserved, as well as the proteins that are involved in the detection, processing, and degradation of these signals. Similarly, the conservation of the proteins responsible for the production and detection of cAMP, polyketides like DIF-1, cyclic di-GMP, cytokinins, ethylene, ammonia, etc. can be assessed. More recently, the evolution of entire families of signal transduction proteins, such as small G-proteins, has been analysed (Forbes et al. 2022).

Comparative analysis of transcription factors across Dictyostelia revealed that evolution of new and more complex morphologies correlated more with change in transcription factor expression patterns than with de novo acquisition, amplification, or domain change (Forbes et al. 2019). Unexpectedly, some transcription factors with roles in *D. discoideum* cell type specification like CudA, GtaC, and STATs are conserved across the dictyostelids and even in the solitary Amoebozoans *Protostelium aurantium* and *Acanthamoeba castellanii* (Hillmann et al. 2018).

The basic components of cAMP signalling (adenylate cyclases, cAMP receptors, phosphodiesterases, PKA) are conserved in all dictyostelids, but their expression pattern has been brought forward in development in group 4. Ancestrally, cAMP likely mediated induction of encystation by environmental stress (Ritchie et al. 2008). Solitary Amoebozoa, like *Acanthamoeba castellanii,* also have sensor histidine kinases, and enzymes like AcrA, RegA, and PKA, which are used in encystation, but no cAMP receptors, which evolved only in the Dictyostelia.

In group 4 dictyostelids, the expression of the aggregation stage cAMP receptor gene *carA* is directed from a distal promoter element and later expression from a proximal element (Louis et al. 1993). The expression from the proximal promoter is probably ancestral and cAR orthologues are expressed at the multicellular stages of group 1–3 species (Alvarez-Curto et al. 2005). Similarly, the promoters of the extracellular phosphodiesterase *pdsA* and the adenylate cyclase *acaA* have acquired distal elements in group 4 that direct early gene expression (Faure et al. 1990; Galardi-Castilla et al. 2010). Loss of cAR1 function in group 4 species blocks aggregation and further development. However, disruption of cAMP signalling by SpcAMPs (a non-hydrolysable cAMP analog) did not affect aggregation in most group 1–3 species but delayed or inhibited their fruiting body formation (Romeralo et al. 2013). Similarly, disruption of the cAR receptor genes *tasA* and *tasB* did not inhibit *H. album* (group 2) aggregation but resulted in disorganised fruiting bodies (Kawabe et al. 2002, 2009). Instead of oblong spores, encapsulated round cells with high similarity to microcysts were formed in the fruiting bodies. This suggests that extracellular cAMP in the multicellular stage is used to coordinate fruiting body morphogenesis and to provide a cue to direct the specification of spores rather than cysts in those species that can form both. This implies that induction of sporulation and coordination of fruiting body morphology are the ancestral roles of extracellular cAMP signalling, whilst its role in aggregation is evolutionary derived in group 4.

Comparative phylogenetic analysis also shows that the adenylate cyclase ACG and its role in osmoregulation of spore germination are conserved in all four taxon groups (Kawabe et al. 2015). Moreover, in species that form microcysts, PKA and ACG, together with AcrA, critically regulate encystation (Ritchie et al. 2008; Kawabe et al. 2015). Microcysts usually form when cells are starved, but high osmolarity triggers encystation independently. For soil amoebas, high osmolarity may signal

Table 2.1 List of available dictyostelid genome and transcriptome assemblies. The table contains the new species names, reference publication (where available), and GenBank Bioproject numbers of the assemblies

Species	Assembly type of assembly	Publication	Genbank # accession
Dictyostelium discoideum	Genome	(Eichinger et al. 2005)	PRJNA201
Dictyostelium discoideum/ Dictyostelium purpureum	Transcriptome	(Parikh et al. 2010)	PRJNA565993
Dictyostelium intermedium	Genome		PRJNA45879
Dictyostelium citrinum	Genome		PRJNA45877
Dictyostelium firmibasis	Genome		PRJNA45875
Dictyostelium purpureum	Genome	(Sucgang et al. 2011)	PRJNA30991
Dictyostelium rosarium	Genome		PRJNA639678
	Transcriptome		PRJNA636272
Dictyostelium brefeldianum	Genome		PRJNA638801
	Transcriptome		PRJNA635606
Dictyostelium capitatum	Genome		PRJNA638827
	Transcriptome		PRJNA636270
Dictyostelium gargantuum	Genome		PRJNA639417
	Transcriptome		PRJNA636271
Polysphondylium violaceum	Genome	(Narita et al. 2020)	PRJNA45881
	Transcriptome	(Narita et al. 2020)	PRJEB34080
Tieghemostelium lacteum	Genome	(Gloeckner et al. 2016)	PRJNA305239
	Transcriptome	(Singh et al. 2016)	PRJEB12908
Tieghemostelium menorah	Transcriptome		PRJNA633982
Hagiwaraea rhizopodium	Transcriptome		PRJNA634000
Raperostelium potamoides	Transcriptome		PRJNA633998
Raperostelium gracile	Transcriptome		PRJNA633972
Raperostelium australe	Transcriptome		PRJNA633971
Speleostelium caveatum	Genome		PRJNA495862
Coremiostelium polycephalum	Genome	(Singh et al. 2016)	PRJEB14640
Heterostelium gloeosporum	Transcriptome		PRJNA634006
Heterostelium multicystogenum	Genome		PRJNA495730
Heterostelium album	Genome	(Heidel et al. 2011)	PRJNA40191
	Transcriptome	(Singh et al. 2016)	PRJEB12907
Heterostelium colligatum	Metagenome		PRJNA454077
Rostrostelium ellipticum	Genome	(Singh et al. 2016)	PRJEB14640
Acytostelium subglobosum	Genome	(Urushihara et al. 2015)	PRJDB1513
Acytostelium leptosomum	Genome	(Singh et al. 2016)	PRJEB14640
Cavenderia parvispora	Transcriptome		PRJNA633780
	Transcriptome		PRJNA630548
Cavenderia bifurcata	Transcriptome		PRJNA630419
Cavenderia multistipes	Transcriptome		PRJNA630548
Cavenderia fasciculata	Genome	(Heidel et al. 2011)	PRJNA40189
	Transcriptome	(Singh et al. 2016)	PRJEB12909
Cavenderia deminutiva	Genome	(Singh et al. 2016)	PRJEB14640
Synstelium polycarpum	Genome	(Singh et al. 2016)	PRJEB14640

approaching drought when the concentration of soil minerals increases. High osmolarity directly activates cAMP synthesis by ACG, which, together with AcrA, activates PKA and initiates the encystation process. Since encystation represents an ancestral mode of stress survival, this implies that the roles of AcrA and PKA in prespore and spore differentiation during multicellular development were evolutionarily derived from homologous roles in encystation (Ritchie et al. 2008; Du et al. 2015).

Comparative genomics also allows us to examine conservation of specific members of the polyketide synthase gene family, since this may lead us to the signal for prestalk differentiation that has remained elusive (Thompson and Kay 2000b; Saito et al. 2008). The three DIF-1 biosynthetic enzymes StlB, DmtA, and ChlA are present in all four dictyostelid taxon groups (Gloeckner et al. 2016). The basal disc inducer DIF-1 is present in the group 4 species, *Dictyostelium mucoroides*, which also has the DIF degrading enzyme, DIF-dechlorinase, and in *Polysphondylium violaceum,* between groups 3 and 4 (Kay et al. 1992). *Raperostelium minutum,* in group 3, can sense DIF-1 and direct changes in gene expression (Van Es et al. 1994). However, group 3 species possess neither DIF-dechlorinase nor basal disc cells, suggesting that DIF-1 induction of basal disc cells is a recent innovation in group 4. However, STATc, the transcription factor that is activated and rapidly accumulates in the nucleus by DIF-1 (Fukuzawa et al. 2001), is conserved in all dictyostelids, and it would be interesting to investigate its function in other taxon groups. The molecular identity of the receptor for DIF-1 has so far remained elusive.

DgcA, the enzyme which catalyses the production of cyclic di-GMP, is conserved across all dictyostelids but remarkably absent in all Amoebozoan genomes investigated so far. The receptor for DGC has not been identified yet, and it remains to be tested if cyclic di-GMP has analogous functions in culmination and stalk differentiation in other dictyostelids.

Ammonia blocks aggregation and causes cells to form microcysts in the group 2 species *H. album* (Lonski and Pesut 1977; Choi and O'Day 1982). However, in other species ammonia has an inhibitory effect on encystation (Romeralo et al. 2013). Effects on later development have not been investigated but are very likely considering the fact that all species produce this catabolite.

Dictyostelids can synthesise at least six different cytokinins and the function of the best-studied one, discadenine, in spore dormancy is restricted to group 4 species. However, IptA, which is involved in the synthesis of discadenine, and its receptor DhkB are conserved in all dictyostelids. Interestingly, IptA and two other enzymes involved in the synthesis of signalling molecules, ChlA and DgcA, as well as the osmolyte sensor DokA have been acquired by horizontal gene transfer from bacteria (Gloeckner et al. 2016).

2.4 Conclusion

The social amoebas with their combination of a unicellular and a multicellular lifestyle are an outstanding model to study the evolution of aggregative multicellularity. Their recently completed molecular phylogeny provides a framework for evolutionary studies and provides information on the core components of developmental pathways and the patterns of gene gain and loss in each of the major groups. This will show how ancestral signalling mechanisms were elaborated on and how developmental robustness was increased by bringing crucial steps in the life cycle under multiple levels of control.

The number of sequenced dictyostelid genomes and transcriptomes has been extended massively at the species level (Table 2.1). This allows for data mining and comparative genomic studies. However, genome mining can only lead so far, and experimental evidence thus far shows that most evolutionary change happened at the level of gene regulation with most genes deeply conserved. This needs to be followed up with functional studies to further consolidate this trend. One of the strengths of the *Dictyostelium* system is that such hypotheses can relatively easily be tested by gene replacement. In addition to the model *D. discoideum*, genetic tractability has also been developed for group 2 species *H. album* and the group 3–4 intermediate *P. violaceum* (Fey and Cox 1997; Narita et al. 2020). The development of a CRISPR/Cas9 system and new transfection methods for dictyostelids now allows functional studies in more species and is helped by the fact that all dictyostelids have haploid genomes (Iriki et al. 2019; Paschke et al. 2019). A restricting factor is the availability of suitable selection markers since many species display resistance to commonly used selection drugs.

Ecology, or rather adaptation to specific habitats, is the ultimate cause for the transition to a multicellular lifestyle and the evolution of multicellular complexity. We can only hope to elucidate how multicellularity evolved if we know what has driven it in the first place. Mapping the factors that characterise the habitats of species to the phylogeny, such as latitude, altitude, (micro)climate, vegetation, and soil type, would provide insights into the factors driving phenotypical change. We can now correlate morphological traits with ecological features (Romeralo et al. 2013). By integrating this information at the gene and genome level together with phenotypic change, we may get closer to understanding why and how multicellular organisms evolved.

Acknowledgements This chapter is based on the previous edition of *Evolution of Signalling and Morphogenesis in the Dictyostelids* in *The Mycota XIV* authored by Pauline Schaap. I would like to express my immense gratitude for her input into the generation of the updated version and her help with the manuscript.

References

Abe H, Uchiyama M, Tanaka Y, Saito H (1976) Structure of discadenine, a spore germination inhibitor from the cellular slime mold. Tetrahedron Lett 17(42): 3807–3810

Adhikari N, Kuburich NA, Hadwiger JA (2020) Mitogen-activated protein kinase regulation of the phosphodiesterase RegA in early Dictyostelium development. Microbiology 166(2):129–140. https://doi.org/10.1099/mic.0.000868

Alvarez-Curto E, Rozen DE, Ritchie AV, Fouquet C, Baldauf SL, Schaap P (2005) Evolutionary origin of cAMP-based chemoattraction in the social amoebae. Proc Natl Acad Sci U S A 102(18):6385–6390

Alvarez-Curto E, Saran S, Meima M, Zobel J, Scott C, Schaap P (2007) cAMP production by adenylyl cyclase G induces prespore differentiation in Dictyostelium slugs. Development 134(5):959–966

Amagai A (1984) Induction by ethylene of macrocyst formation in the cellular slime mould Dictyostelium mucuroides. J Gen Microbiol 130:2961–2965

Anjard C, Loomis WF (2005) Peptide signaling during terminal differentiation of Dictyostelium. Proc Natl Acad Sci U S A 102(21):7607–7611

Anjard C, Loomis WF (2008) Cytokinins induce sporulation in Dictyostelium. Development 135:819–827

Anjard C, Zeng CJ, Loomis WF, Nellen W (1998) Signal transduction pathways leading to spore differentiation in dictyostelium discoideum. Dev Biol 193:146–155

Baldauf SL, Roger AJ, Wenk-Siefert I, Doolittle WF (2000) A kingdom-level phylogeny of eukaryotes based on combined protein data. Science 290(5493): 972–977. https://doi.org/10.1126/science.290.5493.972

Baldauf SL, Romeralo M, Fiz-Palacios O, Heidari N (2018) A deep hidden diversity of Dictyostelia. Protist 169(1):64–78. https://doi.org/10.1016/j.protis.2017.12.005

Bichler G, Weijer CJ (1994) A Dictyostelium anterior-like cell mutant reveals sequential steps in the prespore prestalk differentiation pathway. Development 120(10):2857–2868

Bloomfield G (2019) Sex and macrocyst formation in Dictyostelium. Int J Dev Biol 63:439–446. https://doi.org/10.1387/ijdb.190183gb

Bonner JT (1950) Observations on polarity in the slime mold Dictyostelium discoideum. Biol Bull 99:143–151

Bonner JT, Lamont DS (2005) Behavior of cellular slime molds in the soil. Mycologia 97(1):178–184

Bonner JT, Chiang A, Lee J, Suthers HB (1988) The possible role of ammonia in phototaxis of migrating slugs of Dictyostelium discoideum. Proc Natl Acad Sci U S A 85(11):3885–3887. https://doi.org/10.1073/pnas.85.11.3885

Brazill DT, Gundersen R, Gomer RH (1997) A cell-density sensing factor regulates the lifetime of a chemoattractant-induced Ga-DTP conformation. FEBS Lett 404:100–104

Brown JM, Firtel RA (1999) Regulation of cell-fate determination in Dictyostelium. Dev Biol 216(2):426–441

Castillo DI, Switz GT, Foster KR, Queller DC, Strassmann JE (2005) A cost to chimerism in Dictyostelium discoideum on natural substrates. Evol Ecol Res 7(2): 263–271

Cavallo D, Cervi D, Sands TW, Cotter DA (1999) Differential in vitro activation and deactivation of cysteine proteinases isolated during spore germination and vegetative growth of Dictyostelium discoideum. Eur J Biochem 266(1):132–142

Cavender JC, Vadell EM (2000) The genus Acytostelium. Mycologia 92(5):992–1008

Cavender JC, Worley AC, Raper KB (1981) The yellow-pigmented Dictyostelia. Am J Bot 68:373–382
Cavender JC, Vadell E, Landolt JC, Stephenson SL (2005) New species of small dictyostelids from the Great Smoky Mountains National Park. Mycologia 97(2): 493–512. https://doi.org/10.3852/mycologia.97.2.493
Chang MT, Raper KB, Poff KL (1983) The effect of light on morphogenesis of Dictyostelium mucuroides. Exp Cell Res 143:335–341
Chen ZH, Schaap P (2012) The prokaryote messenger c-di-GMP triggers stalk cell differentiation in Dictyostelium. Nature 488(7413):680–683. https://doi.org/10.1038/nature11313
Chen G, Wang J, Xu X, Wu X, Piao R, Siu CH (2013) TgrC1 mediates cell-cell adhesion by interacting with TgrB1 via mutual IPT/TIG domains during development of Dictyostelium discoideum. Biochem J 452(2): 259–269. https://doi.org/10.1042/BJ20121674
Chen ZH, Singh R, Cole C, Lawal HM, Schilde C, Febrer M, Barton GJ, Schaap P (2017) Adenylate cyclase a acting on PKA mediates induction of stalk formation by cyclic diguanylate at the Dictyostelium organizer. Proc Natl Acad Sci U S A 114(3):516–521. https://doi.org/10.1073/pnas.1608393114
Choi AH, O'Day DH (1982) Ammonia and the induction of microcyst differentiation in wild type and mutant strains of the cellular slime mold Polysphondylium pallidum. Dev Biol 92:356–364
Clarke M, Gomer RH (1995) PSF and CMF, autocrine factors that regulate gene expression during growth and early development of Dictyostelium. Experientia 51:1124–1134
Cotter DA (1977) The effects of osmotic pressure changes on the germination of *Dictyostelium discoideum* spores. Can J Microbiol 23(9):1170–1177
De Wit RJW, Konijn TM (1983a) Identification of the acrasin of *Dictyostelium minutum* as a derivative of folic acid. Cell Differ 12:205–210
De Wit RJW, Konijn TM (1983b) Pterins and folates as extracellular signals for chemotaxis and differentiation in the cellular slime molds. Biochem Clin Aspects Pteridines 2:383–400
Deery WJ, Gao T, Ammann R, Gomer RH (2002) A single cell density-sensing factor stimulates distinct signal transduction pathways through two different receptors. J Biol Chem 277(35):31972–31979
Devreotes PN (1994) G protein-linked signaling pathways control the developmental program of *Dictyostelium*. Neuron 12:235–241
Dormann D, Weijer CJ (2001) Propagating chemoattractant waves coordinate periodic cell movement in *Dictyostelium* slugs. Development 128(22): 4535–4543
Dormann D, Vasiev B, Weijer CJ (1998) Propagating waves control *Dictyostelium discoideum* morphogenesis. Biophys Chem 72:21–35
Du Q, Kawabe Y, Schilde C, Chen ZH, Schaap P (2015) The evolution of aggregative multicellularity and cell-cell communication in the Dictyostelia. J Mol Biol 427(23):3722–3733. https://doi.org/10.1016/j.jmb.2015.08.008
Eichinger L, Pachebat JA, Glockner G, Rajandream MA, Sucgang R, Berriman M, Song J, Olsen R, Szafranski K, Xu Q, Tunggal B, Kummerfeld S, Madera M, Konfortov BA, Rivero F, Bankier AT, Lehmann R, Hamlin N, Davies R, Gaudet P, Fey P, Pilcher K, Chen G, Saunders D, Sodergren E, Davis P, Kerhornou A, Nie X, Hall N, Anjard C, Hemphill L, Bason N, Farbrother P, Desany B, Just E, Morio T, Rost R, Churcher C, Cooper J, Haydock S, van Driessche N, Cronin A, Goodhead I, Muzny D, Mourier T, Pain A, Lu M, Harper D, Lindsay R, Hauser H, James K, Quiles M, Madan Babu M, Saito T, Buchrieser C, Wardroper A, Felder M, Thangavelu M, Johnson D, Knights A, Loulseged H, Mungall K, Oliver K, Price C, Quail MA, Urushihara H, Hernandez J, Rabbinowitsch E, Steffen D, Sanders M, Ma J, Kohara Y, Sharp S, Simmonds M, Spiegler S, Tivey A, Sugano S, White B, Walker D, Woodward J, Winckler T, Tanaka Y, Shaulsky G, Schleicher M, Weinstock G, Rosenthal A, Cox EC, Chisholm RL, Gibbs R, Loomis WF, Platzer M, Kay RR, Williams J, Dear PH, Noegel AA, Barrell B, Kuspa A (2005) The genome of the social amoeba *Dictyostelium discoideum*. Nature 435(7038):43–57
Escalante R, Sastre L (2002) Regulated expression of the MADS-box transcription factor SrfA mediates activation of gene expression by protein kinase A during *Dictyostelium* sporulation. Mech Dev 117(1-2): 201–208
Faure M, Franke J, Hall AL, Podgorski GJ, Kessin RH (1990) The cyclic nucleotide phosphodiesterase gene of *Dictyostelium discoideum* contains three promoters specific for growth, aggregation, and late development. Mol Cell Biol 10:1921–1930
Fey P, Cox EC (1997) Gene trapping with GFP: the isolation of developmental mutants in the slime mold polysphondylium. Curr Biol 7:909–912
Fisher PR, Williams KL (1981) Bidirectional phototaxis by Dictyostelium discoideum slugs. FEMS Microbiol Lett 12:87–89
Forbes G, Chen ZH, Kin K, Lawal HM, Schilde C, Yamada Y, Schaap P (2019) Phylogeny-wide conservation and change in developmental expression, cell-type specificity and functional domains of the transcriptional regulators of social amoebas. BMC Genomics 20(1):890. https://doi.org/10.1186/s12864-019-6239-3
Forbes G, Schilde C, Lawal H, Kin K, Du Q, Chen ZH, Rivero F, Schaap P (2022) Interactome and evolutionary conservation of Dictyostelid small GTPases and their direct regulators. Small GTPases 13(1):239–254. https://doi.org/10.1080/21541248.2021.1984829
Forget M, Adiba S, De Monte S (2021) Social conflicts in Dictyostelium discoideum: a matter of scales. Preprints 2020:2020080554. https://doi.org/10.20944/preprints202008.0554.v2

Fosnaugh KL, Loomis WF (1991) Coordinate regulation of the sport coat genes in Dictyostelium discoideum. Dev Gen 12:123–132

Fujimori T, Nakajima A, Shimada N, Sawai S (2019) Tissue self-organization based on collective cell migration by contact activation of locomotion and chemotaxis. Proc Natl Acad Sci U S A 116(10):4291–4296. https://doi.org/10.1073/pnas.1815063116

Fukuzawa M, Hopper N, Williams J (1997) cudA: a *Dictyostelium* gene with pleiotropic effects on cellular differentiation and slug behaviour. Development 124: 2719–2728

Fukuzawa M, Araki T, Adrian I, Williams JG (2001) Tyrosine phosphorylation-independent nuclear translocation of a dictyostelium STAT in response to DIF signaling. Mol Cell 7(4):779–788

Galardi-Castilla M, Garciandia A, Suarez T, Sastre L (2010) The *Dictyostelium discoideum acaA* gene is transcribed from alternative promoters during aggregation and multicellular development. PLoS One 5(10): e13286. https://doi.org/10.1371/journal.pone.0013286

Gerisch G (1960) Zellfunktionen und Zelfunktionswechsel in der Entwicklung von Dictyostelium discoideum. I. Zellagglutination und Induktion der Fruchtkörperpolarität. Wilhelm Roux Arch Entwickl Mech Org 152:632–654

Gerisch G, Fromm H, Huesgen A, Wick U (1975) Control of cell-contact sites by cyclic AMP pulses in differentiating Dictyostelium cells. Nature 255:547–549

Ghosh R, Chhabra A, Phatale PA, Samrat SK, Sharma J, Gosain A, Mohanty D, Saran S, Gokhale RS (2008) Dissecting the functional role of polyketide synthases in Dictyostelium discoideum: biosynthesis of the differentiation regulating factor 4-methyl-5-pentylbenzene-1,3-diol. J Biol Chem 283(17): 11348–11354. https://doi.org/10.1074/jbc.M709588200

Ginsburg GT, Gollop R, Yu Y, Louis JM, Saxe CL, Kimmel AR (1995) The regulation of Dictyostelium development by transmembrane signalling. J Eukaryot Microbiol 42:200–205

Gloeckner G, Lawal HM, Felder M, Singh R, Singer G, Weijer CJ, Schaap P (2016) The multicellularity genes of dictyostelid social amoebas. Nat Commun 7:12085. https://doi.org/10.1038/ncomms12085

Gomer RH, Jang W, Brazill D (2011) Cell density sensing and size determination. Dev Growth Differ 53(4): 482–494. https://doi.org/10.1111/j.1440-169X.2010.01248.x

Hartley AD, Ward MP, Garrett S (1994) The Yak1 protein kinase of Saccharomyces cerevisiae moderates thermotolerance and inhibits growth by an Sch9 protein kinase-independent mechanism. Genetics 136(2): 465–474

Harwood AJ, Hopper NA, Simon M-N, Driscoll DM, Veron M, Williams JG (1992) Culmination in *Dictyostelium* is regulated by the cAMP-dependent protein kinase. Cell 69:615–624

Heidel AJ, Lawal HM, Felder M, Schilde C, Helps NR, Tunggal B, Rivero F, John U, Schleicher M, Eichinger L, Platzer M, Noegel AA, Schaap P, Glöckner G (2011) Phylogeny-wide analysis of social amoeba genomes highlights ancient origins for complex intercellular communication. Genome Res 21(11): 1882–1891. https://doi.org/10.1101/gr.121137.111

Hillmann F, Forbes G, Novohradska S, Ferling I, Riege K, Groth M, Westermann M, Marz M, Spaller T, Winckler T, Schaap P, Glockner G (2018) Multiple roots of fruiting body formation in Amoebozoa. Genome Biol Evol 10(2):591–606. https://doi.org/10.1093/gbe/evy011

Hirose S, Benabentos R, Ho H-I, Kuspa A, Shaulsky G (2011) Self-recognition in social amoebae is mediated by allelic pairs of tiger genes. Science 333(6041): 467–470. https://doi.org/10.1126/science.1203903

Ho HI, Shaulsky G (2015) Temporal regulation of kin recognition maintains recognition-cue diversity and suppresses cheating. Nat Commun 6:7144. https://doi.org/10.1038/ncomms8144

Ho HI, Hirose S, Kuspa A, Shaulsky G (2013) Kin recognition protects cooperators against cheaters. Curr Biol 23(16):1590–1595. https://doi.org/10.1016/j.cub.2013.06.049

Hopper NA, Harwood AJ, Bouzid S, Véron M, Williams JG (1993) Activation of the prespore and spore cell pathway of *Dictyostelium* differentiation by cAMP-dependent protein kinase and evidence for its upstream regulation by ammonia. EMBO J 12:2459–2466

Huang E, Talukder S, Hughes TR, Curk T, Zupan B, Shaulsky G, Katoh-Kurasawa M (2011) BzpF is a CREB-like transcription factor that regulates spore maturation and stability in Dictyostelium. Dev Biol 358(1):137–146. https://doi.org/10.1016/j.ydbio.2011.07.017

Iriki H, Kawata T, Muramoto T (2019) Generation of deletions and precise point mutations in Dictyostelium discoideum using the CRISPR nickase. PLoS One 14(10):e0224128. https://doi.org/10.1371/journal.pone.0224128

Kawabe Y, Kuwayama H, Morio T, Urushihara H, Tanaka Y (2002) A putative serpentine receptor gene tasA required for normal morphogenesis of primary stalk and branch structure in Polysphondylium pallidum. Gene 285(1-2):291–299. https://doi.org/10.1016/s0378-1119(01)00887-3

Kawabe Y, Morio T, James JL, Prescott AR, Tanaka Y, Schaap P (2009) Activated cAMP receptors switch encystation into sporulation. Proc Natl Acad Sci U S A 106(17):7089–7094. https://doi.org/10.1073/pnas.0901617106

Kawabe Y, Schilde C, Du Q, Schaap P (2015) A conserved signalling pathway for amoebozoan encystation that was co-opted for multicellular development. Sci Rep 5:9644. https://doi.org/10.1038/srep09644

Kay RR, Taylor GW, Jermyn KA, Traynor D (1992) Chlorine-containing compounds produced during Dictyostelium development. Biochem J 281:155–161
Kin K, Forbes G, Cassidy A, Schaap P (2018) Cell-type specific RNA-Seq reveals novel roles and regulatory programs for terminally differentiated Dictyostelium cells. BMC Genomics 19(1):764. https://doi.org/10.1186/s12864-018-5146-3
Kirsten JH, Xiong YH, Dunbar AJ, Rai M, Singleton CK (2005) Ammonium transporter C of Dictyostelium discoideum is required for correct prestalk gene expression and for regulating the choice between slug migration and culmination. Dev Biol 287(1):146–156
Kriebel PW, Parent CA (2004) Adenylyl cyclase expression and regulation during the differentiation of Dictyostelium discoideum. IUBMB Life 56(9):541–546
Kuwayama H, Kubohara Y (2009) Differentiation-inducing factor-1 and -2 function also as modulators for Dictyostelium chemotaxis. PLoS One 4(8):e6658. https://doi.org/10.1371/journal.pone.0006658
Lawal HM, Schilde C, Kin K, Brown MW, James J, Prescott AR, Schaap P (2020) Cold climate adaptation is a plausible cause for evolution of multicellular sporulation in Dictyostelia. Sci Rep 10(1):8797. https://doi.org/10.1038/s41598-020-65709-3
Leach CK, Ashworth JM, Garrod DR (1973) Cell sorting out during the differentiation of mixtures of metabolically distinct populations of Dictyostelium discoideum. J Embryol Exp Morphol 29:647–661
Lonski J (1976) The effect of ammonia on fruiting body size and microcyst formation in the cellular slime molds. Dev Biol 51(1):158–165. https://doi.org/10.1016/0012-1606(76)90131-7
Lonski JU, Pesut N (1977) Induction of encystment of *Polysphondylium pallidum* amoeba. Can J Microbiol 23:518–521
Louis JM, Saxe CL III, Kimmel AR (1993) Two transmembrane signaling mechanisms control expression of the cAMP receptor gene cAR1 during *Dictyostelium* development. Proc Natl Acad Sci U S A 90:5969–5973
Maeda M, Lu S, Shaulsky G, Miyazaki Y, Kuwayama H, Tanaka Y, Kuspa A, Loomis WF (2004) Periodic signaling controlled by an oscillatory circuit that includes protein kinases ERK2 and PKA. Science 304(5672):875–878. https://doi.org/10.1126/science.1094647
Mann SKO, Richardson DL, Lee S, Kimmel AR, Firtel RA (1994) Expression of cAMP-dependent protein kinase in prespore cells is sufficient to induce spore cell differentiation in *Dictyostelium*. Proc Natl Acad Sci U S A 91:10561–10565
McRobbie SJ, Jermyn KA, Duffy K, Blight K, Williams JG (1988) Two DIF-inducible, prestalk-specific mRNAs of Dictyostelium encode extracellular matrix protein of the slug. Development 104:275–284
Meima ME, Schaap P (1999) Fingerprinting of adenylyl cyclase activities during *Dictyostelium* development indicates a dominant role for adenylyl cyclase B in terminal differentiation. Dev Biol 212(1):182–190
Mercer SE, Friedman E (2006) Mirk/Dyrk1B: a multifunctional dual-specificity kinase involved in growth arrest, differentiation, and cell survival. Cell Biochem Biophys 45(3):303–315. https://doi.org/10.1385/CBB:45:3:303
Mesquita A, Cardenal-Muñoz E, Dominguez E, Muñoz-Braceras S, Nuñez-Corcuera B, Phillips BA, Tábara LC, Xiong Q, Coria R, Eichinger L, Golstein P, King JS, Soldati T, Vincent O, Escalante R (2017) Autophagy in Dictyostelium: mechanisms, regulation and disease in a simple biomedical model. Autophagy 13(1):24–40. https://doi.org/10.1080/15548627.2016.1226737
Miermont A, Antolović V, Lenn T, Nichols JME, Millward LJ, Chubb JR (2019) The fate of cells undergoing spontaneous DNA damage during development. Development 146(12):dev174268. https://doi.org/10.1242/dev.174268
Mohanty S, Jermyn KA, Early A, Kawata T, Aubry L, Ceccarelli A, Schaap P, Williams JG, Firtel RA (1999) Evidence that the Dictyostelium Dd-STATa protein is a repressor that regulates commitment to stalk cell differentiation and is also required for efficient chemotaxis. Development 126(15):3391–3405
Mohri K, Kiyota Y, Kuwayama H, Urushihara H (2013) Temporal and non-permanent division of labor during sorocarp formation in the social amoeba Acytostelium subglobosum. Dev Biol 375(2):202–209. https://doi.org/10.1016/j.ydbio.2013.01.003
Mohri K, Tanaka R, Nagano S (2020) Live cell imaging of cell movement and transdifferentiation during regeneration of an amputated multicellular body of the social amoeba Dictyostelium discoideum. Dev Biol 457(1):140–149. https://doi.org/10.1016/j.ydbio.2019.09.014
Mutzel R, Lacombe M-L, Simon M-N, De Gunzburg J, Veron M (1987) Cloning and cDNA sequence of the regulatory subunit of cAMP-dependent protein kinase from *Dictyostelium discoideum*. Proc Natl Acad Sci U S A 84:6–10
Narita TB, Chen ZH, Schaap P, Saito T (2014) The hybrid type polyketide synthase SteelyA is required for cAMP signalling in early Dictyostelium development. PLoS One 9(9):e106634. https://doi.org/10.1371/journal.pone.0106634
Narita TB, Kawabe Y, Kin K, Gibbs RA, Kuspa A, Muzny DM, Richards S, Strassmann JE, Sucgang R, Worley KC, Schaap P (2020) Loss of the polyketide synthase StlB results in stalk cell overproduction in Polysphondylium violaceum. Genome Biol Evol 12(5):674–683. https://doi.org/10.1093/gbe/evaa079
Nichols JME, Antolović V, Reich JD, Brameyer S, Paschke P, Chubb JR (2020) Cell and molecular transitions during efficient dedifferentiation. eLife 9:e55435. https://doi.org/10.7554/eLife.55435
Obata Y, Abe H, Tanaka Y, Yanagisawa K, Uchiyama M (1973) Isolation of a spore germination inhibitor from a cellular slime mold Dictyostelium discoideum. Agr Biol Chem 37:1989–1990

Oehme F, Schuster SC (2001) Osmotic stress-dependent serine phosphorylation of the histidine kinase homologue DokA. BMC Biochem 2(1):2

Otto GP, Wu MY, Kazgan N, Anderson OR, Kessin RH (2003) Macroautophagy is required for multicellular development of the social amoeba Dictyostelium discoideum. J Biol Chem 278(20):17636–17645

Pal DS, Li X, Banerjee T, Miao Y, Devreotes PN (2019) The excitable signal transduction networks: movers and shapers of eukaryotic cell migration. Int J Dev Biol 63:407–416. https://doi.org/10.1387/ijdb.190265pd

Parikh A, Miranda ER, Katoh-Kurasawa M, Fuller D, Rot G, Zagar L, Curk T, Sucgang R, Chen R, Zupan B, Loomis WF, Kuspa A, Shaulsky G (2010) Conserved developmental transcriptomes in evolutionarily divergent species. Genome Biol 11(3):R35. https://doi.org/10.1186/gb-2010-11-3-r35

Paschke P, Knecht DA, Williams TD, Thomason PA, Insall RH, Chubb JR, Kay RR, Veltman DM (2019) Genetic engineering of Dictyostelium discoideum cells based on selection and growth on bacteria. J Vis Exp 143. https://doi.org/10.3791/58981

Peracino B, Borleis J, Jin T, Westphal M, Schwartz JM, Wu L, Bracco E, Gerisch G, Devreotes P, Bozzaro S (1998) G protein beta subunit-null mutants are impaired in phagocytosis and chemotaxis due to inappropriate regulation of the actin cytoskeleton. J Cell Biol 141(7):1529–1537. https://doi.org/10.1083/jcb.141.7.1529

Poff KL, Skokut M (1977) Thermotaxis by pseudoplasmodia of Dictyostelium discoideum. Proc Natl Acad Sci USA 74:2007–2010

Raper KB (1940) Pseudoplasmodium formation and organization in Dictyostelium discoideum. J Elisha Mitchell Sci Soc 56:241–282

Raper KB (1984) The Dictyostelids. Princeton University Press, Princeton, NJ

Raper KB, Fennell DI (1952) Stalk formation in *Dictyostelium*. Bull Torrey Bot Club 79:25–51

Rathi A, Kayman SC, Clarke M (1991) Induction of gene expression in Dictyostelium by prestarvation factor, a factor secreted by growing cells. Dev Gen 12:82–87

Ritchie AV, van Es S, Fouquet C, Schaap P (2008) From drought sensing to developmental control: evolution of cyclic AMP signaling in social amoebas. Mol Biol Evol 25(10):2109–2118. https://doi.org/10.1093/molbev/msn156

Roisin-Bouffay C, Jang W, Caprette DR, Gomer RH (2000) A precise group size in Dictyostelium is generated by a cell-counting factor modulating cell-cell adhesion. Mol Cell 6(4):953–959

Romeralo M, Cavender JC, Landolt JC, Stephenson SL, Baldauf SL (2011) An expanded phylogeny of social amoebas (Dictyostelia) shows increasing diversity and new morphological patterns. BMC Evol Biol 11(1):84. https://doi.org/10.1186/1471-2148-11-84

Romeralo M, Skiba A, Gonzalez-Voyer A, Schilde C, Lawal H, Kedziora S, Cavender JC, Glöckner G, Urushihara H, Schaap P (2013) Analysis of phenotypic evolution in Dictyostelia highlights developmental plasticity as a likely consequence of colonial multicellularity. Proc Biol Sci 280(1764):20130976. https://doi.org/10.1098/rspb.2013.0976

Römling U, Galperin Michael Y, Gomelsky M (2013) Cyclic di-GMP: the first 25 years of a universal bacterial second messenger. Microbiol Mol Biol Rev 77(1): 1–52. https://doi.org/10.1128/MMBR.00043-12

Saito T, Kato A, Kay RR (2008) DIF-1 induces the basal disc of the *Dictyostelium* fruiting body. Dev Biol 317(2):444–453. https://doi.org/10.1016/j.ydbio.2008.02.036

Saran S, Schaap P (2004) Adenylyl cyclase G is activated by an intramolecular osmosensor. Mol Biol Cell 15(3): 1479–1486

Schaap P (1983) Quantitative analysis of the spatial distribution of ultrastructural differentiation markers during development of Dictyostelium discoideum. Roux's Arch Dev Biol 192:86–94

Schaap P (2021) From environmental sensing to developmental control: cognitive evolution in dictyostelid social amoebas. Philos Trans R Soc Lond Ser B Biol Sci 376(1820):20190756. https://doi.org/10.1098/rstb.2019.0756

Schaap P, Van Lookeren Campagne MM, Van Driel R, Spek W, Van Haastert PJ, Pinas J (1986) Postaggregative differentiation induction by cyclic AMP in Dictyostelium: intracellular transduction pathway and requirement for additional stimuli. Dev Biol 118(1):52–63

Schilde C, Skiba A, Schaap P (2014) Evolutionary reconstruction of pattern formation in 98 Dictyostelium species reveals that cell-type specialization by lateral inhibition is a derived trait. EvoDevo 5:34. https://doi.org/10.1186/2041-9139-5-34

Schilde C, Lawal HM, Kin K, Shibano-Hayakawa I, Inouye K, Schaap P (2019) A well supported multi gene phylogeny of 52 dictyostelia. Mol Phylogenet Evol 134:66–73. https://doi.org/10.1016/j.ympev.2019.01.017

Schindler J, Sussman M (1977) Ammonia determines the choice of morphogenetic pathways in Dictyostelium discoideum. J Mol Biol 116:161–169

Schulkes C, Schaap P (1995) cAMP-dependent protein kinase activity is essential for preaggegative gene expression in *Dictyostelium*. FEBS Lett 368:381–384

Schuster SC, Noegel AA, Oehme F, Gerisch G, Simon MI (1996) The hybrid histidine kinase DokA is part of the osmotic response system of *Dictyostelium*. EMBO J 15:3880–3889

Shaulsky G, Fuller D, Loomis WF (1998) A cAMP-phosphodiesterase controls PKA-dependent differentiation. Development 125:691–699

Sheikh S, Gloeckner G, Kuwayama H, Schaap P, Urushihara H, Baldauf SL (2015) Root of Dictyostelia based on 213 universal proteins. Mol Phylogenet Evol 92:53–62. https://doi.org/10.1016/j.ympev.2015.05.017

Sheikh S, Thulin M, Cavender JC, Escalante R, Kawakami SI, Lado C, Landolt JC, Nanjundiah V, Queller DC, Strassmann JE, Spiegel FW, Stephenson SL, Vadell EM, Baldauf SL (2018) A new classification of the Dictyostelids. Protist 169(1):1–28. https://doi.org/10.1016/j.protis.2017.11.001

Shimomura O, Suthers HLB, Bonner JT (1982) Chemical identity of the acrasin of the cellular slime mold *Polysphondylium violaceum*. Proc Natl Acad Sci U S A 79:7376–7379

Singer G, Araki T, Weijer CJ (2019) Oscillatory cAMP cell-cell signalling persists during multicellular Dictyostelium development. Commun Biol 2:139. https://doi.org/10.1038/s42003-019-0371-0

Singh R, Schilde C, Schaap P (2016) A core phylogeny of Dictyostelia inferred from genomes representative of the eight major and minor taxonomic divisions of the group. BMC Evol Biol 16(1):251. https://doi.org/10.1186/s12862-016-0825-7

Singleton CK, Zinda MJ, Mykytka B, Yang P (1998) The histidine kinase dhkC regulates the choice between migrating slugs and terminal differentiation in *Dictyostelium discoideum*. Dev Biol 203:345–357

Singleton CK, Kirsten JH, Dinsmore CJ (2006) Function of ammonium transporter a in the initiation of culmination of development in Dictyostelium discoideum. Eukaryot Cell 5(7):991–996

Smith J, Queller DC, Strassmann JE (2014) Fruiting bodies of the social amoeba Dictyostelium discoideum increase spore transport by drosophila. BMC Evol Biol 14:105. https://doi.org/10.1186/1471-2148-14-105

Soderbom F, Anjard C, Iranfar N, Fuller D, Loomis WF (1999) An adenylyl cyclase that functions during late development of *Dictyostelium*. Development 126: 5463–5471

Souza GM, Lu SJ, Kuspa A (1998) Yaka, a protein kinase required for the transition from growth to development in dictyostelium. Development 125:2291–2302

Souza GM, daSilva AM, Kuspa A (1999) Starvation promotes Dictyostelium development by relieving PufA inhibition of PKA translation through the YakA kinase pathway. Development 126(14):3263–3274

Steinert M (2011) Pathogen-host interactions in Dictyostelium, legionella, mycobacterium and other pathogens. Semin Cell Dev Biol 22(1):70–76. https://doi.org/10.1016/j.semcdb.2010.11.003

Sternfeld J (1998) The anterior-like cells in Dictyostelium are required for the elevation of the spores during culmination. Dev Genes Evol 208:487–494

Sternfeld J, David CN (1981) Cell sorting daring pattern formation in Dictyostelium. Differentiation 20(1-3): 10–21. https://doi.org/10.1111/j.1432-0436.1981.tb01150.x

Sternfeld J, David CN (1982) Fate and regulation of anterior-like cells in Dictyostelium slugs. DevBiol 93: 111–118

Strassmann JE, Zhu Y, Queller DC (2000) Altruism and social cheating in the social amoeba Dictyostelium discoideum. Nature 408(6815):965–967

Sucgang R, Kuo A, Tian X, Salerno W, Parikh A, Feasley CL, Dalin E, Tu H, Huang E, Barry K, Lindquist E, Shapiro H, Bruce D, Schmutz J, Salamov A, Fey P, Gaudet P, Anjard C, Babu MM, Basu S, Bushmanova Y, van der Wel H, Katoh-Kurasawa M, Dinh C, Coutinho PM, Saito T, Elias M, Schaap P, Kay RR, Henrissat B, Eichinger L, Rivero F, Putnam NH, West CM, Loomis WF, Chisholm RL, Shaulsky G, Strassmann JE, Queller DC, Kuspa A, Grigoriev IV (2011) Comparative genomics of the social amoebae Dictyostelium discoideum and Dictyostelium purpureum. Genome Biol 12(2). https://doi.org/10.1186/gb-2011-12-2-r20

Swanson AR, Vadell EM, Cavender JC (1999) Global distribution of forest soil dictyostelids. J Biogeogr 26: 133–148

Thomason PA, Traynor D, Cavet G, Chang W-T, Harwood AJ, Kay RR (1998) An intersection of the cAMP/PKA and two-component signal transduction systems in *Dictyostelium*. EMBO J 17:2838–2845

Thomason PA, Traynor D, Stock JB, Kay RR (1999) The RdeA-RegA system, a eukaryotic phospho-relay controlling cAMP breakdown. J Biol Chem 274(39): 27379–27384

Thompson CR, Kay RR (2000a) Cell-fate choice in dictyostelium: intrinsic biases modulate sensitivity to DIF signaling. Dev Biol 227(1):56–64

Thompson CR, Kay RR (2000b) The role of DIF-1 signaling in Dictyostelium development. Mol Cell 6(6): 1509–1514

Toama MA, Raper KB (1967) Microcysts of the cellular slime mold *Polysphondylium pallidum*. I Factors influencing microcyst formation. J Bacteriol 94(4): 1143–1149

Traub F, Hohl HR, Cavender JC (1981) Cellular slime molds of Switzerland. I. Description of new species. Am J Bot 68:162–171

Urushihara H, Kuwayama H, Fukuhara K, Itoh T, Kagoshima H, Shin-I T, Toyoda A, Ohishi K, Taniguchi T, Noguchi H, Kuroki Y, Hata T, Uchi K, Mohri K, King JS, Insall RH, Kohara Y, Fujiyama A (2015) Comparative genome and transcriptome analyses of the social amoeba Acytostelium subglobosum that accomplishes multicellular development without germ-soma differentiation. BMC Genomics 16(1):80. https://doi.org/10.1186/s12864-015-1278-x

Vadell EM, Cavender JC (2007) Dictyostelids living in the soils of the Atlantic forest, Iguazu region, Misiones, Argentina: description of new species. Mycologia 99(1):112–124. https://doi.org/10.3852/mycologia.99.1.112

Van Es S, Hodgkinson S, Schaap P, Kay RR (1994) Metabolic pathways for differentiation-inducing factor-1 and their regulation are conserved between closely related Dictyostelium species, but not between distant members of the family. Differentiation 58:95–100

Van Es S, Virdy KJ, Pitt GS, Meima M, Sands TW, Devreotes PN, Cotter DA, Schaap P (1996) Adenylyl cyclase G, an osmosensor controlling germination of *Dictyostelium* spores. J Biol Chem 271:23623–23625

Virdy KJ, Sands TW, Kopko SH, van Es S, Meima M, Schaap P, Cotter DA (1999) High cAMP in spores of *Dictyostelium discoideum*: association with spore dormancy and inhibition of germination. Microbiology 145(Pt8):1883–1890

Waddell DR (1982) A predatory slime mould. Nature 298(5873):464–466. https://doi.org/10.1038/298464a0

Wang M, Van Haastert PJM, Schaap P (1986) Multiple effects of differentiation-inducing factor on prespore differentiation and cyclic-AMP signal transduction in Dictyostelium. Differentiation 33:24–28

Wang M, Aerts RJ, Spek W, Schaap P (1988a) Cell cycle phase in Dictyostelium discoideum is correlated with the expression of cyclic AMP production, detection, and degradation. Dev Biol 125:410–416

Wang M, Van Driel R, Schaap P (1988b) Cyclic AMP-phosphodiesterase induces dedifferentiation of prespore cells in Dictyostelium discoideum slugs: evidence that cyclic AMP is the morphogenetic signal for prespore differentiation. Development 103:611–618

Wang N, Soderbom F, Anjard C, Shaulsky G, Loomis WF (1999) SDF-2 induction of terminal differentiation in *Dictyostelium discoideum* is mediated by the membrane-spanning sensor kinase DhkA. Mol Cell Biol 19(7):4750–4756

Weijer CJ, Duschl G, David CN (1984) Dependence of cell-type proportioning and sorting on cell cycle phase in Dictyostelium discoideum. J Cell Sci 70:133–145

West CM (2003) Comparative analysis of spore coat formation, structure, and function in Dictyostelium. Int Rev Cytol 222:237–293

Wilkins MR, Williams KL (1995) The extracellular matrix of the Dictyostelium discoideum slug. Experientia 51: 1189–1196

Yamada Y, Cassidy A, Schaap P (2018) The transcription factor spores absent A is a PKA dependent inducer of Dictyostelium sporulation. Sci Rep 8(1):6643. https://doi.org/10.1038/s41598-018-24915-w

Yamamoto A, Takeuchi I (1983) Vital staining of autophagic vacuoles in differentiating cells of Dictyostelium discoideum. Differentiation 24:83–87

Zinda MJ, Singleton CK (1998) The hybrid histidine kinase dhkB regulates spore germination in *Dictyostelium discoideum*. Dev Biol 196:171–183

Zucko J, Skunca N, Curk T, Zupan B, Long PF, Cullum J, Kessin RH, Hranueli D (2007) Polyketide synthase genes and the natural products potential of Dictyostelium discoideum. Bioinformatics 23(19): 2543–2549. https://doi.org/10.1093/bioinformatics/btm381

The Evolution of Mitochondrial Genomes in Fungi

3

Vassili N. Kouvelis, Alexandra M. Kortsinoglou, and Timothy Y. James

Abstract

Fungal mitochondrial (mt) genomes (mitogenomes) are diverse and highly variable both at the inter- and intra-species level. While they contain a certain set of conserved genes, exceptions with respect to gene content and order may be found among and within species of all phyla across the kingdom of fungi. Phylogenies based on the concatenated matrix of the conserved mitochondrial protein-coding genes are robust and at least as informative as the ones provided by nuclear-based gene matrices, irrelevant to their matrix sizes. The diversity of mitogenomes' size and structure, as well as presence (or not) of accessory elements (introns, ORFs, and plasmids), provides strong information regarding the genome's evolution and to an extent the evolution of the organism. This diversity is also evident in their variable gene order (synteny). Gene shuffling is common among the mt genomes of fungal species, even of closely related ones, like those belonging to the same taxonomic order. Recombination and horizontal gene transfer (HGT) events are among the major mechanisms involved, although there are several cases of gene shuffling which are the result of plasmid integration within the genome, intron mobility, and transposition. Nowadays, analyses of mitogenomes in fungal species provide evidenced insights related to the endosymbiotic event which led to the genesis of the mitochondrion in the proto-eukaryote and its evolution. This is due to the different evolutionary divergence rates that fungal mitogenomes present, given their rates differ from the known respective ones in metazoan and plant mt genomes. In this chapter, all aspects of fungal mitochondrial evolution are described, summarizing the existing knowledge on fungal mitogenomic structure, evolution, and dynamics.

Keywords

Mitochondrial genomes · Synteny · Structure of mitogenomes · Introns · Homing endonuclease genes (HEGs) · Plasmids · mt genetic codes · mtRNA polymerase · Phylogeny · α-proteobacterial endosymbiont

3.1 Introduction

The accumulation of the available whole genome sequencing (WGS) projects of fungal species has

V. N. Kouvelis (✉) · A. M. Kortsinoglou
Section of Genetics and Biotechnology, Department of Biology, National and Kapodistrian University of Athens, Athens, Greece
e-mail: kouvelis@biol.uoa.gr

T. Y. James
Department of Ecology and Evolutionary Biology, University of Michigan, Ann Arbor, MI, USA

© The Author(s), under exclusive license to Springer Nature Switzerland AG 2023

S. Pöggeler, T. James (eds.), *Evolution of Fungi and Fungal-Like Organisms*, The Mycota 14,
https://doi.org/10.1007/978-3-031-29199-9_3

progressed exponentially over this last decade (Smith 2016). Among the sequences created by the analyses of these projects for the species under examination, a small sized genome (compared to the nuclear counterpart) is frequently assembled, the mitochondrial (mt) genome or mitogenome that is located within the mitochondria of the organisms. Even before the WGS projects, analyses of mitogenomes are of great interest to the scientific community for many different reasons (Smith 2016; Rubinoff and Holland 2005). They were preferred firstly because they were easier to handle, since mitogenomes are multi-copied even within a single cell, smaller in size, and more conserved in their gene content than the nuclear chromosomes (Clark-Walker 1992). This is probably the reason why mt genomes were the first non-viral genomes to be sequenced and used in early genetic studies [e.g., for fungal mitogenomes like the ones of *Podospora anserina* (Cummings et al. 1990) and *Hansenula wingei* (Sekito et al. 1995)]. Secondly, many of the genetic exceptions known up to now, such as the high and variable copy number per cell, different genetic codes, self-splicing introns, high A/T content, missing genes, and various mutational and evolutionary rates depending on the organism that carries them (Moritz et al. 1987; Clark-Walker 1992; Christinaki et al. 2022), are found in fungal mitogenomes. Even small genes or regions present differences that may be useful for investigating genetic attributes, as well as evaluating the taxonomic, phylogenetic, and evolutionary status of the organisms that carry them (Hebert et al. 2003; Rubinoff and Holland 2005; Song et al. 2020; Liu and Wang 2021). In fact, mitochondrial genomic regions are among the most widely used and informative genetic markers for population and evolutionary studies (Smith 2016; Wolters et al. 2015; De Chiara et al. 2020).

Fungi present a diversity which is fundamentally informative for studying the evolutionary mechanisms that govern life. Within the kingdom of fungi, more than 3.3 million different species are expected to be discovered (Hawksworth and Lücking 2017), demonstrating species variability which is worth studying, given that fungi inhabit all global ecosystems and exhibit diverse modes of life, including saprotrophism, parasitism, and commensalism. Therefore, their genomic study is imperative for obtaining the required insights to explain evolution among other basic or applied scientific questions.

In this chapter, the acquired knowledge concerning the mitogenomes of fungal species will be presented under the prism of their contribution to evolution. Specifically, general features including genome content and size, gene order (synteny), intron abundance, intergenic regions' variability, genetic code usage, and mutational divergence rates will be discussed in depth, in respect of their contribution in the evolution of mitogenomes, as well as the evolution of the organisms. For reasons of consistency throughout this chapter, mt gene names will follow the nomenclature proposed by Boore (1999).

3.2 General Features of Mitogenomes

Fungal mt genomes present remarkable diversity in size, while their core gene content can be considered generally conserved. Their features are similar to the ones found in metazoa, but enriched with accessory elements like introns, additional genes, usually unidentified open reading frames or pseudogenes, plasmids, and large intergenic regions. The inclusion of these elements usually renders fungal mitogenomes larger than their metazoan counterparts (Lang et al. 1999). There are exceptions within early-diverging species belonging to Microsporidia and Neocallimastigomycota, which do not contain any mt genomes (Bullerwell and Lang 2005; James et al. 2013).

The typically conserved fungal mitogenome consists of ~30–40 genes implicated in the most important function of the cell, i.e., production of ATP through aerobic respiration. Specifically, there are genes that encode proteins functional as subunits for NADH dehydrogenase (*nad*1-*nad*6 and *nad*4L), apocytochrome b (*cob*), cytochrome c oxidase (*cox*1-*cox*3), and ATP synthase (*atp*6, *atp*8, and *atp*9). The conserved package of fungal mitogenomes is completed with 2 genes encoding rRNA subunits, i.e., *rnl* and *rns* for the large (23S) and small (16S) rRNA subunits, respectively, a protein important to the assembly of the small ribosomal subunit, *rps*3 (in the early literature it is also called *var*1), a variable number of ~22–26 tRNA *trn* genes, which usually cover all tRNA needs of the protein synthesis within the organelle, and tRNA processing genes, like *rnp*B (Cummings et al. 1990; Clark-Walker 1992). There are, however, several fungal mitogenomes which lack a varying number of these core genes. For instance, the mt genomes of species belonging to Cryptomycota, Schizosaccharomycetaceae, Saccharomycetaceae, and Saccharomycodaceae lack the *nad* genes (Bullerwell et al. 2003; James et al. 2013; Freel et al. 2015; Quandt et al. 2017; Christinaki et al. 2022). Additionally, the mitogenomes of most Chytridiomycetes lack many *trn* genes (Forget et al. 2002; van de Vossenberg et al. 2018), while several mitogenomes of taxonomically dispersed species lack one or two genes, like *rps*3 (Bullerwell et al. 2000; Korovesi et al. 2018), *atp*8 and *atp*9 (Franco et al. 2017), or *trn* genes, such as *trn*A (for $tRNA^{Ala}$) and *trn*C (for $tRNA^{Cys}$) in *Lecanicillium muscarium* (Kouvelis et al. 2004) (see Table 3.1).

Other genes, which often enrich the mitogenomes' gene content, are the ones encoding for (a) homing endonucleases (HEs) or reverse transcriptases (RTs) needed for horizontal mobility of themselves (and usually of the mt introns which carry them) and (b) even less frequently, DNA and RNA polymerases or remnants of these genes and (c) unidentified ORFs with either an unidentified function or a role in fungal pathogenesis (see corresponding sections below).

Studies of genome sequences show that DNA transfer from organelles to the nucleus occurs at very high rates and that this influx is an ongoing process, occurring since the first endosymbiotic event (Timmis et al. 2004). Still, questions related to the reasons for which the few genes of fungal mitogenomes comprising its main core remained "captive" within the organelle, when the first bacterial endosymbiont turned out to be the mitochondrion (Adams and Palmer 2003), remain unclarified and addressed with suggested hypotheses. These core genes were not even transferred during the much later organellar gene transfer (OGT) incidents, which include the occasional transfer of large DNA genes/fragments from the mitochondrion to the nucleus and, in rare cases, vice versa (Martin 2003). The main arguments that try to explain this hotly debated issue are (a) the protein structure of these genes' products or (b) the mitochondrion's need for redox control. The first theory, called hydrophobicity theory, supports that these genes remain within the mt genome because their products cannot be transferred from the cytoplasm to the inner mitochondrial membrane due to their amino acid content and structure. The hypothesis that these proteins remained due to their transmembrane structure seems plausible and also acceptable for the fungal mt genomes (Adams and Palmer 2003; Daley and Whelan 2005). The second theory supports that these genes had to be maintained and not transferred to the nucleus, due to their implication in the organelle's redox balance (Allen 2003). Thus, this core of genes can be found in all eukaryotes, including fungi and metazoa. Findings show that most of the hydrophobic proteins of both mitochondria and chloroplasts are encoded by the nucleus and transferred to the organelles, a fact that conflicts the hydrophobicity theory (Allen 2003). Mitogenomes, however, are highly compact in metazoa (sizes usually ranging from 17 to 26 kb), whereas in fungi the respective range is from 11,223 bp (*Hanseniaspora pseudoguilliermondii*) to 272.2 kb (*Morchella importuna*) (Liu et al. 2020b). In other words, if the evolution of mitogenomes was studied by focusing only on metazoa, the conclusion would have been that mitogenomes shrink in size during evolution. However, as it has recently been shown in a few studies of fungi and more specifically in species of the subphylum Saccharomycotina of the phylum Ascomycota (Christinaki et al. 2022; Hao 2022), the size of mitogenomes changes without a clear pattern of reduction or expansion, since both of these processes seem to have happened in many different

Table 3.1 Available mitogenomes of fungi in public databases (e.g., NCBI Nucleotide and Organelle databases), the species taxonomy, and other attributes of their mtDNA. The number of fungal genera and their species according to the taxonomic order to which they belong is provided. The range of tRNA (*trn*) genes in each order, as well as "missing" protein encoding genes, i.e., genes of ribosomal protein S3 (*rps*3), ATP synthase (*atp*6, 8, 9), NADH dehydrogenase (*nad*1–6, 4L), cytochrome c oxidases (*cox*1-3), and apocytochrome b (*cob*) subunits, are given as numbers and ticks, respectively. Ribosomal RNA genes are not shown as they are always present in fungal mitogenomes (The table is based on data acquired until May 2022 from the Kortsinoglou A.M., Myridakis A.I., and Kouvelis V.N.—manuscript under preparation)

							"Missing" genes[a]														
						No of	*atp*			*nad*							*cox*				
Phylum	Subphylum	Class	Order	Genera	Species	*trns*	6	8	9	1	2	3	4	4L	5	6	cob	1	2	3	*rps*3
Ascomycota	Pezizomycotina	Arthoniomycetes	Arthroniales	2	2	12,18			✓												
		Dothideomycetes	Botryosphaeriales	1	1	26		✓													
			Cladosporiales	1	3	26–28															✓
			Dothideales	1	1	26															
			Mycosphaerellales	5	6	25–38			✓	✓	✓	✓	✓	✓	✓	✓					✓
			Pleosporales	11	14	21–38		✓										✓			✓
			Venturiales	1	1	N/A[c]															
		Eurotiomycetes	Chaetothyriales	3	5	22–25										✓					✓
			Eurotiales	6	32	10–31		✓			✓			✓						✓	✓
			Onygenales	9	14	25–26															✓
		Lecanoromycetes	Caliciales	1	2	24–28			✓												
			Lecanorales	13	25	29–17		✓													✓
			Ostropales	1	1	26		✓													
			Peltigerales	4	7	21–29		✓	✓							✓					✓
			Pertusariales	1	1	29															
		Leotiomycetes	Erysiphales	4	5	24–26													✓		
			Helotiales	9	13	21–33		✓		✓	✓	✓	✓	✓	✓	✓				✓	✓
			N/C[b]	1	4	27–29															
			Thelebolales	1	1	17		✓		✓	✓	✓	✓	✓	✓	✓					
		Orbiliomycetes	Orbiliales	5	7	18–26															✓
		Pezizomycetes	Pezizales	3	3	14–31															
		Sordariomycetes	Diaporthales	9	4	25–28									✓	✓					
			Glomerellales	2	12	25–29						✓								✓	

			Hypocreales		121	18–30			✓	✓			✓	✓		✓	✓	✓	✓	✓	✓
			Magnaporthales		–	N/A															
			Microascales	5	12	25–31			✓												
			Ophiostomatales	3	6	24–27			✓												✓
			Sordariales	4	6	26–28			✓								✓		✓	✓	✓
			Xylariales	4	4	25–32		✓								✓					✓
	Saccharomycotina	Saccharomycetes	Saccharomycetales	38	245	23–47[a]		✓		✓	✓	✓	✓	✓	✓	✓	✓		✓	✓	✓
	Taphrinomycotina	Pneumocystidomycetes	N/C	1	8	24–28											✓				✓
		Schizosaccharomycetes	Schizosaccharomycetales	1	3	24–25				✓	✓	✓	✓	✓	✓	✓				✓	✓
		Taphrinomycetes	Taphrinales		–	N/A															
Basidiomycota	Agaricomycotina	Agaricomycetes	Agaricales	30	51	14–36				✓								✓			✓
			Auriculariales	1	2	22,23															✓
			Boletales	7	10	25–27															✓
			Cantharellales	3	6	25–27															
			Gomphales	1	1	24															
			Hymenochaetales	9	10	25–30															
			Phallales	2	2	24															
			Polyporales	7	20	25–29															
			Russulales	6	16	24–27												✓			✓
			Sebacinales	–	–	N/A															
			Thelephorales	1	2	25,28															✓
			Tremellales	4	7	16–23															✓
			Trichosporonales	1	3	14–25		✓	✓		✓	✓		✓	✓	✓	✓		✓	✓	
	Pucciniomycotina	Agaricostilbomycetes	Agaricostilbales	1	1	23															✓
		Cystobasidiomycetes	Cystobasidiales	1	1	24															
		Microbotryomycetes	Microbotryales	1	2	24,31															
		Pucciniomycetes	Pucciniales	3	6	23–24		✓									✓				✓
		Microbotryomycetes	Sporidiobolales	2	3	23–28															
	Ustilaginomycotina	Exobasidiomycetes	Microstromatales	1	1	22														✓	
			Tilletiales	1	2	24		✓		✓	✓	✓	✓	✓	✓	✓					✓
		Malasseziomycetes	Malasseziales	1	9	22–25															

(continued)

Table 3.1 (continued)

Phylum	Subphylum	Class	Order	Genera	Species	No of *trns*	"Missing" genes[a]														
							atp			*nad*							cob	*cox*			*rps*3
							6	8	9	1	2	3	4	4L	5	6		1	2	3	
		Ustilaginomycetes	Ustilaginales	4	6	21–22															✓
	N/C	Blastocladiomycetes	Blastocladiales	2	3	25–26		✓	✓												
		Chytridiomycetes	Chytridiales	1	1	7															✓
			Rhizophydiales	1	1	7															
			Spizellomycetales	1	2	8,12								✓							✓
			Synchytriales	1	2	5															✓
		Monoblepharidomycetes	Monoblepharidales	3	4	7–9															✓
		N/C	N/C	2	2	23															
	Glomeromycotina	Glomeromycetes	Diversisporales	1	2	22,25															✓
			Glomerales	2	7	20–26														✓	✓
	Mortierellomycotina	Mortierellomycetes	Mortierellales	1	1	26															
	Mucoromycotina	Mucoromycetes	Mucorales	5	5	24–25															✓
	Entomophthoromycotina	Entomophthoromycetes	Entomophthorales	1	1	23															✓
	Kickxellomycotina	Harpellomycetes	Harpellales	1	1	27															
	Cryptomycota (phylum)	N/C	N/C	2	2	4,23				✓	✓	✓	✓	✓	✓	✓					✓
	TOTAL MITOGENOMES OF DIFFERENT SPECIES:				762																

[a]"missing" = not found in at least one species of the order. Moreover, the lack of all genes within an order (e.g., in Mycosphaerellales) may be the sum of different "missing" genes in different species of the order

[b]N/C = Non-characterized taxonomic group

[c]N/A = Submitted mitogenome BUT Not Annotated

independent events (Christinaki et al. 2022). A similar study, but with only 25 mitogenomes under examination, had reached the same conclusion for the kingdom of fungi (Aguileta et al. 2014). Interestingly enough, size diversity working restlessly toward both directions, i.e., either to expansion or reduction, can be even monitored within the same species, as it was shown with *Saccharomyces cerevisiae* (De Chiara et al. 2020).

3.3 Topology: Structure of Mitogenome

Another intriguing feature of fungal mitogenomes is their morphology. Almost all bilaterian animal mitogenomes are found as circular molecules (Boore 1999), with a small, unique region called D-loop as the epicenter for starting their replication (Falkenberg 2018). In fungi, such a unique region has not been determined, but in *Saccharomyces cerevisiae* and its close relatives, several specific regions have been determined to act as origins of replication (ori), while in other yeasts, different modes of replication have been proposed, i.e., the rolling circle replication followed by template switching. Thus, it is suggested that ori elements are not a universal attribute in mt genomes of yeasts (Chen and Clark-Walker 2018). Therefore, even today, except for *S. cerevisiae* and some of its related species, the mechanism of mt genome replication is not fully and experimentally clarified. Mitogenome structure remains ambiguous regarding circularity or linearity in fungi, with a few experimentally substantiated exceptions (Zardoya 2020). As a common belief, mitochondrial genomes are mostly circular, but findings regarding the prevalence of linear fungal mt genomes are growing exponentially. The notion about fungal mitogenome circularity is still, rather wrongfully, retained for three logically sound facts: (a) beyond any doubt, the circularity of most mt genomes of bilaterian metazoa, like the human mitogenome, which historically was among the first to be fully studied (Menger et al. 2021), (b) the endosymbiotic ancestor of mitochondrion which was a bacterium with a circular genome (Williamson 2002), and (c) the circularity of the restriction maps of fungal mt genomes (Bendich 1993). Exceptions that experimentally verified the linearity of the mitogenome were found in early-diverging fungi, like the chytrids *Hyaloraphidium curvatum* (Forget et al. 2002) and *Synchytrium endobioticum* (van de Vossenberg et al. 2018), and in yeasts, either as multipartite linear array (e.g., *Saccharomyces cerevisiae* and *Candida* spp. Maleszka et al. 1991; Valach et al. 2011) or monomeric linear with inverted terminal repeats (e.g., *Pichia pijperi*—Dinouël et al. 1993; *Hanseniaspora uvarum*—Pramateftaki et al. 2006). Especially in yeasts, full analyses, employing pulse field gel electrophoresis (PFGE) and mapping of termini with exonucleases, provided further knowledge by discriminating the telomeric regions of the linear mitogenomes into two types of terminal structures: type I, with inverted terminal repeats containing a covalently closed single-stranded hairpin (Dinouël et al. 1993), and type II, with the inverted terminal repeats composed of tandem arrays of large repetitive units which end with an incomplete repeat unit possessing a 5′ single-stranded extension of defined length (Nosek et al. 1995). Mitogenome linearity can also be found in other eukaryotes, besides fungi, and it seems to be a trait acquired independently many times throughout evolutionary history assuming genome circularity of the α-bacterial endosymbiont. This is also confirmed by observations that genome structure is not related to species phylogeny, as closely related fungal species may have different mt genome structures (Nosek et al. 1998; Christinaki et al. 2022). The prevailing hypothesis for this morphological change of the mitogenome involves the integration of mobile elements including linear mt plasmids or transposons (Nosek and Tomáska 2003; Fricova et al. 2010).

Regardless of the genome's conformation, mitogenomes are clustered in different numbers in the mitochondria and are packaged and protected within proteins in a structure that is reminiscent of the chromosomal DNA form of bacteria which is in accordance with endosymbiotic theory (Gilkerson et al. 2013). Further experiments would, however, have to be performed in order to decipher the in vivo conformation of fungal mitogenomes, and thus, at the moment, it is prudent to refer to it as a circularly mapped fungal mitogenome when there is no clear experimental evidence for the linearity of the genome examined.

3.4 Synteny (Gene Order) and Genetic Rearrangements

The content of the core genes in fungal mt genomes is generally conserved (with only a few exceptions in which one or more genes are missing, as previously mentioned). The gene order (or synteny) of fungal mt genomes, however, presents extraordinary plasticity (Kouvelis et al. 2004; Freel et al. 2015; Aguileta et al. 2014) (Fig. 3.1). Nowadays, thanks to the WGS projects and the advent of next-generation sequencing (NGS), the known mt genome sequences have increased exponentially, and it is evident that mitogenomic synteny is often so different from one species to another that its study should be considered a necessity if fungal mt genome evolution is to be analyzed. Synteny can be studied with two different approaches, considering (a) only the large sized core genes, i.e., rRNA and protein-coding genes, or (b) with the additional inclusion of smaller genes as the *trn* genes (coding tRNAs). The first approach offers the advantage of retrieving the order of the large, core genes in a simple depiction, which may provide information about ancestral core gene pairs or clusters (Kouvelis et al. 2004), with the additional retrieval of trans-spliced or divided genes, like the *rns* and *cox*1 genes of *Gigaspora rosea* (Nadimi et al. 2012), the *rns* of *Hyaloraphidium curvatum* (Forget et al. 2002), or of pseudogenes and chimeric genes [i.e., fused intronic homing endonuclease gene (HEG) with the preceding exon of the mt gene], like the ones found in the mitogenome of *Sclerotinia borealis* (Mardanov et al. 2014) and *Beauveria brongniartii* (Ghikas et al. 2010), respectively. However, the inclusion of *trn* genes offers more insights into the synteny and evolution of mitogenomes, because *trn* genes can be found either as single units, scattered among the large core genes, or as clusters, which are usually conserved when considering mt genomes of fungi belonging to the same order (Ghikas et al. 2006; Christinaki et al. 2022). Clear syntenic groups and patterns can be found among fungal species belonging to the same order (Fig. 3.1), especially in the case of Ascomycota (Pantou et al. 2008; Christinaki et al. 2022). Both types of distribution provide useful information under the prism of genomics and evolution. Single *trn* genes may play a dual role, either as termination signals of transcription (Schäfer et al. 2005; Varassas and Kouvelis 2022) or as hotspots for recombination events which may contribute to gene shuffling (Pantou et al. 2008). This latter theory has been proposed earlier for mt gene shuffling in metazoa (Saccone et al. 2002), but it must be used cautiously in fungal mitogenomes, since (a) species belonging to early-diverging fungi (EDF), like the ones belonging to Chytridiomycetes, usually contain a very small number of *trn* genes (Forget et al. 2002; Zardoya 2020) and (b) other repetitive elements like the GC islands of yeasts and *Neurospora crassa* may play the role of the target sequences for recombination (Yin et al. 1981; Dieckmann and Gandy 1987).

3.5 Mt Plasmids

The mt plasmids can be defined as sequences with little or no homology to the mitochondrial genome and have an independent evolutionary history that may render them intracellular parasites. They have been found in plants, fungi, and protists (Hausner 2012). Fungal mt plasmids are either circular or linear, usually carrying one or more genes involved in replication and transcription (DNA or RNA polymerases) (Griffiths 1995; Hausner 2012). They can be mainly discriminated into three different types: (a) linear or circular retroplasmids, which usually carry a gene that encodes a reverse transcriptase (Kennell and Cohen 2004), (b) linear plasmids

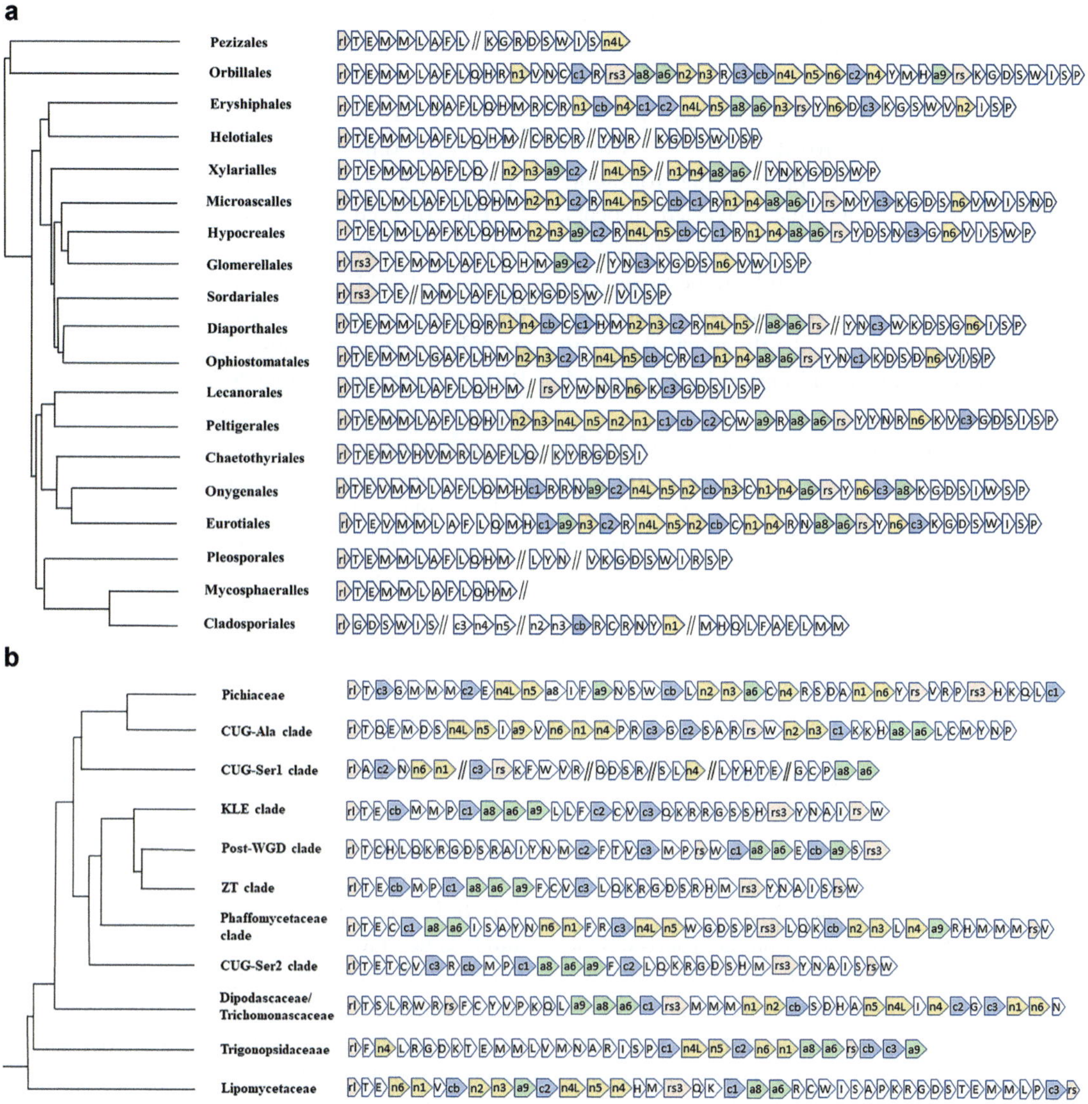

Fig. 3.1 Gene order of mitogenomes of species belonging to different orders of Ascomycota and their phylogenetic relationships: synteny of (**a**) various orders of Pezizomycotina subphylum and (**b**) the major phylogenetic clades of Saccharomycotina subphylum. Due to the variability of syntenic units among all members of each order, the sequence that represents most of the taxa of each order for Pezizomycotina and one representative from each clade of Saccharomycotina presented. Orders that contained less than three strains were not included in the analysis. Each arrow represents a gene. Pink arrows with rl and rs represent *rnl* and *rns* genes, respectively; blue and yellow arrows with cb, c1-c3, n1-n6, and n4L correspond to *cob*, *cox*1-3, *nad*1-6, and *nad*4L, respectively. White arrows with a single capital letter are for the *trn* genes which will encode the tRNAs of the respective one-letter designated amino acid. b) The double slash symbol (//) depicts a high variability of gene order at this part of the sequence. The dendrograms were based on the phylogenetic trees of Li et al. (2021b), Abdollahzadeh et al. (2020), and Christinaki et al. (2022) and represent the phylogenetic relationships of each order without showing distances

with terminal inverted repeats and with genes encoding either a DNA or an RNA polymerase, or both (Nakai et al. 2000; Klassen and Meinhardt 2007), and (c) circular plasmids that encode a DNA polymerase (Griffiths 1995). Linear plasmids, in particular, present the following usual features: (a) terminal inverted repeats, (b) genes for RNA and DNA polymerases of viral origin, and (c) termini with either covalently, hairpin-like closed ends (i.e., "hairpin" plasmids) or proteins which are covalently attached to their 5′ termini (i.e., "inverton" plasmids) (Meinhardt et al. 1997).

Their size varies usually from 5 to 10 kb for linear DNA plasmids, which encode for both DNA-dependent DNA and DNA-dependent RNA polymerases, or from 2.5 to 5 kb circular DNA plasmids that encode either a DNA polymerase or a reverse transcriptase (Cahan and Kennell 2005). It is highly intriguing that mt plasmids can be found either stably integrated in the mt genome, like in *Moniliophthora perniciosa* (Formighieri et al. 2008), or free and occasionally integrated in the mt genome, like in *Neurospora* spp. (Myers et al. 1989) and in *Podospora anserina* (Hermanns et al. 1994).

They have been considered as cryptic mitochondrial elements, since there is no evidence for a change in their host's phenotype (Hausner 2012), but according to several experimentally examined cases, they have also been correlated to senescence, life-span elongation, vegetative incompatibility, and reduced pathogenicity. Specifically, circular retroplasmids, like the *kalilo* or the *Maranhar* plasmids of *Neurospora intermedia*, may be inserted in several different domains of the mt genome and thus disrupt the function of essential genes, leading to growth cessation and aging, a process called senescence (Fox and Kennell 2001). However, in *P. anserina*, insertion of the linear pAL2-1 is correlated with an increased life span, and therefore, this mitochondrial plasmid has been named life-span-prolonging plasmid (Hermanns et al. 1994). Moreover, hairpin plasmids with variable hairpin loops have been correlated to vegetative incompatibility groups of *Rhizoctonia solani* (Katsura et al. 1997), while the circular plasmid pCRY1 of *Cryphonectria parasitica* is an infectious agent that reduces pathogenicity (Monteiro-Vitorello et al. 2000). Finally, in *Candida subhashii*, a linear invertron-type plasmid may have been domesticated to serve as its mitochondrial genome's telomeres (Fricova et al. 2010).

The potential of plasmid integration into their respective mt genome has not been fully resolved for all different types of mt plasmids, but plasmid-like sequences have been found to cluster in domains of mt genomes where low numbers of PstI palindromic sequences exist (Cahan and Kennell 2005). Short sequence homologies of the target regions within the terminal sites of the plasmid are necessary for the integration of the *kalilo* plasmid, whereas other integrated plasmids, like *Maranhar,* do not require a sequence homology for being incorporated in the mt genome. This renders plasmid duplications, horizontal plasmid transfer, or other unknown mechanisms as the proposed explanations for the existence and integration of mt plasmids into the mt genomes (Cahan and Kennell 2005). It is worth mentioning that in a few cases, like the mt genome of *Agrocybe aegerita*, there are two polymerase genes (*pol*B) flanked by two large, inverted repeats longer than 2.4 kb. Each repeat contains an identical copy of *nad*4 gene and, thus, this whole region is of linear-plasmid origin or might even be a remnant of a plasmid (Liu et al. 2020a). In *Moniliophthora perniciosa*, the existence of ORFs between the inverted repeats of the plasmid was considered as an indication of an integration-recombination event (Formighieri et al. 2008), despite the fact that no ORF coding for integrase activity has been found in mt plasmids of any fungal species so far (Cahan and Kennell 2005).

Overall, mt plasmids can be found quite often in fungal mitochondria, mostly in Ascomycetes and Basidiomycetes. Even if they have been integrated into the mt genome, they always carry at least a gene implicated in their function, usually a DNA or RNA polymerase. These genes share a common ancestor with their counterparts in phages (Pöggeler and Kempken 2004), but more recent horizontal gene transfer events cannot be excluded (Kempken 1995). In *M. perniciosa,* the integration of the plasmid in the mt genome was a recent event, since the plasmid sequence is not adapted to the rest of the mt sequence, but the plasmid was not found in its close relative, *M. roreri* (Formighieri et al. 2008). The existence of the mt integrated plasmid might be an evolutionary advantage if the plasmid has been acquired recently and prolongs the life span of the organism carrying it (Formighieri et al. 2008). Similarly, in the mt genomes of *Glomus* spp., numerous DNA polymerase (*dpo*) sequences were retrieved with no similarity to each other, but they seemed to be related to *dpo* plasmid sequences in *Daucus carota*, a plant that may act as a host (Beaudet et al. 2013). This example is an indication that the plasmid integration or even the integration of plasmid-derived genes may occur repeatedly and independently throughout evolution. Additionally, the recent finding that mt plasmids are common in Agaricomycetes (with only the exception of *Cantharellus cibarius,* out of the 12 strains examined), but not in other known classes of Basidiomycota (Himmelstrand et al. 2014), indicates that vertical inheritance of mt plasmids and their genes from the common ancestor of Agaricomycetes may explain their existence with subsequent integration events into the mt genomes at many different independent occasions.

3.6 Nuclear–Mitochondrial Interactions

Nuclear mitochondrial interactions may be studied under the prism of three different approaches: (a) the involvement of nuclear (nc)-encoded proteins in the functions of mitogenomes, like replication, transcription, and DNA repair, (b) the mechanisms and processes that regulate the stoichiometry of nc- and mt-encoded products for efficient mitochondrial homeostasis, and (c) the exchange of nc- and mt-originated sequences harbored in both genomes. In brief, the fungal contribution to the evolution of these interactions to date may be presented as follows:

(a) Mt functions controlled by nc-encoded proteins—the example of mtDNA transcription

Functional processes of fungal mitogenomes rely on nc-encoded genes, as it has been shown in several studies, primarily in fungal model organisms like *Saccharomyces cerevisiae*, *Neurospora crassa*, and *Schizosaccharomyces pombe* (e.g., Burger et al. 1985; Kleidon et al. 2003; Schäfer et al. 2005) and, recently, in other fungi like *Candida albicans* and entomopathogenic hyphomycetes *Beauveria bassiana* and *Metarhizium brunneum* (Kolondra et al. 2015; Varassas and Kouvelis 2022). While many related questions about the evolutionary establishment of these mechanisms remain unanswered, the study of the mt gene transcription in fungi for more than three decades provides certain well-substantiated hypotheses and brings insight into fungal mitochondrial genome evolution.

In specific, the finding of an eubacteria-like multisubunit RNA (msuRNA) polymerase in the mitogenome of a close relative of early-diverging amitochondriate eukaryotes, i.e., the protist *Reclinomonas americana* (Lang et al. 1997), as well as in fission yeasts (Seif et al. 2003) provided credibility to the theory that mt transcription was initially performed by a msuRNA polymerase which was subsequently lost and a nucleus-encoded single-subunit RNA (ssuRNA) polymerase was employed for the transcription of mt genes (Burger and Lang 2003). This replacement must have occurred early in the evolution, as a ssuRNA polymerase gene which resembled a T7 phage RNA polymerase gene has been identified in the early-diverging amoeba *Naegleria fowleri* (Cermakian et al. 1996). An alternative explanation to this is that the ancestral α-proteobacterium, which acted as the progenitor of the mitochondrion, was already infected with a lysogenic T7-like phage that had its genome integrated into the bacterial one and that this phage-originated gene was transferred along with the majority of all the other bacterial genes into the nucleus (Varassas and Kouvelis 2022). The experimentally proven necessity of a sigma-like subunit (Mtf1 gene) for the infallible function of mt RNA polymerase Rpo41 in yeasts (Yang et al. 2015) adds further credibility to the latter theory.

Nevertheless, the transfer of genes from the endosymbiotic bacterial ancestor to the nucleus is beyond any doubt, as shown by the fungal mitogenome contents (see chapter above). The main mechanism proposed for the transfer of these genes to the nucleus is their transfer by mobile elements that employ the non-homologous (NHEJ) recombination for integration (Berg and Kurland 2000; Fonseca et al. 2021).

(b) Mechanisms of mitochondrial homeostasis

The existence of two independently originated protein products in almost all eukaryotic organisms, which have to interact and cooperate in order to achieve vital cell processes—including the production of ATP in mitochondria through oxidative phosphorylation (OXPHOS) and aerobic respiration in general—has recently been elucidated effectively through the study of yeasts and other organismal cells including mammalian ones (Youle 2019). In yeasts, as well as in the majority of other fungi, 14 (at maximum) different mt-encoded proteins have to interact with a median of 80 different nc-encoded counterparts, in order to assemble all the necessary complexes of OXPHOS (for a review, see Barros and McStay 2020 and references therein). Intriguing questions about the mechanisms and processes which are involved in the establishment of functional coordination of both mt- and nc-encoded proteins within the cell have only recently been partially addressed through comparative studies among yeasts and mammals (for reviews, see Isaac et al. 2018; Youle 2019). Therefore, in this subsection, the new advances based on the impact of fungi in deciphering their evolution will be presented in brief.

In a fungal cell there is usually one nucleus, while mitochondria and their mitogenomes are always larger in number (reaching in some cases 100 mitogenomes per cell; Burger et al. 2003). Therefore, a stoichiometry of the mt- and nc-encoded products is imperative. The balance of this stoichiometry is achieved through the coordination of RNA transcription in the nucleus with protein translation in the mitochondrial matrix and by degrading extraneous protein subunits (for a review, see Youle 2019). Proteolysis in the mitochondrial matrix (Liao et al. 2020), mitophagy and mitochondrial biogenesis (Pickles et al. 2018), and mitochondria-derived vesicles which directly target outer mitochondrial membrane proteins to lysosomes (Klecker et al. 2014) may act as additional backup processes. This may be achieved by eliminating excess proteins imported from cytosolic translation of nuclear genes or translated in the matrix from mtDNA-encoded genes that misfold in the absence of interacting partners from the alternate genome (for a review, see Youle 2019 and references therein). Moreover, when hydrophobic inner mitochondrial membrane proteins are overexpressed, they may clog the translocase of the outer membrane (TOM)/translocase of the inner-membrane (TIM) import channels and protect the cell through the unfolded protein response activated by the mistargeting of proteins (UPRam) and mitochondrial precursor overaccumulation stress (mPOS) pathways (Weidberg and Amon 2018). As for mtDNA-encoded proteins, mitochondrial translation machinery and translation itself (but not transcription) are markedly downregulated after clogging (Boos et al. 2019).

Thus, there are several processes which contribute to the maintenance of proteostasis and control of stoichiometry in nc- and mt-encoded products. These mechanisms have also been found in mammals (Youle 2019). Therefore, there is strong indication of a common strategy that may have originated early in eukaryotic evolution and most probably at the stage of the last eukaryotic common ancestor (LECA.) That being said, a few other processes which help in the effective establishment of normal functional nc–mt interactions, like the mitophagy pathway, may be a later evolutionary acquisition (Youle 2019).

(c) NUMTs (NUclear sequences of MiTochondrial origin)

At present, it is well established that nuclear genomes in fungi, as well as in other eukaryotes, contain integrated fragments of mitochondrial DNA called NUMTs (NUclear sequences of MiTochondrial origin) (see, e.g., Kleine et al. 2009; Lafontaine et al. 2004). These usually correspond to pseudogenes or non-coding regions, but in a few cases, they have been incorporated within functional genes as exon patches (Noutsos et al. 2007). A comparative analysis of NUMTs in six different yeast species showed broad diversity in the number, size, and distribution of these elements within the chromosomes (Sacerdot et al. 2008). It is worth mentioning that in the examined yeasts the size of NUMTs was small (<400 bp), in contrast to the ones found in *Neurospora crassa* (the largest 4136 bp and average size of 647 bp—Richly and Leister 2004), and was distributed as either a tight mosaic (NUMTs overlap or are very close) or a loose one (more distant). In both cases, the mechanism of this mtDNA insertion into the nuclear genome may be considered a mutagenic phenomenon that may frequently inactivate a gene (Sacerdot et al. 2008). There are, however, NUMTs which have undergone a positive selection from the evolutionary aspect, since they have produced a new functional gene (Noutsos et al. 2007). Still, the mechanism of transferring NUMTs from the mitochondria to the nucleus is unresolved, but Beaudet et al. (2014) experimentally demonstrated the mobility of the *nad*1-*nad*4 intergenic region from the mitogenome of a *Rhizophagus irregularis* strain to its chromosomal DNA, thus suggesting an ongoing evolutionary process which may differ even among strains of the same species (not found in a second examined strain). Similarly, two *Trichoderma* strains harbored three NUMTs localized within the mitogenomic regions of *nad*5, *nad*6 and an intergenic region, while in the nuclear chromosomes they were found, as expected, in AT-rich regions which do not encode for any protein or RNA. It has been proposed that NUMTs have been implicated in increasing genetic diversity and facilitate genome evolution (Li et al. 2021a). Furthermore, it is experimentally proven in *Saccharomyces cerevisiae* and *Kluyveromyces lactis* that the NUMT acquisition and proliferation is a result of the existing DNA repair mechanism, either with the involvement of double-strand break (DSB) repair or the non-homologous end joining (NHEJ) pathway (Sacerdot et al. 2008).

3.7 Mt Gene-Based Phylogenies

Even from the early stages of molecular phylogenies, mt genes have been used as plausible alternative markers that may provide useful information in order to determine the phylogenetic relationships among fungal species, and to an extent their evolution. For example, studies have employed the genes of ATP synthase subunit 6 (*atp*6) (Kretzer and Bruns 1999) and of the small rRNA subunit (*rns*) (Pantou et al. 2005). When it became obvious that single gene phylogenies may be misleading in determining the phylogeny of whole organisms, these mt gene-based markers were used either in combination with nuclear and other mitochondrial genes as a concatenated matrix (Fahleson et al. 2004; Kouvelis et al. 2008; Theelen et al. 2021) or as a combined dataset of all the conserved mt genes

(Bullerwell et al. 2003; Kouvelis et al. 2004; Nadimi et al. 2016; Christinaki et al. 2022).

Mt-based phylogenies of the major lineages of fungi usually exhibit similar topologies to those produced by nuclear datasets (Bullerwell et al. 2003; Christinaki et al. 2022) with a few major differentiations (Pantou et al. 2008). One such "discrepancy" is the positioning of subphylum Saccharomycotina. Based on nuclear datasets, this subphylum is a sister clade to Pezizomycotina, and Taphrinomycotina are basal to both. Mt concatenated matrices produce phylogenetic trees showing the sisterhood of all yeasts (Saccharomycotina with Taphrinomycotina) and with Pezizomycotina as a sister clade of both (Bullerwell et al. 2003; Pantou et al. 2008; Freel et al. 2015; Christinaki et al. 2022). A plausible explanation for such phylogenetic incongruence is the different divergence rates of proteins and genes of mt vs nc origin. Recently, Christinaki et al. (2022) suggested a different molecular clock governing the evolution of the mt genes when compared with their nc counterparts. While the notion that mt genes evolve faster than nc genes present in metazoa is prevalent (Brown et al. 1982), it has lately been suggested that in plants and fungi, mt genomes exhibit a slower divergence rate (Sandor et al. 2018). However, Christinaki et al. (2022) proposed that fungal mt genes evolved slower than nc genes during the early phases of yeast evolution, and only recently has this changed so that mt genes evolve faster than their nc counterparts.

3.8 Intergenic Regions

As previously mentioned, introns and intergenic regions are some of the major determinants of fungal mt genome diversity. Intergenic regions vary greatly in fungal mt genomes in terms of both their size and content. Mitochondrial genomes of other eukaryotes, such as plants, may harbor larger intergenic regions, but fungi tend to have a very broad range of intergenic region size diversity. These regions in fungi may be as large as 88% of their whole mitogenome, as shown in the case of *Nakaseomyces bacillisporus* (Christinaki et al. 2022), while in certain cases they can also be as small as 1 bp, or non-existent, in cases when gene sequences overlap by 1 bp (Kouvelis et al. 2004). A characteristic example of such non-existing intergenic region is the case of *nad*4L and *nad*5 genes in several species such as *Lecanicillium muscarium*, *Neurospora crassa*, and *Podospora anserina*, where these genes overlap one another by a single base pair (Kouvelis et al. 2004). It has been suggested that intergenic regions may play an important role in genomic rearrangements and gene shuffling (Paquin et al. 2000). They often host repetitive sequences which can act as hot spots for recombination, as well as introns and ORFs with unknown functions (Paquin and Lang 1996; Liu et al. 2020b). This may explain the mobility of certain syntenic units, as there are indications of repeats in the intergenic regions that flank such syntenic units (Yildiz and Ozkilinc 2021; Zhao et al. 2021). It has been previously shown that the number of rearrangement events in fungal mitochondria is significantly correlated with the number of repeats in intergenic regions, the accumulation of which is mainly due to drift and mutation pressure (Aguileta et al. 2014). Similarly, small intergenic regions between genes may favor the formation and stability of syntenic units over time. Such an example can be observed in the case of *atp*6 and *atp*8 genes which are found as a conserved syntenic unit in most of the representative genomes of Pezizomycotina and Saccharomycotina, as well as in some species of Taphrinomycotina (Kouvelis et al. 2004; Christinaki et al. 2022), thus forming an ancestral conserved element. Additionally, small intergenic regions may help in the transcription of mt genes as fungal mt transcripts are initially polycistronic, but immediately after the formation of the polycistronic molecules, they mature to single cistrons as has been already shown in Ascomycetes (Breitenberger et al. 1985; Varassas and Kouvelis 2022). Therefore, repetitive elements in intergenic regions are most probably the result of stochastic duplication and recombinational events However, from the evolutionary prism, they can be involved in a diversity of functionally relevant outcomes such as conversions of the mitogenome from linear to circular forms and vice versa (Valach et al. 2011), duplication of certain genes within the repeats (Theelen et al. 2021), and mobility through transposition (Paquin et al. 2000). The latter is further directly associated with the structure and organization of mitogenomes and, indirectly, with the regulation of the genomes' expression (de Zamaroczy and Bernardi 1986; Koll et al. 1996).

3.9 Introns and Intronic ORFs

Fungal mt genomes harbor various introns, in contrast to metazoa which rarely carry introns (Boore 1999). Based on their RNA secondary structure, sequence conservation, and splicing mechanisms, mt introns can be classified into Group I and Group II introns, but the former are more frequent in fungal mt genomes (Lang et al. 2007; Hausner et al. 2014; Sandor et al. 2018). Both types of introns have the ability to move within the genome by a self-catalyzation mechanism (for a recent review, see Mukhopadhyay and Hausner 2021). In order to do so, self-splicing introns of Group I usually contain homing endonuclease genes (HEGs), coding for homing endonucleases (HEs), i.e., enzymes that recognize site-specific DNA targets (Belfort and Roberts 1997). HEGs have almost exclusively been found in Group I introns of fungal mitogenomes (Megarioti and Kouvelis 2020), but they may also be located within Group II introns or as free-standing genes, a characteristic most commonly seen in early-diverging fungal species (Toor and Zimmerly 2002; Megarioti and Kouvelis 2020). In general, HEs may be classified in one of the following four families based on the amino acid motifs participating in the enzyme active site: GIY-YIG, LAGLIDADG, His-Cys box, and HNH (Stoddard 2014). Fungal mt genomes usually contain the first two types; LAGLIDADG is generally the most frequently found and HNH extremely rare (Mukhopadhyay and Hausner 2021; Toor and Zimmerly 2002).

Group II introns rarely carry a LAGLIDADG endonuclease gene, as mentioned above, and they usually include a reverse transcriptase gene that helps splicing and intron's mobility through an intermediate step of reverse transcription (Novikova and Belfort 2017; Mukhopadhyay and Hausner 2021). However, both Group I and II introns can also take part in several intronic arrangements. One of the most complex intron formations involves introns within other introns, a term called twintron (Copertino and Hallick 1991). Twintrons may be formed by different combinations of Group I and II introns, or other intron types (e.g., tRNA introns), and there is no necessity for them to be of the same type (Hafez and Hausner 2015). Twintrons may be formed by different mechanisms which are based on the mobility of their components. The simplest and most common example is when an intron moves to a critical sequence of another one, preventing its splicing until the inside intron moves again and allows for the second one to become splicing competent. There are also cases where the invading intron interrupts an ORF sequence, so the gene is expressed only after its splicing (Guha and Hausner 2014). A characteristic example is the case of the expression of a functional *rps*3 gene located in an external intron, only after the splicing of the internal intron which contains a LAGLIDADG endonuclease in the mitogenome of *Grosmannia piceiperda* (Rudski and Hausner 2012). Similarly, in the fungus *Chaetomium thermophilum*, an ORFless intron interrupts an external intron that harbors an ORF for a homing endonuclease which can only be expressed upon self-splicing of the invading internal intron (Guha and Hausner 2014). In cases where the sequence is not critical for the splicing, the outside intron has the ability to move independently (Hafez et al. 2013). The role and significance of twintron formations is yet unknown, but their complex and various formations suggest that they could be useful as regulatory elements for gene expression regulation through the various alternative splicing mechanisms they provide (Younis et al. 2013; Hafez and Hausner 2015).

In general, the number, as well as length of introns present in mt genomes, is highly variable, and along with the mitochondrial intergenic regions, introns are major determinants of mt genome size diversity (Hausner 2012; Friedrich et al. 2012; Megarioti and Kouvelis 2020). There are cases, like the fungi *Podospora anserina* and *Phlebia radiata*, where introns account for up to 75% and 80% of the overall size of mtDNA, respectively (Cummings et al. 1990; Salavirta et al. 2014), while in a few other fungal species, i.e., in *Hyaloraphidium curvatum,* no introns were found in their mt genome (Forget et al. 2002). While introns contribute to genome size expansion, intron abundance is not always analogous to fungal mt genome size (Christinaki et al. 2022), and closely related species can vary immensely in the mt genome content of introns. At the same time, intron presence varies among different genes, in respect to both their number and their type. Genes like *cox*1 and *cob* usually contain many more introns than other genes (Freel et al. 2015; Wolters et al. 2015; Christinaki et al. 2022). Accordingly, the various intron types and their HEGs exhibit a preference for certain genes, showing a wide or narrow gene host preference (Hausner et al. 2014; Zardoya 2020; Christinaki et al. 2022).

Introns have a restless evolutionary history. According to the intron-early theory, introns were abundant during the first stages of evolution after the initial endosymbiosis and thereafter had a tendency to disappear. In several cases, an intronic ORF may be fused with a preceding exon in one frame. As suggested in the modified Goddard and Burt model, mutations, which change the target sites for intron acquisition and alternative splicing, may lead to the loss of the intronic ORF fused to the upstream exon (Goddard and Burt 1999; Guha et al. 2018). Recent studies have shown, however, that introns may remain intact at conserved gene locations (Korovesi et al. 2018; Wai et al. 2019), a fact that indicates a form of habituation in certain genes. In reality, introns not only stay intact without disappearing, at least as frequently as the intron-early theory suggests, but may in some instances become more numerous, with new introns being acquired through HGT events or transposition (Wu et al. 2015). Introns and HEGs have a long and co-related evolutionary history. According to the recently proposed "aenaon" model, introns have undergone a complex evolutionary transformation, from an ancestral compact form to a more expanded and complex structure. HEGs tend to invade introns throughout evolution in a perpetual manner, through many recombination, transposition, and horizontal gene transfer events. Ancestral introns still exist, while new introns may be found in Dikarya fungi (Megarioti and Kouvelis 2020). Finally, it has been shown that intron diversity and mobility within the genome play a substantial role in recombination events. Introns tend to move from an intron-containing allele to an intron-minus allele in a homology-dependent gene conversion, a term called intron homing (Dujon 1989). During this process of intron movement, sequences from the intron-containing allele are transferred to the recipient gene, thus shaping genetic diversity through intron homing (Repar and Warnecke 2017; Wu and Hao 2019).

3.10 Unidentified ORFs with Other Function than Intron Splicing and Mobility

In addition to the conserved fungal mt genes or Group I and II intronic ORFs that have an already known role (i.e., HEGs, RT and *rps*3), mitochondria also contain a number of non-conserved ORFs (ncORFs), otherwise called unidentified ORFs. Their presence varies substantially among fungal mt genomes and can either be found in introns or as free-standing genes within intergenic regions (Liu et al. 2020b; Losada et al. 2014; Duò et al. 2012). There have been three main theories regarding their origin: (a) they are derived from existing mt loci that have accumulated point mutations, (b) they are derived from a nuclear gene that was transferred in the mitochondrion, and/or (c) horizontal transfer has occurred from another organism. In addition, mobile elements, such as HEGs, can alter ORFs so they could possibly create unidentified ORFs (Al-Reedy et al. 2012).

Despite being numerous, the majority of them have unknown roles (Hane et al. 2007; Duò et al. 2012). Regardless, it has been suggested that a large number of nuclear ORFs are typically found in large fungal mt genomes, and thus, they may contribute to genome size expansion (Li et al. 2015). Mitochondrial genomes of *Fusarium* spp. were found to contain highly variable regions of 7–9 kb that encode Group I intron associated ORFs, as well as an exceptionally large unidentified ORF which was probably horizontally transmitted before the divergence of *Fusarium* spp. (Al-Reedy et al. 2012). Similarly, free-standing unidentified ORFs were also located in the mt genome of four *Rhynchosporium* spp., among which there was a large unidentified ORF, i.e., Ro_ORF11, which encoded an 807 amino acid in size protein with five transmembrane domains without a recognized function (Torriani et al. 2014). It has also been suggested that non-conserved ORFs with unknown functions may be involved in mitonuclear interactions (Clergeot and Olson 2021). Such interactions can contribute to virulence and host interactions of pathogenic fungi (Hu et al. 2020).

Unidentified ORFs' random and non-conserved presence in fungal mt genomes, in combination with their location within introns or intergenic regions, implies that they contribute largely to mt diversity among different species/taxa. This is in compliance with their role as a rapidly evolving accessory compartment in the mt genome, as their location in non-conserved areas allows them to be the subject of mutations or other genetic rearrangements (Croll and McDonald 2012; Torriani et al. 2014).

3.11 Genetic Codes

The canonical genetic code was established before the LECA. Its essential role in protein synthesis suggested that its alteration is not promoted or favored, as such a thing would cause an alteration in the amino acid sequences and thus possibly in protein structure or function (Crick 1968). As more genome sequences become available, it is now clear that there are numerous deviations throughout different groups of organisms, and indications exist that there is a greater evolvability and flexibility of the code (Knight et al. 2001a, b, c). However, this suggests that the various established genetic codes would require deep structural, functional, and evolutionary changes, which are deep modifications. Still, these have not yet been elucidated upon. Mitochondrial genomes employ variable non-standard genetic codes compared to their nuclear counterparts (Swire et al. 2005). Fungal mt genomes, in particular, employ the genetic codes NCBI 4 (The Mold, Protozoan, and Coelenterate Mitochondrial Code and the Mycoplasma/Spiroplasma Code) and NCBI 3 (The Yeast Mitochondrial Code), while only a few of them employ genetic code NCBI 1 (The Standard Code) and NCBI 16 (Chlorophycean Mitochondrial Code) (Fonseca et al. 2021). Deviations from the canonical code involve reassignments between different amino acids or from amino acids to stop codons and vice versa. Genetic code 4 (UGA codon from Stop to Trp) is the most common change used in fungal mitochondria, while genetic code 3 is mostly used by yeast species (AUA from Ile to Met, CUN from Leu to Thr, and UGA from Stop to Trp). Genetic code 1 is found in species of Mucoromycota and *Schizosaccharomyces pombe* and genetic code NCBI 16 (UAG from stop to Leu) in Chytridiomycota (Laforest et al. 1997).

Sengupta et al. (2007) summarize the reassignments that occurred after the establishment of the canonical code and classify them according to a gain-loss framework. In brief, "loss" is the deletion of the gene for the tRNA originally associated with the codon to be reassigned or the loss of function of this gene due to a mutation or base modification in the anticodon. "Gain" is the addition of a new type of tRNA for the reassigned codon through gene duplication, or the gain of function of an existing tRNA due to a mutation or a base modification, enabling it to pair with the reassigned codon. Therefore, four possible mechanisms for codon reassignments (Sengupta and Higgs 2005; Sengupta et al. 2007) were proposed (Fig. 3.2):

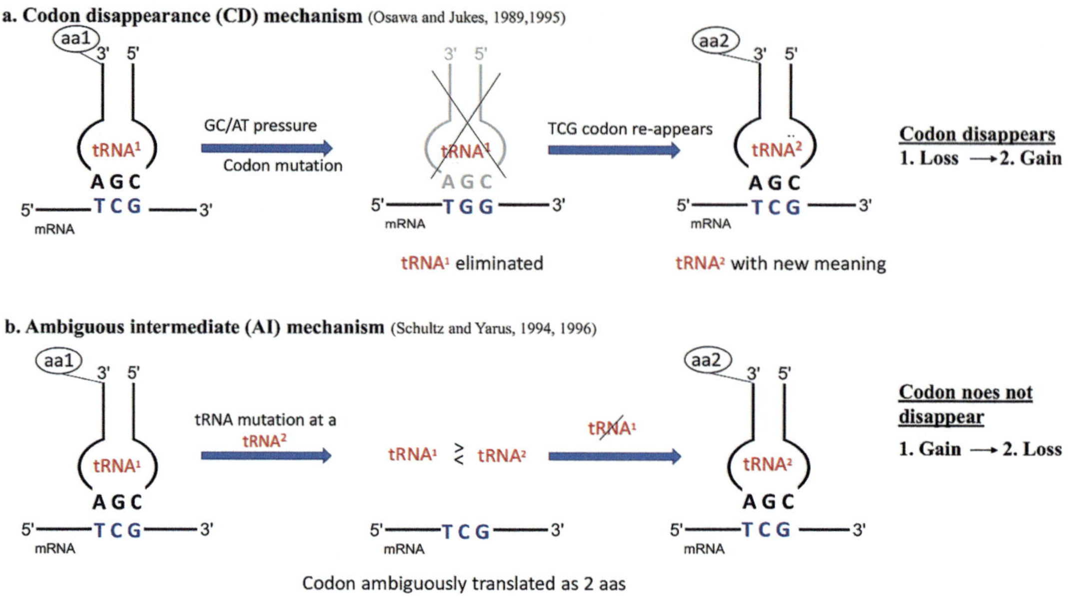

Fig. 3.2 Schematic representation of the two most widely accepted models for codon reassignments

codon disappearance (CD), unassigned codon (UC), ambiguous intermediate (AI), and compensatory change mechanism. CD involves codon loss prior to any other gain and loss events, a disappearance occurring through random drift and directional mutation pressure. It has been previously suggested that variation in GC content is a major determinant in the CD mechanism and influences synonymous codon usage (Knight et al. 2001b). This hypothesis, alternatively stated as the "codon capture" model by Osawa and Jukes, suggested that this mechanism would minimize the possible catastrophic effects of codon alteration before its disappearance (Osawa 1995; Osawa et al. 1992). In both the unassigned codon (UC) and ambiguous intermediate (AI) mechanisms, codons do not disappear prior to the reassignment, but *trn* loss and gain occur first, respectively (Schultz and Yarus 1996). None of the above models can fully interpret all genetic code alterations, and most importantly, they are not mutually exclusive. Sengupta et al. (2007) suggest that CD mechanism may explain stop codons to protein codon changes, such as UGA from stop to Trp (genetic code 3) or UAG to Leu (genetic code 16), as well as some of the other reassignments, because stop codons are rarely found in the mt genome, and thus, are more likely to be lost by chance. In addition, CD mechanism may explain the reassignment of CUN and CGN codons in some yeast species, as it appears that there is a strong bias in their mt genome against C, which leads to codon loss. In contrast, Knight et al. (2001b, c) support the AI mechanism, which acts alone or after the codon disappearance.

All models described above may be used to explain differential appearance of their genetic codes in fungal mitogenomes, but it is important to have in mind that changes may have happened independently many times throughout evolution. In favor of this argument is gene codon usage employed in mitogenomes of early-diverging fungal species and other species from Basidiomycetes and Ascomycetes. Specifically, recently reassigned codons like UGA (a stop codon in the universal genetic code) have only recently been utilized, in mt genomes in Ascomycetes, and randomly in the respective genomes of Basidiomycetes, while they are absent in the respective genomes of early-diverging fungal fungi (Korovesi et al. 2018).

Similarly, the occurrence of alternative codon usage in the mitogenomes of yeasts is variable even within members of the same families (Christinaki et al. 2022), thus supporting the incidental but restless change of genetic codes throughout evolution.

3.12 Major Fungal Evolutionary Transitions Based on Mitogenomes (Conclusions)

The analyses of all the features of fungal mitogenomes, as summarized above, offer many insights pertaining to the evolution of eukaryotes within the kingdom of fungi. Likewise, they provide insights into the evolution of the alpha-proteobacterial endosymbiont which was "adopted" by the archaeon acting as the proto-eukaryote, according to the prevailing endosymbiotic theory (Martin et al. 2015).

From the above-mentioned analyses, it becomes evident that a diverse group of species, like the one belonging to the kingdom of fungi, carries all the conserved ancestral elements, such as genes for the subunits of oxidative phosphorylation, since they cannot be transferred to the nucleus (Adams and Palmer 2003; Daley and Whelan 2005). At the same time, these ancestral elements are differentiated and evolved with the addition of mobile elements like introns and HEGs (Megarioti and Kouvelis 2020; Mukhopadhyay and Hausner 2021), plasmids, and DNA/RNA polymerases (Formighieri et al. 2008; Fricova et al. 2010). There is a restless "back and forth" mechanism in mitochondrial genomes, which is based on recombination and horizontal gene transfer (HGT) events. Even though this is not a deterministic procedure, but a rather stochastic one, it presumably helps the species carrying these genomes to better adapt to their environment. Under this context, studies like the one of Clergeot and Olson (2021) provide evidence that there is a correlation of mt ORFs with pathogenicity, due to their interaction with their nuclear counterparts. NUMTs have revealed an ongoing mechanism of ceaseless interaction between nc chromosomes and mt genomes (Sacerdot et al. 2008; Beaudet et al. 2014), thus offering fungal organisms the ability to evolve with the creation of new genes and pseudogenes or gene hitchhiking (Noutsos et al. 2007; Xiao et al. 2017). Moreover, mt plasmids are, beyond any doubt, involved in processes like senescence and life-span determination (Hermanns et al. 1994; Monteiro-Vitorello et al. 2000) and along with introns, tRNA genes and repetitive sequences may act as hotspots for recombination within the genome and eventually lead to gene shuffling, under a model similar to the one already proposed for the mobility of Group I introns and their HEGs, i.e., the "aenaon" model (Megarioti and Kouvelis 2020). In other words, all mitogenomic studies which tried to decipher the evolution of these genomes and the fungi carrying them revealed that all known mechanisms needed for evolution, including genetic drift, genetic draft (gene hitchhiking), and HGT, also apply in the fungal mitogenomes (Lang et al. 2014; Xiao et al. 2017; Hao 2022; Christinaki et al. 2022). Moreover, unique (to our knowledge) mechanisms in fungal mitochondria, such as the programmed translational jumping in mt genomes of yeasts (Lang et al. 2014), provide evidence that evolution may happen indiscriminately at any time, adding to the plasticity and diversity of mitogenomes. Finally, mitogenomes and studies which try to examine nucleus–mitochondrion interactions in the context of the mitogenome's well-tempered functioning, such as the coevolution of the major proteins implicated in the transcription of mt genes, help in further supporting the proposed hypotheses of mt genome changes through different genetic mechanisms (Varassas and Kouvelis 2022). As mentioned earlier, recent analyses examining fungal mt transcription led to support for the hypothesis that the bacterial endosymbiont was most probably infected with a T7-like phage, and this prophage genome offered the RNA polymerase to the newly formed mitochondrion (Filée and Forterre 2005; Varassas and Kouvelis 2022). Needless to say, other fungal studies, which cannot be overruled, suggest that this phage RNA polymerase, as well as the DNA polymerase, both presenting a phage origin of the mt respective enzymes, might have been acquired in the mitochondrion at a later stage (Burger and Lang 2003).

Irrelevant to which theories are the correct ones, it is evident that the fungal mitogenome diversity offers data needed to resolve such crucial questions such as the origin of the mitochondria and how mitogenomes evolve over time. Therefore, in this era of broad sequencing and WGS projects, the mitogenome analyses of fungal species must be performed in depth.

Acknowledgments VNK wishes to thank Fulbright Greece for its Visiting Scholar Program that funded his visit to TYJ's lab, and thus, VNK had the needed fruitful conversations with TYJ, which led to the writing of this chapter. TYJ is a fellow of CIFAR program Fungal Kingdom: Threats & Opportunities.

References

Abdollahzadeh J, Groenewald JZ, Coetzee MPA, Wingfield MJ, Crous PW (2020) Evolution of lifestyles in Capnodiales. Stud Mycol 95(1):381–414. https://doi.org/10.1016/jsimyco202002004

Adams KL, Palmer JD (2003) Evolution of mitochondrial gene content: gene loss and transfer to the nucleus. Mol Phylogenet Evol 29:380–395. https://doi.org/10.1016/s1055-7903(03)00194-5

Aguileta G, de Vienne DM, Ross ON, Hood ME, Giraud T, Petit E, Gabaldón T (2014) High variability of mitochondrial gene order among fungi. Genome Biol Evol 6:451–465. https://doi.org/10.1093/gbe/evu028

Allen JF (2003) Why chloroplasts and mitochondria contain genomes. Comp Funct Genomics 4:31–36. https://doi.org/10.1002/cfg245

Al-Reedy RM, Malireddy R, Dillman CB, Kennell JC (2012) Comparative analysis of *Fusarium* mitochondrial genomes reveals a highly variable region that encodes an exceptionally large open reading frame. Fungal Genet Biol 49:2–14. https://doi.org/10.1016/jfgb201111008

Barros MH, McStay GP (2020) Modular biogenesis of mitochondrial respiratory complexes. Mitochondrion 50:94–114. https://doi.org/10.1016/jmito201910008

Beaudet D, Nadimi M, Iffis B, Hijri M (2013) Rapid mitochondrial genome evolution through invasion of mobile elements in two closely related species of arbuscular mycorrhizal fungi. PLoS One 8(4):e60768. https://doi.org/10.1371/journalpone0060768

Beaudet D, de la Providencia IE, Labridy M, Roy-Bolduc A, Daubois L, Hijri M (2014) Intraisolate mitochondrial genetic polymorphism and gene variants coexpression in arbuscular mycorrhizal fungi. Genome Biol Evol 7:218–227. https://doi.org/10.1093/gbe/evu275

Belfort M, Roberts RJ (1997) Homing endonucleases: keeping the house in order. Nucleic Acids Res 25(17):3379–3388. https://doi.org/10.1093/nar/25173379

Bendich AJ (1993) Reaching for the ring: the study of mitochondrial genome structure. Curr Genet 24(4): 279–290. https://doi.org/10.1007/BF00336777

Berg OG, Kurland CG (2000) Why mitochondrial genes are most often found in nuclei. Mol Biol Evol 17:951–961. https://doi.org/10.1093/oxfordjournalsmolbeva026376

Boore JL (1999) Animal mitochondrial genomes. Nucleic Acids Res 27(8):1767–1780. https://doi.org/10.1093/nar/2781767

Boos F, Krämer L, Groh C, Jung F, Haberkant P, Stein F, Wollweber F, Gackstatter A, Zöller E, van der Laan M, Savitski MM, Benes V, Herrmann JM (2019) Mitochondrial protein-induced stress triggers a global adaptive transcriptional programme. Nat Cell Biol 21:442–451. https://doi.org/10.1038/s41556-019-0294-5

Breitenberger CA, Browning KS, Alzner-DeWeerd B, RajBhandary UL (1985) RNA processing in *Neurospora crassa* mitochondria: use of transfer RNA sequences as signals. EMBO J 4:185–195. https://doi.org/10.1002/j1460-20751985tb02335x

Brown WM, Prager EM, Wang A, Wilson AC (1982) Mitochondrial DNA sequences of primates: tempo and mode of evolution. J Mol Evol 18:225–239. https://doi.org/10.1007/BF01734101

Bullerwell CE, Lang BF (2005) Fungal evolution: the case of the vanishing mitochondrion. Curr Opin Microbiol 8:362–369. https://doi.org/10.1016/jmib200506009

Bullerwell CE, Burger G, Lang BF (2000) A novel motif for identifying *rps*3 homologs in fungal mitochondrial genomes. TIBS 25(8):363–365. https://doi.org/10.1016/s0968-0004(00)01612-1

Bullerwell CE, Leigh J, Forget L, Lang BF (2003) A comparison of three fission yeast mitochondrial genomes. Nucleic Acids Res 31:759–768. https://doi.org/10.1093/nar/gkg134

Burger G, Lang BF (2003) Parallels in genome evolution in mitochondria and bacterial symbionts. IUBMB Life 55:205–212. https://doi.org/10.1080/1521654031000137380

Burger G, Helmer CM, Nelson MA, Werner S, Macino G (1985) RNA processing in Neurospora crassa mitochondria: transfer RNAs punctuate a large precursor transcript. EMBO J 4(1):197–204. https://doi.org/10.1002/j.1460-2075.1985.tb02

Burger G, Gray MW, Lang BF (2003) Mitochondrial genomes: anything goes. Trends Genet 19:709–716. https://doi.org/10.1016/jtig200310012

Cahan P, Kennell JC (2005) Identification and distribution of sequences having similarity to mitochondrial plasmids in mitochondrial genomes of filamentous

fungi. Mol Genet Genomics 273:462–473. https://doi.org/10.1007/s00438-005-1133-x
Cermakian N, Ikeda TM, Cedergren R, Gray MW (1996) Sequences homologous to yeast mitochondrial and bacteriophage T3 and T7 RNA polymerases are widespread throughout the eukaryotic lineage. Nucleic Acids Res 24:648–654. https://doi.org/10.1093/nar/244648
Chen XJ, Clark-Walker GD (2018) Unveiling the mystery of mitochondrial DNA replication in yeasts. Mitochondrion 38:17–22. https://doi.org/10.1016/jmito201707009
Christinaki AC, Kanellopoulos SG, Kortsinoglou AM, Andrikopoulos MA, Theelen B, Boekhout T, Kouvelis VN (2022) Mitogenomics and mitochondrial gene phylogeny decipher the evolution of Saccharomycotina yeasts. Genome Biol Evol evac073. https://doi.org/10.1093/gbe/evac073
Clark-Walker GD (1992) Evolution of mitochondrial genomes in fungi. In: Welstenholme DR, Jeon KW (eds) Mitochondrial genomes. Academic Press, San Diego, pp 89–127
Clergeot P-H, Olson Å (2021) Mitonuclear genetic interactions in the basidiomycete *Heterobasidion parviporum* involve a non-conserved mitochondrial open reading frame. Front Fungal Biol 2:779337. https://doi.org/10.3389/ffunb2021779337
Copertino DW, Hallick RB (1991) Group II twintron: an intron within an intron in a chloroplast cytochrome b-559 gene. EMBO J 10:433–442. https://doi.org/10.1002/j1460-20751991tb07965x
Crick FH (1968) The origin of the genetic code. J Mol Biol 38:367–379. https://doi.org/10.1016/0022-2836(68)90392-6
Croll D, McDonald BA (2012) The accessory genome as a cradle for adaptive evolution in pathogens. PLoS Pathog 8(4):e1002608. https://doi.org/10.1371/journalppat1002608
Cummings DJ, McNally KL, Domenico JM, Matsuura ET (1990) The complete DNA sequence of the mitochondrial genome of *Podospora anserina*. Curr Genetics 17:375–402. https://doi.org/10.1007/BF00334517
Daley DO, Whelan J (2005) Why genes persist in organelle genomes. Genome Biol 6(5):110. https://doi.org/10.1186/gb-2005-6-5-110
De Chiara M, Friedrich A, Barré B, Breitenbach M, Schacherer J, Liti G (2020) Discordant evolution of mitochondrial and nuclear yeast genomes at population level. BMC Biol 18(49). https://doi.org/10.1186/s12915-020-00786-4
de Zamaroczy M, Bernardi G (1986) The GC clusters of the mitochondrial genome of yeast and their evolutionary origin. Gene 41:1–22. https://doi.org/10.1016/0378-1119(86)90262-3
Dieckmann CL, Gandy B (1987) Preferential recombination between GC clusters in yeast mitochondrial DNA. EMBO J 6:4197–4203. https://doi.org/10.1002/j1460-20751987tb02767x
Dinouël N, Drissi R, Miyakawa I, Sor F, Rousset S, Fukuhara H (1993) Linear mitochondrial DNAs of yeasts: closed-loop structure of the termini and possible linear-circular conversion mechanisms. Mol Cell Biol 13:2315–2323. https://doi.org/10.1128/mcb1342315-23231993
Dujon B (1989) Group I introns as mobile genetic elements: facts and mechanistic speculations: a review. Gene 82:91–114
Duò A, Bruggmann R, Zoller S, Bernt M, Grünig CR (2012) Mitochondrial genome evolution in species belonging to the *Phialocephala fortinii sl* - *Acephala applanata* species complex. BMC Genomics 13:166. https://doi.org/10.1186/1471-2164-13-166
Fahleson J, Hu Q, Dixelius C (2004) Phylogenetic analysis of *Verticillium* species based on nuclear and mitochondrial sequences. Arch Microbiol 181:435–442. https://doi.org/10.1007/s00203-004-0677-1
Falkenberg M (2018) Mitochondrial DNA replication in mammalian cells: overview of the pathway. Essays Biochem 62(3):287–296. https://doi.org/10.1042/EBC20170100
Filée J, Forterre P (2005) Viral proteins functioning in organelles: a cryptic origin? Trends Microbiol 13:510–513. https://doi.org/10.1016/jtim200508012
Fonseca P, De-Paula RB, Araújo DS, Tomé L, Mendes-Pereira T, Rodrigues W, Del-Bem LE, Aguiar E, Góes-Neto A (2021) Global characterization of fungal mitogenomes: new insights on genomic diversity and dynamism of coding genes and accessory elements. Front Microbiol 12:787283. https://doi.org/10.3389/fmicb2021787283
Forget L, Ustinova J, Wang Z, Huss VA, Lang BF (2002) *Hyaloraphidium curvatum*: a linear mitochondrial genome, tRNA editing, and an evolutionary link to lower fungi. Mol Biol Evol 19:310–319. https://doi.org/10.1093/oxfordjournalsmolbeva004084
Formighieri EF, Tiburcio RA, Armas ED, Medrano FJ, Shimo H, Carels N, Góes-Neto A, Cotomacci C, Carazzolle MF, Sardinha-Pinto N, Thomazella DP, Rincones J, Digiampietri L, Carraro DM, Azeredo-Espin AM, Reis SF, Deckmann AC, Gramacho K, Gonçalves MS, Moura Neto JP, Barbosa LV, Meinhardt LW, Cascardo JCM, Pereira GA (2008) The mitochondrial genome of the phytopathogenic basidiomycete *Moniliophthora perniciosa* is 109 kb in size and contains a stable integrated plasmid. Mycol Res 112:1136–1152. https://doi.org/10.1016/jmycres200804014
Fox AN, Kennell JC (2001) Association between variant plasmid formation and senescence in retroplasmid-containing strains of *Neurospora* spp. Curr Genet 39:92–100. https://doi.org/10.1007/s002940100187
Franco M, López S, Medina R, Lucentini CG, Troncozo MI, Pastorino GN, Saparrat M, Balatti PA (2017) The mitochondrial genome of the plant-pathogenic fungus *Stemphylium lycopersici* uncovers a dynamic structure due to repetitive and mobile elements. PLoS One

12(10):e0185545. https://doi.org/10.1371/journalpone0185545
Freel KC, Friedrich A, Schacherer J (2015) Mitochondrial genome evolution in yeasts: an all-encompassing view. FEMS Yeast Res 15(4):fov023. https://doi.org/10.1093/femsyr/fov023
Fricova D, Valach M, Farkas Z, Pfeiffer I, Kucsera J, Tomaska L, Nosek J (2010) The mitochondrial genome of the pathogenic yeast *Candida subhashii*: GC-rich linear DNA with a protein covalently attached to the 5′ termini. Microbiology 156(7):2153–2163. https://doi.org/10.1099/mic0038646-0
Friedrich A, Jung PP, Hou J, Neuveglise C, Schacherer J (2012) Comparative mitochondrial genomics within and among yeast species of the *Lachancea* genus. PLoS One 7:6. https://doi.org/10.1371/journalpone0047834
Ghikas DV, Kouvelis VN, Typas MA (2006) The complete mitochondrial genome of the entomopathogenic fungus *Metarhizium anisopliae* var *anisopliae*: gene order and *trn* gene clusters reveal a common evolutionary course for all Sordariomycetes, while intergenic regions show variation. Arch Microbiol 185:393–401. https://doi.org/10.1007/s00203-006-0104-x
Ghikas DV, Kouvelis VN, Typas MA (2010) Phylogenetic and biogeographic implications inferred by mitochondrial intergenic region analyses and ITS1-5.8S-ITS2 of the entomopathogenic fungi *Beauveria bassiana* and *B. brongniartii*. BMC Microbiol 10:174. https://doi.org/10.1186/1471-2180-10-174
Gilkerson R, Bravo L, Garcia I, Gaytan N, Herrera A, Maldonado A, Quintanilla B (2013) The mitochondrial nucleoid: integrating mitochondrial DNA into cellular homeostasis. Cold Spring Harb Perspect Biol 5(5): a011080. https://doi.org/10.1101/cshperspecta011080
Goddard MR, Burt A (1999) Recurrent invasion and extinction of a selfish gene. PNAS 96:13880–13885. https://doi.org/10.1073/pnas962413880
Griffiths AJ (1995) Natural plasmids of filamentous fungi. Microbiol Rev 59:673–685. https://doi.org/10.1128/mr594673-6851995
Guha TK, Hausner GA (2014) Homing endonuclease with a switch: characterization of a twintron encoded homing endonuclease. Fungal Genet Biol 65:57–68. https://doi.org/10.1016/jfgb201401004
Guha TK, Wai A, Mullineux ST, Hausner G (2018) The intron landscape of the mtDNA *cyt*b gene among the Ascomycota: introns and intron-encoded open reading frames. Mitochondrial DNA A DNA Mapp Seq Anal 29:1015–1024. https://doi.org/10.1080/2470139420171404042
Hafez M, Hausner G (2015) Convergent evolution of twintron-like configurations: one is never enough. RNA Biol 12(12):1275–1288. https://doi.org/10.1080/1547628620151103427
Hafez M, Majer A, Sethuraman J, Rudski SM, Michel F, Hausner G (2013) The mtDNA *rns* gene landscape in the Ophiostomatales and other fungal taxa: twintrons, introns, and intron-encoded proteins. Fungal Genet Biol 53:71–83. https://doi.org/10.1016/jfgb201301005
Hane JK, Lowe RG, Solomon PS, Tan KC, Schoch CL, Spatafora JW, Crous PW, Kodira C, Birren BW, Galagan JE, Torriani SF, McDonald BA, Oliver RP (2007) Dothideomycete plant interactions illuminated by genome sequencing and EST analysis of the wheat pathogen *Stagonospora nodorum*. Plant Cell 19:3347–3368. https://doi.org/10.1105/tpc107052829
Hao W (2022) From genome variation to molecular mechanisms: what we have learned from yeast mitochondrial genomes? Front Microbiol 13:806575. https://doi.org/10.3389/fmicb2022806575
Hausner G (2012) Introns, Mobile elements, and plasmids. In: Bullerwell CE (ed) Organelle genetics. Springer, Berlin, pp 329–357
Hausner G, Hafez M, Edgell DR (2014) Bacterial group I introns: mobile RNA catalysts. Mob DNA 5(1):8–12. https://doi.org/10.1186/1759-8753-5-8
Hawksworth DL, Lücking R (2017) Fungal diversity revisited: 22 to 38 million species. Microbiol Spectr 5:79–95. https://doi.org/10.1128/microbiolspecFUNK-0052-2016
Hebert PD, Cywinska A, Ball SL, deWaard JR (2003) Biological identifications through DNA barcodes. Proc Biol Sci 270:313–321. https://doi.org/10.1098/rspb20022218
Hermanns J, Asseburg A, Osiewacz HD (1994) Evidence for a life span-prolonging effect of a linear plasmid in a longevity mutant of *Podospora anserina*. Mol Gen Genet 243:297–307. https://doi.org/10.1007/BF00301065
Himmelstrand K, Olson A, Brandström Durling M, Karlsson M, Stenlid J (2014) Intronic and plasmid-derived regions contribute to the large mitochondrial genome sizes of Agaricomycetes. Curr Genet 60: 303–313. https://doi.org/10.1007/s00294-014-0436-z
Hu Y, Elfstrand M, Stenlid J, Durling MB, Olson Å (2020) The conifer root rot pathogens *Heterobasidion irregulare* and *Heterobasidion occidentale* employ different strategies to infect Norway spruce. Sci Rep 10:5884. https://doi.org/10.1038/s41598-020-62521-x
Isaac RS, McShane E, Churchman LS (2018) The multiple levels of mitonuclear coregulation. Annu Rev Genet 52:511–533. https://doi.org/10.1146/annurev-genet-120417-031709
James TY, Pelin A, Bonen L, Ahrendt S, Sain D, Corradi N, Stajich JE (2013) Shared signatures of parasitism and phylogenomics unite Cryptomycota and microsporidia. Curr Biol 23:1548–1553. https://doi.org/10.1016/jcub201306057
Katsura K, Suzuki F, Miyashita SI, Nishi T, Hirochika H, Hashiba T (1997) The complete nucleotide sequence and characterization of the linear DNA plasmid pRS64-2 from the plant pathogenic fungus *Rhizoctonia solani*. Curr Genet 32:431–435. https://doi.org/10.1007/s002940050299

Kempken F (1995) Horizontal transfer of a mitochondrial plasmid. Mol Gen Genet 248:89–94. https://doi.org/10.1007/BF02456617

Kennell JC, Cohen SM (2004) Fungal mitochondria: genomes, genetic elements and gene expression. In: Arora DK (ed) The handbook of fungal biotechnology, 2nd edn. Marcel Dekker, New York, pp 131–143

Klassen R, Meinhardt F (2007) Linear protein primed replicating plasmids in eukaryotic microbes. In: Meinhardt F, Klassen R (eds) Microbial linear plasmids. Springer, Berlin, pp 188–226

Klecker T, Böckler S, Westermann B (2014) Making connections: interorganelle contacts orchestrate mitochondrial behavior. Trends Cell Biol 24:537–545. https://doi.org/10.1016/jtcb201404004

Kleidon J, Plesofsky N, Brambl R (2003) Transcripts and transcript binding proteins in mitochondria of *Neurospora crassa*. Mitochondrion 2:345–360. https://doi.org/10.1016/S15677249(03)00002-3

Kleine T, Maier UG, Leister D (2009) DNA transfer from organelles to the nucleus: the idiosyncratic genetics of endosymbiosis. Annu Rev Plant Biol 60:115–138. https://doi.org/10.1146/annurevarplant043008092119

Knight RD, Freeland SJ, Landweber LF (2001a) A simple model based on mutation and selection explains trends in codon and amino-acid usage and GC composition within and across genomes. Genome Biol 2: research0010. https://doi.org/10.1186/gb-2001-2-4-research0010

Knight RD, Landweber LF, Yarus M (2001b) How mitochondria redefine the code. J Mol Evol 53:299–313. https://doi.org/10.1007/s002390010220

Knight R, Freeland S, Landweber L (2001c) Rewiring the keyboard: evolvability of the genetic code. Nat Rev Genet 2:49–58. https://doi.org/10.1038/35047500

Koll F, Boulay J, Belcour L, d'Aubenton-Carafa Y (1996) Contribution of ultra-short invasive elements to the evolution of the mitochondrial genome in the genus *Podospora*. Nucleic Acids Res 24:1734–1741. https://doi.org/10.1093/nar/2491734

Kolondra A, Labedzka-Dmoch K, Wenda JM, Drzewicka K, Golik P (2015) The transcriptome of *Candida albicans* mitochondria and the evolution of organellar transcription units in yeasts. BMC Genomics 16:827. https://doi.org/10.1186/s12864-015-2078-z

Korovesi AG, Ntertilis M, Kouvelis VN (2018) Mt-*rps*3 is an ancient gene which provides insight into the evolution of fungal mitochondrial genomes. Mol Phylogenet Evol 127:74–86. https://doi.org/10.1016/jympev201804037

Kouvelis VN, Ghikas DV, Typas MA (2004) The analysis of the complete mitochondrial genome of *Lecanicillium muscarium* (synonym *Verticillium lecanii*) suggests a minimum common gene organization in mtDNAs of Sordariomycetes: phylogenetic implications. Fungal Genet Biol 41:930–940. https://doi.org/10.1016/jfgb200407003

Kouvelis VN, Sialakouma A, Typas MA (2008) Mitochondrial gene sequences alone or combined with ITS region sequences provide firm molecular criteria for the classification of *Lecanicillium* species. Mycol Res 112:829–844. https://doi.org/10.1016/jmycres200801016

Kretzer AM, Bruns TD (1999) Use of *atp*6 in fungal phylogenetics: an example from the boletales. Mol Phylogenet Evol 13:483–492. https://doi.org/10.1006/mpev19990680

Lafontaine I, Fischer G, Talla E, Dujon B (2004) Gene relics in the genome of the yeast *Saccharomyces cerevisiae*. Gene 335:1–17. https://doi.org/10.1016/jgene200403028

Laforest MJ, Roewer I, Lang BF (1997) Mitochondrial tRNAs in the lower fungus Spizellomyces punctatus: tRNA editing and UAG 'stop' codons recognized as leucine. Nucleic Acids Res 25(3):626–632. https://doi.org/10.1093/nar/253626

Lang BF, Burger G, O'Kelly CJ, Cedergren R, Golding GB, Lemieux C, Sankoff D, Turmel M, Gray MW (1997) An ancestral mitochondrial DNA resembling a eubacterial genome in miniature. Nature 387:493–497. https://doi.org/10.1038/387493a0

Lang BF, Gray MW, Burger G (1999) Mitochondrial genome evolution and the origin of eukaryotes. Annu Rev Genet 33:351–397. https://doi.org/10.1146/annurevgenet331351

Lang BF, Laforest MJ, Burger G (2007) Mitochondrial introns: a critical view. Trends Genet 23:119–125. https://doi.org/10.1016/jtig200701006

Lang BF, Jakubkova M, Hegedusova E, Daoud R, Forget L, Brejova B, Vinar T, Kosa P, Fricova D, Nebohacova M, Griac P, Tomaska L, Burger G, Nosek J (2014) Massive programmed translational jumping in mitochondria. PNAS 111:5926–5931. https://doi.org/10.1073/pnas1322190111

Li Y, Hu XD, Yang RH, Hsiang T, Wang K, Liang DQ, Liang F, Cao DM, Zhou F, Wen G, Yao YJ (2015) Complete mitochondrial genome of the medicinal fungus *Ophiocordyceps sinensis*. Sci Rep 5:13892. https://doi.org/10.1038/srep13892

Li WC, Lin TC, Chen CL, Liu HC, Lin HN, Chao JL, Hsieh CH, Ni HF, Chen RS, Wang TF (2021a) Complete genome sequences and genome-wide characterization of *Trichoderma* biocontrol agents provide new insights into their evolution and variation in genome organization, sexual development, and fungal-plant interactions. Microbiol Spectr 9:e0066321. https://doi.org/10.1128/Spectrum00663-21

Li Y, Steenwyk JL, Chang Y, Wang Y, James TY, Stajich JE, Rokas A (2021b) A genome-scale phylogeny of the kingdom fungi. Curr Biol 31(8):1653–1665.e5. https://doi.org/10.1016/jcub202101074

Liao PC, Wolken D, Serrano E, Srivastava P, Pon LA (2020) Mitochondria-associated degradation pathway (MAD) function beyond the outer membrane. Cell Rep 32:107902. https://doi.org/10.1016/jcelrep2020107902

Liu Q, Wang X (2021) Characterization and phylogenetic analysis of the complete mitochondrial genome of pathogen *Trichosporon inkin* (Trichosporonales: Trichosporonaceae). Mitochondrial DNA B Resour 6(3):803–805. https://doi.org/10.1080/23802359202111882912

Liu X, Wu X, Tan H, Xie B, Deng Y (2020a) Large inverted repeats identified by intra-specific comparison of mitochondrial genomes provide insights into the evolution of *Agrocybe aegerita*. Comput Struct Biotechnol J 18:2424–2437. https://doi.org/10.1016/jcsbj202008022

Liu W, Cai Y, Zhang Q, Chen L, Shu F, Ma X, Bian Y (2020b) The mitochondrial genome of *Morchella importuna* (2722 kb) is the largest among fungi and contains numerous introns, mitochondrial non-conserved open reading frames and repetitive sequences. Int J Biol Macromol 143:373–381. https://doi.org/10.1016/jijbiomac201912056

Losada L, Pakala SB, Fedorova ND, Joardar V, Shabalina SA, Hostetler J, Pakala SM, Zafar N, Thomas E, Rodriguez-Carres M, Dean R, Vilgalys R, Nierman WC, Cubeta MA (2014) Mobile elements and mitochondrial genome expansion in the soil fungus and potato pathogen *Rhizoctonia solani* AG-3. FEMS Microbiol Lett 352:165–173. https://doi.org/10.1111/1574-696812387

Maleszka R, Skelly PJ, Clark-Walker GD (1991) Rolling circle replication of DNA in yeast mitochondria. EMBO J 10:3923–3929. https://doi.org/10.1002/j1460-20751991tb04962x

Mardanov AV, Beletsky AV, Kadnikov VV, Ignatov AN, Ravin NV (2014) The 203 kbp mitochondrial genome of the phytopathogenic fungus *Sclerotinia borealis* reveals multiple invasions of introns and genomic duplications. PLoS One 9(9):e107536. https://doi.org/10.1371/journalpone0107536

Martin W (2003) Gene transfer from organelles to the nucleus: frequent and in big chunks. PNAS 100: 8612–8614. https://doi.org/10.1073/pnas1633606100

Martin WF, Garg S, Zimorski V (2015) Endosymbiotic theories for eukaryote origin. Philos Trans R Soc Lond Ser B Biol Sci 370:20140330. https://doi.org/10.1098/rstb20140330

Megarioti AH, Kouvelis VN (2020) The coevolution of fungal mitochondrial introns and their homing endonucleases (GIY-YIG and LAGLIDADG). Genome Biol Evol 12(8):1337–1354. https://doi.org/10.1093/gbe/evaa126

Meinhardt F, Schaffrath R, Larsen M (1997) Microbial linear plasmids. Appl Microbiol Biotechnol 47:329–336. https://doi.org/10.1007/s002530050936

Menger KE, Rodríguez-Luis A, Chapman J, Nicholls TJ (2021) Controlling the topology of mammalian mitochondrial DNA. Open Biol 11(9):210168. https://doi.org/10.1098/rsob210168

Monteiro-Vitorello CB, Baidyaroy D, Bell JA, Hausner G, Fulbright DW, Bertrand H (2000) A circular mitochondrial plasmid incites hypovirulence in some strains of *Cryphonectria parasitica*. Curr Genet 37:242–256. https://doi.org/10.1007/s002940050526

Moritz C, Dowling TE, Brown WM (1987) Evolution of animal mitochondrial DNA: relevance for population biology and systematics. Annu Rev Ecol Evol 18:269–292. https://doi.org/10.1146/annureves18110187001413

Mukhopadhyay J, Hausner G (2021) Organellar introns in fungi, algae, and plants. Cells 10:2001. https://doi.org/10.3390/cells10082001

Myers CJ, Griffiths AJ, Bertrand H (1989) Linear kalilo DNA is a *Neurospora* mitochondrial plasmid that integrates into the mitochondrial DNA. Mol Gen Genet 220:113–120. https://doi.org/10.1007/BF00260864

Nadimi M, Beaudet D, Forget L, Hijri M, Lang BF (2012) Group I intron-mediated trans-splicing in mitochondria of *Gigaspora rosea* and a robust phylogenetic affiliation of arbuscular mycorrhizal fungi with Mortierellales. Mol Biol Evol 29(9):2199–2210. https://doi.org/10.1093/molbev/mss088

Nadimi M, Daubois L, Hijri M (2016) Mitochondrial comparative genomics and phylogenetic signal assessment of mtDNA among arbuscular mycorrhizal fungi. Mol Phylogenet Evol 98:74–83. https://doi.org/10.1016/jympev201601009

Nakai R, Sen K, Kurosawa S, Shibai H (2000) Basidiomycetous fungus *Flammulina velutipes* harbors two linear mitochondrial plasmids encoding DNA and RNA polymerases. FEMS Microbiol Lett 190(1): 99–102. https://doi.org/10.1111/j1574-69682000tb09269x

Nosek J, Tomáska L (2003) Mitochondrial genome diversity: evolution of the molecular architecture and replication strategy. Curr Genet 44:73–84. https://doi.org/10.1007/s00294-003-0426-z

Nosek J, Dinouël N, Kovac L, Fukuhara H (1995) Linear mitochondrial DNAs from yeasts: telomeres with large tandem repetitions. Mol Gen Genet 247(1):61–72. https://doi.org/10.1007/BF00425822

Nosek J, Tomáska L, Fukuhara H, Suyama Y, Kovác L (1998) Linear mitochondrial genomes: 30 years down the line. Trends Genet 14:184–188. https://doi.org/10.1016/s0168-9525(98)01443-7

Noutsos C, Kleine T, Armbruster U, DalCorso G, Leister D (2007) Nuclear insertions of organellar DNA can create novel patches of functional exon sequences. Trends Genet 23:597–601. https://doi.org/10.1016/jtig200708016

Novikova O, Belfort M (2017) Mobile group II introns as ancestral eukaryotic elements. Trends Genet 33:773–783. https://doi.org/10.1016/jtig201707009

Osawa S (1995) Evolution of the genetic code. Oxford University Press, Oxford

Osawa S, Jukes TH, Watanabe K, Muto A (1992) Recent evidence for evolution of the genetic code. Microbiol Rev 56:229–264. https://doi.org/10.1128/mr561229-2641992

Pantou MP, Strunnikova OK, Shakhnazarova VY, Vishnevskaya NA, Papalouka VG, Typas MA (2005) Molecular and immunochemical phylogeny of *Verticillium* species. Mycol Res 109:889–902. https://doi.org/10.1017/s0953756205003345
Pantou MP, Kouvelis VN, Typas MA (2008) The complete mitochondrial genome of *Fusarium oxysporum*: insights into fungal mitochondrial evolution. Gene 419(1–2):7–15. https://doi.org/10.1016/jgene200804009
Paquin B, Lang BF (1996) The mitochondrial DNA of *Allomyces macrogynus*: the complete genomic sequence from an ancestral fungus. J Mol Biol 255: 688–701. https://doi.org/10.1006/jmbi19960056
Paquin B, Laforest MJ, Lang BF (2000) Double-hairpin elements in the mitochondrial DNA of allomyces: evidence for mobility. Mol Biol Evol 17:1760–1768. https://doi.org/10.1093/oxfordjournalsmolbeva026274
Pickles S, Vigié P, Youle RJ (2018) Mitophagy and quality control mechanisms in mitochondrial maintenance. Curr Biol 28:R170–R185. https://doi.org/10.1016/jcub201801004
Pöggeler S, Kempken F (2004) Mobile genetic elements in mycelial fungi. In: Kück U (ed) The mycota II: genetics and biotechnology, 2nd edn. Springer, Berlin, pp 165–197
Pramateftaki PV, Kouvelis VN, Lanaridis P, Typas MA (2006) The mitochondrial genome of the wine yeast *Hanseniaspora uvarum*: a unique genome organization among yeast/fungal counterparts. FEMS Yeast Res 6(1):77–90. https://doi.org/10.1111/j1567-1364200500018x
Quandt CA, Beaudet D, Corsaro D, Walochnik J, Michel R, Corradi N, James TY (2017) The genome of an intranuclear parasite, *Paramicrosporidium saccamoebae*, reveals alternative adaptations to obligate intracellular parasitism. eLife 6:e29594. https://doi.org/10.7554/eLife29594
Repar J, Warnecke T (2017) Mobile introns shape the genetic diversity of their host genes. Genetics 205(4): 1641–1648. https://doi.org/10.1534/genetics116199059
Richly E, Leister D (2004) NUMTs in sequenced eukaryotic genomes. Mol Biol Evol 21(6):1081–1084. https://doi.org/10.1093/molbev/msh110
Rubinoff D, Holland BS (2005) Between two extremes: mitochondrial DNA is neither the panacea nor the nemesis of phylogenetic and taxonomic inference. Syst Biol 54:952–961. https://doi.org/10.1080/10635150500234674
Rudski SM, Hausner G (2012) The mtDNA *rps*3 locus has been invaded by a group I intron in some species of *Grosmannia*. Mycoscience 53:471–475. https://doi.org/10.1007/S10267-012-0183-2
Saccone C, Gissi C, Reyes A, Larizza A, Sbisà E, Pesole G (2002) Mitochondrial DNA in metazoa: degree of freedom in a frozen event. Gene 286:3–12. https://doi.org/10.1016/s0378-1119(01)00807-1
Sacerdot C, Casaregola S, Lafontaine I, Tekaia F, Dujon B, Ozier-Kalogeropoulos O (2008) Promiscuous DNA in the nuclear genomes of hemiascomycetous yeasts. FEMS Yeast Res 8:846–857. https://doi.org/10.1111/j1567-1364200800409x
Salavirta H, Oksanen I, Kuuskeri J, Mäkelä M, Laine P, Paulin L, Lundell T (2014) Mitochondrial genome of *Phlebia radiata* is the second largest (156 kbp) among fungi and features signs of genome flexibility and recent recombination events. PLoS One 9:e97141. https://doi.org/10.1371/journalpone0097141
Sandor S, Zhang Y, Xu J (2018) Fungal mitochondrial genomes and genetic polymorphisms. Appl Microbiol Biotechnol 102:9433–9448. https://doi.org/10.1007/s00253-018-9350-5
Schäfer B, Hansen M, Lang BF (2005) Transcription and RNA-processing in fission yeast mitochondria. RNA 11:785–795. https://doi.org/10.1261/rna7252205
Schultz DW, Yarus M (1996) On malleability in the genetic code. J Mol Evol 42:597–601. https://doi.org/10.1007/BF02352290
Seif ER, Forget L, Martin NC, Lang BF (2003) Mitochondrial RNase P RNAs in ascomycete fungi: lineage-specific variations in RNA secondary structure. RNA 9:1073–1083. https://doi.org/10.1261/rna5880403
Sekito T, Okamoto K, Kitano H, Yoshida K (1995) The complete mitochondrial DNA sequence of *Hansenula wingei* reveals new characteristics of yeast mitochondria. Curr Genet 28:39–53. https://doi.org/10.1007/BF00311880
Sengupta S, Higgs PG (2005) A unified model of codon reassignment in alternative genetic codes. Genetics 170:831–840. https://doi.org/10.1534/genetics104037887
Sengupta S, Yang X, Higgs PG (2007) The mechanisms of codon reassignments in mitochondrial genetic codes. J Mol Evol 64:662–688. https://doi.org/10.1007/s00239-006-0284-7
Smith DR (2016) The past, present and future of mitochondrial genomics: have we sequenced enough mtDNAs? Brief Funct Genomics 15:47–54. https://doi.org/10.1093/bfgp/elv027
Song N, Geng Y, Li X (2020) The mitochondrial genome of the phytopathogenic fungus *Bipolaris sorokiniana* and the utility of mitochondrial genome to infer phylogeny of Dothideomycetes. Front Microbiol 11:863. https://doi.org/10.3389/fmicb202000863
Stoddard BL (2014) Homing endonucleases from mobile group I introns: discovery to genome engineering. Mob DNA 5(1):7. https://doi.org/10.1186/1759-8753-5-7
Swire J, Judson OP, Burt A (2005) Mitochondrial genetic codes evolve to match amino acid requirements of proteins. J Mol Evol 60:128–139. https://doi.org/10.1007/s00239-004-0077-9
Theelen B, Christinaki AC, Dawson TL, Boekhout T, Kouvelis VN (2021) Comparative analysis of *Malassezia furfur* mitogenomes and the development of a mitochondria-based typing approach. FEMS Yeast

Res 21:foab051. https://doi.org/10.1093/femsyr/foab051

Timmis JN, Ayliffe MA, Huang CY, Martin W (2004) Endosymbiotic gene transfer: organelle genomes forge eukaryotic chromosomes. Nat Rev Genet 5:123–135. https://doi.org/10.1038/nrg1271

Toor N, Zimmerly S (2002) Identification of a family of group II introns encoding LAGLIDADG ORFs typical of group I introns. RNA 8(11):1373–1377. https://doi.org/10.1017/s1355838202023087

Torriani SF, Penselin D, Knogge W, Felder M, Taudien S, Platzer M, McDonald BA, Brunner PC (2014) Comparative analysis of mitochondrial genomes from closely related *Rhynchosporium* species reveals extensive intron invasion. Fungal Genet Biol 62:34–42. https://doi.org/10.1016/jfgb201311001

Valach M, Farkas Z, Fricova D, Kovac J, Brejova B, Vinar T, Pfeiffer I, Kucsera J, Tomaska L, Lang BF, Nosek J (2011) Evolution of linear chromosomes and multipartite genomes in yeast mitochondria. Nucleic Acids Res 39:4202–4219. https://doi.org/10.1093/nar/gkq1345

van de Vossenberg B, Brankovics B, Nguyen H, van Gent-Pelzer M, Smith D, Dadej K, Przetakiewicz J, Kreuze JF, Boerma M, van Leeuwen G, André Lévesque C, van der Lee T (2018) The linear mitochondrial genome of the quarantine chytrid *Synchytrium endobioticum*; insights into the evolution and recent history of an obligate biotrophic plant pathogen. BMC Evol Biol 18:136. https://doi.org/10.1186/s12862-018-1246-6

Varassas SP, Kouvelis VN (2022) Mitochondrial transcription of entomopathogenic fungi reveals evolutionary aspects of mitogenomes. Front Microbiol 13: 821638. https://doi.org/10.3389/fmicb2022821638

Wai A, Shen C, Carta A, Dansen A, Crous PW, Hausner G (2019) Intron-encoded ribosomal proteins and N-acetyltransferases within the mitochondrial genomes of fungi: here today, gone tomorrow? Mitochondrial DNA 30(3):573–584. https://doi.org/10.1080/2470139420191580272

Weidberg H, Amon A (2018) MitoCPR-A surveillance pathway that protects mitochondria in response to protein import stress. Science 360:eaan4146. https://doi.org/10.1126/scienceaan4146

Williamson D (2002) The curious history of yeast mitochondrial DNA. Nat Rev Genet 3(6):475–481. https://doi.org/10.1038/nrg814

Wolters JF, Chiu K, Fiumera HL (2015) Population structure of mitochondrial genomes in *Saccharomyces cerevisiae*. BMC Genomics 16:281

Wu B, Hao W (2019) Mitochondrial-encoded endonucleases drive recombination of protein-coding genes in yeast. Environ Microbiol 21:4233–4240. https://doi.org/10.1111/1462-292014783

Wu B, Buljic A, Hao W (2015) Extensive horizontal transfer and homologous recombination generate highly chimeric mitochondrial genomes in yeast. Mol Biol Evol 32:2559–2570. https://doi.org/10.1093/molbev/msv127

Xiao S, Nguyen DT, Wu B, Hao W (2017) Genetic drift and indel mutation in the evolution of yeast mitochondrial genome size. Genome Biol Evol 9:3088–3099. https://doi.org/10.1093/gbe/evx232

Yang X, Chang HR, Yin YW (2015) Yeast mitochondrial transcription factor Mtf1 determines the precision of promoter-directed initiation of RNA polymerase Rpo41. PLoS One 10:e0136879. https://doi.org/10.1371/journalpone0136879

Yildiz G, Ozkilinc H (2021) Pan-Mitogenomics approach discovers diversity and dynamism in the prominent Brown rot fungal pathogens. Front Microbiol 12: 647989. https://doi.org/10.3389/fmicb2021647989

Yin S, Heckman J, RajBhandary UL (1981) Highly conserved GC-rich palindromic DNA sequences flank tRNA genes in *Neurospora crassa* mitochondria. Cell 26:325–332. https://doi.org/10.1016/0092-8674(81)90201-4

Youle RJ (2019) Mitochondria-striking a balance between host and endosymbiont. Science 365:eaaw9855. https://doi.org/10.1126/scienceaaw9855

Younis I, Dittmar K, Wang W, Foley SW, Berg MG, Hu KY, Wei Z, Wan L, Dreyfuss G (2013) Minor introns are embedded molecular switches regulated by highly unstable U6atac snRNA. Elife 2:e00780. https://doi.org/10.7554/eLife0078

Zardoya R (2020) Recent advances in understanding mitochondrial genome diversity. F1000Research 9:27. https://doi.org/10.12688/f1000research.21490.1

Zhao Z, Zhu K, Tang D, Wang Y, Wang Y, Zhang,G, Geng Y, & Yu H (2021) Comparative analysis of mitochondrial genome features among four *Clonostachys* species and insight into their systematic positions in the order hypocreales. Int J Mol Sci 22: 5530. https://doi.org/10.3390/ijms22115530

Part II

Evolution of Pathogenic Strategies

Dimorphism and Pathogenesis in *Mucor* Species

4

Alexis Garcia, Courtney P. Smith, and Soo Chan Lee

Abstract

Many pathogenic fungi are dimorphic, in which they switch the morphology between yeast and filamentous forms. This dimorphism has been observed in the Dikarya. Interestingly, the genus *Mucor* is the only group of fungi among the early diverged fungi outside of the Dikarya that exhibits the yeast–hyphae transition. Their morphogenic switch is controlled by environmental factors such as low oxygen and high carbon dioxide concentrations. Genetically, the genes encoding G-protein coupled receptors (GPCR), a serine/threonine phosphatase calcineurin, and subunits of protein kinase A (PKA) all have been documented in the regulation of the dimorphic transitions in *Mucor*. *Mucor circinelloides* is one of causative agents of the deadly opportunistic fungal infection mucormycosis. Similar to other known dimorphic fungi (e.g. *Candida albicans* and *Coccidioides* species), the morphology directly contributes to the virulence of this fungus. Upon entering a host, it grows as filamentous hyphae and invades the host tissue. This chapter further highlights the recent findings on how genes and the environment play a critical role in dimorphism and the virulence in *Mucor* and discusses how these findings can serve as a platform for new therapeutic interventions.

Keywords

Mucor · Dimorphism · Pathogenesis · Calcineurin · Protein kinase A

4.1 Introduction

There is a dramatic variety of cellular morphologies in fungi that can be observed. Two of the most common are the yeast and hyphal morphologies. The ability to shift between two morphologies is referred to as dimorphism. The yeast morphology is characterized as a single cell with a nucleus that reproduces via budding or fission (Sudbery et al. 2004). Typically, this morphology has a spherical or ellipsoid shape that can span a few micrometers in length (Sudbery et al. 2004). Other morphotypes of yeast cells do exist, such as the syncytium, which consists of a single cytoplasm with multiple nuclei (Knop 2011). The hyphal morphology consists of long segments called filaments. These filaments can be 1–30 μm in diameter and can grow from a few microns to meters in length (Islam et al. 2017). Many fungi can transition between yeast and filamentous forms; the dimorphic transition

A. Garcia · C. P. Smith · S. C. Lee (✉)
Department of Molecular Microbiology and Immunology, South Texas Center for Emerging Infectious Diseases (STCEID), The University of Texas at San Antonio, San Antonio, TX, USA
e-mail: soochan.lee@utsa.edu

© The Author(s), under exclusive license to Springer Nature Switzerland AG 2023

S. Pöggeler, T. James (eds.), *Evolution of Fungi and Fungal-Like Organisms*, The Mycota 14,
https://doi.org/10.1007/978-3-031-29199-9_4

between these two morphologies depends on factors such as pH, temperature, and aerobic/anaerobic environmental conditions (Orlowski 1991; Mitchell 1998; Kirkland and Fierer 2018).

Dimorphism shows convergent evolution across the tree of fungi. Most of these fungi belong to the phyla Ascomycota or Basidiomycota, such as *Candida* spp. and *Cryptococcus* spp. (Fig. 4.1a). Many studies indicate that the yeast morphology is only conserved in the Dikarya (Ascomycota and Basidiomycota) and that fungi in the basal lineages do not undergo the yeast form (Laundon et al. 2020). However, yeast growth is also observed in fungi in the phylum Mucoromycota. Mucoromycota has not been included in a yeast-forming group, which is due to only species found in the genus *Mucor* having the ability to grow as yeast (Fig. 4.1b) (Nagy et al. 2017; Laundon et al. 2020). There are three scenarios with regard to the evolution of the yeast morphology. In the first scenario, yeast may have emerged early in the evolution of the fungal kingdom, followed by a loss of the morphology that may have occurred in most basal lineages and some Dikarya clades (Prostak et al. 2021). In the second scenario, the yeast morphology may have emerged prior to the emergence of Mucoromycota and Dikarya. If so, then a loss of the morphology could have occurred in most of the lineages in Mucoromycota and some lineages in Dikarya. In the final scenario, multiple independent convergent evolutions may have occurred in distinct, unrelated clades.

4.2 Dimorphic Transitions in *Mucor* Species

4.2.1 Environmental Factors

Mucor species have been studied longer than most of the world's fungi. They were first described in 1665 in the publication *Micrographia* by Robert Hooke, who described the morphology of a *Mucor* species that grew on his leather belt (Hooke 1667). As time passed and research on Mucorales delved deeper, it was discovered that fungi in the *Mucor* genus are dimorphic (Orlowski 1991; Lee et al. 2013). Interestingly, the hyphal growth of *Mucor* was thought to result from a transmutation of *Saccharomyces* species (Bartnicki-García 1963). However, Louis Pasteur wrote in his book *Études Sur La Bière* that *Mucor* is a filamentous fungus but grows as a multi-budded yeast in conditions with low O_2 and high CO_2 concentrations (Fig. 4.1b), similar to the conditions in the bottom of wine or beer barrels (Pasteur 1876). Bartnicki-Garcia and his colleagues later rediscovered that the major inducer of *Mucor* yeast growth is CO_2 (Bartnicki-Garcia and Nickerson 1962a, b).

Although *Mucor* species exhibit a dimorphic transition in response to various environments, the primary factors that affect the type of growth are the concentration of O_2 and CO_2 and carbon sources (Orlowski 1991). High O_2 concentrations trigger hyphal growth even when fungi are exposed to high concentrations of CO_2 (Fig. 4.1b). This indicates that respiration and mitochondrial function is involved in the dimorphic transition. In fact, chemicals that inhibit mitochondrial function, such as potassium cyanide and antimycin A, which are known to block electron transport, or oligomycin and phenyl alcohol, which inhibit oxidative phosphorylation, can induce yeast growth even in aerobic conditions (Schulz et al. 1974; Gordon et al. 1972; Friedenthal et al. 1974). In addition, chloramphenicol, which inhibits mitochondrial components, results in yeast growth in *Mucor* (Clark-Walker 1973; Rogers et al. 1974). The lipid metabolism inhibitor cerulenin and the S-adenosylmethionine synthetase inhibitor cycloleucine both inhibit the formation of hyphal growth from yeast in aerobic conditions as well (Aoki and Ito-Kuwa 1982; Garcia et al. 1980). Importantly, the addition of bicarbonate that activates adenylyl cyclase or cAMP to the culture media results in yeast growth of *Mucor* (Linz and Orlowski 1991; Lee et al. 2013), implying that protein kinase A (PKA) plays an important role in yeast growth.

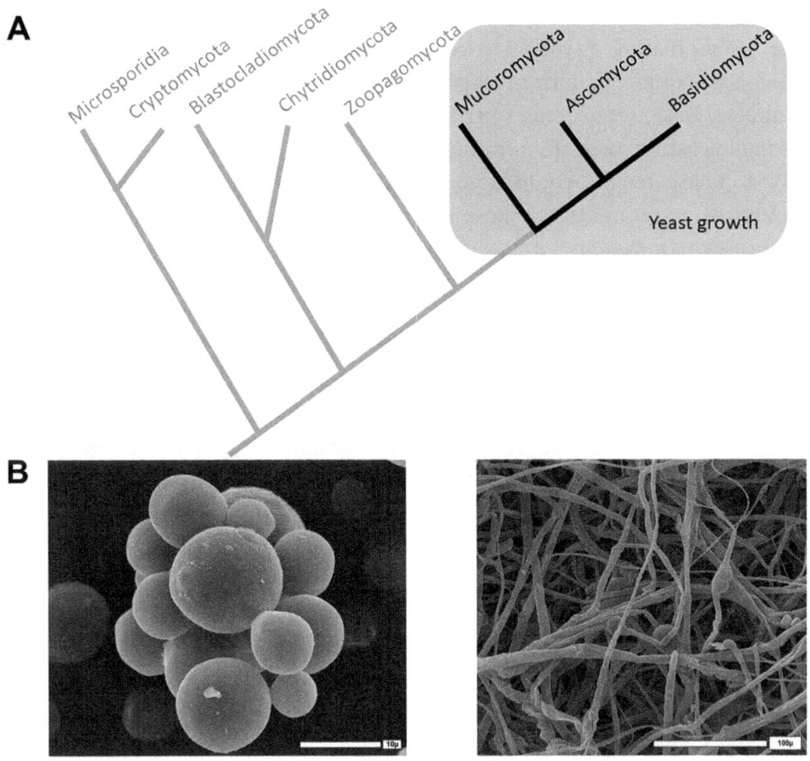

Fig. 4.1 Phylogeny figure. (**a**) Phylogenetic tree of different phyla of fungi. The yeast morphology is found throughout the Dikarya (Ascomycete and Basidiomycota). Fungi in the basal lineages, except for Mucoromycota, do not grow in the yeast morphology or at least have not been observed. (**b**) Two morphologies of *Mucor*. *Mucor* is a filamentous fungus but grows as a multi-budded yeast in conditions with low O_2 and high CO_2 concentrations (left). High O_2 concentrations trigger hyphal growth (right)

4.2.2 Genetics

Dimorphism in *Mucor* species has been documented to be controlled by environmental and genetic cues (Lee et al. 2015; Nadal et al. 2008; Valle-Maldonado et al. 2020). Genetic changes in fungi, such as mutations, generate various detectable signals; the dimorphic fungi receive these signals and may undergo a morphological shift. Currently, there are several documented signaling pathways in the regulation of the dimorphic transitions in *Mucor* - those related to G-protein coupled receptors (GPCR), calcineurin, and subunits of PKA.

Phylogenetic analysis of heterotrimeric G-proteins in fungal species has gained traction in the past two decades. Heterotrimeric G-proteins are abundant across eukaryotic cells; these proteins are fundamental in biological processes, including but not limited to growth, cell differentiation, pathogenesis, and signaling transduction (Li et al. 2007). The heterotrimeric G-proteins are composed of three subunits: Gα, Gβ, and Gγ. Most fungal genomes consist of three Gα, one Gβ, and one Gγ subunits that are conserved (Li et al. 2007). These subunits are attached to the GPCR and sense various ligands (Beckerman 2005; Li et al. 2007). The signaling cascade of the G-proteins is initiated when guanosine triphosphate (GTP) binds to the Gα subunit. The heterotrimeric G-protein dissociates into two signaling components: the Gα–GTP complex and the Gβ–Gγ dimer. Both of these signaling components may regulate downstream effectors. The hydrolysis of the GTP by the Gα subunit results in the reassociation of the heterotrimeric complex (Li et al. 2007).

It has been documented that *Mucor* possesses twelve Gα (Gpa 1–12), three Gβ (Gpb 1–3), and three Gγ subunits (Gpg 1–3), which is the largest collection of heterotrimeric G-proteins in a fungal species (Valle-Maldonado et al. 2015). Phylogenetic and mRNA quantitation analysis of these genes during dimorphic shifting showed that the expression of *gpb1* specifically was upregulated during mycelial growth when compared to spore or yeast growth, which may indicate that it plays a role in regulating mycelial growth. *gpb3* and *gpg2* are expressed during mycelial growth; in contrast, *gpa1*, *gpb2*, and *gpg2* are expressed during yeast growth (Valle-Maldonado et al. 2015). These results could together outline the biological importance of these genes and the dimorphism in *Mucor*.

Another pathway is the calcium-signaling pathway. Calcineurin is a Ca^{2+}/calmodulin-dependent, serine/threonine-specific protein phosphatase and is a conserved virulence factor in many pathogenic fungi (Vellanki et al. 2020). *Mucor* encodes for one regulatory B subunit (CnbR) and three calcineurin catalytic A subunits (CnaA, CnaB, and CnaC) (Lee et al. 2013). Some nonsynonymous mutations that occur in the genes *cnaA* or *cnbR* result in hyphal growth in the presence of calcineurin inhibitors such as FK506 (tacrolimus) or cyclosporine A (CsA), which indicates that the mutations confer resistance to the drugs. Also, loss-of-function mutations that occurred in the *cnbR* gene, which is necessary for calcineurin activity, result in yeast-locked cells (Lee et al. 2013; Garcia et al. 2017) (Fig. 4.2b). In addition, epimutations in the gene encoding FKBP12, the cellular receptor for FK506, also result in hyphal growth in the presence of FK506 (Calo et al. 2014). These combined findings strongly indicate that calcineurin plays a central role in the morphological transition of *Mucor*.

Additionally, it has been found that when *Mucor* is yeast locked, either genetically or environmentally, PKA activity is elevated. This implies that there is a link between calcineurin and PKA (Lee et al. 2013). The cyclic AMP (cAMP)-dependent PKA signaling pathway is conserved and controls many cellular processes in numerous fungi, such as development, growth, and virulence (D'Souza and Heitman 2001; Kronstad et al. 1998; Taylor et al. 1992). The central components of this signaling pathway have been defined for various fungal species (e.g. *Cryptococcus neoformans*, *Ustilago maydis*, and *Candida* spp.), including the model yeast system *Saccharomyces cerevisiae* (Thevelein and De Winde 1999). The *Mucor* genome harbors multiple copies of the genes encoding two subunits of the cAMP-dependent PKA: regulatory PKA (PKAR) and catalytic PKA (PKAC). cAMP serves as a secondary messenger that regulates PKA activity through the binding and interaction with the regulatory subunit. As a result of this interaction, the catalytic subunit is released, initiating the phosphorylation cascade that causes a dimorphic shift (Taylor et al. 1992). The role of this signaling pathway in regulating morphology and filamentation/branching has been documented using genetic approaches in various fungal species, including *Aspergillus niger* and *Mucor* (Saudohar et al. 2002; Wolff et al. 2002). The regulatory and catalytic subunits are encoded by *pkaR* (regulatory subunit) and *pkaC* (catalytic subunit) genes. Both genes are upregulated under yeast growing conditions. However, only *pkaR* expression is increased when shifting from yeast to filamentous growth. Additionally, the overexpression of *pkaR* results in the multi-branched phenotype. These results together indicate that the PKAR subunit plays a major role in filamentation in *Mucor* (Lübbehüsen et al. 2004; Wolff et al. 2002; Valle-Maldonado et al. 2020).

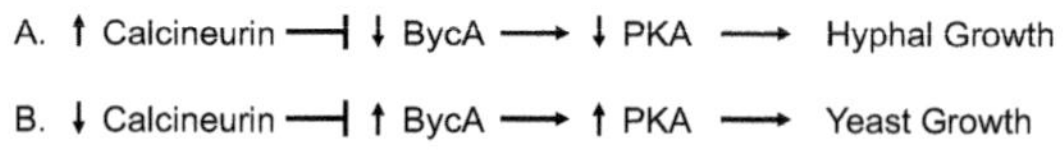

Fig. 4.2 How calcineurin, BycA, and PKA are expressed in the morphogenesis of *Mucor*. (**a**) Calcineurin positively regulates hyphal growth through suppressing *bycA* expression and consequently preventing PKA activity from increasing. (**b**) Under conditions when calcineurin is not functional (either through loss-of-function mutations or drug inhibition), *bycA* gene expression is elevated. BycA then activates PKA, promoting yeast growth in *Mucor*

Notably, one study revealed that calcineurin activity is inversely correlated with PKA activity (Lee et al. 2013). When calcineurin is active, PKA activity is lowered and *Mucor* exhibits hyphal growth (Fig. 4.2a). On the other hand, when calcineurin is inhibited by drugs or the calcineurin gene has loss-of-function mutations PKA activity is elevated, and the fungus is forced to grow in the yeast morphology. Another study found that the *bycA* gene encoding an amino acid permease serves as a link between calcineurin and

PKA (Vellanki et al. 2020). Calcineurin negatively regulates the expression of the *bycA* gene, and the *bycA* gene products function as a positive regulator of PKA activity (Fig. 4.2b).

4.3 Dimorphism and Virulence in Pathogenic Fungi

4.3.1 *Candida* Species

In many pathogenic fungal systems, dimorphic transitions play a central role in pathogenesis. The commensal and opportunistic fungus *Candida albicans* has the ability to transition between individual yeast cells and filamentous hyphae (Sudbery et al. 2004) (Fig. 4.3a). As noted above, environmental conditions can initiate the dimorphic transition in different fungi. In *C. albicans*, filamentation can be stimulated when the organism is grown at 37 °C either in serum or in a neutral pH (Mitchell 1998). Dimorphism and filamentation have been known to be a major virulence factor in *C. albicans*. Interestingly, it is known that these filaments adhere well to different surfaces, such as the tubing used in catheters or even mammalian cells. This allows *C. albicans* to form biofilms that can cause chronic infections in immunocompromised patients or patients with medically implanted/indwelling devices (Gulati and Nobile 2016). Another advantage of dimorphism in *C. albicans* can be observed in the evasion of macrophages (Mitchell 1998; Cutler 1991). Some studies have found that yeast cells taken up by macrophages can switch to the hyphal form and lyse the macrophages from within, leading to the successful evasion of the host immune system (Cutler 1991). Additionally, a yeast-locked mutant of *C. albicans* was used to infect mice and was compared to a filamenting strain. In this study, it was observed that the filamentous strain of *C. albicans* was more virulent than the yeast-locked mutant (Cutler 1991).

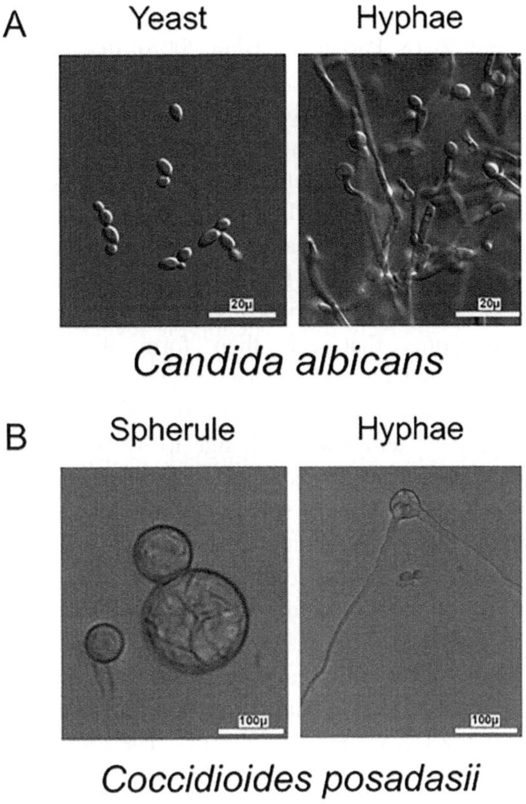

Fig. 4.3 Varying types of morphologies presented by different dimorphic fungi. (**a**) Varying morphologies of *Candida albicans*. *C. albicans* has three morphologies, with the most two common being the yeast and hyphal form; the third is called pseudohyphae (not depicted), which is an intermittent morphology that exhibits characteristics of both yeast and true hyphae. (**b**) Different morphologies of *Coccidioides* spp. The two major morphologies of *Coccidioides* spp. are the spherules and the hyphae forming long filaments that protrude from the spherules

4.3.2 *Coccidioides* Species

The *Coccidioides* spp. of the dimorphic fungi can also transition between the yeast and hyphal forms. *C. immitis* and *C. posadasii* are the two highly pathogenic fungal species endemic to the arid regions of the western USA (Kirkland and Fierer 2018). These species of *Coccidioides* are known to grow as filamentous mold in the soil and can form parasitic spherules (Kirkland and Fierer 2018; Parish and Blair 2008). In the wild, the *Coccidioides* spp. form mycelium in the soil (Fig. 4.3b) and make arthroconidia. Once the arthroconidia are released, they can further propagate if the spores land in the soil (Kirkland and Fierer 2018; Cole and Sun 1985). Unfortunately, if the spores are inhaled, the arthroconidia can then form spherules (Fig. 4.3b). These spherules will begin to swell and convert into hundreds of endospores that can then form their own spherules (Cole and Sun 1985). The environmental cues that trigger dimorphism in *Coccidioides* spp. are a temperature change to 37 °C and an increase in CO_2, making animals a perfect host for the replication of spherules (Parish and Blair 2008). Interestingly, it has been found that when the arthroconidia encounter neutrophils they can begin to form spherules as well. This is further backed by evidence showing that in an organ lacking neutrophils, such as the lungs, an attempt to reverse the hyphal form can be observed (Muñoz-Hernández et al. 2014). The dimorphic transition of different fungi can play major roles in their survival in various environments, including an infected host. It is important to understand dimorphism since it can be an excellent target for therapeutic treatments against disease-causing dimorphic fungi.

4.4 Dimorphism and Virulence in *Mucor* Species

Mucor is typically found in the wild as a mold, which poses little threat to healthy individuals. In recent years, *Mucor* has rapidly emerged as a causative agent of the infection mucormycosis (Smith and Lee 2022). Patients who lack a functional immune system are at the highest risk of this infection that has unacceptably high mortality rates of 90–100% in disseminated cases (Reid et al. 2020). When *Mucor* infects an individual, it grows as filamentous hyphae and has been known to invade the host tissue via angiogenesis (Vellanki et al. 2020). The biggest issue related to the mortality of these infections is that Mucorales fungi are intrinsically resistant to most antifungal treatments, making it difficult to contain the infection (Challa 2019). One possible treatment for mucormycosis is to target the dimorphic transition that these fungi make in stressful environments. Studies have found that the yeast-locked phenotype of *Mucor* is avirulent, making dimorphism an excellent target for treating mucormycosis infections (Lee et al. 2013; Lee et al. 2015).

Dimorphism in *Mucor* is governed by calcineurin, and calcineurin inhibitors block the formation of invasive hyphal growth in *Mucor*. Thus, further investigation into *Mucor* dimorphism can provide a platform of understanding the invasive filamentous growth of Mucorales fungi and aid in the development of anti-mucormycosis drugs. Calcineurin inhibitors such as FK506 and CsA inhibit hyphal growth in *Mucor* and other Mucorales fungi (Juvvadi et al. 2017; Haider et al. 2019; Schwarz et al. 2019; Vellanki et al. 2020). However, these calcineurin inhibitors cannot be used to treat fungal infections in humans because calcineurin is highly conserved in fungi and humans. In humans, calcineurin is required for T-cell function and is a popular target to immunosuppress patients (Williams and Gooch 2012). A recent study suggests that a gene downstream of calcineurin can be targeted to inhibit the calcineurin pathway without the direct inhibition of calcineurin itself (Vellanki et al. 2020). The *bycA* gene is downstream from calcineurin, and its overexpression potentially recapitulates the inhibition of calcineurin seen when calcineurin inhibitors are used. Importantly, this amino acid permease is highly diverged and unique to Mucorales. Therefore, any molecules that directly affect the expression of *bycA* will be excellent calcineurin pathway inhibitors without compromising calcineurin and therefore will have no effect on human calcineurin function.

Among the fungi in the phylum Mucoromycota, *Mucor* species in the Mucorales order are the only ones to exhibit dimorphism between yeast and hyphal growth. This is the only dimorphism observed in the basal fungal lineages. The yeast phase of fungi is an evolutionary conundrum in which evolutionarily distinct species show both yeast and hyphal growths. In contrast, in other cases of closely related species only one species exhibits dimorphism, but others do not. For example, *C. albicans* is dimorphic and another fungus in the *Candida* clade *Candida lusitaniae* is not (Shapiro and Cowen 2012). The evolution of dimorphism in the *Mucor* lineage is quite interesting. Further comparative genomics followed by genetic manipulations will provide clues to understanding the evolution of dimorphism in *Mucor* and even in the fungal kingdom.

4.5 Conclusion and Future Directions

In the last two decades, many studies have unveiled how genes and the environment play a critical role in the virulence of Mucorales. These studies have shown promising developments in host–*Mucor* interactions and new treatments. Among the recent studies, it was demonstrated that iron uptake systems also play a role in virulence and dimorphism (Navarro-Mendoza et al. 2018), indicating the presence of regulatory factors orchestrating both processes. Therefore, these factors could represent a target for new treatment therapies. Additionally, other studies are currently using antifungal compounds, like FK506, and gene silencing targeting calcineurin to research dimorphism and virulence. Unfortunately, compounds like FK506 do not have a clinical application as an antifungal because of its strong immunosuppressive effect on patients (Hooks 1994). In spite of that, new compounds are needed to target the calcineurin pathway and in tandem with gene silencing could prove to be a promising avenue for new therapies.

It is of interest that only *Mucor* species exhibit yeast-hyphal transition, even when considering other fungi in Mucoromycota and other basal lineages. Has dimorphism only evolved in *Mucor* species? Or is the extent of dimorphism in other basal lineages simply yet to be found? More attention needs to be paid to the basal fungal lineages to address these questions.

References

Aoki S, Ito-Kuwa S (1982) Respiration of Candida albicans in relation to its morphogenesis. Plant Cell Physiol 23(4):721–726

Bartnicki-García S (1963) III. Mold-yeast dimorphism of mucor. Bacteriol Rev 27(3):293–304

Bartnicki-Garcia S, Nickerson WJ (1962a) Isolation, composition, and structure of cell walls of filamentous and yeast-like forms of Mucor rouxii. Biochim Biophys Acta 58(1):102–119

Bartnicki-Garcia S, Nickerson WJ (1962b) Nutrition, growth, and morphogenesis of Mucor rouxii. J Bacteriol 84(4):841–858

Beckerman M (2005) Signaling in the endocrine and nervous systems through GPCRs. Mol Cell Signal 2005: 275–303

Calo S, Shertz-Wall C, Lee SC, Bastidas RJ, Nicolás FE, Granek JA, Mieczkowski P, Torres-Martínez S, Ruiz-Vázquez RM, Cardenas ME, Heitman J (2014) Antifungal drug resistance evoked via RNAi-dependent epimutations. Nature 513(7519):555–558. https://doi.org/10.1038/nature13575

Challa S (2019) Mucormycosis: pathogenesis and pathology. Curr Fungal Infect Rep 13(1):11–20

Clark-Walker G (1973) Relationship between dimorphology and respiration in Mucor genevensis studied with chloramphenicol. J Bacteriol 116(2): 972–980

Cole GT, Sun S (1985) Arthroconidium-spherule-endospore transformation in Coccidioides immitis. In: Fungal dimorphism. Springer, New York, pp 281–333

Cutler JE (1991) Putative virulence factors of Candida albicans. Annu Rev Microbiol 45:187–218

D'Souza CA, Heitman J (2001) Conserved cAMP signaling cascades regulate fungal development and virulence. FEMS Microbiol Rev 25(3):349–364

Friedenthal M, Epstein A, Passeron S (1974) Effect of potassium cyanide, glucose and anaerobiosis on morphogenesis of Mucor rouxii. Microbiology 82(1): 15–24

Garcia J, Hiatt W, Peters J, Sypherd P (1980) S-Adenosylmethionine levels and protein methylation during morphogenesis of Mucor racemosus. J Bacteriol 142(1):196–201

Garcia A, Adedoyin G, Heitman J, Lee SC (2017) Construction of a recyclable genetic marker and serial gene deletions in the human pathogenic Mucorales *Mucor circinelloides*. G3 (Bethesda) 7(7):2047–2054. https://doi.org/10.1534/g3.117.041095

Gordon P, Lowdon M, Stewart P (1972) Effects of chloramphenicol isomers and erythromycin on enzyme and lipid synthesis induced by oxygen in wild-type and petite yeast. J Bacteriol 110(2):504–510

Gulati M, Nobile CJ (2016) Candida albicans biofilms: development, regulation, and molecular mechanisms. Microbes Infect 18(5):310–321

Haider M, Dambuza IM, Asamaphan P, Stappers M, Reid D, Yamasaki S, Brown GD, Gow NA, Erwig

LP (2019) The pattern recognition receptors dectin-2, mincle, and Far impact the dynamics of phagocytosis of Candida, Saccharomyces, Malassezia, and Mucor species. PLoS One 14(8):e0220867
Hooke R (1667) Micrographia. Royal Society, London
Hooks MA (1994) Tacrolimus, a new immunosuppressant: a review of the literature. Ann Pharmacother 28(4):501–511. https://doi.org/10.1177/106002809402800414
Islam MR, Tudryn G, Bucinell R, Schadler L, Picu RC (2017) Morphology and mechanics of fungal mycelium. Sci Rep 7(1):13070. https://doi.org/10.1038/s41598-017-13295-2
Juvvadi PR, Lee SC, Heitman J, Steinbach WJ (2017) Calcineurin in fungal virulence and drug resistance: prospects for harnessing targeted inhibition of calcineurin for an antifungal therapeutic approach. Virulence 8(2):186–197
Kirkland TN, Fierer J (2018) Coccidioides immitis and posadasii; a review of their biology, genomics, pathogenesis, and host immunity. Virulence 9(1):1426–1435
Knop M (2011) Yeast cell morphology and sexual reproduction: a short overview and some considerations. C R Biol 334(8–9):599–606
Kronstad J, De Maria A, Funnell D, Laidlaw RD, Lee N, de Sá MM, Ramesh M (1998) Signaling via cAMP in fungi: interconnections with mitogen-activated protein kinase pathways. Arch Microbiol 170(6):395–404
Laundon D, Chrismas N, Wheeler G, Cunliffe M (2020) Chytrid rhizoid morphogenesis resembles hyphal development in multicellular fungi and is adaptive to resource availability. Proc R Soc B 287(1928): 20200433
Lee SC, Li A, Calo S, Heitman J (2013) Calcineurin plays key roles in the dimorphic transition and virulence of the human pathogenic zygomycete Mucor circinelloides. PLoS Pathog 9(9):e1003625. https://doi.org/10.1371/journal.ppat.1003625
Lee SC, Li A, Calo S, Inoue M, Tonthat NK, Bain JM, Louw J, Shinohara ML, Erwig LP, Schumacher MA, Ko DC, Heitman J (2015) Calcineurin orchestrates dimorphic transitions, antifungal drug responses and host–pathogen interactions of the pathogenic mucoralean fungus Mucor circinelloides. Mol Microbiol 97(5):844–865. https://doi.org/10.1111/mmi.13071
Li L, Wright SJ, Krystofova S, Park G, Borkovich KA (2007) Heterotrimeric G protein signaling in filamentous fungi. Annu Rev Microbiol 61:423–452. https://doi.org/10.1146/annurev.micro.61.080706.093432
Linz JE, Orlowski M (1991) Differential gene expression and evidence of selective translation during anaerobic germination of Mucor racemosus sporangiospores. Microbiology 137(4):827–835
Lübbehüsen T, Polo VG, Rossi S, Nielsen J, Moreno S, McIntyre M, Arnau J (2004) Protein kinase A is involved in the control of morphology and branching during aerobic growth of Mucor circinelloides. Microbiology 150(1):143–150. https://doi.org/10.1099/mic.0.26708-0
Mitchell AP (1998) Dimorphism and virulence in Candida albicans. Curr Opin Microbiol 1(6):687–692
Muñoz-Hernández B, Palma-Cortés G, Cabello-Gutiérrez C, Martínez-Rivera MA (2014) Parasitic polymorphism of Coccidioides spp. BMC Infect Dis 14(1):1–9
Nadal M, García-Pedrajas MD, Gold SE (2008) Dimorphism in fungal plant pathogens. FEMS Microbiol Lett 284(2):127–134. https://doi.org/10.1111/j.1574-6968.2008.01173.x
Nagy LG, Tóth R, Kiss E, Slot J, Gácser A, Kovács GM (2017) Six key traits of fungi: their evolutionary origins and genetic bases. Microbiol Spectr 5(4). https://doi.org/10.1128/microbiolspec.FUNK-0036-2016
Navarro-Mendoza MI, Pérez-Arques C, Murcia L, Martínez-García P, Lax C, Sanchis M, Capilla J, Nicolás FE, Garre V (2018) Components of a new gene family of ferroxidases involved in virulence are functionally specialized in fungal dimorphism. Sci Rep 8(1):7660. https://doi.org/10.1038/s41598-018-26051-x
Orlowski M (1991) Mucor dimorphism. Microbiol Rev 55(2):234–258. https://doi.org/10.1128/mr.55.2.234-258.1991
Parish JM, Blair JE (2008) Coccidioidomycosis. In: Mayo Clinic proceedings, vol 3. Elsevier, New York, pp 343–349
Pasteur L (1876) Études sur la bière, ses maladies, causes qui les provoquent, procédé pour la rendre inaltérable: avec une théorie nouvelle de la fermentation. Gauthier-Villars, Paris
Prostak SM, Robinson KA, Titus MA, Fritz-Laylin LK (2021) The actin networks of chytrid fungi reveal evolutionary loss of cytoskeletal complexity in the fungal kingdom. Curr Biol 31(6):1192–1205.e1196
Reid G, Lynch JP III, Fishbein MC, Clark NM (2020) Mucormycosis. In: Seminars in respiratory and critical care medicine, vol 01. Thieme Medical Publishers, New York, pp 099–114
Rogers P, Clark-Walker G, Stewart P (1974) Effects of oxygen and glucose on energy metabolism and dimorphism of Mucor genevensis grown in continuous culture: reversibility of yeast-mycelium conversion. J Bacteriol 119(1):282–293
Saudohar M, Bencina M, van de Vondervoort PJ, Panneman H, Legisa M, Visser J, Ruijter GJ (2002) Cyclic AMP-dependent protein kinase is involved in morphogenesis of *Aspergillus niger*[aa]The EMBL accession number for the sequence reported in this paper is AJ296317. Microbiology 148(8):2635–2645
Schulz BE, Kraepelin G, Hinkelmann W (1974) Factors affecting dimorphism in Mycotypha (Mucorales): a correlation with the fermentation/respiration equilibrium. Microbiology 82(1):1–13
Schwarz P, Schwarz PV, Felske-Zech H, Dannaoui E (2019) In vitro interactions between isavuconazole

and tacrolimus, cyclosporin A or sirolimus against Mucorales. J Antimicrob Chemother 74(7):1921–1927
Shapiro RS, Cowen LE (2012) Uncovering cellular circuitry controlling temperature-dependent fungal morphogenesis. Virulence 3(4):400–404
Smith C, Lee SC (2022) Current treatments against mucormycosis and future directions. PLoS Pathog 18(10):e1010858. https://doi.org/10.1371/journal.ppat.1010858
Sudbery P, Gow N, Berman J (2004) The distinct morphogenic states of Candida albicans. Trends Microbiol 12(7):317–324
Taylor SS, Knighton DR, Zheng J, Ten Eyck LF, Sowadski JM (1992) Structural framework for the protein kinase family. Annu Rev Cell Biol 8:429–462
Thevelein JM, De Winde JH (1999) Novel sensing mechanisms and targets for the cAMP–protein kinase A pathway in the yeast Saccharomyces cerevisiae. Mol Microbiol 33(5):904–918. https://doi.org/10.1046/j.1365-2958.1999.01538.x
Valle-Maldonado MI, Jácome-Galarza IE, Díaz-Pérez AL, Martínez-Cadena G, Campos-García J, Ramírez-Díaz MI, Reyes-De la Cruz H, Riveros-Rosas H, Díaz-Pérez C, Meza-Carmen V (2015) Phylogenetic analysis of fungal heterotrimeric G protein-encoding genes and their expression during dimorphism in Mucor circinelloides. Fungal Biol 119(12):1179–1193. https://doi.org/10.1016/j.funbio.2015.08.009
Valle-Maldonado MI, Patiño-Medina JA, Pérez-Arques C, Reyes-Mares NY, Jácome-Galarza IE, Ortíz-Alvarado R, Vellanki S, Ramírez-Díaz MI, Lee SC, Garre V, Meza-Carmen V (2020) The heterotrimeric G-protein beta subunit Gpb1 controls hyphal growth under low oxygen conditions through the protein kinase A pathway and is essential for virulence in the fungus Mucor circinelloides. Cell Microbiol 22(10): e13236. https://doi.org/10.1111/cmi.13236
Vellanki S, Billmyre RB, Lorenzen A, Campbell M, Turner B, Huh EY, Heitman J, Lee SC (2020) A novel resistance pathway for calcineurin inhibitors in the human-pathogenic Mucorales mucor circinelloides. mBio 11(1). https://doi.org/10.1128/mBio.02949-19
Williams CR, Gooch JL (2012) Calcineurin inhibitors and immunosuppression: a tale of two isoforms. Expert Rev Mol Med 14
Wolff AM, Appel KF, Petersen JB, Poulsen U, Arnau J (2002) Identification and analysis of genes involved in the control of dimorphism in Mucor circinelloides (syn. racemosus). FEMS Yeast Res 2(2):203–213. https://doi.org/10.1111/j.1567-1364.2002.tb00085.x

Genome Evolution in Fungal Plant Pathogens: From Populations to Kingdom-Wide Dynamics

5

Ursula Oggenfuss, Alice Feurtey, Claudia Sarai Reyes-Avila, Emile Gluck-Thaler, Guido Puccetti, Hanna Maren Glad, Leen Nanchira Abraham, Luzia Stalder, Sabina Moser Tralamazza, Sandra Milena González-Sáyer, and Daniel Croll

Abstract

Plant pathogenic fungi have emerged as major concerns for plant health in agriculture and occupy important niches in natural ecosystems. Pathogen species are often highly diverse and respond rapidly to external factors. Our understanding of how this diverse group of fungi interacts with hosts and the environment has been critically advanced by analyses of fungal genomes. Here, we provide an overview of recent research into surveys of genomic diversity within species and review drivers of diversification including transposable elements. Analyses of epigenetic modifications of the genome have revealed important links to pathogenicity factors and genomic defense mechanisms of the genome. We also review population genomics analyses focused on short-term dynamics in plant pathogens such as recombination, genetic exchange, host adaptation, and the genetic basis of phenotypic traits.

Keywords

Plant pathogens · Agriculture · Genome analyses · Structural variation · Transposable elements · Population genetics · Phenotypic traits · Genome-wide association mapping

Abbreviations

5mC	5-methylcytosine
GWAS	Genome wide association studies
QTL	Quantitative trait locus
RIP	Repeat-induced silencing
RNAi	RNA interference

U. Oggenfuss
Laboratory of Evolutionary Genetics, Institute of Biology, University of Neuchâtel, Neuchâtel, Switzerland

Department of Microbiology and Immunology, University of Minnesota, Minneapolis, MN, USA

A. Feurtey
Laboratory of Evolutionary Genetics, Institute of Biology, University of Neuchâtel, Neuchâtel, Switzerland

Plant Pathology, D-USYS, ETH Zurich, Zurich, Switzerland

C. S. Reyes-Avila · E. Gluck-Thaler · H. M. Glad · L. N. Abraham · L. Stalder · S. M. Tralamazza · D. Croll (✉)
Laboratory of Evolutionary Genetics, Institute of Biology, University of Neuchâtel, Neuchâtel, Switzerland
e-mail: daniel.croll@unine.ch

G. Puccetti
Laboratory of Evolutionary Genetics, Institute of Biology, University of Neuchâtel, Neuchâtel, Switzerland

Syngenta Crop Protection AG, Stein, Switzerland

S. M. González-Sáyer
Laboratory of Evolutionary Genetics, Institute of Biology, University of Neuchâtel, Neuchâtel, Switzerland

Laboratory of Molecular Characterization, Biotechnology Institute - Universidad Nacional de Colombia, Bogotá, DC, Colombia

© The Author(s), under exclusive license to Springer Nature Switzerland AG 2023

S. Pöggeler, T. James (eds.), *Evolution of Fungi and Fungal-Like Organisms*, The Mycota 14,
https://doi.org/10.1007/978-3-031-29199-9_5

SIS	Sex-induced silencing
SV	Structural variation
TE	Transposable element

5.1 Introduction

Fungal plant pathogens play key roles in ecosystems and threaten global food security (Fisher et al. 2012, 2018). A key focus of research over the past several decades has been on genetic factors governing the interaction of fungal pathogens with their environment including host organisms. The discovery of effectors and secondary metabolites governing such interactions were major achievements (Lo Presti et al. 2015). Advances in our understanding of functions encoded by pathogen genomes allowed researchers to further analyze the chromosomal locations of genes underlying important phenotypes. In plant pathogens, effector genes tend to be encoded in more repetitive and dynamic regions of the genome; genes underlying antifungal drug resistance are found to evolve rapidly through duplications or insertions into regulatory regions. Comparative analyses of genome sequences have additionally illuminated the evolutionary processes underlying pathogen adaptation. In this chapter, we review recent advances in the fields related to plant pathogen evolutionary genomics. We provide an overview of variation in genome sequences among individuals and how such variation can be linked to adaptation to the host and environment. We focus on transposable elements (TEs) as key drivers of variation in genomes over different evolutionary time frames. We review the emergence of defenses against TEs including epigenetic changes of pathogen genomes. Finally, we discuss how variation in genomes can be used to study within-species processes such as hybridization and gene flow, as well as how genetic variation can be linked to variation in phenotypic traits. Such population genomics analyses provide powerful insights into the recent adaptation of pathogens to new hosts, fungicides, and the environment.

5.2 Variation in Pathogen Genome Structure

Genetic variation is at the origin of all evolutionary innovation. Genetic variation ranges from single nucleotide differences among individuals to extensive variation in the content and organization of genomes called structural variants (SVs) (Ho et al. 2020; Priest et al. 2020). Compared with mutations at the single nucleotide level, SVs often have substantial implications for genome stability and affect inheritance and linkage patterns across the genome. SVs can impact large chromosomal regions in a single mutational step, thus affecting the expression of several phenotypic traits at once. TEs are major factors in initiating SV mutations due to their repetitive nature and are thought to contribute to the evolvability of lineages (Grandaubert et al. 2014; Yeaman et al. 2016; Hartmann 2022). Identifying the mechanisms generating SVs and associated changes in phenotypic traits have been highly fruitful to understand the process of pathogen adaptation at the genetic level.

5.2.1 Origins of SVs and Their Impact on Pathogenicity Evolution

SVs arise through distinct mechanisms, and characterizing their origins and effects on phenotypic traits is important for understanding their role in pathogen evolution. In broad terms, SVs can be either "balanced" or "unbalanced." SVs in either category span orders of magnitude in length, ranging from tens to hundreds of thousands of base pairs (Fig. 5.1) (Hartmann 2022). Balanced SVs, such as inversions, translocations, and chromosomal fissions and fusions, lead to the reorganization of genomic sequences but no net sequence gain or loss. Although balanced SVs do not change the sequence content, such SVs can destabilize chromosomal integrity or affect gene order (Pepper 2003). In outcrossing species, balanced mutations tend to decrease recombination rates between loci, thus increasing linkage disequilibrium. In

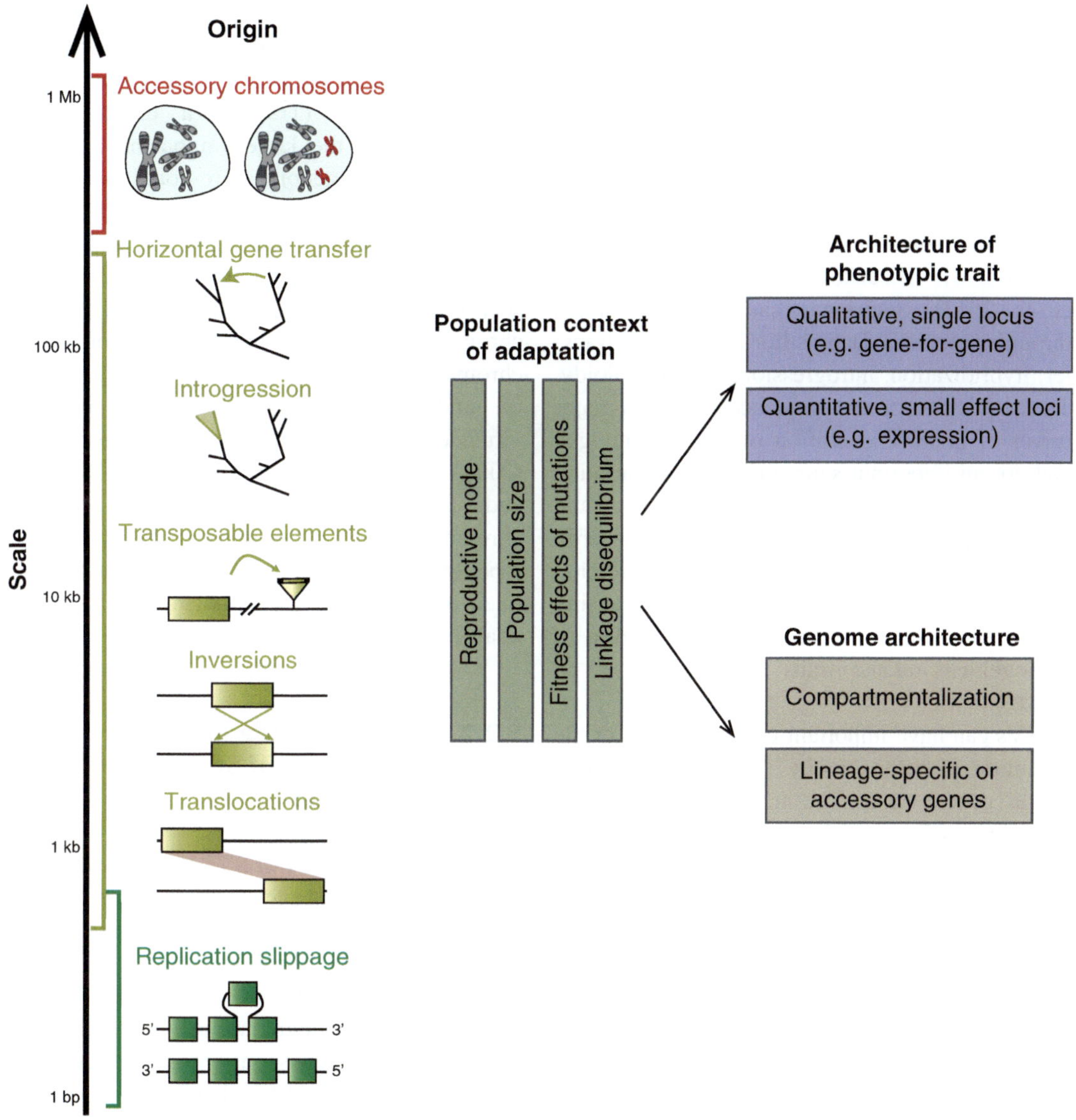

Fig. 5.1 Origin, fate, and consequences of structural variants among fungal plant pathogen genomes. Depending on the trigger and mechanism, structural variation can affect only a few base pairs to several million base pairs. Genetic variation can impact phenotypic traits, influence genome architecture, and have long-term consequences for populations and species

general, higher linkage disequilibrium decreases the efficacy of selection, but can have the positive effect of preserving associations if a region encodes loci with co-adapted alleles (Hill and Robertson 2008; Faria et al. 2019). For example, two inversion polymorphisms in *Microbotryum* anther-smut fungi show high degrees of divergence among host-specialized populations and contain genes likely involved in ecological adaptation with evidence for selective sweeps (Hartmann et al. 2020a). Preserving associations between co-adapted alleles may be particularly beneficial under conditions of high gene flow from maladaptive populations, as these may occur during bouts of local adaptation (Yeaman 2022).

In contrast to balanced SVs, unbalanced SVs result in either a net gain or loss of sequences. Sequence gains and losses often take the form of duplications that arise from polymerase replication slippage, unequal crossing-over, and non-homologous recombination (Bzymek and Lovett 2001; Mieczkowski et al. 2006; Vogan et al. 2013). Sequences can also be gained through processes such as horizontal gene transfer, hybridization, introgression, and polyploidy. The acquisition of new sequences can be a major driver of genome evolution of fungal pathogen and are an important source of genetic variation facilitating adaptation to new hosts and environments (Gardiner et al. 2012; Menardo et al. 2016). For instance, repeated introgression among lineages of fungal pathogens causing Dutch elm disease has driven the evolution of pathogenicity-related traits (Hessenauer et al. 2020).

SVs can have important consequences for the evolution of pathogenicity even at the level of single gene events. The transfer of the *ToxA* gene responsible for the production of a host-specific toxin from the wheat pathogen *Parastagonospora nodorum* to *Pyrenophora tritici-repentis* resulted in the emergence of the tan spot disease in the 1940s (Friesen et al. 2006). Non-homologous recombination deleted a gene encoding a recognized effector located between copies of TE in the wheat pathogen *Zymoseptoria tritici* likely enabling the escape from recognition by specific wheat cultivars (Hartmann et al. 2017). Such effector genes and other loci underpinning "gene-for-gene" interactions with the host are particularly prone to favor gains or losses as these genes can mediate host specificity (Franceschetti et al. 2017). For example, *Leptosphaeria maculans*, the causative agent of the stem cancer disease in *Brassica napa*, can infect specific host cultivars depending on the effector complement a strain encodes in the genome (Jiquel et al. 2021).

SVs spanning multiple genes can affect more complex phenotypes such as the production of metabolites. The frequent gain and loss of metabolic gene clusters generates variation in secondary metabolite production in multiple pathogens, including the *Fusarium graminearum* species complex (Tralamazza et al. 2021). Finally, among the largest unbalanced SVs in fungal pathogen genomes are conditionally dispensable or accessory chromosomes, which are frequently gained, transferred, and lost among individuals (Croll and McDonald 2012). Accessory chromosomes can be among the smallest chromosomes (a few hundred kb) or largest chromosomes of an organism carrying upward of hundreds of protein-coding genes, including virulence factors that determine host range (Akagi et al. 2009; Bertazzoni et al. 2018; Habig and Stukenbrock 2020). Many accessory chromosomes are a cradle for TEs and other repetitive sequences, yet are often not being fixed or can be lost in pathogen species (Yang et al. 2020).

5.2.2 Transposable Elements as Drivers of Structural Variation

An important category of SVs are TEs, which are DNA sequences that encode mechanisms for their own excision, replication, and re-integration into genomic sequence, thus generating either balanced or unbalanced SVs through excision and translocation or through duplication (Wicker et al. 2007). A notable consequence of TE activity is the potential to generate adaptive genetic variation at higher rates than expected from point mutations. Rates of SVs generated by TE activity in fungal pathogen genomes have been shown to increase under conditions of stress during infection (Fouché et al. 2020). The activity of class II cut-and-paste transposable elements, such as *Tc1* mariner and MITEs, may result in both balanced and unbalanced SVs depending on successful re-integration at a new locus. In contrast, class I copy-and-paste TEs, such as *Copia* or LINEs, result in the generation of unbalanced SVs through sequence duplication.

Several families of TEs have significant potential to shape pathogen genome evolution by carrying genes necessary for their own replication and act as cargo carriers capable of mobilizing host genes implicated in pathogen adaptation. For example, the recently discovered class of giant *Starship* TEs can carry virulence factors implicated in pathogen adaptation (McDonald et al. 2019; Gluck-Thaler et al. 2022; Urquhart et al. 2022). Similarly, a smaller class II DNA transposon identified in the wheat pathogen *Z. tritici* carries a gene, which negatively impacts pathogen reproduction (Wang et al. 2021). Types of cargo-carrying TEs such as Helitrons and Mutator-like elements were found in non-pathogenic fungi as well (Castanera et al. 2014; Dallaire et al. 2021), yet the impact of TEs on fungal lifestyle evolution remains largely unexplored. Beyond contributing directly to the generation of SVs, TEs also potentiate the frequency of other events creating SVs (Faino et al. 2016; Seidl et al. 2020). Especially in the case of copy-and-paste TEs, the creation of duplicated sequences at multiple loci increases opportunities for non-homologous recombination (i.e., affecting distinct loci in the genome) triggering gene deletions (Hartmann et al. 2017).

5.2.3 The Fate of SVs and TEs in Pathogen Populations

Deletions, insertions, and rearrangements have a high likelihood of affecting chromosomal integrity and being deleterious under most circumstances. As a consequence, most novel SVs will not remain in populations. In particular, large SVs affecting the faithful transmission of chromosomes during meiosis are likely to be under strong purifying selection (Petrov et al. 2003). How likely SVs are generated at each generation stems from multiple factors as discussed above including the activity of TEs and the repetitiveness of the genome. SVs with weak effects on fitness can evolve neutrally in populations and escape counter-selection. Furthermore, how efficient selection acts on SVs is a function of effective population size, rates of sex, and recombination rates. Sexual reproduction breaks up associations between beneficial alleles through recombination (Agrawal 2006). Asexual reproduction constrains selection against detrimental SV. TEs in particular may be more active drivers of diversification in clonal species compared to species with regular sexual cycles (e.g., *Colletotrichum higginsianum* (Tsushima et al. 2019)). Hence, the extent of SVs (and TEs) present in a pathogen species is a complex product of both population genetic factors and properties of a species' genome.

5.3 Compartmentalization of Pathogen Genomes

In many fungal plant pathogen genomes, most SVs cluster into compartments with low gene content. SV-rich compartments consist mostly of TEs and other repetitive elements. Depending on the pathogen, compartments can be large and include entire chromosomes or occur as a fine-grained mosaic along chromosomes. The few genes located in TE-rich compartments are often involved in the interaction with the plant host, and include effectors, which are secreted proteins manipulating the host to escape recognition and repress immune responses by the host (Lo Presti et al. 2015). Effectors may also act as avirulence factors if the host retains the ability to recognize the proteins (Rouxel and Balesdent 2010). In the generalist fungal pathogen *Verticillium dahliae*, the fungal avirulence factor *Ave1* is encoded in a TE-rich compartment and was independently lost in strains infecting host plants that recognized Ave1 (Faino et al. 2016). Relaxed selection, insertion of TEs, and higher rates of chromosomal rearrangement in TE-rich compartments are thought to allow effector genes to mutate more rapidly in order to escape recognition by the host (Rouxel and Balesdent 2010; Schmidt et al. 2013). In some pathogens, defense mechanisms against TEs also contribute to higher mutation rates in TE-rich compartments (Wang et al. 2020). In contrast to the gene-poor compartments, gene-rich compartments encode most essential genes and are under strong purifying selection with few newly inserted TEs and SVs observed at the population level (Croll and McDonald 2012).

The differentiation of genome compartments and the association of particular compartments with pathogenicity is referred to as the two-speed genome of plant pathogens (Raffaele and Kamoun 2012; Croll and McDonald 2012; Dong et al. 2015). However, two-speed genomes are found in fungi with non-pathogenic lifestyles as well and are also not a shared feature of all fungal plant pathogens (Torres et al. 2020). In the wheat powdery mildew *Blumeria graminis* f. sp. *hordei*, mutation rates and gene as well as TE densities are fairly uniform along the genome (Spanu et al. 2010; Frantzeskakis et al. 2018). Abundant and likely still active TEs likely contributed to the observed polymorphisms and rapid turnover in effector genes (Müller et al. 2019). Such a genome was referred to as a de facto one-speed genome (Frantzeskakis et al. 2018). Other fungal pathogens were proposed to carry a so-called multi-speed genome, including *Fusarium oxysporum* f. sp. *lycopersici*. In multi-speed genomes, more than two distinct compartments were recognized with differing levels of polymorphisms and associations with pathogenicity factors (Vlaardingerbroek et al. 2016; Frantzeskakis et al. 2019). The use of the term "compartment" has been suggested over "speed," stressing the importance of structural variation itself and avoiding some of the challenging concepts of assessing and comparing evolutionary rates between different genomic regions (Frantzeskakis et al. 2019; Torres et al. 2020).

5.4 Transposable Element Activity and Defenses in Fungal Genomes

Pathogen genomes vary strongly in their TE composition and content even between closely related species, consistent with recent activity (Raffaele and Kamoun 2012). Beyond affecting gene functions through disruption or changes in regulation, activation of specific TEs can underpin TE content and genome size variation within pathogen species such as *Z. tritici* (Oggenfuss et al. 2021). In some species such as *Pseudocercospora ulei* and *Cenococcum geophilum*, single TE families were likely responsible for most recent insertions (Peter et al. 2016; González-Sayer et al. 2022). In genomes with high TE content, it is more likely for an active TE to insert into regions already populated by TEs; thus, most TE insertions do not have a strong deleterious impact. Consequently, ongoing TE activity can lead to further genome size increases. While some short-term benefits are apparent for fungal pathogens (see above), the ongoing proliferation of TEs may incur long-term costs. Genome compartmentalization may allow fungal plant pathogens to retain TE activity to a certain degree through compartmentalization and epigenetic control (Fouché et al. 2022). Nevertheless, TEs may regain activity due, e.g., to stress caused by host plant defenses or fungicides (Chen et al. 2015; Fouché et al. 2020). The tenuous relationship of pathogens with TEs in their genomes has recently been proposed as a devil's bargain, where the presence and activity of TEs is adding potential benefits and disadvantages at the same time (Fouché et al. 2022).

5.4.1 Repeat-Induced Point Mutations as a Defense Against TEs

All organisms including fungi have evolved mechanisms to defend their genomes against the deleterious effects of TEs, with most of these mechanisms having deep evolutionary origins in the defense against viruses (Slotkin and Martienssen 2007; Jordan and Miller 2008). Of particular interest are the fungal-specific methyltransferases DIM-2 and RID, which are suspected to be involved in a major genome defense pathway called repeat-induced point mutations (RIP). RIP is exclusively found in ascomycete fungi and is thought to occur each generation between fertilization and meiosis (Selker 1990). RIP induces mutations in any duplicated region such as TEs and duplicated genes (Watters et al. 1999). Smaller repeats, including the non-autonomous MITEs, seem to be able to escape RIP (Pereira et al. 2020b). RIP mutations mostly follow a CpA to TpA pattern, but questions remain about the conservation of this mutation pattern (Horns et al. 2012). The impact of RIP is highly variable, even among closely related lineages (van Wyk et al. 2021) and may be mediated by the presence of the different methyltransferases (Bewick et al. 2019). Experiments in *Neurospora crassa* showed reduced but not abolished RIP activity when either RID or DIM-2 had been deactivated (Freitag et al. 2002). Some species such as *Z. tritici* show RIP in the absence of a functional copy of *dim2*, indicating that RID- and DIM-2-mediated RIP pathways may be individually dispensable (Gladyshev 2017; van Wyk et al. 2021). TE-rich compartments provide long stretches of RIP targets, and leakage of RIP into neighboring loci may contribute to the rapid diversification of effectors (van de Wouw et al. 2010; van Wyk et al. 2021). A loss of RIP may lead to an increase in TE activity, followed by higher diversification rates as it was shown for the recent loss of *dim2* in *Z. tritici* (Habig et al. 2021; Lorrain et al. 2021). Populations of the pathogen across the global distribution range show uneven patterns of RIP activity. Populations with reduced signatures of RIP also tend to carry higher TE loads, which may be a direct consequence of reduced RIP efficacy (Feurtey et al. 2023).

5.4.2 RNAi-Mediated Defenses

Proteins of the Argonaute family are at the basis of all RNA interference (RNAi)-based defenses and utilize small RNAs to guide the RNA-induced silencing complex to target RNAs (Rivas et al. 2005). RNAi may induce post-transcriptional silencing of repetitive sequences through quelling, meiotic silencing, and sex-induced silencing (Wang et al. 2010; Dang et al. 2011). In some host–pathogen interactions, small RNAs are exchanged either unidirectionally or bidirectionally between the host and the pathogen. *Botrytis cinerea* was shown to export small RNAs capable of silencing immunity genes of its host *Arabidopsis thaliana* (Weiberg et al. 2013). In parallel, hosts are also capable of using small RNAs to target pathogenicity genes upon infection (Zhang et al. 2016). Overall, small RNA cross-talk during infection may result in different evolutionary pressures on the RNAi pathways in fungal pathogens as compared to non-pathogenic lineages.

5.5 Epigenetic Landscapes and Control of Transcription

Repeat-rich compartments in the genome are often associated with specific epigenetic signatures in fungi. Epigenetic signatures include DNA or histone methylation, and both are associated with transcriptional silencing of TEs in *Magnaporthe oryzae*, and *N. crassa*, respectively (Selker et al. 2003; Jeon et al. 2015). Global reprogramming of DNA and histone methylation serves various additional purposes including the regulation of fungal growth, the expression of virulence factors, or the development of morphology (Jeon et al. 2015; Ramos-García et al. 2022). Epigenetic modifications at the level of chromatin conformation are formed by the nucleosome, an octamer with two copies each of histone protein (i.e., H2A, H2B, H3, and H4) with 146 base pairs of DNA wrapped around histones (Luger et al. 1997). Chemical modifications of nucleosome components invoke significant chromatin spatial rearrangements and changes to DNA accessibility (Fig. 5.2). The joint effect of nucleosomes may modulate or stabilize gene expression during the cell cycle (Klemm et al. 2019). Cytosine DNA methylation is a covalent modification involving the transfer of a methyl group onto the C5 position of cytosine sites (i.e., 5-methylcytosine, 5mC) by a DNA methyltransferase enzyme (Martienssen and Colot 2001). DNA methylation occurs mostly in intergenic and repeated regions of fungal plant pathogen genomes. A major role of DNA methylation is to transcriptionally silence transposable elements and other repetitive elements (Bewick et al. 2019). After the inactivation of a DNA methyltransferase in the wheat

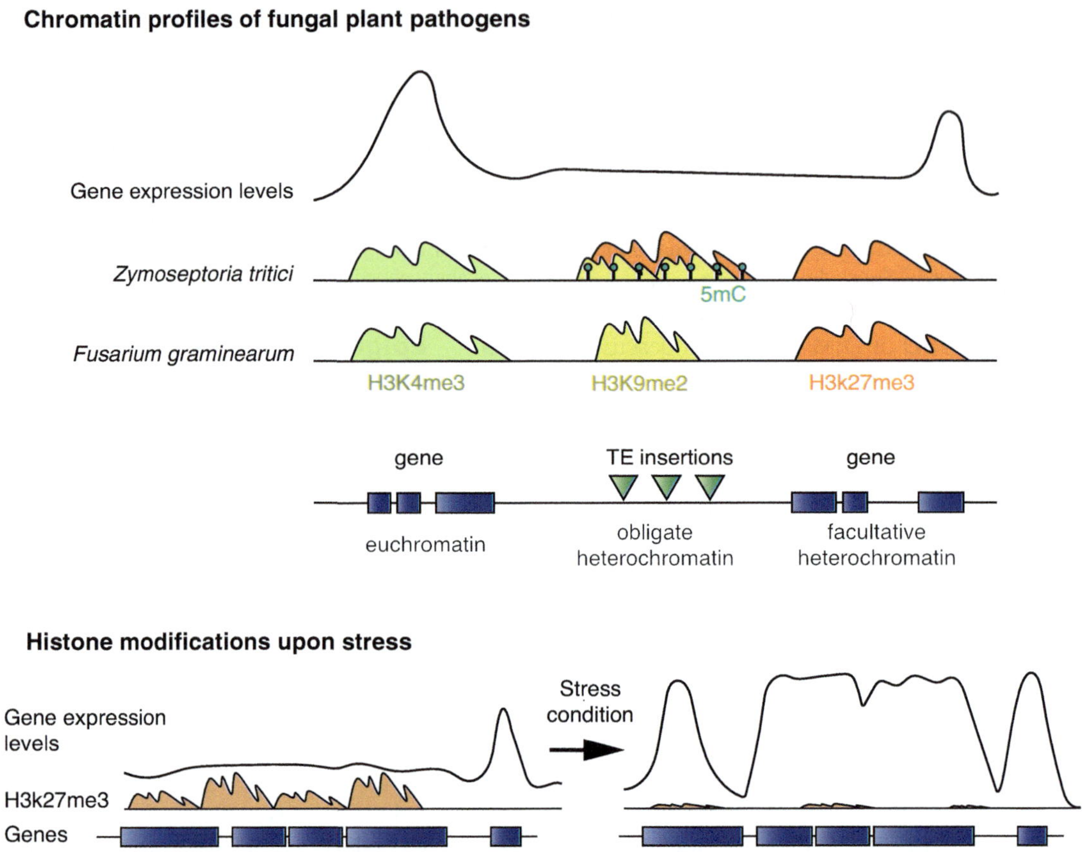

Fig. 5.2 Epigenetic changes in plant pathogen genomes based on histone modifications. Species vary in their specific mechanisms how regions of euchromatin and obligate and facultative heterochromatin are determined based on histone methylations and 5mC methylation. De-repression of facultative heterochromatin upon stress (e.g., experienced during host infection) can upregulate otherwise silenced genes

pathogen *Z. tritici*, 5mC levels had decreased, and TE activity subsequently increased (Möller et al. 2021; Kramer et al. 2022).

Histone modifications have a wide range of impacts on the epigenetic landscape of a genome (Bannister and Kouzarides 2011). For fungal plant pathogens, methylation of the lysine residue of histone 3 has been the focus of most studies so far. Histone methylation patterns in the genome show a high degree of conservation among pathogen species and underpin spatial and regulatory defined genomic regions (Freitag 2017). Transcriptionally active euchromatin is frequently enriched in histones that are methylated at lysine 4 (H3K4me) (Gates et al. 2017). Deletion of histone methyltransferases in *F. graminearum*, *Colletotrichum fructicola*, and *M. oryzae* led to the loss of H3K4me2/me3 (i.e., fourth lysine with di- or tri-methylation), defects in growth, reduced spore production, decreased metabolism, and virulence (Liu et al. 2015; Zhou et al. 2021; Gao et al. 2022). H3K9me3 is frequently found in transcriptionally silent heterochromatin such as the centromeres and repeat-rich compartments and its disruption is associated with changes in the development and regulation of metabolism in fungal plant pathogens, including *L. maculans*, *Fusarium verticillioides, Fusarium mangiferae,* or *V. dahliae* (Soyer et al. 2014; Gu et al. 2017; Gates et al. 2017; Atanasoff-Kardjalieff et al. 2021; Kramer et al. 2022).

5.5.1 Epigenetic Regulation of Pathogenicity Factors

H3K27me3 modifications underpin facultative heterochromatin and gene expression regulation and are frequently associated with secondary metabolism gene clusters and effectors modulated during the infection cycle (Connolly et al. 2013; Studt et al. 2016; Meile et al. 2020; Tralamazza et al. 2021; Kramer et al. 2022). Genomic regions rich in repetitive elements and effector genes are often marked by H3K9me3 and H3K27me3 (Meile et al. 2020; Soyer et al. 2021). In *L. maculans*, H3K9me3 is shown to be involved in the epigenetic regulation of effector genes enriched in AT isochores of the genome (Soyer et al. 2014, 2021; Clairet et al. 2021). The expression of the highly diverse effector gene *Avr3D1* among isolates of *Z. tritici* is likely influenced by chromatin changes and sequence rearrangements (Meile et al. 2020). In *M. oryzae*, genes important for host infection tend to be enriched in H3K27me3 repressing transcription outside of the host but triggering transcription during the stressful plant infection period (Zhang et al. 2021). Differences in virulence among *V. dahliae* strains were suggested to be governed by variation in DNA methylation profiles (Ramírez-Tejero et al. 2020). Furthermore, specific chromatin profiles define virulence-associated genomic regions highlighting how different epigenetic factors contribute to the organization of such regions (Cook et al. 2020). Secondary metabolites contribute to fungal pathogenicity in many species (Scharf et al. 2014) with epigenetic regulation playing a major role in the activation of the underlying gene clusters (Pfannenstiel and Keller 2019). The maize pathogen *F. verticillioides* produces the mycotoxin fumonisin regulated by histone-acetylation modifications (Visentin et al. 2012). In *F. graminearum*, a third of all genes are marked with H3K27me3 and de-repression of marked genes mainly regulates secondary metabolite production (Connolly et al. 2013).

5.6 Evolutionary Histories of Pathogen Species Through the Lens of Genomics

Understanding the evolutionary history of pathogen species, such as the origin, migration routes, or the co-evolutionary interaction with their host, is a major challenge. In many cases, host and pathogen evolution are tightly linked, from short-term local adaptation to co-speciation. *Microbotryum* anther-smut pathogens show a highly similar population structure compared to their *Silene* host species, with local adaptation of the host for pathogen resistance (Feurtey et al. 2016; Hartmann et al. 2020b). Similarly, the apple scab fungus *Venturia inaequalis* shows geographical clustering depending on the co-evolved wild apple species (Gladieux et al. 2010). Co-evolution between host and pathogen can go beyond population differentiation and lead to co-speciation. *Z. tritici* was domesticated alongside its wheat host and further co-evolves with different wheat cultivars, resulting in complex adaptation of the pathogen toward different host genotypes (Stukenbrock et al. 2007; Karisto et al. 2018). *Hymenoscyphus* species co-evolved with their local ash tree host species giving rise to non-pathogenic endophytes (Cleary et al. 2016). In contrast, the introduction of Asian *Hymenoscyphus* species into Europe led to a dramatic ash dieback disease outbreak and a strong decline in native ash species (Gross et al. 2014). Beyond striking examples of host–pathogen co-evolution, the history of many pathogen species and genera is marked by host jumps and, thus, decoupled evolutionary paths of the hosts and pathogens (De Vienne et al. 2013; Thines 2019).

5.6.1 The Evolution of Genetic Diversity and Pathogen Emergence

Fungal pathogen populations display a wide range of genomic diversity from extremely diverse pathogen populations to globally distributed quasi-clonal species. In *Z. tritici*, a population from a single field can contain millions of genotypes and upward of 37 SNPs per kilobase making the pathogen one of the most polymorphic species known (Singh et al. 2021; McDonald et al. 2022). Sexual reproduction of *Z. tritici* is a key aspect generating diversity in the form of recombinant genotypes. However, the lack of meaningful population bottlenecks throughout the seasons constitutes an even stronger factor maintaining diversity for the pathogen (Feurtey et al. 2023). High genetic diversity was also observed in *V. inaequalis* and *P. nodorum* (Le Cam et al. 2019; Richards et al. 2019) and to a lesser extent in the globally distributed rust *Austropuccinia psidii* with strong geographic effects on diversity (Stewart et al. 2018). Recent pathogen emergence is often associated with a dramatic loss of genetic diversity and a rapid spread of clonal lineages. In *F. oxysporum* f. sp. *cubense*, the causative agent of the Panama disease in bananas, a single clonal lineage led to severe disease on the most widely used banana cultivar in the 1960s (Ploetz 2008). A recently emerged clonal lineage is now threatening new banana cultivars resistant to the previous clonal lineage (Magdama et al. 2020). A clonal strain of *Cryphonectria parasitica*, the causative agent of the chestnut blight, caused a devastating invasion in southeastern Europe (Stauber et al. 2021). However, how long such aggressive clones remain successful is not known. Aggressive clones may go extinct as either the host plant is decimated or gains resistance leading to co-evolutionary dynamics.

Following devastating clonal plant pathogen invasions, the diversification of the lineage through the acquisition of variability seems to a be a general feature. Genetic variability can be acquired through mutations, gene flow between populations, hybridization with closely related species, horizontal gene transfer, or introgression. Such gene flow among pathogen populations or species depends on the distribution of the hosts. Pathogens colonizing well-connected host populations tend to form well-connected pathogen populations as well. The main route for many pathogens to disperse is through wind or rain splash dispersed spores (Karisto et al. 2021). Spores can also be transported through the atmosphere and have thus the potential to be spread globally (Golan and Pringle 2017). Human activity is another important factor in the distribution of fungal plant pathogens, and often, fungal material can hitchhike on plant material including seeds, fruits, or grass to become established in a new area (Fones et al. 2020). The introduction of chestnut blight caused by *C. parasitica* into North America and later into Europe in the twentieth century dramatically decimated chestnut (Bissegger et al. 1997). More recently, the spread of wheat blast from South America to Bangladesh and Zambia was facilitated by the shipment of infected plant material (Latorre et al. 2022). Both pathogen invasions were also likely facilitated by the introduction of pathogen genotypes, which were able to create new variants favorable for further spread (i.e., the establishment of a bridgehead). For *C. parasitica,* the bridgehead was likely formed in Europe early during the invasion helping to generate a genotype adapted to the new environment, including the host (Stauber et al. 2021). Wheat blast exists as a highly diverse pathogen population on South American wheat helping create a globally invasive lineage that twice independently colonized African and South Asian wheat (Latorre et al. 2022). In contrast to factors facilitating their spread, obligate asexual and clonal pathogens may face more significant challenges to invade new areas with the clear exception of global rust lineages (Singh et al. 2011; Karisto et al. 2021). Overall, many sexual pathogens display clear geographic patterns indicating that modest gene flow occurs among regions (Mohd-Assaad et al. 2018; Pereira et al. 2020a; Feurtey et al. 2023).

Hybridization and re-establishment of gene flow after long periods of isolation have been suggested as an important driver of plant pathogen evolution in particular for crop pathogens. Hybridization is defined as reproduction between members of genetically distinct populations, producing offspring of mixed ancestry (Abbott et al. 2013; Stukenbrock 2016). The hybridization between two *B. graminis* formae speciales causing mildews on wheat and rye gave rise to a novel mildew lineage that can infect triticale, which incidentally is itself an artificial hybrid between the two host species (Menardo et al. 2016). In contrast to hybridization, introgression is the process where individuals interbreed and the resulting hybrids undergo backcrossing with one of the progenitor species (Stukenbrock 2016). The backcrossing introduces a restricted amount of genetic material from one species into the species and can drive the success of invasive species via the rapid acquisition of adaptive mutations. For example, the Dutch elm disease pandemic was caused by three fungal lineages with permeable reproductive barriers: *Ophiostoma ulmi*, *O. novo-ulmi* subspecies *novo-ulmi*, and *O. novo-ulmi* subspecies *americana* (Hessenauer et al. 2020). Introgression was the main driver creating genetic diversity in the species and had important effects on fitness-related traits, including growth rates at high temperatures and necrosis sizes (Hessenauer et al. 2020).

5.6.2 Plant Pathogens in the Face of Changing Environments

Plant pathogens are exposed to numerous environmental challenges in particular during the colonization of new environments. Crop pathogens are exposed to the impact of human activity, e.g., through the application of fungicides, the introduction of more resistant cultivars, or changes to irrigation (Fig. 5.3). Furthermore, climate change will have a strong impact on future distributions and selection pressures on pathogen populations (Chaloner et al. 2021). To investigate consequences of environmental stress on pathogen populations requires an understanding of phenotypic trait variation and genes governing such variation. The main approaches to link phenotypic trait variation to polymorphism are quantitative trait locus (QTL) mapping and genome-wide association studies (GWAS). QTL mapping requires the analysis of progeny populations and has been used to investigate evidence for pleiotropy between azole sensitivity and melanization (Lendenmann et al. 2015; Zheng et al. 2018). In contrast, GWAS are typically based on population samples of local, regional, or global sets of isolates (Weigel and Nordborg 2015; Sánchez-Vallet et al. 2018). Both GWAS and QTL mapping can identify genomic regions or even individual genes associated with a specific phenotypic trait. Mapping outcomes can be followed up with molecular genetics approaches recapitulating the genetic variants associated with a phenotype in different genetic backgrounds. GWAS can also be complemented with gene co-expression studies as it has been used to identify soybean resistance factors to multiple fungal pathogens (Gao et al. 2016; Almeida-Silva and Venancio 2021). On the pathogen side, GWAS enabled among others the mapping of effectors using panels of only 100–150 isolates coupled with molecular genetics validation (Zhong et al. 2017; Hartmann et al. 2017).

A further application of GWAS in plant pathogens is to identify mechanisms of fungicide resistance. Fungicides are a strong selection pressure for crop pathogens and shape the evolutionary course of many species in human-dominated environments (Hawkins and Fraaije 2018; Rhodes et al. 2022). Fungicide application in agriculture varied widely across countries and time resulting in different resistance levels along clines helping map mutations linked to resistance (Croll and McDonald 2017). Modern selective fungicides are site-specific, designed to target specific cellular processes and bind to specific protein targets. However, the exact mutations leading to resistance can vary across the globe for the same pathogen (Mohd-Assaad et al. 2016; Pereira et al. 2020b; Hartmann et al. 2021). Beyond mutations modifying the gene encoding the fungicide target protein, additional resistance mechanisms related to, e.g., detoxification were discovered. A striking example includes the multiple insertions of TEs upstream of a gene encoding a major facilitator superfamily gene in *Z. tritici* with resistant strains showing different complements of inserted TEs upregulating gene expression (Omrane et al. 2015, 2017). Gains in resistance were found to be associated with costs for resistant strains in the absence of fungicides (Croll and McDonald 2017). For example, higher

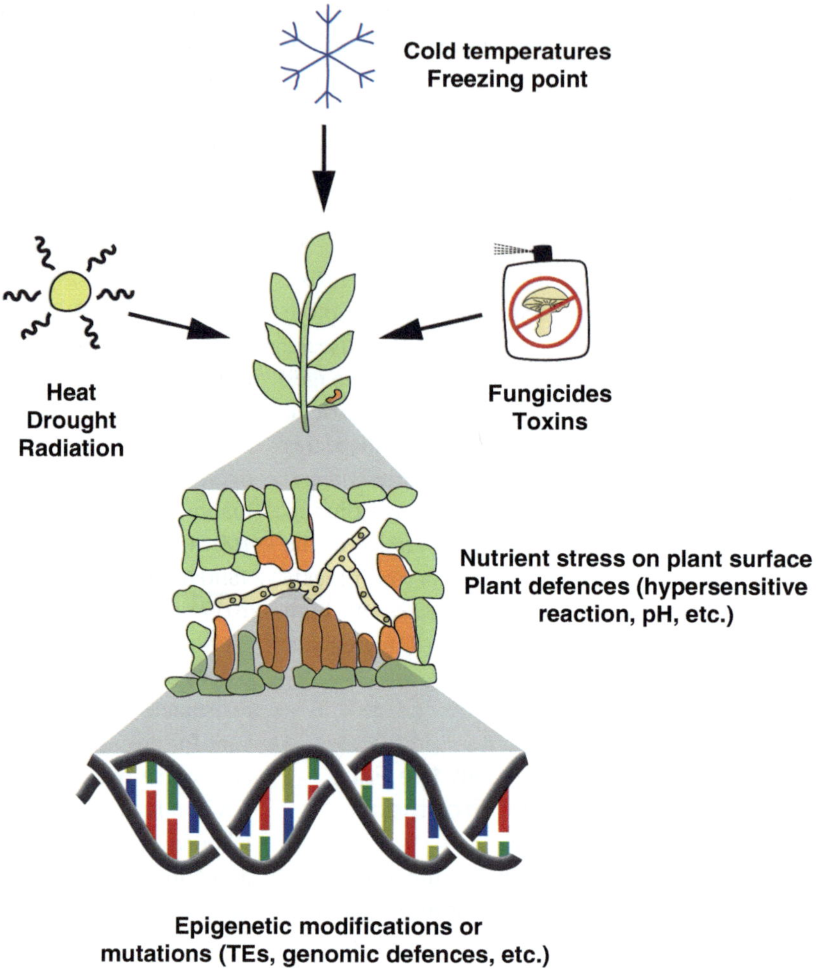

Fig. 5.3 Stress factors for fungal plant pathogens. Different stressors include abiotic factors such as heat stress, cold temperatures, or the application of fungicides. Upon arrival on the plant surface, pathogens typically suffer from nutrient stress and may be exposed to plant defenses such as reduced pH and enzymes. Genetic and epigenetic modifications can be triggered during stress

azole resistance in *Z. tritici* correlates with slower colony growth possibly mediated by melanin production (Lendenmann et al. 2014, 2015). In a systematic analysis of trade-offs among traits expressed by *Z. tritici* revealed multiple instances of distinct traits being either positively or negatively correlated at the phenotypic level but also showing possible pleiotropic effects (Dutta et al. 2021). The study used GWAS to identify loci contributing to the expression of dozens of individual traits related to fungicide resistance or virulence. A number of trait pairs of virulence and fungicide resistance showed negative genetic correlations meaning that mutations contributing positively to one trait on average affected negatively the other trait. Such hints at genetic constraints to evolve both high virulence and high resistance may open the possibility to exploit these constraints to limit pathogen evolution.

5.7 Conclusions and Open Questions

The advent of genome sequencing technologies has been particularly fruitful in investigating the evolution and ecology of fungal plant pathogens. The typically compact genomes, the ability to culture many species, and the imperative to improve crop protection against new pathogen threats have helped build scientific communities with a common goal to generate reference genome sequences early on. Over the past two decades, high-quality genomes have been established for all major crop pathogens. More recently, community efforts to build additional resources for pathogens outside of agricultural concern have emerged. This included efforts to broadly sequence fungal diversity by establishing reference genomes (e.g., the 1K fungal genome project; Grigoriev et al. 2011). In parallel, falling costs of sequencing enabled genome sequencing efforts at the within-species levels. For an increasing number of crop pathogens, extensive sets of field samples were sequenced across the globe. In combination, the genomic resources available for fungal pathogens paved the way for a broad set of scientific discoveries from identifying genetic factors of pathogenicity, processes of rapid adaptation to the environment, epigenetic regulation, TE dynamics, and genome organization. The coupling of research focused on addressing agricultural concerns with basic research has created multidisciplinary and dynamic scientific communities.

Future research on pathogen genomes will likely pursue the parallel tracks of applied research addressing direct needs for managing pathogen damage and basic research tackling broader topics of genome evolution. Highlights of such future research could include the systematic discovery of pathogenicity factors using high-throughput mutant analyses and GWAS. The lack of knowledge around pathogenicity mechanisms and resistance breakdown often complicates resistance breeding efforts. Similarly, predicting evolutionary paths to fungicide resistance before large field applications may prevent rapid failures in efficacy. Standing variation in pathogen populations could be exploited to understand how well individual crop pathogens will be able to adapt to new chemicals. On the basic research side, recent analyses of the epigenetic landscape of pathogen genomes have provided fascinating insights into how different genomic elements including TEs modulate chromosomal structure, gene functions, and regulation. The rapid changes at effector gene loci will further improve our understanding of how gene functions evolve in general. Finally, the global population genomic surveys of multiple pathogen species are among the most extensive datasets for any group of organisms. These public sequence datasets offer vast opportunities to investigate mechanisms of rapid adaptation. Environmental gradients and variation in host genotypes provide powerful gradients across continents to assess how pathogens may have adapted along such environmental factors. In summary, fungal plant pathogens will continue to provide highly fertile grounds for interdisciplinary and impactful research.

References

Abbott R, Albach D, Ansell S et al (2013) Hybridization and speciation. J Evol Biol 26:229–246. https://doi.org/10.1111/j.1420-9101.2012.02599.x

Agrawal AF (2006) Evolution of sex: why do organisms shuffle their genotypes? Curr Biol 16(17):R696–R704. https://doi.org/10.1016/j.cub.2006.07.063

Akagi Y, Akamatsu H, Otani H, Kodama M (2009) Horizontal chromosome transfer, a mechanism for the evolution and differentiation of a plant-pathogenic fungus. Eukaryot Cell 8:1732–1738. https://doi.org/10.1128/EC.00135-09

Almeida-Silva F, Venancio TM (2021) Integration of genome-wide association studies and gene coexpression networks unveils promising soybean resistance genes against five common fungal pathogens. Sci Rep 11:1–10. https://doi.org/10.1038/s41598-021-03864-x

Atanasoff-Kardjalieff AK, Lünne F, Kalinina S et al (2021) Biosynthesis of Fusapyrone depends on the H3K9 methyltransferase, FmKmt1, in Fusarium mangiferae. Front Fungal Biol 2. https://doi.org/10.3389/ffunb.2021.671796

Bannister AJ, Kouzarides T (2011) Regulation of chromatin by histone modifications. Cell Res 21:381–395. https://doi.org/10.1038/cr.2011.22

Bertazzoni S, Williams AH, Jones DA et al (2018) Accessories make the outfit: accessory chromosomes and other dispensable DNA regions in plant-pathogenic fungi. Mol Plant Microbe Interact 31: 779–788. https://doi.org/10.1094/MPMI-06-17-0135-FI

Bewick AJ, Hofmeister BT, Powers RA et al (2019) Diversity of cytosine methylation across the fungal tree of life. Nat Ecol Evol 3:479–490. https://doi.org/10.1038/s41559-019-0810-9

Bissegger M, Rigling D, Heiniger U (1997) Population structure and disease development of Cryphonectria parasitica in European chestnut forests in the presence of natural hypovirulence. Phytopathology 87:50–59. https://doi.org/10.1094/PHYTO.1997.87.1.50

Bzymek M, Lovett ST (2001) Instability of repetitive DNA sequences: the role of replication in multiple

mechanisms. Proc Natl Acad Sci U S A 98:8319–8325. https://doi.org/10.1073/pnas.111008398
Castanera R, Pérez G, López L et al (2014) Highly expressed captured genes and cross-kingdom domains present in Helitrons create novel diversity in pleurotus ostreatus and other fungi. BMC Genomics 15:1–17. https://doi.org/10.1186/1471-2164-15-1071
Chaloner TM, Gurr SJ, Bebber DP (2021) Plant pathogen infection risk tracks global crop yields under climate change. Nat Clim Chang 11:710–715. https://doi.org/10.1038/s41558-021-01104-8
Chen F, Everhart SE, Bryson PK et al (2015) Fungicide-induced transposon movement in Monilinia fructicola. Fungal Genet Biol 85:38–44. https://doi.org/10.1016/j.fgb.2015.10.006
Clairet C, Gay EJ, Porquier A et al (2021) Regulation of effector gene expression as concerted waves in Leptosphaeria maculans: a two-players game. bioRxiv. https://doi.org/10.1101/2021.12.15.472773
Cleary M, Nguyen D, Marciulyniene D et al (2016) Friend or foe? Biological and ecological traits of the European ash dieback pathogen Hymenoscyphus fraxineus in its native environment. Sci Rep 6. https://doi.org/10.1038/srep21895
Connolly LR, Smith KM, Freitag M (2013) The fusarium graminearum histone H3 K27 methyltransferase KMT6 regulates development and expression of secondary metabolite gene clusters. PLoS Genet 9. https://doi.org/10.1371/journal.pgen.1003916
Cook DE, Kramer HM, Torres DE et al (2020) A unique chromatin profile defines adaptive genomic regions in a fungal plant pathogen. Elife 9:1–32. https://doi.org/10.7554/ELIFE.62208
Croll D, McDonald BA (2012) The accessory genome as a cradle for adaptive evolution in pathogens. PLoS Pathog 8:e1002608. https://doi.org/10.1371/journal.ppat.1002608
Croll D, McDonald BA (2017) The genetic basis of local adaptation for pathogenic fungi in agricultural ecosystems. Mol Ecol 26:2027–2040. https://doi.org/10.1111/mec.13870
Dallaire A, Manley BF, Wilkens M et al (2021) Transcriptional activity and epigenetic regulation of transposable elements in the symbiotic fungus Rhizophagus irregularis. Genome Res 31:2290–2302. https://doi.org/10.1101/gr.275752.121
Dang Y, Yang Q, Xue Z, Liu Y (2011) RNA interference in fungi: pathways, functions, and applications. Eukaryot Cell 10:1148–1155. https://doi.org/10.1128/EC.05109-11
De Vienne DM, Refrégier G, López-Villavicencio M et al (2013) Cospeciation vs host-shift speciation: methods for testing, evidence from natural associations and relation to coevolution. New Phytol 198:347–385. https://doi.org/10.1111/nph.12150
Dong S, Raffaele S, Kamoun S (2015) The two-speed genomes of filamentous pathogens: waltz with plants. Curr Opin Genet Dev 35:57–65. https://doi.org/10.1016/j.gde.2015.09.001
Dutta A, Croll D, McDonald BA, Barrett LG (2021) Maintenance of variation in virulence and reproduction in populations of an agricultural plant pathogen. Evol Appl 14:335–347. https://doi.org/10.1111/eva.13117
Faino L, Seidl MF, Shi-Kunne X et al (2016) Transposons passively and actively contribute to evolution of the two-speed genome of a fungal pathogen. Genome Res 26:1091–1100. https://doi.org/10.1101/gr.204974.116
Faria R, Johannesson K, Butlin RK, Westram AM (2019) Evolving inversions. Trends Ecol Evol 34:239–248. https://doi.org/10.1016/j.tree.2018.12.005
Feurtey A, Gladieux P, Hood ME et al (2016) Strong phylogeographic co-structure between the anther-smut fungus and its white campion host. New Phytol 212:668–679. https://doi.org/10.1111/nph.14125
Feurtey, A., Lorrain, C., McDonald, M.C. et al. A thousand-genome panel retraces the global spread and adaptation of a major fungal crop pathogen. Nat Commun 14, 1059 (2023). https://doi.org/10.1038/s41467-023-36674-y
Fisher MC, Henk DA, Briggs CJ et al (2012) Emerging fungal threats to animal, plant and ecosystem health. Nature 484:186–194. https://doi.org/10.1038/nature10947
Fisher MC, Hawkins NJ, Sanglard D, Gurr SJ (2018) Worldwide emergence of resistance to antifungal drugs challenges human health and food security. Science 80:739–742. https://doi.org/10.1126/science.aap7999
Fones HN, Bebber DP, Chaloner TM et al (2020) Threats to global food security from emerging fungal and oomycete crop pathogens. Nat Food 1:332–342. https://doi.org/10.1038/s43016-020-0075-0
Fouché S, Badet T, Oggenfuss U et al (2020) Stress-driven transposable element de-repression dynamics and virulence evolution in a fungal pathogen. Mol Biol Evol 37:221–239. https://doi.org/10.1093/molbev/msz216
Fouché S, Oggenfuss U, Chanclud E, Croll D (2022) A devil's bargain with transposable elements in plant pathogens. Trends Genet 38:222–230. https://doi.org/10.1016/j.tig.2021.08.005
Franceschetti M, Maqbool A, Jiménez-Dalmaroni MJ et al (2017) Effectors of filamentous plant pathogens: commonalities amid diversity. Microbiol Mol Biol Rev 81. https://doi.org/10.1128/mmbr.00066-16
Frantzeskakis L, Kracher B, Kusch S et al (2018) Signatures of host specialization and a recent transposable element burst in the dynamic one-speed genome of the fungal barley powdery mildew pathogen. BMC Genomics 19:1–23. https://doi.org/10.1186/s12864-018-4750-6
Frantzeskakis L, Kusch S, Panstruga R (2019) The need for speed: compartmentalized genome evolution in filamentous phytopathogens. Mol Plant Pathol 20:3–7. https://doi.org/10.1111/mpp.12738
Freitag M (2017) Histone methylation by SET domain proteins in fungi. Annu Rev Microbiol 71:413–439. https://doi.org/10.1146/annurev-micro-102215-095757

Freitag M, Williams RL, Kothe GO, Selker EU (2002) A cytosine methyltransferase homologue is essential for repeat-induced point mutation in Neurospora crassa. Proc Natl Acad Sci U S A 99:8802–8807. https://doi.org/10.1073/pnas.132212899

Friesen TL, Stukenbrock EH, Liu Z et al (2006) Emergence of a new disease as a result of interspecific virulence gene transfer. Nat Genet 38:953–956. https://doi.org/10.1038/ng1839

Gao Y, Liu Z, Faris JD et al (2016) Validation of genome-wide association studies as a tool to identify virulence factors in Parastagonospora nodorum. Phytopathology 106:1177–1185. https://doi.org/10.1094/PHYTO-02-16-0113-FI

Gao Y, Zhang S, Li H (2022) H3K4 methyltransferase CfSet1 is required for development and pathogenesis in Colletotrichum fructicola. J Fungi 8:3–9. https://doi.org/10.3390/jof8040363

Gardiner DM, McDonald MC, Covarelli L et al (2012) Comparative pathogenomics reveals horizontally acquired novel virulence genes in fungi infecting cereal hosts. PLoS Pathog 8. https://doi.org/10.1371/journal.ppat.1002952

Gates LA, Foulds CE, O'Malley BW (2017) Histone marks in the 'driver's seat': functional roles in steering the transcription cycle. Trends Biochem Sci 42:977–989. https://doi.org/10.1016/j.tibs.2017.10.004

Gladieux P, Zhang XG, Róldan-Ruiz I et al (2010) Evolution of the population structure of Venturia inaequalis, the apple scab fungus, associated with the domestication of its host. Mol Ecol 19:658–674. https://doi.org/10.1111/j.1365-294X.2009.04498.x

Gladyshev E (2017) Repeat-induced point mutation (RIP) and other genome Defense mechanisms in fungi. Microbiol Spectr 5. https://doi.org/10.1016/j.physbeh.2017.03.040

Gluck-Thaler E, Ralston T, Konkel Z et al (2022) Giant starship elements mobilize accessory genes in fungal genomes. Mol Biol Evol 39. https://doi.org/10.1093/molbev/msac109

Golan JJ, Pringle A (2017) Long-distance dispersal of fungi. In: The fungal kingdom. ASM Press, Washington, DC, pp 309–333

González-Sayer S, Oggenfuss U, García I et al (2022) High-quality genome assembly of Pseudocercospora ulei the main threat to natural rubber trees. Genet Mol Biol 45:1–5. https://doi.org/10.1590/1678-4685-gmb-2021-0051

Grandaubert J, Lowe RGT, Soyer JL et al (2014) Transposable element-assisted evolution and adaptation to host plant within the Leptosphaeria maculans-Leptosphaeria biglobosa species complex of fungal pathogens. BMC Genomics 15:1–27. https://doi.org/10.1186/1471-2164-15-891

Grigoriev IV, Cullen D, Goodwin SB, Hibbett D, Jeffries TW, Kubicek CP, Kuske C, Magnuson JK, Martin F, Spatafora JW et al (2011) Fueling the future with fungal genomics. Mycology 2:192–209

Gross A, Hosoya T, Queloz V (2014) Population structure of the invasive forest pathogen Hymenoscyphus pseudoalbidus. Mol Ecol 23:2943–2960. https://doi.org/10.1111/mec.12792

Gu Q, Ji T, Sun X et al (2017) Histone H3 lysine 9 methyltransferase FvDim5 regulates fungal development, pathogenicity and osmotic stress responses in fusarium verticillioides. FEMS Microbiol Lett 364:1–8. https://doi.org/10.1093/femsle/fnx184

Habig M, Stukenbrock EH (2020) Origin, function, and transmission of accessory chromosomes. In: Genetics and biotechnology. Springer, Cham, pp 25–47

Habig M, Schotanus K, Hufnagel K et al (2021) Ago1 affects the virulence of the fungal plant pathogen Zymoseptoria tritici. Genes (Basel) 12. https://doi.org/10.3390/genes12071011

Hartmann FE (2022) Using structural variants to understand the ecological and evolutionary dynamics of fungal plant pathogens. New Phytol 234:43–49. https://doi.org/10.1111/nph.17907

Hartmann FE, Sánchez-Vallet A, McDonald BA, Croll D (2017) A fungal wheat pathogen evolved host specialization by extensive chromosomal rearrangements. ISME J 11:1189–1204. https://doi.org/10.1038/ismej.2016.196

Hartmann FE, Rodríguez de la Vega RC, Gladieux P et al (2020a) Higher gene flow in sex-related chromosomes than in autosomes during fungal divergence. Mol Biol Evol 37:668–682. https://doi.org/10.1093/molbev/msz252

Hartmann FE, Snirc A, Cornille A et al (2020b) Congruent population genetic structures and divergence histories in anther-smut fungi and their host plants Silene italica and the Silene nutans species complex. Mol Ecol 29:1154–1172. https://doi.org/10.1111/mec.15387

Hartmann FE, Ament-Velásquez SL, Vogan AA et al (2021) Size variation of the nonrecombining region on the mating-type chromosomes in the fungal Podospora anserina species complex. Mol Biol Evol 38:2475–2492. https://doi.org/10.1093/molbev/msab040

Hawkins NJ, Fraaije BA (2018) Fitness penalties in the evolution of fungicide resistance. Annu Rev Phytopathol 56:339–360. https://doi.org/10.1146/annurev-phyto-080417-050012

Hessenauer P, Fijarczyk A, Martin H et al (2020) Hybridization and introgression drive genome evolution of Dutch elm disease pathogens. Nat Ecol Evol 4:626–638. https://doi.org/10.1038/s41559-020-1133-6

Hill WG, Robertson A (2008) The effect of linkage on limits to artificial selection. Genet Res (Camb) 89:311–336. https://doi.org/10.1017/S001667230800949X

Ho S, Alexander U, Mills R (2020) Structural variation in the sequencing era: comprehensive discovery and integration. Nat Rev Genet 21:171–189. https://doi.org/10.1038/s41576-019-0180-9.Structural

Horns F, Petit E, Yockteng R, Hood ME (2012) Patterns of repeat-induced point mutation in transposable elements

of basidiomycete fungi. Genome Biol Evol 4:240–247. https://doi.org/10.1093/gbe/evs005
Jeon J, Choi J, Lee GW et al (2015) Genome-wide profiling of DNA methylation provides insights into epigenetic regulation of fungal development in a plant pathogenic fungus, Magnaporthe oryzae. Sci Rep 5:1–11. https://doi.org/10.1038/srep08567
Jiquel A, Gervais J, Geistodt-Kiener A et al (2021) A gene-for-gene interaction involving a 'late' effector contributes to quantitative resistance to the stem canker disease in Brassica napus. New Phytol 231:1510–1524. https://doi.org/10.1111/nph.17292
Jordan IK, Miller WJ (2008) Genome Defense against transposable elements and the origins of regulatory RNA. In: Genome dynamics and stability. Springer, New York, pp 77–94
Karisto P, Hund A, Yu K et al (2018) Ranking quantitative resistance to Septoria tritici blotch in elite wheat cultivars using automated image analysis. Phytopathology 108:568–581. https://doi.org/10.1094/PHYTO-04-17-0163-R
Karisto P, Suffert F, Mikaberidze A (2021) Measuring splash-dispersal of a major wheat pathogen in the field. PhytoFrontier. https://doi.org/10.1094/phytofr-05-21-0039-r
Klemm SL, Shipony Z, Greenleaf WJ (2019) Chromatin accessibility and the regulatory epigenome. Nat Rev Genet 20:207–220. https://doi.org/10.1038/s41576-018-0089-8
Kramer HM, Seidl MF, Thomma BP, Cook DE (2022) Local rather than global H3K27me3 dynamics are associated with differential gene expression in verticillium dahliae. MBio 13. https://doi.org/10.1128/mbio.03566-21
Latorre SM, Were VM, Foster AJ et al (2022) A pandemic clonal lineage of the wheat blast fungus. bioRxiv. https://doi.org/10.1101/2022.06.06.494979
Le Cam B, Sargent D, Gouzy J et al (2019) Population genome sequencing of the scab fungal species Venturia inaequalis, Venturia pirina, Venturia aucupariae and Venturia asperata. G3 Genes Genomes Genetics 9:2405–2414. https://doi.org/10.1534/g3.119.400047
Lendenmann MH, Croll D, Stewart EL, McDonald BA (2014) Quantitative trait locus mapping of melanization in the plant pathogenic fungus Zymoseptoria tritici. G3 Genes Genomes Genetics 4:2519–2533. https://doi.org/10.1534/g3.114.015289
Lendenmann MH, Croll D, McDonald BA (2015) QTL mapping of fungicide sensitivity reveals novel genes and pleiotropy with melanization in the pathogen Zymoseptoria tritici. Fungal Genet Biol 80:53–67. https://doi.org/10.1016/j.fgb.2015.05.001
Liu Y, Liu N, Yin Y et al (2015) Histone H3K4 methylation regulates hyphal growth, secondary metabolism and multiple stress responses in Fusarium graminearum. Environ Microbiol 17:4615–4630. https://doi.org/10.1111/1462-2920.12993
Lo Presti L, Lanver D, Schweizer G et al (2015) Fungal effectors and plant susceptibility. Annu Rev Plant Biol 66:513–545. https://doi.org/10.1146/annurev-arplant-043014-114623
Lorrain C, Feurtey A, Möller M et al (2021) Dynamics of transposable elements in recently diverged fungal pathogens: lineage-specific transposable element content and efficiency of genome defenses. G3 Genes Genomes Genetics 11:0–12. https://doi.org/10.1093/g3journal/jkab068
Luger K, Mäder AW, Richmond RK et al (1997) Crystal structure of the nucleosome core particle at 2.8 Å resolution. Nature 389:251–260. https://doi.org/10.1038/38444
Magdama F, Monserrate-Maggi L, Serrano L et al (2020) Genetic diversity of Fusarium oxysporum f. sp. cubense, the fusarium wilt pathogen of Banana, in Ecuador. Plants 9:1133. https://doi.org/10.3390/plants9091133
Martienssen RA, Colot V (2001) DNA methylation and epigenetic inheritance in plants and filamentous fungi. Science 80:1070–1074. https://doi.org/10.1126/science.293.5532.1070
McDonald MC, Taranto AP, Hill E et al (2019) Transposon-mediated horizontal transfer of the host-specific virulence protein ToxA between three fungal wheat pathogens. mBio 10. https://doi.org/10.1128/mBio.01515-19
McDonald BA, Suffert F, Bernasconi A, Mikaberidze A (2022) How large and diverse are field populations of fungal plant pathogens? The case of Zymoseptoria tritici. bioRxiv. https://doi.org/10.1101/2022.03.13.484150
Meile L, Peter J, Puccetti G et al (2020) Chromatin dynamics contribute to the spatiotemporal expression pattern of virulence genes in a fungal plant pathogen. mBio 11:1–18. https://doi.org/10.1128/mBio.02343-20
Menardo F, Praz CR, Wyder S et al (2016) Hybridization of powdery mildew strains gives rise to pathogens on novel agricultural crop species. Nat Genet 48:201–205. https://doi.org/10.1038/ng.3485
Mieczkowski PA, Lemoine FJ, Petes TD (2006) Recombination between retrotransposons as a source of chromosome rearrangements in the yeast Saccharomyces cerevisiae. DNA Repair (Amst) 5:1010–1020. https://doi.org/10.1016/j.dnarep.2006.05.027
Mohd-Assaad N, McDonald BA, Croll D (2016) Multilocus resistance evolution to azole fungicides in fungal plant pathogen populations. Mol Ecol 25:6124–6142. https://doi.org/10.1111/mec.13916
Mohd-Assaad N, McDonald BA, Croll D (2018) Genome-wide detection of genes under positive selection in worldwide populations of the barley scald pathogen. Genome Biol Evol 10:1315–1332. https://doi.org/10.1093/gbe/evy087
Möller M, Habig M, Lorrain C et al (2021) Recent loss of the Dim2 DNA methyltransferase decreases mutation rate in repeats and changes evolutionary trajectory in a fungal pathogen. PLoS Genet 17:1–27. https://doi.org/10.1371/journal.pgen.1009448

Müller MC, Praz CR, Sotiropoulos AG et al (2019) A chromosome-scale genome assembly reveals a highly dynamic effector repertoire of wheat powdery mildew. New Phytol 221:2176–2189. https://doi.org/10.1111/nph.15529
Oggenfuss U, Badet T, Wicker T et al (2021) A population-level invasion by transposable elements triggers genome expansion in a fungal pathogen. Elife 10:1–25. https://doi.org/10.7554/eLife.69249
Omrane S, Sghyer H, Audeon C et al (2015) Fungicide efflux and the MgMFS1 transporter contribute to the multidrug resistance phenotype in Zymoseptoria tritici field isolates. Environ Microbiol 17:2805–2823. https://doi.org/10.1111/1462-2920.12781
Omrane S, Audéon C, Ignace A et al (2017) Plasticity of the MFS1 promoter leads to multidrug resistance in the wheat pathogen Zymoseptoria tritici. mSphere 2:1–42. https://doi.org/10.1128/mSphere.00393-17
Pepper JW (2003) The evolution of evolvability in genetic linkage patterns. Biosystems 69:115–126. https://doi.org/10.1016/S0303-2647(02)00134-X
Pereira D, McDonald BA, Croll D (2020a) The genetic architecture of emerging fungicide resistance in populations of a global wheat pathogen. Genome Biol Evol 12:2231–2244. https://doi.org/10.1093/gbe/evaa203
Pereira D, Oggenfuss U, McDonald BA, Croll D (2020b) The population genomics of transposable element activation in the highly repressive genome of an agricultural pathogen. bioRxiv. https://doi.org/10.1101/2020.11.12.379651
Peter M, Kohler A, Ohm RA et al (2016) Ectomycorrhizal ecology is imprinted in the genome of the dominant symbiotic fungus Cenococcum geophilum. Nat Commun 7:1–15. https://doi.org/10.1038/ncomms12662
Petrov DA, Aminetzach YT, Davis JC et al (2003) Size matters: non-LTR retrotransposable elements and ectopic recombination in drosophila. Mol Biol Evol 20:880–892. https://doi.org/10.1093/molbev/msg102
Pfannenstiel BT, Keller NP (2019) On top of biosynthetic gene clusters: how epigenetic machinery influences secondary metabolism in fungi. Biotechnol Adv 37:107345. https://doi.org/10.1016/j.biotechadv.2019.02.001
Ploetz RC (2008) Panama disease: a classic and destructive disease of banana. Agric Sci 1876:1–7. https://doi.org/10.1094/PHP-2000-1204-01-HM.Panama
Priest SJ, Yadav V, Heitman J (2020) Advances in understanding the evolution of fungal genome architecture. F1000Research 9:776. https://doi.org/10.12688/f1000research.25424.1
Raffaele S, Kamoun S (2012) Genome evolution in filamentous plant pathogens: why bigger can be better. Nat Rev Microbiol 10:417–430. https://doi.org/10.1038/nrmicro2790
Ramírez-Tejero JA, Cabanás CG-L, Valverde-Corredor A et al (2020) Epigenetic regulation of Verticillium dahliae virulence: does DNA methylation level play a role? Int J Mol Sci 21:5197. https://doi.org/10.3390/ijms21155197
Ramos-García VH, Villota-Salazar NA, González-Prieto JM, Cortés-Espinosa DV (2022) Different histone deacetylase inhibitors reduce growth, virulence as well as changes in the morphology of the fungus Macrophomina phaseolina (Tassi) Goid. World J Microbiol Biotechnol 38:1–16. https://doi.org/10.1007/s11274-022-03249-w
Rhodes J, Abdolrasouli A, Dunne K et al (2022) Population genomics confirms acquisition of drug-resistant Aspergillus fumigatus infection by humans from the environment. Nat Microbiol 7:663–674. https://doi.org/10.1038/s41564-022-01091-2
Richards JK, Stukenbrock EH, Carpenter J et al (2019) Local adaptation drives the diversification of effectors in the fungal wheat pathogen Parastagonospora nodorum in the United States. PLoS Genet 15:e1008223. https://doi.org/10.1371/journal.pgen.1008223
Rivas FV, Tolia NH, Song JJ et al (2005) Purified Argonaute2 and an siRNA form recombinant human RISC. Nat Struct Mol Biol 12:340–349. https://doi.org/10.1038/nsmb918
Rouxel T, Balesdent M (2010) Avirulence genes. Wiley, Chichester, UK
Sánchez-Vallet A, Hartmann FE, Marcel TC, Croll D (2018) Nature's genetic screens: using genome-wide association studies for effector discovery. Mol Plant Pathol 19:3–6. https://doi.org/10.1111/mpp.12592
Scharf DH, Heinekamp T, Brakhage AA (2014) Human and plant fungal pathogens: the role of secondary metabolites. PLoS Pathog 10:10–12. https://doi.org/10.1371/journal.ppat.1003859
Schmidt SM, Houterman PM, Schreiver I et al (2013) MITEs in the promoters of effector genes allow prediction of novel virulence genes in Fusarium oxysporum. BMC Genomics 14:119. https://doi.org/10.1186/1471-2164-14-119
Seidl MF, Kramer HM, Cook DE et al (2020) Repetitive elements contribute to the diversity and evolution of centromeres in the fungal genus verticillium. MBio 11:1–22. https://doi.org/10.1128/mBio.01714-20
Selker EU (1990) Premeiotic instability of repeated sequences in neurospora crassa. Annu Rev Genet 24:579–613. https://doi.org/10.1146/annurev.ge.24.120190.003051
Selker EU, Tountas NA, Cross SH et al (2003) The methylated component of the neurospora crassa genome. Nature 422:893–897. https://doi.org/10.1038/nature01564
Singh RP, Hodson DP, Huerta-Espino J et al (2011) The emergence of Ug99 races of the stem rust fungus is a threat to world wheat production. Annu Rev Phytopathol 49:465–481. https://doi.org/10.1146/annurev-phyto-072910-095423
Singh NK, Karisto P, Croll D (2021) Population-level deep sequencing reveals the interplay of clonal and sexual reproduction in the fungal wheat pathogen

Zymoseptoria tritici. Microb Genomics 7. https://doi.org/10.1099/MGEN.0.000678
Slotkin RK, Martienssen R (2007) Transposable elements and the epigenetic regulation of the genome. Nat Rev Genet 8:272–285. https://doi.org/10.1038/nrg2072
Soyer JL, El Ghalid M, Glaser N et al (2014) Epigenetic control of effector gene expression in the plant pathogenic fungus Leptosphaeria maculans. PLoS Genet 10. https://doi.org/10.1371/journal.pgen.1004227
Soyer JL, Clairet C, Gay EJ et al (2021) Genome-wide mapping of histone modifications during axenic growth in two species of Leptosphaeria maculans showing contrasting genomic organization. Chromosom Res 29:219–236. https://doi.org/10.1007/s10577-021-09658-1
Spanu PD, Abbott JC, Amselem J et al (2010) Genome expansion and gene loss in. Science 80:1543–1546
Stauber L, Badet T, Feurtey A et al (2021) Emergence and diversification of a highly invasive chestnut pathogen lineage across southeastern Europe. Elife 10:1–27. https://doi.org/10.7554/eLife.56279
Stewart J, Ross-Davis A, Gra?a R et al (2018) Genetic diversity of the myrtle rust pathogen (Austropuccinia psidii) in the Americas and Hawaii: global implications for invasive threat assessments. For Pathol 48:1–13. https://doi.org/10.1111/efp.12378
Studt L, Rösler SM, Burkhardt I et al (2016) Knock-down of the methyltransferase Kmt6 relieves H3K27me3 and results in induction of cryptic and otherwise silent secondary metabolite gene clusters in Fusarium fujikuroi. Environ Microbiol 18:4037–4054. https://doi.org/10.1111/1462-2920.13427
Stukenbrock EH (2016) The role of hybridization in the evolution and emergence of new fungal plant pathogens. Phytopathology 106:104–112. https://doi.org/10.1094/PHYTO-08-15-0184-RVW
Stukenbrock EH, Banke S, Javan-Nikkhah M, McDonald BA (2007) Origin and domestication of the fungal wheat pathogen Mycosphaerella graminicola via sympatric speciation. Mol Biol Evol 24:398–411
Thines M (2019) An evolutionary framework for host shifts – jumping ships for survival. New Phytol 224: 605–617. https://doi.org/10.1111/nph.16092
Torres DE, Oggenfuss U, Croll D, Seidl MF (2020) Genome evolution in fungal plant pathogens: looking beyond the two-speed genome model. Fungal Biol Rev 34:136–143. https://doi.org/10.1016/j.fbr.2020.07.001
Tralamazza SM, Abraham LN, Reyes-Avila CS et al (2021) Histone H3K27 methylation perturbs transcriptional robustness and underpins dispensability of highly conserved genes in fungi. Mol Biol Evol 39. https://doi.org/10.1093/molbev/msab323
Tsushima A, Gan P, Kumakura N et al (2019) Genomic plasticity mediated by transposable elements in the plant pathogenic fungus Colletotrichum higginsianum. Genome Biol Evol 11:1487–1500. https://doi.org/10.1093/gbe/evz087
Urquhart AS, Chong NF, Yang Y, Idnurm A (2022) A large transposable element mediates metal resistance in the fungus Paecilomyces variotii. Curr Biol 32:937–950.e5. https://doi.org/10.1016/j.cub.2021.12.048
van de Wouw AP, Cozijnsen AJ, Hane JK et al (2010) Evolution of linked avirulence effectors in Leptosphaeria maculans is affected by genomic environment and exposure to resistance genes in host plants. PLoS Pathog 6. https://doi.org/10.1371/journal.ppat.1001180
van Wyk S, Wingfield BD, De Vos L et al (2021) Genome-wide analyses of repeat-induced point mutations in the Ascomycota. Front Microbiol 11. https://doi.org/10.3389/fmicb.2020.622368
Visentin I, Montis V, Döll K et al (2012) Transcription of genes in the biosynthetic pathway for fumonisin mycotoxins is epigenetically and differentially regulated in the fungal maize pathogen fusarium verticillioides. Eukaryot Cell 11:252–259. https://doi.org/10.1128/EC.05159-11
Vlaardingerbroek I, Beerens B, Schmidt SM et al (2016) Dispensable chromosomes in fusarium oxysporum f. Sp. lycopersici. Mol Plant Pathol 17:1455–1466. https://doi.org/10.1111/mpp.12440
Vogan AA, Khankhet J, Xu J (2013) Evidence for mitotic recombination within the basidia of a hybrid Cross of Cryptococcus neoformans. PLoS One 8. https://doi.org/10.1371/journal.pone.0062790
Wang X, Hsueh YP, Li W et al (2010) Sex-induced silencing defends the genome of Cryptococcus neoformans via RNAi. Genes Dev 24:2566–2582. https://doi.org/10.1101/gad.1970910
Wang L, Sun Y, Sun X et al (2020) Repeat-induced point mutation in neurospora crassa causes the highest known mutation rate and mutational burden of any cellular life. Genome Biol 21:1–23. https://doi.org/10.1186/s13059-020-02060-w
Wang C, Milgate AW, Solomon PS, McDonald MC (2021) The identification of a transposon affecting the asexual reproduction of the wheat pathogen Zymoseptoria tritici. Mol Plant Pathol 22:800–816. https://doi.org/10.1111/mpp.13064
Watters MK, Randall TA, Margolin BS et al (1999) Action of repeat-induced point mutation on both strands of a duplex and on tandem duplications of various sizes in neurospora. Genetics 153:705–714
Weiberg A, Wang M, Lin FM et al (2013) Fungal small RNAs suppress plant immunity by hijacking host RNA interference pathways. Science 342:118–123. https://doi.org/10.1126/science.1239705
Weigel D, Nordborg M (2015) Population genomics for understanding adaptation in wild plant species. Annu Rev Genet 49:315–338. https://doi.org/10.1146/annurev-genet-120213-092110
Wicker T, Sabot F, Hua-Van A et al (2007) A unified classification system for eukaryotic transposable elements. Nat Rev Genet 8:973–982. https://doi.org/10.1038/nrg2165
Yang H, Yu H, Ma LJ (2020) Accessory chromosomes in fusarium oxysporum. Phytopathology 110:1488–1496. https://doi.org/10.1094/PHYTO-03-20-0069-IA

Yeaman S (2022) Evolution of polygenic traits under global vs local adaptation. Genetics 220. https://doi.org/10.1093/genetics/iyab134

Yeaman S, Aeschbacher S, Bürger R (2016) The evolution of genomic islands by increased establishment probability of linked alleles. Mol Ecol 25:2542–2558. https://doi.org/10.1111/mec.13611

Zhang T, Zhao YL, Zhao JH et al (2016) Cotton plants export microRNAs to inhibit virulence gene expression in a fungal pathogen. Nat Plants 2:1–6. https://doi.org/10.1038/nplants.2016.153

Zhang W, Huang J, Cook DE (2021) Histone modification dynamics at H3K27 are associated with altered transcription of in planta induced genes in Magnaporthe oryzae. PLoS Genet 17:1–29. https://doi.org/10.1371/JOURNAL.PGEN.1009376

Zheng H, Chen J, Mu C et al (2018) Combined linkage and association mapping reveal QTL for host plant resistance to common rust (Puccinia sorghi) in tropical maize. BMC Plant Biol 18:1–14. https://doi.org/10.1186/s12870-018-1520-1

Zhong Z, Marcel TC, Hartmann FE et al (2017) A small secreted protein in Zymoseptoria tritici is responsible for avirulence on wheat cultivars carrying the Stb6 resistance gene. New Phytol 214:619–631. https://doi.org/10.1111/nph.14434

Zhou S, Liu X, Sun W et al (2021) The COMPASS-like complex modulates fungal development and pathogenesis by regulating H3K4me3-mediated targeted gene expression in Magnaporthe oryzae. Mol Plant Pathol 22:422–439. https://doi.org/10.1111/mpp.13035

6 Host Switching and Geographic Expansions in (Hemi)biotrophic Plant Pathogens

Marco Thines, Hanno Seebens, M. Catherine Aime, Roger G. Shivas, Levente Kiss, Franz Essl, and Anna Schertler

Abstract

While it is increasingly well understood how plants and animals spread around the world, and how they diversify and occupy new niches, such knowledge is fairly limited for fungi and oomycetes. As is true for animals and plants, many plant pathogenic fungi have been spread anthropogenically, but, in contrast to them, only rarely as a deliberate introduction. In addition, the occupation of new niches for plant pathogenic fungi is less defined by the abiotic environment, but more by the biotic environment, as the interaction with the host plant is the major ecological determinant of fitness, especially for biotrophic and hemibiotrophic pathogens. Thus, host switches are a major driver in the diversification of pathogens, which can have an effect similar to the arrival of animals or plants on a previously uninhabited archipelago. This chapter summarises the current state of knowledge on range expansions and host jumps in plant pathogens, focusing on (hemi-) biotrophic fungi and oomycetes.

Keywords

Anthropogenic spread · Evolution · Host jumps · *Erysiphaceae* · *Oomycota* · *Pucciniales* · *Ustilaginomycotina*

M. Thines (✉)
Goethe Universität Frankfurt am Main, Frankfurt, Germany

Senckenberg Biodiversity and Climate Research Centre, Frankfurt, Germany
e-mail: m.thines@thines-lab.eu; Thines@bio.uni-frankfurt.de

H. Seebens
Senckenberg Biodiversity and Climate Research Centre, Frankfurt, Germany

M. C. Aime
Department of Botany and Plant Pathology, Purdue University, West Lafayette, IN, USA

R. G. Shivas · L. Kiss
Centre for Crop Health, University of Southern Queensland, Toowoomba, QLD, Australia

F. Essl · A. Schertler
Division of Bioinvasions, Global Change, Macroecology, Department of Botany and Biodiversity Research, University of Vienna, Vienna, Austria

6.1 How Plant Pathogenic Alien Fungi and Oomycetes Spread Around the World

6.1.1 Biological Invasions and Plant Pathogens

Alien species (syn. introduced, non-native) are those that occur outside their natural range due to human-mediated introduction, intentionally or accidentally. Although most will not find themselves in a suitable new environment, some thrive and form self-sustaining populations (i.e. *established alien species*) and may ultimately

© The Author(s), under exclusive license to Springer Nature Switzerland AG 2023

S. Pöggeler, T. James (eds.), *Evolution of Fungi and Fungal-Like Organisms*, The Mycota 14,
https://doi.org/10.1007/978-3-031-29199-9_6

start spreading and impacting native biota and ecosystems, human well-being, and the economy (i.e. *invasive alien species*) (Convention on Biological Diversity 2014). Nowadays invasive alien species are thought to be one of the main drivers of biodiversity loss, and they are associated with enormous economic costs worldwide (Diagne et al. 2021; IPBES 2019; Pimentel et al. 2005).

Fungi and fungus-like organisms were long neglected in invasion science (Desprez-Loustau et al. 2007) due to several obstacles. Amongst these are their often inconspicuous life-cycle stages, small size, and unsettled taxonomy (Lücking et al. 2020). But as for other taxonomic groups, their spread has accelerated in recent decades (Santini et al. 2013; Seebens et al. 2017) and frequently resulted in devastating impacts on human livelihoods and native biota—often with cascading effects on ecosystem services (Fisher et al. 2012; Newcombe and Dugan 2010). This is especially the case for plant pathogens, i.e. defined hereafter as fungi and fungal-like organisms that are able to cause disease in a plant host, excluding other plant pests such as arthropods and nematodes. There are several prominent examples in natural ecosystems (e.g. Dutch elm disease, ash dieback, chestnut blight), and managed agricultural or plantation-forestry systems, where they cause crop yield losses worth billions of dollars annually and sometimes even threaten food security (e.g. wheat stem rust, potato late blight, and Panama disease) (Fisher et al. 2012; Fones et al. 2017; Gurr et al. 2011; Sikes et al. 2018). A few notable recently emerging pathogens introduced by anthropogenic activity are shown in Fig. 6.1.

6.1.2 Historical Changes in the Introduction Frequency

Although the human-mediated transportation of organisms to new areas is as old as human movement itself, a new order of magnitude was reached after the onset of large-scale intercontinental movement of humans and goods in the fifteenth century [i.e. *Columbian Exchange*; Crosby (2003)] (Santini et al. 2018). While concrete information on introduction frequency prior to the eighteenth century is largely lacking, it can be assumed that a considerable number of plant pathogens were also transported between continents (Kreisel and Scholler 1994). With the unparalleled expansion of the global trade network associated with increases in trade and travel volumes, introductions of fungal and fungal-like plant pathogens have been continuously increasing worldwide (Fig. 6.2). Particularly steep increases took place in recent decades, e.g. in Europe (Desprez-Loustau et al. 2010; Santini et al. 2013), New Zealand (Sikes et al. 2018), North America (Aukema et al. 2010; Nealis et al. 2016), South America (Fuentes et al. 2020), and Australia (Nahrung and Carnegie 2020).

6.1.3 Major Routes of Introduction

A commonly identified major driver of this development is the intensification and expansion of trade (Desprez-Loustau et al. 2010). Plant pathogens are mostly introduced unintentionally, often as contaminants of their hosts. Therefore, the import of living plants (plants-for-planting, i.e. for agriculture, forestry, horticulture) is a well-recognised introduction pathway for alien plant pathogens, e.g., in infected asymptomatic plants (Anderson et al. 2004; Santini et al. 2013, 2018). This issue led to the stepwise implementation of phytosanitary measures and plant quarantine, resulting in multilateral treaties such as the International Plant Protection Convention (IPPC; https://www.ippc.int) which introduced International Standards for Phytosanitary Measures (ISPMs). The implementation and effectiveness of such measures varies amongst countries, but for those with very strict biosecurity measures (e.g. Australia and New Zealand) the main pathway for plant pathogens nowadays is believed to be contaminated goods other than plants (Santini

Fig. 6.1 Notable plant pathogens for which evidence suggests that they have spread via anthropogenic activities onto new, economically important crops. (**a**) *Peronospora belbahrii*, the downy mildew pathogen of basil, introduced from Africa from an unknown original host and now distributed worldwide. (**b**) *Hemileia vastatrix*, the causal agent of coffee leaf rust, which was able to colonise all major coffee growing regions of the world within 150 years. (**c**) *Erysiphe neolycopersici*, one of those three powdery mildew species that infect tomato, started to spread in the 1980s in Europe and has become widespread in the European and North American tomato production by the early 2000s. (**d**) *Sporisorium scitamineum* on cultivated sugarcane, which apparently originated from wild progenitors of sugarcane in India and has spread to almost all sugarcane-producing countries throughout the past five decades

et al. 2018). Indeed, in New Zealand, the arrival rate of new crop- and pasture-associated plant pathogens decreased in the last few decades, coinciding with the implementation of strict agricultural quarantine policies (Sikes et al. 2018). Apart from living plants, other important pathways of plant pathogens are contaminated seeds, soil (e.g. potting soil), timber and wood or associated insects, as well as machinery (Santini et al. 2013; Wingfield et al. 2015). Contaminated wood was identified as the second most important pathway of forest pathogens to Europe (Santini et al. 2013). A rather unexpected pathway is the unintentional introduction of plant pathogens during war activities through allied army supplies, such as the rot fungus *Heterobasidion irregulare* to Italy in WWII (Santini et al. 2018).

6.1.4 Major Sources and Sinks of Plant Pathogens

The highest numbers of alien plant pathogens globally are currently known in New Zealand, Europe (e.g. Austria, France, Italy, Switzerland), Australia, the USA, and China (Fig. 6.3). Generally, alien plant pathogen richness patterns should be treated with caution, as there is a considerable spatial bias in sampling frequency, monitoring effort, and research capacity (Bebber et al. 2019). Together with agricultural production, research efforts can largely explain crop pest numbers across countries, exemplifying the bias in reported pathogen richness (Bebber et al. 2019). Global studies on the plant pathogenic oomycete genus *Phytophthora*, which contains many invasive alien species, showed that many countries are likely to have an underreported

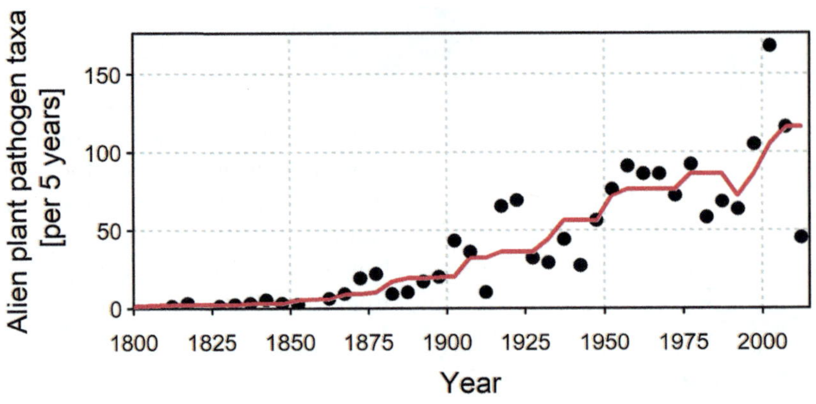

Fig. 6.2 Number of global first records of alien plant pathogens (fungi and oomycetes) in 5-year periods between 1800 and 2015. A global first record is defined as the first available report on the occurrence of a specific pathogen in an area outside its native range (alien or cryptogenic, i.e. likely outside its native range sensu Essl et al. 2018), and first records are based on a range of published sources (Abasova et al. 2018; Arechavaleta et al. 2005, 2009; Barwell et al. 2020; Bradshaw et al. 2021; Bufford et al. 2016; CAB International 2021; Desprez-Loustau et al. 2010; Deutsche Gesellschaft für Mykologie DGfM 2020; Dgebuadze et al. 2018; European and Mediterranean Plant Protection Organization EPPO 2020; Farr and Rossman 2021; Fortini et al. 2019; Fuentes et al. 2020; Graziosi et al. 2020; Gryzenhout et al. 2006; Heluta et al. 2019; IUCN Invasive Species Specialist Group ISSG 2019a, b; LeBlanc et al. 2018; McCook 2006; MyCoPortal 2020; Nahrung and Carnegie 2020; Nielsen et al. 2020; Santini et al. 2013; Seebens et al. 2017; Simpson et al. 2018; Strand et al. 2019; SwissFungi 2021; Talhinhas et al. 2019; Tomoshevich 2019; Voglmayr et al. 2022; Westerdijk Fungal Biodiversity Institute 2020; Wood 2017; Xu and Qiang 2018). Note that the number of records in recent periods might represent an underestimation due to lags in recording and reporting (Seebens et al. 2017) and that the number of new records per region is likely much higher

pathogen diversity, especially in Asia (Scott et al. 2019) and tropical regions worldwide (Bebber et al. 2019).

Alien plant pathogens in Europe mostly originate from temperate North America and Asia (Desprez-Loustau et al. 2010; Santini et al. 2013; Voglmayr et al. 2022). Nevertheless, for a significant amount of newly recorded plant pathogens the origin is unknown but likely alien (i.e. *cryptogenic*, Essl et al. 2018).

6.1.5 Knowledge Gaps and the Way Forward

Besides fungi and oomycetes being understudied in invasion science in general, plant pathogens of economically used plants have received the bulk of attention, while the consequences for natural plant communities are far less well known (Roy et al. 2017). Also, historical information on fungal introductions before the eighteenth century is largely missing, with some exceptions as the introduction of the *Armillaria* root rot pathogen in the 1600s to South Africa by early European settlers (Graziosi et al. 2020). Due to a lack of baseline data on their natural distribution, the biogeographic status of newly recorded fungi can often only be inferred from their hosts or through population genetic studies (Newcombe and Dugan 2010). Methodological improvements such as the application of microscopy in the late nineteenth century and the development of molecular phylogenetics and environmental DNA sampling in the last few decades aided the description and detection of plant pathogens (Hawksworth and Lücking 2017) also resulting in an increased number of first records and better knowledge of the distribution, such as for the oomycete genus *Phytophthora* (Scott et al. 2019). Despite the significant advances made, unresolved taxonomy, unknown and cryptic

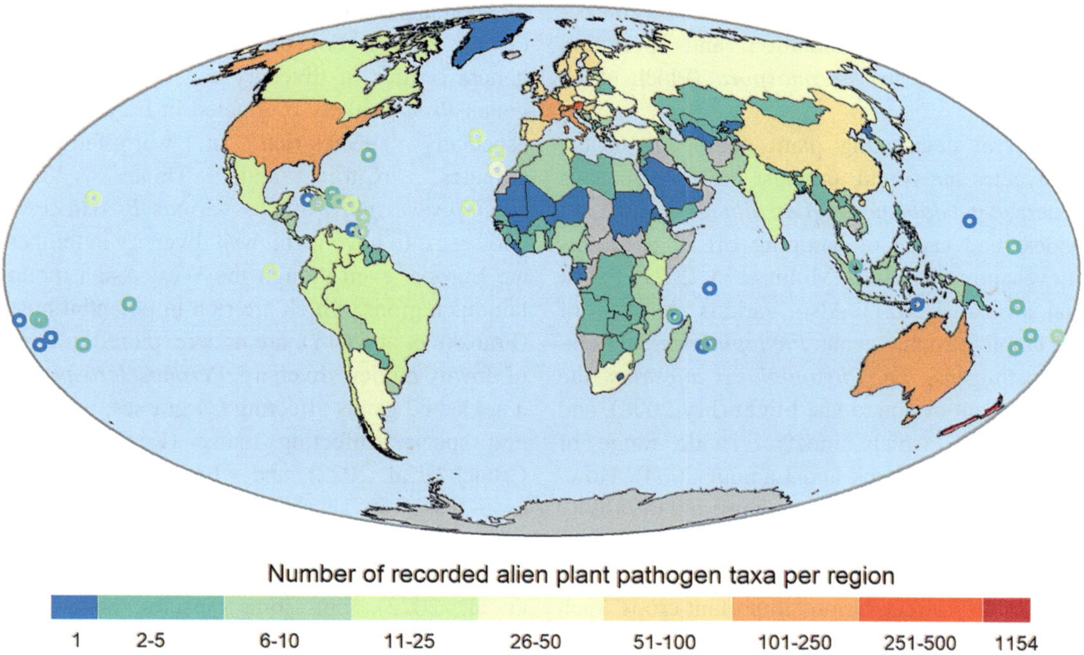

Fig. 6.3 Global patterns of known alien and cryptogenic plant pathogen (fungi and oomycetes) richness. In total 183 mainland and island regions (the latter represented as circles) are coloured after the number of recorded plant pathogen taxa (species or infraspecific rank) per region based on a range of published sources (see references listed in Fig. 6.2)[1]. Colours represent either the minimum or maximum or an interval and grey colour indicates that no data were available for a region

species, unclear biogeographic origin, and strong regional differences in research and recording intensity still hamper the way forward. Even for better studied taxa simple baseline knowledge is often missing (Desprez-Loustau et al. 2007).

A more interdisciplinary approach, integrating fungal invasion biology with plant pathology and other relevant disciplines as well as better integration into international policies, can help to improve our global understanding of fungal biological invasions and minimise resulting economic costs and damage to natural ecosystems (Fisher et al. 2020; Paap et al. 2022; Roy et al. 2017; Santini et al. 2018).

6.2 *Oomycota*

6.2.1 General Introduction

So far, about 1800 species of oomycetes have been described (Beakes and Thines 2017; Wijayawardene et al. 2022). Amongst these, more than 1000 are plant pathogens (Thines 2014). Within the plant pathogenic lineages, various levels of host specificity have been observed. Generally, opportunistic pathogens, e.g. in the genera *Pythium* and *Globisporangium*, have broad host spectra (Middleton 1943), while the obligate biotrophic downy mildew and white blister pathogens have rather narrow host spectra (Ploch et al. 2010; Thines and Choi 2016). However, there are some exceptions, e.g. *Albugo candida*, which parasitises a wide range of *Brassicales* (Choi et al. 2009), and *Pseudoperonospora cubensis*, which parasitises several dozen species of the *Cucurbitaceae* (Runge et al. 2011). In these cases, it is likely that a genetic innovation has led to a host jump and subsequent radiation (Thines 2019), which has not yet resulted in speciation. Like in other groups, the process of host jumping, radiation, and subsequent speciation has driven the diversity of at least biotrophic oomycetes (Choi and Thines

2015; Thines 2019), as also evidenced by consecutive host jumping, radiation, and speciation found in the genus *Peronospora* (Ploch et al. 2022).

Several devastating plant pathogens in the oomycetes are found amongst the pythiacaeous genera, e.g. *Globisporangium ultimum*, which is a widespread cause of damping off of seedlings (Lévésque et al. 2010; Molin et al. 2021; Pánek and Střížková 2021). Also, various members of the hemibiotrophic genus *Phytophthora* are notable pathogens, e.g. *Phytophthora infestans*, the causal agent of potato late blight (Fry 2008), and *Ph. capsici*, which infects a wide range of vegetables (Hausbeck and Lamour 2004). However, due to their seed-borne nature and the ability to cause latent systemic infections, downy mildew pathogens are equally threatening agriculture, as they infect various important crops, such as maize (Shaw 1978; Suharjo et al. 2020), soybean (Pathak et al. 1978), grape (Gessler et al. 2011), lettuce (Lebeda et al. 2008), spinach (Feng et al. 2018), and sunflower (Viranyi et al. 2015), as well as culinary herbs, e.g. basil, which is parasitised by *Peronospora belbahrii* (Thines et al. 2009), and ornamentals, e.g. balsamines, which are parasitised by *Plasmopara destructor* and *Pl. velutina* (Görg et al. 2017).

6.2.2 Centres of Diversity for Different Groups

For the opportunistic and hemibiotrophic species of oomycetes, centres of diversity cannot be defined at present, as their diversity is poorly explored and many species seem to have a cosmopolitan distribution (Diéguez-Uribeondo et al. 2007; Tojo et al. 2012). Also, for the white blister pathogens an assessment on their origin and centre of diversity is difficult, even though it is noteworthy that most species of the genus *Albugo* have been described from temperate to subtropical regions of Eurasia (Choi and Priest 1995; Choi et al. 2011). Downy mildews seem to have originated in the Paleotropics, probably as descendants of obligate biotrophic grass parasites (Thines 2009). For the major downy mildew genera centres of diversity can be defined. The genus *Bremia* likely originated in Eurasia and is especially species-rich in warm-temperate climates (Choi and Thines 2015). *Hyaloperonospora* affects various *Brassicaceae* and seems to have its highest diversity in temperate Europe, even though the West Asian mountainous regions, which are rich in potential hosts (Franzke et al. 2011), are underexplored in terms of downy mildew diversity. *Peronosclerospora* is a neglected genus affecting C4 grasses, with several species infecting maize (Kenneth 1981; Crouch et al. 2022), and which leads to severe crop losses in several countries of East Asia (Suharjo et al. 2020). The majority of its species have been reported from Australasia (Crouch et al. 2022), but some species have been introduced into other regions along with their hosts, e.g. *Pscl. sorghi* (Prom et al. 2015). *Peronospora* is the most speciose genus of oomycetes, with more than 400 described species. It is common throughout the Holarctic, and its highest diversity seems to be in temperate and warm-temperate Eurasia, from where also most species were described (Gäumann 1923; Gustavsson 1959; Constantinescu 1991). The genus *Plasmopara* is widespread throughout North America and Eurasia, but the highest diversity of affected hosts is in North America (Thines, unpublished). Given the high degree of host specialisation that has been observed in the genus (Görg et al. 2017; Salgado-Salazar and Thines 2022), it seems likely that North America is the centre of diversity for this genus.

6.2.3 Anthropogenic Spread

Anthropogenic spread of oomycete pathogens is with infested soil, seeds, and plants. Due to the increase in global trade, new downy mildew diseases have been reported at a high frequency throughout the past two decades, some of which are causing notable diseases, such as the downy mildews of basil (Belbahri et al. 2005), poppy (Voglmayr et al. 2014), balsamines (Görg et al. 2017), and columbines (Denton et al. 2015; Thines et al. 2019, 2020). Especially the trade with infected seeds is an important source of pathogens, as seed infection has been shown in various downy mildew pathogens (Inaba et al. 1983; Adenle and Cardwell 2000; Ioos et al. 2007) and is difficult to detect if infections are not severe. Thus, utmost care should be taken to avoid import of seeds from areas in which downy mildew has been reported on a specific host plant (Thines and Choi 2016).

Infested soil and plant parts are the most important cause of anthropogenic spread of hemibiotrophic and opportunistic pathogens, and can have potentially devastating effects, such as in case of *Ph. infestans*, which triggered the Great Irish Famine in the mid-nineteenth century after having been introduced from southern North America (Yoshida et al. 2013; Fry et al. 2015; Leesutthiphonchai et al. 2018), or severe ecological effects, such as in case of *Ph. ramorum* and *Ph. cinnamomi*, which led to sudden oak death in the western USA (Rizzo et al. 2002) and to Jarrah dieback in Australia (Hardham and Blackman 2018), respectively.

6.2.4 Most Notable Host Jumps

As mentioned earlier, host jumps are a driving force in the diversification of oomycetes. However, there are a few host jumps that are notable, as they are across orders or even classes. Probably the widest host jump has happened in the genus *Peronospora*—here a host jump from dicots to monocots took place and led to the evolution of *Pe. destructor*, an economically important pathogen of onions (Kofoet et al. 1990). However, this host jump to *Alliaceae* so far has not led to a wide radiation of the pathogen, with only rather few species being affected. Contrary, the host jump from hops to *Cucurbitaceae*, leading to the two lineages of *Ps. cubensis*, has proven to be rather successful, with a wide range of hosts reported to be colonised by the pathogen under natural and laboratory conditions (Thomas et al. 1987; Colucci 2008; Runge and Thines 2011). Two further noteworthy host jumps have occurred in the white blister pathogens—one from dicots to *Piperaceae* (Ploch et al. 2018), and one from *Brassicaceae* to *Fabaceae* (Choi et al. 2012). In both cases, these host jumps have not led to new radiations, and it can be assumed that most host shifts will not be evolutionarily successful, due to the difficulties in sustaining infections in the initial phase of suboptimal interaction (Thines 2019). Along these lines, the host shift of *Ps. cubensis* from *Cucurbitaceae* to *Balsaminaceae*, observed in Africa (Voglmayr et al. 2009), is remarkable, and it seems to be warranted to monitor if similar host shifts occur in the future, as they might herald a new epidemic, if the pathogen became able to sustain infections on cultivated balsamines.

6.2.5 Most Notable Range Expansions

Apart from a few hemibiotrophic species, such as *Ph. infestans,* which was already mentioned above, the most notable range expansions occurred in downy mildew pathogens. Historic examples for this are *Pl. viticola*, which was restricted to North America and is now present in almost all regions in which grapes are cultivated (Gobbin et al. 2006; Rouxel et al. 2014), and *Pl. halstedii*, a pathogen of cultivated sunflower, which, after its introduction into Europe, has spread globally to all sunflower-producing areas (Gulya et al. 1991; Spring 2019), except for Australia (Constantinescu and Thines 2010). In the past two decades, most range expansions were due to an insufficient knowledge regarding the diversity of downy mildews, associated with the erroneous assumption that species of downy mildews could often infect a whole host family (Yerkes and Shaw 1959). This led to an insufficient action against *Pe. belbahrii*, a disease that was first attributed to the widespread species *Pe. lamii* or referred to as *Peronospora* sp. (Lefort et al. 2003; Coosemans 2004; Garibaldi et al. 2004; Thines et al. 2009). Likewise, *Pe. somniferi* did not receive adequate attention, as it was initially thought to be the same species as *Pe. arborescens* on weedy poppy (Landa et al. 2005; Voglmayr et al. 2014). Also *Pl. destructor* was not opposed with strong phytosanitary measures, as it was first thought to be conspecific with a pathogen of wild impatiens, *Pl. obducens* (Lane et al. 2005; Cunnington et al. 2008a; Görg et al. 2017). All of these pathogens are now present in most regions in which their hosts are commercially grown, and it seems unlikely that it will be possible to fully defeat them.

6.2.6 Future Directions

As no general testing methods are in use and resistance to fungicides is common (Ziogas et al. 2006; Bi et al. 2014), the spread of pathogens can often only be delayed, such as in the case of *Pe. aquilegiicola* (Denton et al. 2015; Thines et al. 2019), which was first restricted to England and Wales after its import from East Asia, but now has crossed the channel to the European mainland, where several cases have been reported in Germany (Thines et al. 2020). Thus, it will be of great importance to enhance testing for and monitoring of new oomycete diseases. To achieve this, it seems necessary that the great knowledge gaps regarding the diversity and ecology of oomycetes are being filled. It is estimated that less than 10% of oomycete species are presently known (Beakes and Thines 2017), and for most pathogens, natural and potential host ranges are poorly known. In addition, species occurring in the tropics are notoriously understudied, which is exemplified by graminicolous downy mildews, which pose a substantial threat to maize farming, yet major lineages affecting maize are insufficiently known (Suharjo et al. 2020). Thus, it will be important in the coming years to gain a better understanding of oomycete diversity on crops and related species to enable an informed risk assessment and countermeasures against future pathogen spread.

6.3 *Pucciniales*

6.3.1 General Introduction

Pucciniales, known as the rust fungi, comprise the largest group of plant pathogens. Close to 8000 species have been described in ca. 180 currently accepted genera (Aime and McTaggart 2020). As a group the rust fungi have evolved a suite of characters that are otherwise unique or rare in fungi. All rust fungi are obligate biotrophs. Species undergo alternation of generations with separate gametothalli and sporothalli that are usually, though not exclusively, produced on different host plant species. Additionally, rust species produce up to five different spore stages that are morphologically and functionally distinct, and, as a result, the rusts are believed to have the most complex life cycles in fungi. Some of the largest genomes known in fungi are of rusts; for example, *Gymnosporangium confusum* has a reported genome size of 893 Mb, and *Austropuccinia psidii* has a reported haploid genome size of over 1 Gb, which is one of the largest assembled fungal genomes to date (Tavares et al. 2014; Tobias et al. 2021).

Rust fungal species infect a wide range of flowering plants that include most important agricultural crops such as wheat, coffee, and soybean. Wheat stem rust, for example, caused by the fungus *Puccinia graminis*, was known as a scourge of crops to early Greeks and Romans, the latter developing an annual ritual, known as the Robigalia, in attempts to placate the gods and reduce wheat crop losses. *Puccinia graminis*, which alternates between wheat and barberry (*Berberis* spp.), is also an important organism for advancing the disciplines of plant pathology and agriculture. The first agricultural bans, against barberry, were instigated to mitigate wheat stem rust outbreaks, and the modern green revolution was sparked by the effort to breed durable resistance against *P. graminis* into wheat.

A rust fungus that alternates between two host species to complete its life cycle is known as heteroecious; one that can complete its life cycle on a single host is termed autoecious. Rust species that cycle through all five spore stages are termed macrocyclic; those that eliminate one or more spore stages can be demicyclic, microcyclic, or endocyclic. The different, unrelated, host species necessary for completion of the life cycle of a heteroecious rust are termed alternate hosts, for example, barberry and wheat are alternate hosts of *P. graminis*. In contrast, alternative hosts are related species that can be utilised by a rust species by either the gametothallus or sporothallus. For examples, Asian soybean rust, caused by *Phakopsora pachyrhizi*, can utilise other alternative leguminous host species, in addition to soybean, for development of the sporothallus. While, in terms of species richness, *Pucciniales* can be considered a very successful lineage, the complexities of the life cycle, and incomplete data on geographic and host species ranges for most species have hindered our ability to understand the drivers of this diversity. While forces such as co-speciation (or biological specialisation) and host jumping (or biogenic radiation) have been variously used to explain rust species diversification, modern analyses are highlighting how the heteroecious nature of *Pucciniales* may have made it possible for both forces to act, but on different stages of the life cycle, leading to the tremendous extant diversity in this group (Aime et al. 2018).

6.3.2 Centres of Diversity for Different Groups

The centre of diversity for most rust fungal lineages has not been explored in depth. The rust fungi likely evolved during the middle Jurassic, with two major diversification events—that of the *Raveneliineae*, which occurred during the early Cretaceous, and that of the *Uredinineae*—the lineage to which the majority of extant rust species belong—which occurred during the middle Cretaceous (Aime et al. 2018; Aime and McTaggart 2020). The earliest extant rusts produce gametothalli that are usually associated with Gymnosperms in temperate climates and, given that most tropical rust species appear to be derived (Aime 2006), a temperate origin of *Pucciniales* appears most likely. Some rust genera are known to have radiated in regions where their host genera have also undergone radiations, especially for the autoecious rusts, for example, *Endoraecium* species on *Acacia* hosts in Australia, and *Dasyspora* species on *Xylopia* hosts in the neotropics (McTaggart et al. 2015; Beenken et al. 2012).

6.3.3 Anthropogenic Spread

The common name, "rust" fungi, comes from the typical orange-brown colour of urediniospores—one of the five spore stages produced by *Pucciniales* and the stage responsible for most epidemic diseases—that *en masse* impart a rusty colour to infected hosts. Urediniospore production can be so dense in cropping systems that a person walking through an infected area can emerge with a powdery coating of spores on clothing, skin, and machinery. Because urediniospores may remain viable for long periods of time—more than 300 days in some systems (Barua et al. 2018)—the potential for human-mediated transmission is immense. Additionally, as other spore stages can be more cryptic, rust fungi are frequently and unwittingly transported in infected plant material (Helfer 2013). For example, the origins of the coffee rust pathogen, *Hemileia vastatrix*, are unknown, but likely North African (Ramírez-Camejo et al. 2022a). Humans are believed to have mediated the global spread of this rust, and detailed epidemiological and genotypic analyses of its most recent incursion into Hawaii implicate human-mediated, and not natural, spread (Ramírez-Camejo et al. 2022b). Other anthropogenic factors, such as climate change and host range expansions, whether intentional or unintentional, also impact the spread of rust fungal disease (Helfer 2013).

6.3.4 Most Notable Host Jumps

Host jumps, especially for the sporothallus, are believed to be one of the driving forces in rust diversification, similar to other pathogen groups (Thines 2019), yet a lack of data on natural host ranges for most rust species have limited empirical testing of this hypothesis. Nonetheless, a few instances of host jumps have been documented in the rust fungi, and while these have not been linked, yet, to species diversification, they can be devastatingly consequential to ecosystems. In modern times, host jumps are most frequently mediated by anthropogenic factors. One well-documented case is that of *Cronartium ribicola*—a rust fungus that alternates between species of *Ribes* (sporothallus) and pines (gametothallus). *Cronartium ribicola* is a native of China that was accidentally introduced into the USA on nursery seedlings and has since jumped to several native North American pine species with devastating results to both natural and managed forest ecosystems (Kinloch 2001). *Austropuccinia psidii*, a native of *Myrtaceae* species in South America, jumped to introduced *Eucalyptus* species on that continent (Coutinho et al. 1998), thereby facilitating one of the most important recent range expansions of a rust fungus.

6.3.5 Most Notable Range Expansions

Austropuccinia psidii is an autoecious rust fungi, first reported from guava (*Myrtaceae*) in Brazil (Winter 1884), which over the last two decades has undergone a tremendous host and range expansion, infecting more than 400 species of *Myrtaceae* across the globe. Although the first reported host jumps, to introduced *Eucalyptus* plantations in Brazil, were recorded in the 1940s, the fungus did not begin its expansive range and host expansions until much more recently (Almeida et al. 2021). *Austropuccinia psidii* is now considered a threat to natural ecosystems in the South Pacific and Australia where it may well cause the extinction of numerous native species (Carnegie et al. 2016; McTaggart et al. 2016a). The reason for this massive host and range expansion is unknown, although evolution of a new race has been implicated (Granados et al. 2017), as has the increased transposable element content—estimated at 91% of the genome—which may favour rapid adaptation to new hosts (Tobias et al. 2021).

6.3.6 Future Directions

As is true for so many groups of fungi, much more data are needed to fully elucidate the causes and consequences of host and range expansions in *Pucciniales*. In contrast to most other fungal groups, the heteroecism of most rust species further complicates both the ability to work out entire life cycles for species and the ability to meaningfully analyse host/rust interaction data. Nonetheless, some general patterns are beginning to emerge. Heteroecious rust species tend to conserve gametothallus hosts—the site of fertilisation—and patterns consistent with co-evolution or co-speciation are evident at this stage; however, evidence for co-speciation is much less evident between heteroecious rusts and their sporothallus hosts, and host jumping may be the driving force behind speciation at this stage (Aime et al. 2018). However, diversification in autoecious rusts may also be driven by co-speciation (e.g. McTaggart et al. 2015). Additionally, rust speciation can be driven by localised host extinctions that can drive the rapid speciation of alternate life cycles, such as has been shown in some microcylic rusts (Scholler et al. 2019).

6.4 *Erysiphaceae*

6.4.1 General Introduction

Powdery mildews (*Erysiphaceae*, *Helotiales*, *Ascomycota*) are conspicuous plant pathogens that infect altogether more than 10,000 dicot and monocot species globally (Braun and Cook 2012). Non-flowering plants are not infected by powdery mildews. The powdery mildews are obligate biotrophs as they take up nutrients exclusively from the living cells of their host plants through specialised intracellular structures known as haustoria (Hückelhoven and Panstruga 2011; Zheng et al. 2013). Consequently, powdery mildews cannot grow and reproduce without being structurally and functionally connected to living host plant tissues. There are over 900 powdery mildew species in total, included in 19 genera (Kiss et al. 2020), which belong to a large monophyletic lineage within the order *Helotiales* based on recent genome-scale phylogenetic analyses (Johnston et al. 2019; Vaghefi et al. 2022). These results make the order *Erysiphales* redundant; currently, the family *Erysiphaceae* is the highest taxonomic group that only includes powdery mildews (Vaghefi et al. 2022).

Powdery mildew is also the name of the plant disease caused by these obligate biotrophic fungi. Symptoms of powdery mildew are easy to recognise and appear as whitish patches localised exclusively on aerial plant surfaces, mainly on leaves and stems, and sometimes also on flowers and young fruits, wherever epiphytic mycelium develops. In addition, powdery mildews may also cause chlorotic spots, leaf distortions, yellowing, and premature fall or death of the infected plant organs, and may also kill entire plants, especially at the seedling stage. Some powdery mildew species cause economically important diseases by reducing the yield and quality of agricultural and horticultural crops, including wheat, barley, grapevine, and other fruit, and vegetable species (Fuller et al. 2014; Kelly et al. 2021; Kusch et al. 2020; Zhang et al. 2022). Others are forest pathogens (Marçais and Desprez-Loustau 2014; Demeter et al. 2021). Some are known as invasive plant pathogens in different parts of the world (Kiss 2005; Desprez-Loustau et al. 2010, 2018; Kiss et al. 2020; Choi et al. 2022). *Blumeria* spp. on cereals and *Erysiphe necator* on grapevine have become model species of powdery mildews in plant pathology research (Gadoury et al. 2012; Bindschedler et al. 2016). Another species, *Podosphaera plantaginis* that infects a weedy host, *Plantago lanceolata*, has long been in focus in the study of wild plant pathosystems (e.g. Halliday et al. 2020).

6.4.2 Centres of Diversity of Tree-Pathogenic and Herb-Pathogenic Powdery Mildews

There are only speculations on the centre(s) of diversity of the *Erysiphaceae*. These were based on the estimated age and evolutionary radiation of the angiosperms as powdery mildews are unable to infect other plants, and molecular clock calculations that were built on the 18S and the 28S regions of the nuclear ribosomal DNA (nrDNA) of powdery mildews. Altogether, these data indicated that the first powdery mildews appeared approximately 100 million years ago (Berbee and Taylor 2001; Takamatsu and Matsuda 2004; Takamatsu 2013a), about 30–60 million years after the first angiosperms had emerged (Soltis et al. 2008; Coiro et al. 2019). Takamatsu (2013a, b) hypothesised that during the evolution of the *Erysiphaceae* the most ancestral powdery mildews may have emerged as obligate biotrophic pathogens of woody hosts, especially broad-leaved deciduous trees, while species infecting herbaceous hosts appeared later. This hypothesis has been supported by all phylogenetic analyses to date (Frantzeskakis et al. 2019; Bradshaw and Tobin 2020; Kiss et al. 2020; Kusch et al. 2022; Vaghefi et al. 2022) and helps decipher the possible centres of diversity of powdery mildew lineages.

Phylogenies and morphological analyses of powdery mildews infecting *Nothofagus* in South America (Niinomi et al. 2008) and another study on the co-evolution of *Golovinomyces* spp. and their asteraceous hosts (Takamatsu et al. 2006) have both indicated that powdery mildews first emerged in the Northern Hemisphere. The distribution of those angiosperm families that are known as hosts of the *Erysiphaceae* has also revealed that powdery mildews are more common in the Northern Hemisphere and occur less frequently on tropical hosts and plant families that are confined to the Southern Hemisphere (Takamatsu 2013a). When the number of powdery mildew species that infect herbaceous plants in Europe, North America, Central and West Asia, and East Asia was compared to those on woody hosts, i.e. tree and shrub species, based on the data compiled by Amano (1986), it was found that the highest number of powdery mildew species on trees and shrubs is native to East Asia. This indicated that East Asia could be the centre of diversity of the *Erysiphaceae* on woody hosts. Those species are considered ancestral; thus, East Asia may also be the centre of origin of all powdery mildews (Takamatsu 2013a, b). However, the number of *Golovinomyces* spp. pathogenic on herbaceous plants is much higher in Europe than Asia, which may indicate that *Golovinomyces* has diversified in Europe rather than East Asia (Takamatsu 2013a).

6.4.3 Anthropogenic Spread

The introduction of *Erysiphe necator*, the causal agent of grapevine powdery mildew, from North America to Europe in the nineteenth century, has been widely presented as a textbook example of human-mediated dispersal of a crop disease through intercontinental movement of living plant materials, such as cuttings and rootstocks (Weltzien 1978; Brewer and Milgroom 2010; Gadoury et al. 2012). Grape powdery mildew was also introduced to Australia, New Zealand, South Africa, South America, and other grapevine-growing regions of the world (Csikós et al. 2020), possibly from North America, the presumed centre of origin of *E. necator* based on comprehensive genetic analyses (Brewer and Milgroom 2010; Gadoury et al. 2012; Frenkel et al. 2012). A recent study on wild and traditional vines in the Middle East has also revealed the genetic fingerprints of the North American populations of *E. necator* there and, in addition, has identified a distinct population that may have been introduced to Israel from Asia (Gur et al. 2021). Although the centre of origin of *E. necator* is still debated (Brewer and Milgroom 2010; Gadoury et al. 2012; Gur et al. 2021), there is little doubt that its recent introduction to different grapevine-growing regions around the globe and the economic damage that it has caused in the new areas (Scholefield and Morison 2010; Fuller et al. 2014) are an unwanted outcome of global horticultural trade.

From the early 1900s, the emergence and epidemic spread of *Podosphaera mors-uvae* on gooseberry across Europe may have also been the result of horticultural trade (Weltzien 1978). This new crop disease was named 'American gooseberry mildew' in Europe shortly after it has emerged because it had been known in the USA since the nineteenth century. There are no detailed genetic and genomic analyses of diverse *P. mors-uvae* populations; therefore, the origin of American gooseberry mildew in Europe is still unknown.

Surprisingly, the first comprehensive inventory of all powdery mildew taxa in Australia found that the total number of species is very low compared to other continents (Kiss et al. 2020). Moreover, all powdery mildew taxa identified in this Australian study were known overseas from the same hosts or other plant species. Only a few powdery mildew infections were observed on Australian native plants, and the causal agents were always polyphagous species that infect other plant species both overseas and in Australia. The conclusion of this study was that Australia is probably a continent without native powdery mildews, and most, if not all, species have been introduced inadvertently since 1788, the beginning of the European colonisation (Kiss et al. 2020). This hypothesis implies that the Australian land vegetation has been exposed recently to a massive human-mediated introduction and dispersal of an entire group of plant pathogens that have not co-evolved with the local flora. Sporadic powdery mildew infections of Australian native plants could be examples of rapid host range expansions of polyphagous species (see below). The hypothesis aligns with the molecular clock estimates of the first emergence of the *Erysiphaceae* (ca. 100 million years ago) in East Asia (Takamatsu 2013a). This was also about the time when the Australian and the Antarctic continents separated from India, which had all separated from the western half of Gondwanaland about 180 million years ago (Williams et al. 2013). The Australian terrestrial flora has continued to evolve in isolation, without exposure to plant pathogenic fungi from overseas until the arrival of the first European settlers in 1788 (Kiss et al. 2020).

One of the best documented examples of anthropogenic dispersal of a powdery mildew species is a study on the global spread of wheat powdery mildew caused by *Blumeria graminis* (Sotiropoulos et al. 2022). An analysis of 172 *B. graminis* genomes indicated that this fungus emerged in the Fertile Crescent during wheat domestication. After its spread throughout Eurasian wheat fields, the European colonisation had brought it to North America, where hybridisation events may have occurred with unknown *Blumeria* lineages on wild grasses. US populations of wheat powdery mildew were then introduced to Japan, and European ones to China. Apparently, in both China and Japan, the introduced lineages hybridised with local populations. This comprehensive study has documented how human activities have driven the global spread and evolution of an important crop pathogen and has also highlighted rapid evolution events through hybridisation of diverse *Blumeria* lineages (Sotiropoulos et al. 2022).

6.4.4 Most Notable Host Jumps

In the 1960s, triticale (×*Triticosecale*), an artificial intergeneric hybrid of wheat and rye, was released for commercial use in Europe and elsewhere (Walker et al. 2011; Troch et al. 2012). By the late 1990s, triticale had become a widely grown cereal in Europe due to its favourable agronomic traits that included complete resistance to powdery mildew. However, in 2001, powdery mildew was observed in triticale fields across Europe, where it has since become a major limiting factor of production. Initial analyses of a limited number of DNA markers concluded that wheat-pathogenic *B. graminis* strains have acquired virulence factors that enabled those lineages to cause disease on triticale. This process was interpreted as a host range expansion of some *B. graminis* lineages infecting wheat (Walker et al. 2011; Troch et al. 2012). However, comparative analyses of the genomes of 46 powdery mildew isolates indicated that the strains that infect triticale were, in fact, hybrids between wheat-pathogenic and rye-pathogenic *Blumeria* strains, infecting triticale, an artificial hybrid created by plant breeders (Menardo et al. 2016). The role of hybridisation in the rapid evolution of *Blumeria graminis* lineages has also been documented in a global study of wheat powdery mildew isolates (Sotiropoulos et al. 2022) mentioned above.

Podosphaera xanthii is another common powdery mildew fungus mostly known from cucurbits (Pérez-García et al. 2009) and fabaceous hosts (Kelly et al. 2021). It has been reported as a pathogen of many diverse, only distantly related plants (Hirata et al. 2000), which indicates that *P. xanthii* is a polyphagous species. In Australia, where *P. xanthii* was unintentionally introduced during European colonisation, several host jumps onto native Australian plants were reported (Kiss et al. 2020). *Podosphaera xanthii* was found on *Cephalotus follicularis*, an insectivorous plant endemic to Western Australia (Cunnington et al. 2008b); rainforest spinach (*Elatostema reticulatum*) that often grows along rainforest creeks in Queensland and New South Wales (Kiss and Vaghefi 2021); and *Trema tomentosa*, a native rainforest tree species (Kiss et al. 2020). In other parts of the world, *P. xanthii* caused unusual and unique infections on diverse other plants, such as a serious disease of jellyfish tree (*Medusagyne oppositifolia*) imported from the Seychelles to the UK (Pettitt et al. 2010), and an infection of water poppy (*Hydrocleys nymphoides*), a monocot grown in an urban area in Korea (Cho et al. 2018).

6.4.5 Most Notable Geographic Range Expansions

In the late 1980s, an apparently new powdery mildew disease appeared in tomato production in Europe. It was first reported from the

Netherlands (Noordeloos and Loerakker 1989), then shortly after from almost every other European country (Kiss 1996; Whipps et al. 1998; Jones et al. 2001) as well as Canada and the USA (Kiss et al. 2005). The pathogen that has spread so rapidly across Europe and North America was recognised as a new species, *Erysiphe neolycopersici* (Kiss et al. 2001; Hsiao et al. 2022).

Many other powdery mildews are known to rapidly expand their geographic ranges if their host plants are available in the new environments and the climatic conditions are conducive. Most crop pathogenic powdery mildews, such as *Blumeria graminis* on cereals (Sotiropoulos et al. 2022), *E. necator* on grapes (Gadoury et al. 2012), and *P. xanthii* on cucurbits (Pérez-García et al. 2009), have quickly spread around the world causing serious epidemics in the newly colonised regions. Accidental introductions of North American powdery mildews to Europe, and their quick spread in new regions, have been well documented. For example, *E. flexuosa* infecting horse chestnut (*Aesculus* spp.) and *E. elevata* infecting Indian bean (*Catalpa bignonioides*) are both considered as North American species. The two powdery mildews have been established in Europe since the early 2000s and spread rapidly from one European country to another causing serious epidemics on their hosts planted as ornamentals in urban areas (Kiss 2005).

It is often difficult to explain how obligate biotrophic plant pathogens reach new and/or remote geographic regions, particularly if their hosts are weeds or other little studied native plants (Fontaine et al. 2013; Kiss et al. 2018). Powdery mildews are not seed-borne; their airborne conidia are short-lived; and their long-distance dispersal by natural means is debated (Glawe 2008). Some powdery mildews regularly produce sexual morphs (chasmothecia), and their ascospores enclosed in chasmothecia may remain viable for a few years (Jankovics et al. 2015; Vági et al. 2016). The chasmothecia of *Blumeria* spp. may have been spread between continents by humans, e.g., in straw bales transported overseas during military or other missions (especially if horses were taken from one continent to another) provided that at the destination there were young host plants available for infection by ascospores released from chasmothecia.

Such potential long-distance dispersal and mostly human-mediated spread is not possible for many powdery mildew species because most of them do not produce chasmothecia (and ascospores). Most likely, powdery mildews reach new geographic areas as living mycelia on their living host plants that are transported to new destinations.

6.4.6 Future Directions

It may be surprising that amongst all the known plant pathogenic fungi, some powdery mildews are amongst the most rapidly spreading species. By 2050, many powdery mildews are predicted to spread to all those countries where their hosts are currently present (Bebber et al. 2014, 2019; Bebber and Gurr 2015).

In addition to their highly successful dispersal, powdery mildews are also capable of rapid evolution (Thines 2019; Frantzeskakis et al. 2020). Host range expansions are part of their lifestyle (Vági et al. 2007). Most probably, their unique, large genomes drive their rapid and versatile evolutionary processes (Wicker et al. 2013; Frantzeskakis et al. 2018; Müller et al. 2019; Sotiropoulos et al. 2022).

The Australian continent offers a special environment to study the rapid evolutionary processes of powdery mildews in the field. Apparently, all species of the *Erysiphaceae* that are currently found in Australia have reached this continent inadvertently, not more than 200–230 years ago (Kiss et al. 2020). Since then, microevolutionary processes, such as radiation through host jumps, may have started in some populations. Experimental and genome-level studies are needed to decipher the mechanisms of host range expansions in this highly successful group of plant pathogens.

6.5 *Ustilaginomycotina*

6.5.1 General Introduction

The *Ustilaginomycotina* comprises four classes (*Exobasidiomycetes*, *Malasseziomycetes*, *Moniliellomycetes*, and *Ustilaginomycetes*) (Begerow et al. 2018; James et al. 2020). The *Exobasidiomycetes* and *Ustilaginomycetes* include mostly plant pathogens (smut fungi), which are characterised by a saprobic, monokaryotic yeast or yeast-like stage and a plant pathogenic, dikaryotic filamentous stage (Begerow et al. 2014; Spatafora et al. 2017). *Fereydounia khargensis* is an unusual ustilaginomycetous yeast as it is an opportunistic pathogen of humans (Nasr et al. 2014).

The *Malasseziomycetes* and *Moniliellomycetes* are asexual fungi known only as yeasts or yeast-like fungi (Wang et al. 2014), with some, especially *Malassezia* spp., causing diseases in humans and animals (Das et al. 2021). Of evolutionary interest are the four known species of *Cintractiella* (*Cintractiellales*), which share a most recent common ancestor with the *Malasseziomycetes*, and do not share a most recent common ancestor with other orders of smut fungi (McTaggart et al. 2020). *Cintractiella* spp. are unusual smut fungi that infect some tropical sedges (*Cyperaceae*, subfamily *Mapaniodeae*) and produce black spore masses in adventitious spikelets that form on infected leaves (McTaggart et al. 2020).

Prior to the application of molecular phylogenies to the classification of fungi, the ustilaginomycetous yeasts were classified in a largely artificial taxonomic system that was often incompatible with that of the smut fungi (Wang et al. 2015). Traditionally, ustilaginomycetous yeasts were classified by physiological traits in culture, and smut fungi were classified by morphology and host. Now many ustilaginomycetous yeast species have been transferred to corresponding genera of smut fungi based on strongly supported molecular phylogenetic affinities (Wang et al. 2015; Li et al. 2021). The smut fungi take their name from the often conspicuous and copious amounts of dikaryotic teliospores produced in fungal structures called sori (sing. sorus) that are produced in plant tissues. The sori of grass-infecting smut fungi are often located in the ovaries, flowers, or the entire inflorescences of grasses (McTaggart et al. 2012a, c; Vánky 2012), which includes the domesticated cereal crops. The smut fungi are classified in about 50 genera (Vánky 2013) with about 1650 species (Vánky 2012).

6.5.2 Centres of Diversity for Different Groups

The most current and widely accepted model of plant pathogen evolution proposes that diversity is driven by host jumps, followed by radiation, specialisation and speciation, which occurs rapidly, within centuries and millennia rather than over millions of years (Thines 2019). Species of smut fungi are highly host specific with most restricted to a single host species, and very few of the remainder having more than five host species (Begerow et al. 2004; Vánky 2012). Divergence time estimates indicate that *Ustilaginomycotina* originated about 450 MYA before the age of angiosperms and coincided with the major expansion of gymnosperms (Riess et al. 2016). Consequently, Riess et al. (2016) proposed that early smut fungi must have been saprobic organisms as hosts plants did not exist at that time.

Most smut fungi parasitise grasses (*Poaceae*) and other monocots (Bauer et al. 2001; Vánky 2012). To find insights into the possible origins (and centres of diversity) of smut fungi, one must look to the distribution and nature of grasslands worldwide. Grasslands are one of the largest biomes on earth and dominate the landscape, possibly covering up to 70% of the world's land surface (Gibson 2009; Conant 2010; Chapin et al. 2013). Unfortunately, many of the world's grasslands have been degraded by farming, overgrazing, fires, invasive species, and climate change. Further, smut fungi have been under-collected and their diversity is likely to be far greater than currently known in comparison to the diversity of grasses that have about 12,000 species in 780 genera (Christenhusz and Byng 2016).

Many of the smut fungi on dicots belong to the family *Melanotaeniaceae* (*Ustilaginomycetes*) (Begerow et al. 2018). There is still much to discover about the evolution of members in the *Melanotaeniaceae*. For example, *Yelsemia* contains four species, which each occur on hosts in different plant families, specifically *Y. arthropodii* on *Arthropodium minus* (*Asparagaceae*), *Y. droserae* on *Drosera burmannii* (*Droseraceae*), *Y. lowrieana* on *Byblis rorida* (*Byblidaceae*), and *Y. speculariae* on *Specularia perfoliata* (*Campanulaceae*) (Vánky 2012). It is curious that *Y. droserae* and *Y. lowrieana* can co-occur on carnivorous host plants in different families (Shivas et al. 2014).

6.5.3 Anthropogenic Spread

Smut fungi have followed cultivated grass hosts (including barley, bamboo, corn, couch turf grass, oats, millet, wheat) around the world with their spread often assisted by human activity that moves infected soil, plants, or seed. Biosecurity regulations have restricted or slowed the spread of some serious cereal smut fungi, e.g. *Tilletia indica* (the cause of karnal bunt in wheat) (Bishnoi et al. 2020). Many smut fungi are spread by wind dispersed spores that infect highly susceptible sentinel crops established in regions/countries previously free of disease. For example, *Sporisorium scitamineum* (sugarcane smut) was reported for the first time in a remote agricultural region in north-western Australia in 1998 (Croft and Braithwaite 2006) before its spread to the major sugarcane regions of eastern Australia as well as to Papua New Guinea in 2016 (Bhuiyan et al. 2021).

6.5.4 Most Notable Host Jumps

The diversification of plant pathogens, whether facultative or obligate biotrophic pathogens, is largely driven by host jumps followed by rapid radiation and speciation (Choi and Thines 2015; McTaggart et al. 2016b; Thines 2019). Two notable examples of smut fungi that have been shown by molecular phylogenetic studies to have made host jumps are (i) *Melanopsichium pennsylvanicum*, which causes gall smut of several species of *Persicaria* (*Polygonaceae*, which is eudicotyledonous), and represents a host jump from an unknown grass (*Poaceae*, which is monocotyledonous) (Sharma et al. 2014), and (ii) *Pattersoniomyces tillandsiae*, which produces sori in the flowers of some *Tillandisia* ssp. (*Bromeliaceae*), and represents a host jump from either a sedge (*Cyperaceae*) or grass, without further radiation on other bromeliad genera (Piątek et al. 2017). Explosive radiation of species following a rapid succession of host jumps is seen in *Entyloma* (*Exobasidiales*), which contains smut fungi that mostly parasitise the leaves and stems of species in the eudicot plant families, *Apiaceae* and *Asteraceae* (Vánky 2012; Begerow et al. 2014). Evidence of rapid radiation and speciation in smut fungi is seen in the phylogenetic analyses of the grass-infecting *Ustilago s.l.* that reveal monophyletic clades of closely related species restricted to individual host genera or closely related host genera, some of which have been reclassified in genera such as *Langdonia* (on *Aristida* spp.), *Stollia* (on tribe *Andropogoneae*), and *Triodiomyces* (on *Triodia*) (McTaggart et al. 2012b).

6.5.5 Most Notable Range Expansions

In general, epidemic plant diseases are often caused by asexual (clonal) stages of fungi that disperse and build inocula from successful genotypes (Drenth et al. 2019). However, the smut fungi are strict outbreeders that become pathogenic only after dikaryotisation (Bakkeren et al. 2008). Smut fungi require monocultures of susceptible host plants to cause epidemics. For example, *Quambalaria pitereka*, which causes leaf and shoot blight on *Corymbia* (*Myrtaceae*), had greatest impact when susceptible hosts were planted in large monocultures in eastern Australia (McTaggart et al. 2022). Australia is considered the centre of origin for *Quambalaria* spp., as all occur on tree species native to Australia (de Beer et al. 2006; Pegg et al. 2008). *Quambalaria* spp. have spread to many parts of the world where eucalypt plantations have been established (Bragança et al. 2016).

Species of *Restiosporium* (*Ustilaginomycetes*) infect host plants in the *Restionaceae* in Australia and New Zealand (Vánky 2013). *Restiosporium* spp. are not found in South Africa, where the *Restionaceae* are most diverse. It appears that *Restiosporium* has originated in Australia from a jump from an unknown host after the southern continents split from Gondwana about 65 MYA (Bremer 2002).

6.5.6 Future Directions

There are several genera of smut fungi with unknown host origins that may be explained by host jumps, for example, *Centrolepidosporium, Ceraceosorus, Cintractiella, Entyloma, Eriocaulago, Exobasidium, Graphiola, Melanopsichium, Pattersoniomyces, Pericladium, Restiosporium, Uleiella, Websdanea,* and *Yelsemia*. Elucidation of the evolutionary journey of these genera awaits further specimen collection and molecular phylogenetic analyses.

6.6 Conclusions

Facilitated by anthropogenic spread, plant pathogens have colonised a wide range of new hosts and regions. The speed of this process seems to be still accelerating, as global trade with plants still increases, and generally, there is no saturation of alien species colonising new territories in sight (Seebens et al. 2017). As range expansions by the invasion of new territories and host jumps can be regarded as a main driver of pathogen diversity (Thines 2019), it seems likely that the anthropogenic impact on the evolution of plant pathogens is high, with yet unpredictable outcomes and implications for global health and food security. The spread of pathogens and the colonisation of new hosts must be closely monitored, so that potential threats can be identified early, to enable timely countermeasures and to devise mitigation strategies. To halt the future spread of plant pathogens, the trade with seeds and plants should be done with utmost caution, especially between geographically divergent regions. In addition, the introduction of, e.g., new exotic ornamental plants should be carefully weighed against the risk of introducing pathogens that could have a major impact on crops or native ecosystems.

Acknowledgements MT is supported by the German Federal Ministry BLE and the React-EU program of the European Union. FE and AS appreciate funding by the Austrian Science Foundation FWF (project: Alien Fungi, P 34688-B). RGS would like to thank Dr. Alistair R. McTaggart (University of Queensland, Australia) for his insightful comments and suggestions.

References

Abasova L, Aghayeva D, Takamatsu S (2018) Notes on powdery mildews of the genus *Erysiphe* from Azerbaijan. Curr Res Environ Appl Mycol 8(1):30–53

Adenle VO, Cardwell KF (2000) Seed transmission of maize downy mildew (*Peronosclerospora sorghi*) in Nigeria. Plant Pathol 49(5):628–634

Aime MC (2006) Toward resolving family-level relationships in rust fungi (*Uredinales*). Mycoscience 47:112–122

Aime MC, McTaggart AG (2020) A higher-rank classification for rust fungi, with notes on genera. Fungal Syst Evol 7:21–47

Aime MC, Bell C, Wilson AW (2018) Deconstructing the evolutionary complexity of rust fungi (*Pucciniales*) and their hosts. Stud Mycol 89:143–152

Almeida RF, Machado PS, Damacena MB, Santos SA, Guimarães LMS, Klopfenstein NB, Alfenas AC (2021) A new, highly aggressive race of *Austropuccinia psidii* infects a widely planted myrtle rust-resistant, eucalypt genotype in Brazil. Forest Pathol 51:e12679

Amano K (1986) Host range and geographical distribution of the powdery mildew fungi. Japan Scientific Societies Press, Tokyo

Anderson PK, Cunningham AA, Patel NG, Morales FJ, Epstein PR, Daszak P (2004) Emerging infectious diseases of plants: pathogen pollution, climate change and agrotechnology drivers. Trends Ecol Evol 19(10): 535–544

Arechavaleta M, Zurita N, Marrero MC, Martín JL (eds) (2005) Lista preliminar de especies silvestres de Cabo Verde. Hongos, plantas y animales terrestres. Consejería de Medio Ambiente y Ordenación Territorial. Gobierno de Canarias, p 155

Arechavaleta M, Rodríguez S, Zurita N, García A (2009) Lista de especies silvestres de Canarias. Hongos, plantas y animales terrestres. Gobierno de Canarias, p 579

Aukema JE, McCullough DG, Von Holle B, Liebhold AM, Britton K, Frankel SJ (2010) Historical accumulation of nonindigenous forest pests in the continental United States. BioScience 60(11):886–897

Bakkeren G, Kämper J, Schirawski J (2008) Sex in smut fungi: structure, function and evolution of mating-type complexes. Fungal Genet Biol 45:S15–S21

Barua P, You MP, Bayliss KL, Lanoiselet V, Barbetti MJ (2018) Extended survival of *Puccinia graminis* f. sp. *tritici* urediniospores: implications for biosecurity and on-farm management. Plant Pathol 67:799–809

Barwell LJ, Perez-Sierra A, Henricot B, Harris A, Burgess TI, Hardy G, Scott P, Williams N, Cooke DEL, Green S, Chapman DS, Purse BV (2020) Evolutionary trait-based approaches for predicting future global impacts of plant pathogens in the genus *Phytophthora*. J Appl Ecol 00:1–13

Bauer R, Begerow D, Oberwinkler F, Piepenbring M, Berbee ML (2001) Ustilaginomycetes. In: McLaughlin DJ, McLaughlin EG, Lemke PA (eds) The Mycota, Volume 7, systematics and evolution. Springer, Heidelberg, pp 57–83

Beakes GW, Thines M (2017) *Hyphochytriomycota* and *Oomycota*. In: Archibald J, Simpson A, Slamovits C (eds) Handbook of the protists, 2nd edn. Cham, Springer, pp 435–505

Bebber DP, Gurr SJ (2015) Crop-destroying fungal and oomycete pathogens challenge food security. Fungal Genet Biol 74:62–64

Bebber DP, Holmes T, Gurr SJ (2014) The global spread of crop pests and pathogens. Glob Ecol Biogeogr 23: 1398–1407

Bebber DP, Field E, Gui H, Mortimer P, Holmes T, Gurr SJ (2019) Many unreported crop pests and pathogens are probably already present. Glob Chang Biol 25(8): 2703–2713

Beenken L, Zoller S, Berndt R (2012) Rust fungi on *Annonaceae* II: the genus *Dasyspora* Berk. & M.A. Curtis. Mycologia 104:659–681

Begerow D, Göker M, Lutz M, Stoll M (2004) On the evolution of smut fungi on their hosts. In: Frontiers in basidiomycote mycology. IHW-Verlag & Verlagsbuchhandlung, Eching, Germany, pp 81–98

Begerow D, Schäfer AM, Kellner R, Yurkov A, Kemler M, Oberwinkler F, Bauer R (2014) Ustilaginomycotina. In: McLaughlin DJ, Spatafora JW (eds) The Mycota, Volume 7A, systematics and evolution. Springer, Heidelberg, pp 295–329

Begerow D, McTaggart AR, Agerer R (2018) Syllabus of plant families - A, Engler's Syllabus der Pflanzenfamilien Part 1/3. (Frey W. ed). Schweizerbart'sche Verlagsbuchhandlung, Stuttgart, Germany

Belbahri L, Calmin G, Pawlowski J, Lefort F (2005) Phylogenetic analysis and real time PCR detection of a presumably undescribed *Peronospora* species on sweet basil and sage. Mycol Res 109(11):1276–1287

Berbee ML, Taylor JW (2001) Fungal molecular evolution: gene trees and geologic time. In: McLaughlin DJ, Mclaughlin EG, Lemke PA (eds) The Mycota VII. Systematics and evolution, part B. Springer, Berlin, pp 229–245

Bhuiyan SA, Magarey RC, McNeil MD, Aitken KS (2021) Sugarcane smut, caused by *Sporisorium scitamineum*, a major disease of sugarcane: a contemporary review. Phytopathology 111(11):1905–1917

Bi Y, Chen L, Cai M, Zhu S, Pang Z, Liu X (2014) Two non-target recessive genes confer resistance to the anti-oomycete microtubule inhibitor zoxamide in *Phytophthora capsici*. PLoS One 9(2):e89336

Bindschedler LV, Panstruga R, Spanu PD (2016) Mildew-omics: how global analyses aid the understanding of life and evolution of powdery mildews. Front Plant Sci 7:123

Bishnoi SK, He X, Phuke RM, Kashyap PL, Alakonya A, Chhokar V, Singh RP, Singh PK (2020) Karnal bunt: a re-emerging old foe of wheat. Front Plant Sci 1486

Bradshaw M, Tobin P (2020) Sequencing herbarium specimens of a common detrimental plant disease (powdery mildew). Phytopathology 110:1248–1254

Bradshaw M, Braun U, Meeboon J, Tobin P (2021) Phylogeny and taxonomy of powdery mildew caused by *Erysiphe* species on *Corylus* hosts. Mycologia 2021:1–17

Bragança H, Diogo ELF, Neves L, Valente C, Araujo C, Bonifácio L, Phillips AJL (2016) *Quambalaria eucalypti* a pathogen of *Eucalyptus globulus* newly reported in Portugal and in Europe. For Pathol 46(1): 67–75

Braun U, Cook RTA (2012) Taxonomic manual of the *Erysiphales* (powdery mildews). CBS-KNAW Fungal Biodiversity Centre, Utrecht

Bremer K (2002) Gondwanan evolution of the grass Alliance of families (*Poales*). Evolution 56(7):1374–1387

Brewer MT, Milgroom MG (2010) Phylogeography and population structure of the grape powdery mildew fungus, *Erysiphe necator*, from diverse Vitis species. BMC Ecol Evol 10:268

Bufford JL, Hulme PE, Sikes BA, Cooper JA, Johnston PR, Duncan RP (2016) Taxonomic similarity, more than contact opportunity, explains novel plant-pathogen associations between native and alien taxa. New Phytol 212(3):657–667

CAB International (2021) Invasive species compendium ISC. CAB International, Wallingford, UK

Carnegie AJ, Kathuria A, Pegg GS, Entwistle P, Nagel M, Giblin FR (2016) Impact of the invasive rust *Puccinia psidii* (myrtle rust) on native *Myrtaceae* in natural ecosystems in Australia. Biol Invasions 18:127–144

Chapin FS, Sala OE, Huber-Sannwald E (2013) Global biodiversity in a changing environment: scenarios for the 21st century. Springer, New York

Cho S, Han K, Choi I, Shin HD (2018) First report of powdery mildew caused by *Podosphaera xanthii* on *Hydrocleys nymphoides* in Korea. Plant Dis 102:247

Choi D, Priest MJ (1995) A key to the genus Albugo. Mycotaxon 53:261–272

Choi YJ, Thines M (2015) Host jumps and radiation, not co-divergence drives diversification of obligate pathogens. A case study in downy mildews and *Asteraceae*. PLoS One 10(7):e0133655

Choi YJ, Shin HD, Thines M (2009) The host range of *Albugo candida* extends from *Brassicaceae* through *Cleomaceae* to *Capparaceae*. Mycol Prog 8(4): 329–335

Choi YJ, Shin HD, Ploch S, Thines M (2011) Three new phylogenetic lineages are the closest relatives of the widespread species *Albugo candida*. Fungal Biol 115(7):598–607

Choi YJ, Thines M, Tek MP, Shin HD (2012) Morphological evidence supports the existence of multiple species in Pustula (Albuginaceae, Oomycota). Nova Hedwigia 94:181–192

Choi IY, Abasova L, Park JH, Shin HD (2022) First report of *Podosphaera mors-uvae* causing powdery mildew on *Ribes fasciculatum* var *chinense* in Korea. For Pathol 2022:e12746

Christenhusz MJM, Byng JW (2016) The number of known plants species in the world and its annual increase. Phytotaxa 261(3):201–217

Coiro M, Doyle JA, Hilton J (2019) How deep is the conflict between molecular and fossil evidence on the age of angiosperms? New Phytol 223(1):83–99

Colucci SJ (2008) Host range, fungicide resistance and management of *Pseudoperonospora cubensis*, causal agent of cucurbit downy mildew

Conant RT (2010) Challenges and opportunities for carbon sequestration in grassland systems: a technical report on grassland management and climate change mitigation. FAO

Constantinescu O (1991) An annotated list of *Peronospora* names. Thunbergia 15:1–110

Constantinescu O, Thines M (2010) *Plasmopara halstedii* is absent from Australia and New Zealand. Pol Bot J 55(2):293–298

Convention on Biological Diversity (2014) Pathways of introduction of invasive species, their prioritization and management. Convention on Biological Diversity, June 2014, p 18

Coosemans J (2004) First report of *Peronospora lamii*, downy mildew on basil (*Ocimum basilicum*) in Belgium. Parasitica 60(1–2):27–27

Coutinho TA, Wingfield JJ, Alfenas AC, Crous PW (1998) *Eucalyptus* rust: a disease with the potential for serious international implications. Plant Dis 82:819–825

Croft BJ, Braithwaite KS (2006) Management of an incursion of sugarcane smut in Australia. Australas Plant Pathol 35:113–122

Crosby AJ (2003) The Columbian exchange: biological and cultural consequences of 1492. Greenwood Publishing Group, Westport, CT

Crouch JA, Davis WJ, Shishkoff N, Castroagudín VL, Martin F, Michelmore R, Thines M (2022) *Peronosporaceae* species causing downy mildew diseases of *Poaceae*, including nomenclature revisions and diagnostic resources. Fungal Syst Evol 9(1):43–86

Csikós A, Németh MZ, Frenkel O, Kiss L, Váczy KZ (2020) A fresh look at grape powdery mildew (*Erysiphe necator*) A and B genotypes revealed frequent mixed infections and only B genotypes in flag shoot samples. Plan Theory 9:1156

Cunnington JH, Aldaoud R, Loh M, Washington WS, Irvine G (2008a) First record of *Plasmopara obducens* (downy mildew) on impatiens in Australia. Plant Pathol 57(2)

Cunnington JH, Jones RH, De Alwis SK (2008b) First record of powdery mildew on the *Cephalotaceae*. Aust Plant Dis Notes 3:51–52

Das SP, Ahmed SM, Naik B, Laha S, Bejai V (2021) The human fungal pathogen *Malassezia* and its role in cancer. Fungal Biol Rev 38:9–24

de Beer ZW, Begerow D, Bauer R, Pegg GS, Crous PW, Wingfield MJ (2006) Phylogeny of the *Quambalariaceae* fam. nov., including important *eucalyptus* pathogens in South Africa and Australia. Stud Mycol 55:289–298

Demeter L, Molnár ÁP, Öllerer K, Csóka G, Kiš A, Vadász C et al (2021) Rethinking the natural regeneration failure of pedunculate oak: the pathogen mildew hypothesis. Biol Conserv 253:108928

Denton GJ, Beal EJ, Kilty A, Denton JO, Choi YJ, Thines M (2015) Characterisation and risk assessment of the emerging *Peronospora* disease on *Aquilegia*. Mycol Prog 14(9):1–10

Desprez-Loustau ML, Robin C, Buée M, Courtecuisse R, Garbaye J, Suffert F, Sache I, Rizzo DM (2007) The fungal dimension of biological invasions. Trends Ecol Evol 22(9):472–480

Desprez-Loustau ML, Courtecuisse R, Robin C, Husson C, Moreau PA, Blancard D, Selosse MA, Lung-Escarmant B, Piou D, Sache I (2010) Species diversity and drivers of spread of alien fungi (*sensu lato*) in Europe with a particular focus on France. Biol Invasions 12(1):157–172

Desprez-Loustau M-L, Massot M, Toïgo M, Fort T, Kaya AGA, Boberg J et al (2018) From leaf to continent: the multi-scale distribution of an invasive cryptic pathogen complex on oak. Fungal Ecol 36:39–50

Deutsche Gesellschaft für Mykologie DGfM (2020) Datenbank der Pilze Deutschlands [Database of the fungi of Germany]. http://www.pilze-deutschland.de

Dgebuadze YY, Petrosyan VG, Khlyap LA (2018) The most dangerous invasive species in Russia (TOP-100). KMK Scientific Press, Moscow

Diagne C, Leroy B, Vaissière AC, Gozlan RE, Roiz D, Jarić I, Salles JM, Bradshaw CJA, Courchamp F (2021) High and rising economic costs of biological invasions worldwide. Nature 592(7855):571–576

Diéguez-Uribeondo J, Fregeneda-Grandes JM, Cerenius L, Pérez-Iniesta E, Aller-Gancedo JM, Tellería MT et al (2007) Re-evaluation of the enigmatic species complex *Saprolegnia diclina-Saprolegnia parasitica* based on morphological, physiological and molecular data. Fungal Genet Biol 44(7): 585–601

Drenth A, McTaggart AR, Wingfield BD (2019) Fungal clones win the battle, but recombination wins the war. IMA Fungus 10:18

Essl F, Bacher S, Genovesi P, Hulme PE, Jeschke JM, Katsanevakis S, Kowarik I, Kühn I, Pyšek P, Rabitsch W, Schindler S, Van Kleunen M, Vilà M, Wilson JRU, Richardson DM (2018) Which taxa are alien? Criteria, applications, and uncertainties. BioScience 68(7):496–509. https://doi.org/10.1093/biosci/biy057

European and Mediterranean Plant Protection Organization EPPO (2020). EPPO Global Database (online)

Farr DF, Rossman AY (2021) Fungal databases. U.-S. National Fungus Collections, ARS, USDA

Feng C, Saito K, Liu B, Manley A, Kammeijer K, Mauzey SJ et al (2018) New races and novel strains of the spinach downy mildew pathogen *Peronospora effusa*. Plant Dis 102(3):613–618

Fisher MC, Henk DA, Briggs CJ, Brownstein JS, Madoff LC, McCraw SL, Gurr SJ (2012) Emerging fungal threats to animal, plant and ecosystem health. Nature 484(7393):186–194

Fisher MC, Gurr SJ, Cuomo CA, Blehert DS, Jin H, Stukenbrock EH, Stajich JE, Kahmann R, Boone C, Denning DW, Gow NAR, Klein BS, Kronstad JW, Sheppard DC, Taylor JW, Wright GD, Heitman J, Casadevall A, Coweni L (2020) Threats posed by the fungal kingdom to humans, wildlife, and agriculture. MBio 11(3):1–17

Fones HN, Fisher MC, Gurr SJ (2017) Emerging fungal threats to plants and animals challenge agriculture and ecosystem resilience. Microbiol Spectr. https://doi.org/10.1128/microbiolspec.FUNK-0027-2016

Fontaine MC, Gladieux P, Hood ME, Giraud T (2013) History of the invasion of the anther smut pathogen on *Silene latifolia* in North America. New Phytol 198: 946–956

Fortini LB, Kaiser LR, Keith LM, Price J, Hughes RF, Jacobi JD, Friday JB (2019) The evolving threat of rapid ʻŌhiʻa Death (ROD) to Hawaiʻi's native ecosystems and rare plant species. For Ecol Manag 448(6):376–385

Frantzeskakis L, Kracher B, Kusch S, Yoshikawa-Maekawa M, Bauer S, Pedersen C et al (2018) Signatures of host specialization and a recent transposable element burst in the dynamic one-speed genome of the fungal barley powdery mildew pathogen. BMC Genomics 19(1):1–23

Frantzeskakis L, Németh MZ, Barsoum M, Kusch S, Kiss L, Takamatsu S, Panstruga R (2019) The *Parauncinula polyspora* draft genome provides insights into patterns of gene erosion and genome expansion in powdery mildew fungi. mBio 10(5): e01692–e01619

Frantzeskakis L, Di Pietro A, Rep M, Schirawski J, Wu CH, Panstruga R (2020) Rapid evolution in plant-microbe interactions-a molecular genomics perspective. New Phytol 225(3):1134–1142

Franzke A, Lysak MA, Al-Shehbaz IA, Koch MA, Mummenhoff K (2011) Cabbage family affairs: the evolutionary history of *Brassicaceae*. Trends Plant Sci 16(2):108–116

Frenkel O, Portillo I, Brewer MT, Peros JP, Cadle-Davidson L, Milgroom MG (2012) Development of microsatellite markers from the transcriptome of *Erysiphe necator* for analysing population structure in North America and Europe. Plant Pathol 61(1): 106–119

Fry W (2008) *Phytophthora infestans*: the plant (and R gene) destroyer. Mol Plant Pathol 9(3):385–402

Fry WE, Birch PRJ, Judelson HS, Grünwald NJ, Danies G, Everts KL et al (2015) Five reasons to consider *Phytophthora infestans* a reemerging pathogen. Phytopathology 105(7):966–981

Fuentes N, Marticorena A, Saldaña A, Jerez V, Carlos Ortiz J, Victoriano P, Moreno RA, Larraín J, Villaseñor-Parada C, Palfner G, Sánchez P, Pauchard A (2020) Multi-taxa inventory of naturalized species in Chile. NeoBiota 60:25–41

Fuller KB, Alston JM, Sambucci OS (2014) The value of powdery mildew resistance in grapes: evidence from California. Wine Econ Policy 3:90–107

Gadoury DM, Cadle-Davidson L, Wilcox WF, Dry IB, Seem RC, Milgroom MG (2012) Grapevine powdery mildew (*Erysiphe necator*): a fascinating system for the study of the biology, ecology and epidemiology of an obligate biotroph. Mol Plant Pathol 13(1):1–16

Garibaldi A, Minuto A, Minuto G, Gullino ML (2004) First report of downy mildew on basil (*Ocimum basilicum*) in Italy. Plant Dis 88(3):312–312

Gäumann E (1923) Beiträge zu einer Monographie der Gattung *Peronospora* Corda. Beiträge zur Kryptogamenflora der Schweiz 5:1–360

Gessler C, Pertot I, Perazzolli M (2011) Plasmopara viticola: a review of knowledge on downy mildew of grapevine and effective disease management. Phytopathol Mediterr 50(1):3–44

Gibson DJ (2009) Grasses and grassland ecology. Oxford University Press, New York

Glawe DA (2008) The powdery mildews: a review of the world's most familiar (yet poorly known) plant pathogens. Annu Rev Phytopathol 46:27–51

Gobbin D, Rumbou A, Linde CC, Gessler C (2006) Population genetic structure of *Plasmopara viticola* after

125 years of colonization in European vineyards. Mol Plant Pathol 7(6):519–531

Görg M, Ploch S, Kruse J, Kummer V, Runge F, Choi YJ, Thines M (2017) Revision of *Plasmopara* (*Oomycota*, *Peronosporales*) parasitic to impatiens. Mycol Prog 16(8):791–799

Granados GM, McTaggart AR, Barnes I, Roodas CA, Roux J, Wingfield MJ (2017) The pandemic biotype of *Austropuccinia psidii* discovered in South America. Australas Plant Pathol 46:267–275

Graziosi I, Tembo M, Kuate J, Muchugi A (2020) Pests and diseases of trees in Africa: a growing continental emergency. Plants People Planet 2(1):14–28

Gryzenhout M, Rodas CA, Mena Portales J, Clegg P, Wingfield BD, Wingfield MJ (2006) Novel hosts of the *Eucalyptus* canker pathogen *Chrysoporthe cubensis* and a new *Chrysoporthe* species from Colombia. Mycol Res 110(7):833–845

Gulya TJ, Sackston WE, Viranyi F, Masirevic S, Rashid KY (1991) New races of the sunflower downy mildew pathogen (*Plasmopara halstedii*) in Europe and North and South America. J Phytopathol 132(4):303–311

Gur L, Reuveni M, Cohen Y, Cadle-Davidson L, Kisselstein B, Ovadia S, Frenkel O (2021) Population structure of *Erysiphe necator* on domesticated and wild vines in the Middle East raises questions on the origin of the grapevine powdery mildew pathogen. Environ Microbiol 23(10):6019–6037

Gurr S, Samalova M, Fisher M (2011) The rise and rise of emerging infectious fungi challenges food security and ecosystem health. Fungal Biol Rev 25(4):181–188

Gustavsson A (1959) Studies on Nordic Peronosporas. I. Taxonomic part. Opera Bot 3(1): 1–271

Halliday FW, Penczykowski RM, Barrès B, Eck JJ, Numminen E, Laine A-L (2020) Facilitative priority effects drive parasite assembly under coinfection. Nat Ecol Evol 4:1510–1521

Hardham AR, Blackman LM (2018) Phytophthora cinnamomi. Mol Plant Pathol 19(2):260–285

Hausbeck MK, Lamour KH (2004) *Phytophthora capsici* on vegetable crops: research progress and management challenges. Plant Dis 88(12):1292–1303

Hawksworth D, Lücking R (2017) Fungal diversity revisited: 2.2 to 3.8 million species. Microbiol Spectr. https://doi.org/10.1128/microbiolspec.FUNK-0052-2016

Helfer S (2013) Rust fungi and global change. New Phytol 201:770–780

Heluta VP, Makarenko NV, Al-Maali GA (2019) First records of *Erysiphe corylacearum* (*Erysiphales*, *Ascomycota*) on *Corylus avellana* in Ukraine. Mycol Rec 76(3):252–259

Hirata T, Cunnington JH, Paksiri U, Limkaisang S, Shishkoff N, Grigaliunaite B et al (2000) Evolutionary analysis of subsection *Magnicellulatae* of *Podosphaera* section *Sphaerotheca* (*Erysiphales*) based on the rDNA internal transcribed spacer sequences with special reference to host plants. Can J Bot 78:1521–1530

Hsiao HY, Ariyawansa HA, Hsu CC, Wang CJ, Shen YM (2022) New records of powdery mildews from Taiwan: *Erysiphe ipomoeae* comb. nov., *E.* aff. *betae* on buckwheat, and *E. neolycopersici* comb. nov. on *Cardiospermum halicacabum*. Diversity 14(3):204

Hückelhoven R, Panstruga R (2011) Cell biology of the plant-powdery mildew interaction. Curr Opin Plant Biol 14(6):738–746

Inaba T, Takahashi K, Morinaka T (1983) Seed transmission of spinach downy mildew. Plant Dis 67(10): 1139–1141

Ioos R, Laugustin L, Rose S, Tourvieille J, Tourvieille de Labrouhe D (2007) Development of a PCR test to detect the downy mildew causal agent *Plasmopara halstedii* in sunflower seeds. Plant Pathol 56(2): 209–218

IPBES (2019) Summary for policymakers of the global assessment report on biodiversity and ecosystem services of the Intergovernmental Science-Policy Platform on Biodiversity and Ecosystem Services. IPBES Secretariat. https://ipbes.net/system/tdf/ipbes_global_assessment_report_summary_for_policymakers.pdf?file=1&type=node&id=35329

IUCN Invasive Species Specialist Group ISSG (2019a) Global Register of Introduced and Invasive Species GRIIS - Data Download. http://www.griis.org/

IUCN Invasive Species Specialist Group ISSG (2019b) The Global Invasive Species Database. Version 2015.1. http://www.iucngisd.org/gisd/

James TY, Stajich JE, Hittinger CT, Rokas A (2020) Toward a fully resolved fungal tree of life. Annu Rev Microbiol 74:291–313

Jankovics T, Komáromi J, Fábián A, Jager K, Vida G, Kiss L (2015) New insights into the life cycle of the wheat powdery mildew: direct observation of ascosporic infection in *Blumeria graminis* f. sp. *tritici*. Phytopathology 105(6):797–804

Johnston PR, Quijada L, Smith CA, Baral HO, Hosoya T, Baschien C et al (2019) A multigene phylogeny toward a new phylogenetic classification of *Leotiomycetes*. IMA Fungus 10:1

Jones H, Whipps JM, Gurr SJ (2001) The tomato powdery mildew fungus *oidium neolycopersici*. Mol Plant Pathol 2(6):303–309

Kelly LA, Vaghefi N, Bransgrove K, Fechner NA, Stuart K et al (2021) One crop disease, how many pathogens? *Podosphaera xanthii* and *Erysiphe vignae* sp. nov. identified as the two species that cause powdery mildew of mungbean (*Vigna radiata*) and black gram (*V. mungo*) in Australia. Phytopathology 111(7): 1193–1206

Kenneth RG (1981) Downy mildews of graminaceous crops. In: Spencer DM (ed) The Downy Mildews. Academic Press, London, New York, San Francisco, pp 367–394

Kinloch BB (2001) White pine blister rust in North America: past and prognosis. Phytopathology 93:1044–1047

Kiss L (1996) Occurrence of a new powdery mildew fungus (*Erysiphe* sp.) on tomatoes in Hungary. Plant Dis 80:224

Kiss L (2005) Powdery mildews as invasive plant pathogens: new epidemics caused by two North American species in Europe. Mycol Res 109:259–260

Kiss L, Vaghefi N (2021) First report of powdery mildew of rainforest spinach (*Elatostema reticulatum*), native to Australia, caused by *Podosphaera xanthii*. Aust Plant Dis Notes 16:8

Kiss L, Cook RTA, Saenz GS, Cunnington JH, Pascoe I, Bardin M et al (2001) Identification of two powdery mildew fungi, *Oidium neolycopersici* sp. nov. and *O. lycopersici*, infecting tomato in different parts of the world. Mycol Res 105(6):684–697

Kiss L, Takamatsu S, Cunnington JH (2005) Molecular identification of *Oidium neolycopersici* as the causal agent of the recent tomato powdery mildew epidemics in North America. Plant Dis 89(5):491–496

Kiss L, Kovács GM, Bóka K, Bohár G, Bohárné KV, Németh MZ et al (2018) Deciphering the biology of *Cryptophyllachora eurasiatica* gen. et sp. nov., an often cryptic pathogen of an allergenic weed, *Ambrosia artemisiifolia*. Sci Rep 8:10806

Kiss L, Vaghefi N, Bransgrove K, Dearnaley JDW, Takamatsu S, Tan YP et al (2020) Australia: a continent without native powdery mildews? The first comprehensive catalogue indicates recent introductions and multiple host range expansion events, and leads to the re-discovery of *Salmonomyces* as a new lineage of the *Erysiphales*. Front Microbiol 11:1571

Kofoet A, Kik C, Wietsma WA, De Vries JN (1990) Inheritance of resistance to downy mildew (*Peronospora destructor* [Berk.] Casp.) from Allium roylei Stearn in the backcross *Allium cepa* L.× (*A. roylei* × *A. cepa*). Plant Breed 105(2):144–149

Kreisel H, Scholler M (1994) Chronology of Phytoparasitic Fungi introduced to Germany and adjacent countries. Bot Acta 107(6):387–392

Kusch S, Németh MZ, Vaghefi N, Ibrahim HM, Panstruga R, Kiss L (2020) A short-read genome assembly resource for *Leveillula taurica* causing powdery mildew disease of sweet pepper (*Capsicum annuum*). Mol Plant-Microbe Interact 33(6):782–786

Kusch S, Vaghefi N, Takamatsu S, Liu S-Y, Németh MZ, Seress D et al (2022) First draft genome assemblies of *Pleochaeta shiraiana* and *Phyllactinia moricola*, two tree-parasitic powdery mildew fungi with hemiendophytic mycelia. Phytopathology 112(4): 961–967

Landa BB, Montes-Borrego M, Muñoz-Ledesma FJ, Alcaliber SA, Jiménez-Díaz RM (2005) First report of downy mildew of opium poppy caused by *Peronospora arborescens* in Spain. Plant Dis 89(3): 338–338

Lane CR, Beales PA, O'neill TM, McPherson GM, Finlay AR, David J et al (2005) First report of impatiens downy mildew (*Plasmopara obducens*) in the UK. Plant Pathol 54(2)

Lebeda A, Sedlářová M, Petřivalský M, Prokopová J (2008) Diversity of defence mechanisms in plant-oomycete interactions: a case study of *Lactuca* spp. and *Bremia lactucae*. Eur J Plant Pathol 122(1):71–89

LeBlanc N, Salgado-Salazar C, Crouch JA (2018) Boxwood blight: an ongoing threat to ornamental and native boxwood. Appl Microbiol Biotechnol 102(10): 4371–4380

Leesutthiphonchai W, Vu AL, Ah-Fong AM, Judelson HS (2018) How does *Phytophthora infestans* evade control efforts? Modern insight into the late blight disease. Phytopathology 108(8):916–924

Lefort F, Gigon V, Amos B (2003) Premier cas de *Peronospora lamii*, responsable du mildiou du basilic dans le bassin lémanique. Assemblée d'Automne de la Société Suisse de Phytiatrie, EIC, Nyons- Changins, 2 octobre 2003

Lévésque CA, Brouwer H, Cano L, Hamilton JP, Holt C, Huitema E et al (2010) Genome sequence of the necrotrophic plant pathogen *Pythium ultimum* reveals original pathogenicity mechanisms and effector repertoire. Genome Biol 11(7):1–22

Li YY, Wang MM, Groenewald M, Li AH, Guo YT, Wu F, Zhang BQ, Tanaka E, Wang QM, Bai FY, Begerow D (2021) Proposal of two new combinations, twenty new species, four new genera, one new family, and one new order for the anamorphic basidiomycetous yeast species in *Ustilaginomycotina*. Front Microbiol 12:777338–777338

Lücking R, Aime MC, Robbertse B, Miller AN, Ariyawansa HA, Aoki T et al (2020) Unambiguous identification of fungi: where do we stand and how accurate and precise is fungal DNA barcoding? IMA Fungus 11(1):1–32

Marçais B, Desprez-Loustau M-L (2014) European oak powdery mildew: impact on trees, effects of environmental factors, and potential effects of climate change. Ann For Sci 71:633–642

McCook S (2006) Global rust belt: Hemileia vastatrix and the ecological integration of world coffee production since 1850. J Glob Hist 1(2):177–195

McTaggart AR, Shivas RG, Geering ADW, Vanky K, Scharaschkin T (2012a) A review of the *Ustilago-Sporisorium-Macalpinomyces* complex. Persoonia 29(1):55–62

McTaggart AR, Shivas RG, Geering ADW, Callaghan B, Vanky K, Scharaschkin T (2012b) Taxonomic revision of *Ustilago*, *Sporisorium* and *Macalpinomyces*. Persoonia 29(1):116–132

McTaggart AR, Shivas RG, Geering ADW, Callaghan B, Vanky K, Scharaschkin T (2012c) Soral synapomorphies are significant for the systematics of the *Ustilago-Sporisorium-Macalpinomyces* complex (*Ustilaginaceae*). Persoonia 29(1):63–77

McTaggart AR, Doungsa-ard C, Geering ADW, Aime MC, Shivas RG (2015) A co-evolutionary relationship exists between *Endoraecium* (*Pucciniales*) and its *Acacia* hosts in Australia. Persoonia 35:50–62

McTaggart AR, Roux J, Granados GM, Gafur A, Tarrigan M, Santhakumar P, Wingfield MJ (2016a) Rust (*Puccinia psidii*) recorded in Indonesia poses a threat to forests and forestry in South-East Asia. Australas Plant Pathol 45:83–89

McTaggart AR, Shivas RG, van der Nest MA, Roux J, Wingfield BD, Wingfield MJ (2016b) Host jumps shaped the diversity of extant rust fungi (*Pucciniales*). New Phytol 209:1149–1158

McTaggart AR, Prychid CJ, Bruhl JJ, Shivas RG (2020) The PhyloCode applied to *Cintractiellales*, a new order of smut fungi with unresolved phylogenetic relationships in the *Ustilaginomycotina*. Fungal Syst Evol 6:55–64

McTaggart AR, Drenth A, Wingfield MJ, O'Dwyer C, Shuey LS, Pegg GS (2022) Epidemic spread of smut fungi (*Quambalaria*) by sexual reproduction in a native pathosystem. Eur J Plant Pathol 163:341–349

Menardo F, Praz CR, Wyder S, Ben-David R, Bourras S, Matsumae H et al (2016) Hybridization of powdery mildew strains gives rise to pathogens on novel agricultural crop species. Nat Genet 48(2):201–205

Middleton JT (1943) The taxonomy, host range and geographic distribution of the genus *Pythium*. Mem Torrey Bot Club 20(1):1–171

Molin C, Ribeiro NR, Matsumoto MN, Fernando Giasson N, Brollo J, Zanardo B et al (2021) Damping-off of soybean in southern Brazil can be associated with different species of *Globisporangium* spp. and *Pythium* spp. Plant Pathol 70(7):1686–1694

Müller MC, Praz CR, Sotiropoulos AG, Menardo F, Kunz L, Schudel S et al (2019) A chromosome-scale genome assembly reveals a highly dynamic effector repertoire of wheat powdery mildew. New Phytol 221(4):2176–2189

MyCoPortal (2020) Mycology collections data portal. http://mycoportal.org/portal/index.php

Nahrung HF, Carnegie AJ (2020) Non-native forest insects and pathogens in Australia: establishment, spread, and impact. Front For Global Change 3(37): 1–12

Nasr S, Soudi MR, Fazeli SAS, Nguyen HD, Lutz M, Piątek M (2014) Expanding evolutionary diversity in the *Ustilaginomycotina*: *Fereydouniaceae* fam. nov. and *Fereydounia* gen. nov., the first urocystidalean yeast lineage. Mycol Prog 13:1217–1226

Nealis VG, Demerchant I, Langor D, Noseworthy MK, Pohl G, Porter K, Shanks E, Turnquist R, Waring V (2016) Historical occurrence of alien arthropods and pathogens on trees in Canada. Can J For Res 46(2):1–9

Newcombe G, Dugan FM (2010) Fungal pathogens of plants in the homogocene. In: Molecular identification of fungi. Springer, New York. https://doi.org/10.1007/978-3-642-05042-8_1

Nielsen KN, Gopalakrishnan S, Korneliussen TS, Skovrind M (2020) Population genomics of the emerging forest pathogen *Neonectria neomacrospora*. BioRxiv 2020:1–29. https://doi.org/10.1101/2020.12.07.407155

Niinomi S, Takamatsu S, Havrylenko M (2008) Molecular data do not support a southern hemisphere base of *Nothofagus* powdery mildews. Mycologia 100(5): 716–726

Noordeloos ME, Loerakker WM (1989) Studies in plant pathogenic fungi - II. On some powdery mildews (*Erysiphales*) recently recorded from The Netherlands. Persoonia 14:51–60

Paap T, Wingfield MJ, Burgess TI, Wilson JRU, Richardson DM, Santini A (2022) Invasion frameworks: a forest pathogen perspective. Curr For Rep 8(1):74–89

Pánek M, Střížková I (2021) A comparison of the virulence of selected *Pythium*, *Globisporangium*, *Phytopythium* and *Phytophthora* species against strawberry plants. J Plant Dis Prot 128(6):1447–1458

Pathak VK, Mathur SB, Neergaard P (1978) Detection of *Peronospora manshurica* (Naum.) Syd. in seeds of soybean, Glycine max. EPPO Bull 8(1):21–28

Pegg GS, O'Dwyer C, Carnegie AJ, Burgess TI, Wingfield MJ, Drenth A (2008) *Quambalaria* species associated with plantation and native eucalypts in Australia. Plant Pathol 57(4):702–714

Pérez-García A, Romero D, Fernández-Ortuño D, López-Ruiz F, De Vicente A, Tores JA (2009) The powdery mildew fungus *Podosphaera fusca* (synonym *Podosphaera xanthii*), a constant threat to cucurbits. Mol Plant Pathol 10(2):153–160

Pettitt T, Henricot B, Matatiken D, Cook RTA (2010) First record of *oidium* anamorph of *Podosphaera xanthii* on *Medusagyne oppositifolia*. Plant Pathol 59:1168

Piątek M, Lutz M, Sousa FMP, Santos ARO, Félix CR, Landell MF, Gomes FCO, Rosa CA (2017) *Pattersoniomyces tillandsiae* gen. et comb. nov.: linking sexual and asexual morphs of the only known smut fungus associated with *Bromeliaceae*. Org Divers Evol 17:531–543

Pimentel D, Zuniga R, Morrison D (2005) Update on the environmental and economic costs associated with alien-invasive species in the United States. Ecol Econ 52(3):273–288

Ploch S, Choi YJ, Rost C, Shin HD, Schilling E, Thines M (2010) Evolution of diversity in *Albugo* is driven by high host specificity and multiple speciation events on closely related *Brassicaceae*. Mol Phylogenet Evol 57(2):812–820

Ploch S, Choi YJ, Thines M (2018) The only known white blister rust on a basal angiosperm is a member of the genus *Albugo*. Org Divers Evol 18(1):63–69

Ploch S, Kruse J, Choi YJ, Thiel H, Thines M (2022) Ancestral state reconstruction in *Peronospora* provides further evidence for host jumping as a key element in the diversification of obligate parasites. Mol Phylogenet Evol 166:107321

Prom LK, Perumal R, Montes-Garcia N, Isakeit T, Odvody GN, Rooney WL et al (2015) Evaluation of Gambian and Malian sorghum germplasm against downy mildew pathogen, *Peronosclerospora sorghi*, in Mexico and the USA. J Gen Plant Pathol 81(1): 24–31

Ramírez-Camejo LA, Eamvijarn A, Díaz-Valderrama JR, Karlsen-Ayala E, Koch RA, Johnson E, Pruvot-Woehl-S, Meija LC, Montagnon C, Maldonado-Fuentes C, Aime MC (2022a) Global analysis of *Hemileia vastatrix* populations shows clonal reproduction for the coffee leaf rust pathogen throughout most of its range. Phytopathology 112:643–652

Ramírez-Camejo LA, Keith LM, Matsumoto T, Sugiyama L, Fukada M, Brann M, Moffitt A, Liu JRA, Aime MC (2022b) Coffee leaf rust (*Hemileia vastatrix*) from the recent invasion into Hawaii shares a genotypic relationship with Latin-American populations. J Fungi 8:189

Riess K, Schön ME, Lutz M, Butin H, Oberwinkler F, Garnica S (2016) On the evolutionary history of *Uleiella chilensis*, a smut fungus parasite of *Araucaria Araucana* in South America: *Uleiellales* ord. nov. in *Ustilaginomycetes*. PLoS One 11(1):e0147107

Rizzo DM, Garbelotto M, Davidson JM, Slaughter GW, Koike ST (2002) *Phytophthora ramorum* as the cause of extensive mortality of *Quercus* spp. and *Lithocarpus densiflorus* in California. Plant Dis 86(3):205–214

Rouxel M, Mestre P, Baudoin A, Carisse O, Delière L, Ellis MA et al (2014) Geographic distribution of cryptic species of *Plasmopara viticola* causing downy mildew on wild and cultivated grape in eastern North America. Phytopathology 104(7):692–701

Roy HE, Hesketh H, Purse BV, Eilenberg J, Santini A, Scalera R, Stentiford GD, Adriaens T, Bacela-Spychalska K, Bass D, Beckmann KM, Bessell P, Bojko J, Booy O, Cardoso AC, Essl F, Groom Q, Harrower C, Kleespies R et al (2017) Alien pathogens on the horizon: opportunities for predicting their threat to wildlife. Conserv Lett 10(4):476–483

Runge F, Thines M (2011) Host matrix has major impact on the morphology of *Pseudoperonospora cubensis*. Eur J Plant Pathol 129(2):147–156

Runge F, Choi YJ, Thines M (2011) Phylogenetic investigations in the genus *Pseudoperonospora* reveal overlooked species and cryptic diversity in the *P. cubensis* species cluster. Eur J Plant Pathol 129(2): 135–146

Salgado-Salazar C, Thines M (2022) Two new species of *Plasmopara* affecting wild grapes in the USA. Mycol Prog 21(7):1–15

Santini A, Ghelardini L, De Pace C, Desprez-Loustau ML, Capretti P, Chandelier A, Cech T, Chira D, Diamandis S, Gaitniekis T, Hantula J, Holdenrieder O, Jankovsky L, Jung T, Jurc D, Kirisits T, Kunca A, Lygis V, Malecka M et al (2013) Biogeographical patterns and determinants of invasion by forest pathogens in Europe. New Phytol 197(1):238–250

Santini A, Liebhold A, Migliorini D, Woodward S (2018) Tracing the role of human civilization in the globalization of plant pathogens. ISME J 12(3):647–652

Scholefield P, Morison J (2010) GWRDC project GWR 08/04: assessment of economic cost of endemic pests and diseases on the Australian grape and wine industry. Grape and Wine Research and Development Corporation, Australia

Scholler M, Lutz M, Aime MC (2019) Repeated formation of correlated species in *Tranzschelia* (*Pucciniales*). Mycol Prog 18:295–303

Scott P, Bader MKF, Burgess T, Hardy G, Williams N (2019) Global biogeography and invasion risk of the plant pathogen genus *Phytophthora*. Environ Sci Policy 101(September):175–182

Seebens H, Blackburn TM, Dyer EE, Genovesi P, Hulme PE, Jeschke JM, Pagad S, Pyšek P, Winter M, Arianoutsou M, Bacher S, Blasius B, Brundu G, Capinha C, Celesti-Grapow L, Dawson W, Dullinger S, Fuentes N, Jäger H et al (2017) No saturation in the accumulation of alien species worldwide. Nat Commun 8:1–9

Sharma R, Mishra B, Runge F, Thines M (2014) Gene loss rather than gene gain is associated with a host jump from monocots to dicots in the smut fungus *Melanopsichium pennsylvanicum*. Genome Biol Evol 6(8):2034–2049

Shaw CG (1978) *Peronosclerospora* species and other downy mildews of the Gramineae. Mycologia 70(3): 594–604

Shivas RG, Beasley DR, McTaggart AR (2014) Online identification guides for Australian smut fungi (*Ustilaginomycotina*) and rust fungi (*Pucciniales*). IMA Fungus 5:195–202

Sikes BA, Bufford JL, Hulme PE, Cooper JA, Johnston PR, Duncan RP (2018) Import volumes and biosecurity interventions shape the arrival rate of fungal pathogens. PLoS Biol 16(5):1–16

Simpson A, Eyler MC, Sikes D, Bowser M, Sellers E (2018) A comprehensive list of non-native species established in three major regions of the United States. U.S. Geological Survey Data Release

Soltis DE, Bell CD, Kim S, Soltis PS (2008) Origin and early evolution of angiosperms. Ann N Y Acad Sci 1133(1):3–25

Sotiropoulos AG, Arango-Isaza E, Ban T, Barbieri C, Bourras S, Cowger C et al (2022) Global genomic analyses of wheat powdery mildew reveal association of pathogen spread with historical human migration and trade. Nat Commun 13(1):1–14

Spatafora JW, Aime MC, Grigoriev IV, Martin F, Stajich JE, Blackwell M (2017) The fungal tree of life: from molecular systematics to genome-scale phylogenies. Microbiol Spectr 5:5

Spring O (2019) Spreading and global pathogenic diversity of sunflower downy mildew-review. Plant Prot Sci 55(3):149–158

Strand M, Aronsson M, Svensson M (2019) Klassificering av främmande arters effekter på biologisk mångfald i Sverige - ArtDatabankens risklista [classification of the effects of alien species on biodiversity in Sweden - the

Swedish species data Bank's risk list], ArtDatabanken Rapporterar 21. ArtDatabanken SLU, Uppsala

Suharjo R, Swibawa G, Prasetyo J, Fitriana Y, Danaatmadja Y, Budiawan A, Roberts S, Noorhajati N, Amad M, Thines M (2020) Peronosclerospora australiensis is a synonym of P. maydis, which is widespread on Sumatra, and distinct from the most prevalent Java maize downy mildew pathogen. Mycol Prog 19:1309–1315

SwissFungi (2021) Liste der Neomyceten der Schweiz [List of neomycetes of Switzerland]. https://swissfungi.wsl.ch/de/neomyceten/neomyceten-der-schweiz/liste-der-neomyceten.html

Takamatsu S (2013a) Origin and evolution of the powdery mildews (*Ascomycota*, *Erysiphales*). Mycoscience 54(1):75–86

Takamatsu S (2013b) Molecular phylogeny reveals phenotypic evolution of powdery mildews (*Erysiphales*, *Ascomycota*). J Gen Plant Pathol 79:218–226

Takamatsu S, Matsuda S (2004) Estimation of molecular clocks for ITS and 28S rDNA in *Erysiphales*. Mycoscience 45(5):340–344

Takamatsu S, Matsuda S, Niinomi S, Havrylenko M (2006) Molecular phylogeny supports a northern hemisphere origin of *Golovinomyces* (*Ascomycota*: *Erysiphales*). Mycol Res 110(9):1093–1101

Talhinhas P, Carvalho R, Figueira R, Ramos AP (2019) An annotated checklist of rust fungi (*Pucciniales*) occurring in Portugal. Sydowia 71(6):65–84

Tavares S, Ramos AP, Pires AS, Azinheira HG, Caldeirinha P, Link T, Abranches R, Silva MDC, Voegele RT, Loureiro J, Talhinhas P (2014) Genome size analyses of *Pucciniales* reveal the largest fungal genomes. Front Plant Sci 5:422

Thines M (2009) Bridging the gulf: *Phytophthora* and downy mildews are connected by rare grass parasites. PLoS One 4(3):e4790

Thines M (2014) Phylogeny and evolution of plant pathogenic oomycetes—a global overview. Eur J Plant Pathol 138(3):431–447

Thines M (2019) An evolutionary framework for host shifts-jumping ships for survival. New Phytol 224(2): 605–617

Thines M, Choi YJ (2016) Evolution, diversity, and taxonomy of the *Peronosporaceae*, with focus on the genus *Peronospora*. Phytopathology 106(1):6–18

Thines M, Telle S, Ploch S, Runge F (2009) Identity of the downy mildew pathogens of basil, coleus, and sage with implications for quarantine measures. Mycol Res 113(5):532–540

Thines M, Denton GJ, Beal EJ, Kilty A, Denton JO, Shin HD, Choi YJ (2019) *Peronospora aquilegiicola* sp. nov., the downy mildew affecting columbines in the UK is an invasive species from East Asia. Eur J Plant Pathol 155(2):515–525

Thines M, Buaya A, Ali T, Brand T (2020) *Peronospora aquilegiicola* made its way to Germany: the start of a new pandemic? Mycol Prog 19(8):791–798

Thomas CE, Inaba T, Cohen Y (1987) Physiological specialization in Pseudoperonospora cubensis. Phytopathology 77(12):1621–1624

Tobias PA, Schwessinger B, Deng CH, Wu C, Dong C, Sperschneider J, Jones A, Luo Z, Zhang P, Sandhu K, Smith GR, Tibbits J, Chagné D, Park RF (2021) *Austropuccinia psidii*, causing myrtle rust, has a gigabase-sized genome shaped by transposable elements. G3 Genes Genomes Genet 11(3):jkaa015

Tojo M, Van West P, Hoshino T, Kida K, Fujii H, Hakoda A et al (2012) Pythium polare, a new heterothallic oomycete causing brown discolouration of Sanionia uncinata in the Arctic and Antarctic. Fungal Biol 116(7):756–768

Tomoshevich MA (2019) Interrelations between alien and native foliar fungal pathogens and woody plants in Siberia. Contemp Probl Ecol 12(6):642–657

Troch V, Audenaert K, Bekaert B, Höfte M, Haesaert G (2012) Phylogeography and virulence structure of the powdery mildew population on its 'new' host triticale. BMC Ecol Evol 12:76

Vaghefi N, Kusch S, Németh MZ, Seress D, Braun U, Takamatsu S et al (2022) Beyond nuclear ribosomal DNA sequences: evolution, taxonomy, and closest known saprobic relatives of powdery mildew fungi (*Erysiphaceae*) inferred from their first comprehensive genome-scale phylogenetic analyses. Front Microbiol 13:903024

Vági P, Kovács GM, Kiss L (2007) Host range expansion in a powdery mildew fungus (*Golovinomyces* sp.) infecting *Arabidopsis thaliana: Torenia fournieri* as a new host. Eur J Plant Pathol 117(1):89–93

Vági P, Caffi T, Váczy KZ, Kiss L (2016) Refining a method for ascospore viability testing in overwintering chasmothecia of *Erysiphe necator*. Eur J Plant Pathol 144(4):799–802

Vánky K (2012) Smut fungi of the world. American Phytopathological Society (APS Press), St. Paul, MN

Vánky K (2013) Illustrated genera of smut fungi, 3rd edn. American Phytopathological Society (APS Press), St. Paul, MN

Viranyi F, Gulya TJ, Tourvieille DL (2015) Recent changes in the pathogenic variability of *Plasmopara halstedii* (sunflower downy mildew) populations from different continents. Helia 38(63):149–162

Voglmayr H, Piątek M, Mossebo DC (2009) *Pseudoperonospora cubensis* causing downy mildew disease on *Impatiens irvingii* in Cameroon: a new host for the pathogen. Plant Pathol 58(2)

Voglmayr H, Montes-Borrego M, Landa BB (2014) Disentangling *Peronospora* on *Papaver*: phylogenetics, taxonomy, nomenclature and host range of downy mildew of opium poppy (*Papaver somniferum*) and related species. PLoS One 9(5): e96838

Voglmayr H, Schertler A, Essl F, Krisai-Greilhuber I (2022) Alien and cryptogenic fungi and oomycetes in Austria: an annotated checklist (2nd edition). Biol Invasions. https://doi.org/10.1007/s10530-022-02896-2

Walker AS, Bouguennec A, Confais J, Morgant G, Leroux P (2011) Evidence of host-range expansion from new powdery mildew (*Blumeria graminis*) infections of triticale (×*Triticosecale*) in France. Plant Pathol 60: 207–220

Wang QM, Theelen B, Groenewald M, Bai FY, Boekhout T (2014) *Moniliellomycetes* and *Malasseziomycetes*, two new classes in *Ustilaginomycotina*. Persoonia 33: 41–47

Wang QM, Begerow D, Groenewald M, Liu XZ, Theelen B, Bai FY, Boekhout T (2015) Multigene phylogeny and taxonomic revision of yeasts and related fungi in the *Ustilaginomycotina*. Stud Mycol 81(1):149–189

Weltzien HC (1978) Geographical distribution of powdery mildews. In: Spencer DM (ed) The powdery mildews. Academic Press, London, pp 39–49

Westerdijk Fungal Biodiversity Institute (2020) Fungal collection. CBS Strain Database, Westerdijk Fungal Biodiversity Institute

Whipps JM, Budge SP, Fenlon JS (1998) Characteristics and host range of tomato powdery mildew. Plant Pathol 47:36–48

Wicker T, Oberhaensli S, Parlange F, Buchmann JP, Shatalina M, Roffler S et al (2013) The wheat powdery mildew genome shows the unique evolution of an obligate biotroph. Nat Genet 45(9):1092–1096

Wijayawardene NN, Hyde KD, Dai DQ, Sánchez-García M, Goto BT, Saxena RK et al (2022) Outline of Fungi and fungus-like taxa-2021. Mycosphere 13(1):53–453

Williams SE, Whittaker JM, Granot R, Müller DR (2013) Early India-Australia spreading history revealed by newly detected Mesozoic magnetic anomalies in the Perth abyssal plain. J Geophys Res Solid Earth 118(7): 3275–3284

Wingfield MJ, Brockerhoff EG, Wingfield BD, Slippers B (2015) Planted forest health: the need for a global strategy. Science 349(6250):832–836

Winter G (1884) Repertorium. Rabenhorstii fungi europaei et extraeuropaei exsiccati cura Dr. G. Winter, Centuria XXXI et XXXII. Hedwigia 23:164–172

Wood AR (2017) Fungi and invasions in South Africa. Bothalia 47(2):2124

Xu HG, Qiang S (2018) China's invasive alien species (revised). Science Press

Yerkes WD, Shaw CG (1959) Taxonomy of the Peronospora species on Cruciferae and Chenopodiaceae. Phytopathology 49:499–507

Yoshida K, Schuenemann VJ, Cano LM, Pais M, Mishra B, Sharma R et al (2013) The rise and fall of the *Phytophthora infestans* lineage that triggered the Irish potato famine. eLife 2:e00731

Zhang Q, Men X, Hui C, Ge F, Ouyang F (2022) Wheat yield losses from pests and pathogens in China. Agric Ecosyst Environ 326:107821

Zheng Z, Nonomura T, Bóka K, Matsuda Y, Visser RG, Toyoda H, Kiss L, Bai Y (2013) Detection and quantification of *Leveillula taurica* growth in pepper leaves. Phytopathology 103(6):623–632

Ziogas BN, Markoglou AN, Theodosiou DI, Anagnostou A, Boutopoulou S (2006) A high multi-drug resistance to chemically unrelated oomycete fungicides in Phytophthora infestans. Eur J Plant Pathol 115(3):283–292

Part III

Evolution of Mutualistic Interactions

Mycoviruses 7

Jillian M. Myers and Timothy Y. James

Abstract

Viruses are tiny genetic elements that can have enormous impacts on living systems. Relative to viruses of prokaryotes and many other eukaryotes, mycoviruses have received less attention. Recent work, however, has shed light on how mycoviruses modify host phenotypes, the molecular basis for mycovirus–host interactions, and potential applications of mycoviruses for biocontrol in agriculture.

Known from most, but not all, phyla, mycoviruses are still underexplored in many fungal lineages. Mycovirus diversity is classified into 22 families of predominantly RNA viruses. Phenotypic effects vary and examples of mycovirus–host relationships span the parasitism–mutualism continuum, though mycovirus infection often appears asymptomatic in lab culture. However, either the literature is skewed toward or mycoviruses tend toward negative host effects, such as growth rate reduction or inducing hypovirulence. These phenotypic changes are typically associated with altered host gene expression induced by both virus and host. Hosts may respond to infection through the innate immune function of RNA silencing pathways. It is known that mycoviruses may also be able to manipulate third parties (i.e., plant or animal hosts) via their affiliation with fungi. Many mycoviruses have become endogenized into their host's genome, and this process may be involved in facilitating horizontal gene transfer between fungi.

Keywords

Mycovirus · CHV1 · Killer virus · RNAi · Horizontal gene transfer

7.1 Introduction

Viruses are reputed for having an outsized impact on their host species, changing host population sizes, behaviors, and forcing the evolution of complex defensive mechanisms (e.g., CRISPR, the prokaryotic viral defense system derived from phage DNA). Fungi also play host to viruses, but by and large, mycologists have overlooked the presence and impact viruses may be having on their organisms of interest. Fungal viruses (mycoviruses) have been under study for nearly 60 years, gradually increasing in popularity as a study system (Fig. 7.1). The past 20 years of research have been especially fruitful and have put mycoviruses on the map as hyperparasites of ecologically and agriculturally important pathogens, symbionts and, potentially, tools for combating diseases and genetic engineering.

J. M. Myers (✉) · T. Y. James
Department of Ecology and Evolutionary Biology, University of Michigan, Ann Arbor, MI, USA
e-mail: jimyers@umich.edu

© The Author(s), under exclusive license to Springer Nature Switzerland AG 2023

S. Pöggeler, T. James (eds.), *Evolution of Fungi and Fungal-Like Organisms*, The Mycota 14,
https://doi.org/10.1007/978-3-031-29199-9_7

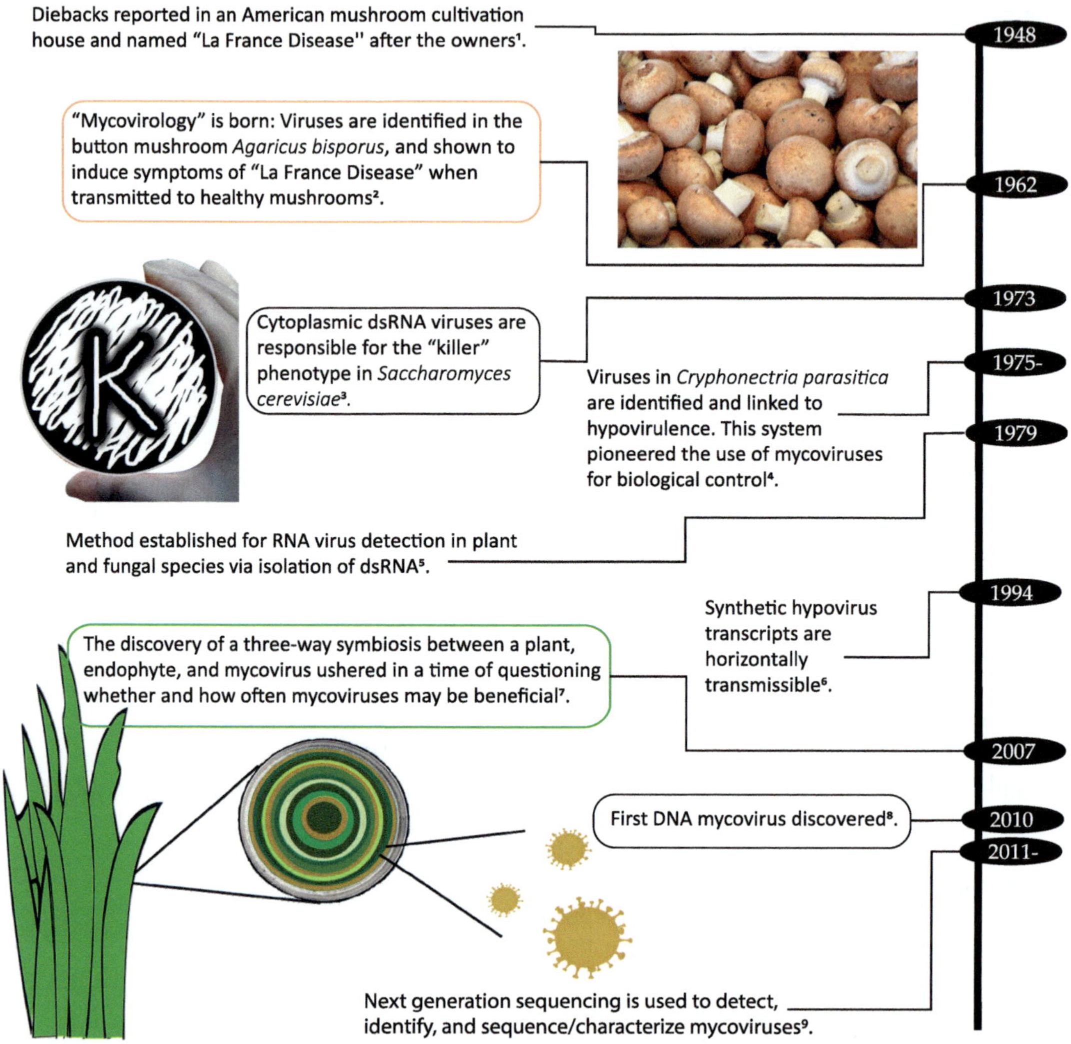

Fig. 7.1. A brief history of mycovirology including selected notable discoveries. (1) Sinden and Hauser 1950 (2) Hollings 1962 (3) Bevan et al. 1973 (4) Anagnostakis 1982; Heiniger and Rigling 1994; Van Alfen et al. 1975 (5) Morris and Dodds 1979 (6) Chen et al. 1994 (7) Márquez et al. 2007 (8) Yu et al. 2010 (9) Al Rwahnih et al. 2011; Marzano et al. 2016; Nerva et al. 2016

Mycoviruses generally have properties that distinguish them from the traditional view of deadly pathogens. This relates to their mode of transmission which may be primarily vertical, a phenomenon that is expected to select for more benign, even mutualistic, interactions between host and virus. Indeed some of the best studied mycoviral systems are known to be cases where the virus can be considered a mutualist, such as the killer viruses of yeast. It is estimated that perhaps as many as 20% of all fungal cultures contain viruses, yet some strains are known to host many (>10!) viruses (Khan et al. 2021b; Myers et al. 2020). In this chapter, we summarize the current knowledge on the diversity of viruses, their impacts on hosts, their mode of transmission, and impacts on host evolution. Numerous excellent reviews have already been published on the subject of mycoviruses (Ghabrial et al. 2015; Kondo et al. 2022; Pearson et al. 2009; Son et al.

2015). Here we take special care to write for the mycologist who may have little or no background knowledge of viruses or methods common in virology.

7.2 Mycoviral Diversity and Taxonomy

Viruses are the most genetically diverse entities on the planet. The known diversity of mycoviruses can only be but a minute fraction of what exists because mycovirology, as a discipline, is in its infancy. And yet, viruses with great variation in morphology and genetic features have been described. At the most fundamental level of genetic diversity, mycovirus genomes can be composed of single-stranded RNA (ssRNA), double-stranded RNA (dsRNA), or single-stranded DNA (ssDNA). Most known mycoviruses have RNA genomes, though this is likely a result of biased sampling since a screening protocol for RNA viruses was developed relatively early on (Morris and Dodds 1979). The first DNA virus was not discovered until 2010 (Yu et al. 2010), and double-stranded DNA mycoviruses have not yet been described.

Mycoviruses with RNA genomes can be naked nucleic acids or form virions that have capsids of various shapes and symmetries assembled into a protective coat around the genome. Most are found in the cytoplasm, but some RNA mycoviruses specialize in the mitochondria. RNA mycovirus genomes are linear, on one or many segments. Relative to some viruses that infect other hosts, the genomes tend to be relatively small, ranging from 2.5 to 23 kb and encoding just one to twelve genes. Although there are no common genes shared by all viruses, all RNA viruses do have one gene in common: the RNA-dependent RNA polymerase (RdRp). This gene is critical to our understanding of viral relatedness and taxonomy. Mycoviruses that are encapsidated encode a coat protein (CP). Though many mycoviruses contain open reading frames (ORFs) in addition to the RdRp and CP, their functions are often unknown.

The most common method for identifying RNA viruses in a culture relies on the specific binding of dsRNA during cellulose chromatography, which captures viruses with both dsRNA and ssRNA genomes since a double-stranded intermediate of ssRNA virus genomes is made during replication (Morris and Dodds 1979). This method is still commonly used, but is limited to detecting RNA viruses only, and may miss some viruses in low titer. The advent of high-throughput sequencing has changed this, as it has much of biology. Numerous mycoviruses have been identified through searching genomic and transcriptomic data (Gilbert et al. 2019; Marzano et al. 2016; Myers et al. 2020; Nerva et al. 2016). However, the ends of viruses detected by data-mining are often lost; to achieve the complete sequence of a virus one must usually perform RNA ligase-mediated rapid amplification of cDNA ends (RLM-RACE) (Coutts and Livieratos 2003; Liu and Gorovsky 1993). For visualization, viruses must be purified by ultracentrifugation and viewed by transmission electron microscopy.

To date, the only DNA mycoviruses known to actively infect fungi are circular rep-encoding single-stranded (CRESS) DNA viruses. As the name describes, these viruses are composed of ssDNA forming a very small (1.3–2.4 kb) circular genome typically encoding just two proteins: Rep, an enzyme which initiates replication by rolling-circle amplification, and a capsid. Virions are 20–22 nm in diameter and are replicated in the nucleus (Zhao et al. 2019). The first of these DNA viruses was identified in 2010 in the phytopathogen *Sclerotinia sclerotiorum* (Yu et al. 2010), and just a few more have been characterized since then (Du et al. 2014; Khalifa and MacDiarmid 2021; Li et al. 2020). Though the first discovery was over a decade ago, still very little is known about the prevalence or diversity of CRESS DNA viruses in fungi. However, recent findings of endogenous CRESS virus-like sequences in fungal genomes suggest a frequency of past, if not contemporary, infection (Liu et al. 2011; Zhao et al. 2021).

Together, mycovirus diversity has been classified into 22 viral families, in at least 6 different phyla. Traditionally, virus taxonomy was based on phenotypic features including viral particle size and symmetry, host range, and disease symptoms. A viral taxonomy based on genetic data began to gradually develop more recently, empowered by improved sequencing technologies. With special thanks to metagenomics and metatranscriptomics we now have the ability to identify and classify viruses without first isolating them. This has created a boom of viral data necessitating the switch to a genetics-based taxonomy.

As of 2020, viruses can be classified into 15 taxonomic ranks, reflecting those used for cellular life (Gorbalenya et al. 2020). Viral realm is the highest classification, of which there are currently six. Realm is analogous to "domain" in cellular life but is markedly different in that not all viruses have common ancestry. Each of the six realms represents at least one novel origin whereby a virus came into existence. Thus, these are broadly defined groups that are decidedly different from one another. For example, the realm *Riboviria* includes all viruses with RNA genomes that replicate with RNA-dependent RNA polymerases (RdRps). *Adnaviria* includes all viruses with dsDNA genomes, filamentous virions, and a particular capsid protein structure.

At fine scale resolutions phylogeny is typically used to inform taxonomy, though supplementary phenotypic characteristics are also sometimes included in demarcation criteria. For example, mycovirus species within the genus *Betachrysovirus* are demarcated based on sequence similarity of RdRP and capsid protein (CP) genes whereby different species share less than 70% and 53% similarity, respectively (Walker et al. 2022). Other criteria can be used to supplement these data including length of genome segments and length of the 5'UTR. Criteria for species demarcation are often genus-specific, and the rules for a particular genus are available through the International Committee on the Taxonomy of Viruses (ICTV).

Importantly, viral taxonomy is not set in stone, and is very often revised based on new data. The ICTV is in charge of establishing, defining, and implementing viral taxonomy and nomenclature. ICTV is a non-profit global organization made up of many virologist volunteers elected to serve terms in specialized Study Groups. Collectively, the ICTV produces reports documenting current information on viral taxonomy and properties of each viral group/taxon and maintains a web-based searchable viral taxonomy (https://ictv.global/taxonomy).

7.3 The Prevalence of Infection

Although the majority of mycovirus research has focused on plant pathogens, mycoviruses exist across the fungal tree of life and arguably there could be at least one virus for every fungal species. The only fungal phylum for which a mycovirus is not known is Cryptomycota, almost surely due to the lack of sampling. Detectable infection prevalence across the Fungi is highly variable, and some clades appear to be more highly infected than others, but more research is needed (Gilbert et al. 2019; Myers et al. 2020). Particularly, a geographically controlled and taxonomically broad and even analysis is needed to reliably determine whether certain phyla or subphlya are more susceptible to mycoviral infection than others.

Both Myers et al. (2020) and Gilbert et al. (2019) found viral prevalence between subphyla to vary greatly; the Glomeromycotina (Mucoromycota) had the highest infection rate at 88.9%, though the sample size was small ($n = 9$). Glomeromycotina are also often host to endosymbiotic bacteria and it is possible that viruses populate these fungi through mechanisms similar to those used by other cytoplasmic symbionts (Bonfante and Desirò 2017). The high rate of mycoviral infection in this group, in combination with prevalent endohyphal bacteria, is interesting to consider in light of the hypothesis that communities of microbial symbionts are more diverse than communities of macrobial hosts (Feldman et al. 2012). Research is needed to corroborate the idea, but this may be an interesting system in which it could be explored. Entomophthoromycotina (Zoopagomycota), which includes many insect pathogens and symbionts, also had a high infection prevalence at 54.5% (Myers et al. 2020). On the other hand, genome sizes of these host-associated fungi are also large and genome sequences difficult to assemble, making it also possible that the viruses are endogenous elements that have inserted into the genome. Endogenous elements are transmitted along with the nuclear genome and would appear as high prevalence viruses.

Between species there is also great variability. *Beauveria bassiana* has infection prevalence ranging from 20% to 50% in nature (Herrero et al. 2012; Kotta-Loizou and Coutts 2017a), while dozens of isolates of certain species have been screened without a single positive (Webb et al. 2022). Notable human pathogens *Cryptococcus neoformans, Candida albicans*, and *Candida glabrata* are curiously lacking reported mycoviruses.

Whether infection prevalence varies by habitat or host ecology is an open question, as is what conditions lead to one isolate being infected by multiple unique viruses. Coinfection of multiple mycoviruses within a single host is common and can sometimes be extreme. A single isolate of *Fusarium mangiferae* was host to 11 viruses (Khan et al. 2021b). Because soil environments support highly diverse fungal communities, we expect that these habitats should be particularly rich with mycoviruses (Sutela et al. 2019). Data suggests that soils have greater viral species richness than even marine environments, which are far better studied, though these data are largely based on bacterial viruses (Williamson et al. 2017). As mycovirus identification and sequencing continues it is reasonable to think that more unknown sequences in databases will be attributed to mycoviruses (Marzano and Domier 2016). Overall, it is tempting to conclude that there are phylogenetic and ecological components to the likelihood of viral presence, though deliberate examination is needed.

7.4 Mycoviral Transmission and Evolution

Over the past 60 years of mycovirus research, the dominant idea regarding transmission has been that mycoviruses do not possess an extracellular phase to their replication cycles, and instead are passed vertically via spores as well as horizontally to compatible individuals via hyphal fusion (Ghabrial 1998). Although multiple lines of evidence are in agreement with this, other modes of transmission are possible, have occurred under laboratory conditions, and occur in nature at least sometimes.

The extent to which transmission to spores occurs seems to be highly variable and without a yet detectable pattern. In some species, such as *Cryphonectria parasitica,* sexual reproduction can limit or restrict viral transmission, but in general both sexual and asexual spores frequently possess viruses, with the rates varying by species from 0 to 100% (Anagnostakis 1988; Chu et al. 2004; Ihrmark et al. 2002; Pearson et al. 2009; Vainio et al. 2015; Varga et al. 2005). However, the prevailing dogma is that many viruses are not transmitted through sexual spores (Ghabrial et al. 2015; Nuss 2005). Vertical transmission implies stable infection over time. This is often true in the laboratory, where many virally infected cultures can apparently be passaged indefinitely without loss of virus. While this is interesting evolutionarily, it can be quite frustrating in the laboratory when trying to obtain an uninfected strain with the same genetic background. In nature, it might be expected for spores that by chance are uninfected to outcompete those with virus and subsequently for viruses to be diminished in a population over time. However, evidence suggests that mycovirus infections can be stable in nature for years (Sutela and Vainio 2020), though this likely depends on the species-specific factors such as efficiency of transmission via spores (Hillman and Suzuki 2004).

In plants, most virus genomes include movement proteins that modulate viral spread within an individual by altering plasmodesmata to allow viruses to pass between cells. In fungi, it is unknown whether an analogous system exists that would facilitate viral spread. Instead, it is presumed that viruses are passively carried along between cells with the cytoplasm. Though this might suggest an accumulation of virus at the hyphal tips, this does not always appear to be so, and in fact hyphal tip fragmentation is one method that can be used to achieve a virus-free strain, albeit with varied success rates depending on the species and strain. Viral accumulation at hyphal tips could facilitate horizontal transmission between vegetatively compatible individuals and, in concept, possibly also through the failed or incomplete reactions of vegetatively incompatible individuals. Indeed, there is an abundance of evidence that horizontal transmission can occur through anastomosis of compatible individuals in the laboratory, though with some variability (Dalzoto et al. 2006; Ihrmark et al. 2002; Smit et al. 1996; Suzaki et al. 2005). There is also some evidence of more limited transmission between incompatible individuals (Cortesi et al. 2001; Suzaki et al. 2005). In a transmission experiment with monokaryotic strains of *Helicobasidium mompa,* a totivirus conferring hypovirulence was transmitted from the donor strain to 2 of 12 recipient strains that were vegetatively incompatible with the donor and comprising 11 different compatibility groups (Suzaki et al. 2005). Those two recipients were then used as new donor strains, and successfully infected an additional 7 monokaryotic strains, demonstrating that horizontal transmission between vegetatively incompatible hosts can occur and that monokaryons have a role in viral transmission. A recent study also showed that viruses can readily transmit between different *Beauveria* spp. (Ning et al. 2022).

Evidence for horizontal transmission in nature, between species as well as incompatible conspecifics, is mixed. One study examined viruses in numerous hosts within a single forest stand highly infected with *Heterobasidion* spp., for which the *Heterobasidion* viruses have been thoroughly studied (Vainio et al. 2017). Interestingly, they did find instances of *Heterobasidion* viruses infecting two other fungal hosts, though it was rare. Another study looking at population structure of closely related species of *Lactarius* in two forests found no cross-species transmission (Sutela and Vainio 2020). Viruses of *Cryphonectria* spp. have been shown to cross the species barrier (Liu et al. 2003). For a given species pair, whether horizontal transmission occurs and to what extent may depend on factors specific to host biology, ecology, and/or habitat.

Phylogenetic evidence supports some amount of cross-species transmission, if limited. That many fungi are infected by multiple, very distantly related viruses implies horizontal transmission, if in the distant past (Arjona-Lopez et al. 2018; Thapa and Roossinck 2019). Tests of cophylogeny indicate that mycovirus families mostly speciate by duplication and cospeciation with the host, but at greater phylogenetic depths there are recurrent host-shifts (Göker et al. 2011; Myers and James 2022). Host-switching across kingdoms seems a substantial hurdle but given the often intimate association between plants and fungi it could be expected to occur at least sometimes. Indeed, many fungal and plant viruses are closely related, and fungal viruses have been shown capable of replicating in plant cells, and vice-versa (Janda and Ahlquist 1993; Nerva et al. 2017; Panavas and Nagy 2003). Remarkably, a natural infection of *Rhizoctonia solani* with cucumber mosaic virus was recently reported, the first direct evidence of cross-kingdom viral transmission from plant to fungus (Andika et al. 2017). Further, in the lab the fungus could become infected with the virus through interactions with the plant, and could then transmit the virus to uninfected plants. Soon after, this group reported widespread, though transient, infection by plant viruses in filamentous fungal strains isolated from plant leaves (Cao et al. 2022b). These recent findings of cross-kingdom viral transmission imply that plants and fungi play a role in the evolution and ecologies of the other's viral infections, perhaps serving as vectors or reservoirs.

Horizontal and environmental transmission is thought to be hindered by the fungal cell wall, though the fungal mechanism of endocytosis provides at least a conceptual entry point (Commer and Shaw 2021). In laboratory settings, protoplasting can be used to facilitate horizontal transmission in some species and sometimes between species, though whether this could occur in nature is unknown (Kanematsu et al. 2010; Lee et al. 2011; van Diepeningen et al. 2000, 1998). It is interesting to consider whether horizontal transmission between individuals of the same or different species might be more frequent among fungi with life cycles that include a zoosporic stage which lack a cell wall.

Although a couple of studies have hinted that insects may also play a role in mycovirus transmission, the consensus remains that transmission is generally limited (Bouneb et al. 2016; Petrzik et al. 2016). Perhaps this idea is changing. It is also possible that the limitation to transmission applies more frequently to RNA mycoviruses, but so few DNA mycoviruses are known it is difficult to say. One of the most exciting recent discoveries in mycovirology is Sclerotinia sclerotiorum hypovirulence-associated DNA virus 1 (SsHAD-V1), a ssDNA virus that is transmissible horizontally to vegetatively incompatible strains (Yu et al. 2010), extracellularly (Yu et al. 2013), and also vectored by a mycophageous insect (Liu et al. 2016). It remains an open, and very interesting, question whether other DNA mycoviruses transmit similarly, and whether RNA mycoviruses may also be environmentally- or vector-transmitted more frequently than currently expected.

7.5 Effects on Host Phenotype

Mycoviruses are notorious for their negative effects on hosts, and these impacts include reduced growth rates, sporulation, and virulence. Some of the best studied examples of mycoviruses include the CHV-1 (*Hypoviridae*) which leads to lower sporulation, reduced growth rate, as well as hypovirulence of the chestnut blight pathogen *Cryphonectria parasitica* (Prospero et al. 2021), and the ssDNA pathogen SsHADV-1 which leads to reduced growth and virulence of *Sclerotinia sclerotiorum* (Yu et al. 2010). Yet the paradigm of mycovirus phenotypic effects is that the majority of viruses have no phenotypic effect on their hosts (Ghabrial et al. 2015). Nonetheless, when phenotypic effects occur, they need not necessarily be detrimental to their host, and numerous instances where viruses promote growth or invasion of their fungal host are known (Applen Clancey et al. 2020; Kotta-Loizou and Coutts 2017a, 2017b; Thapa et al. 2022; Tran et al. 2019). In fact, because mycoviruses are largely vertically transmitted, it stands to reason that they should in general evolve to improve the fitness or virulence of their hosts so that they may themselves increase their own chances of reproducing (Lerer and Shlezinger 2022). To study recent trends in the phenotypic effects uncovered in mycovirus research, we reviewed the recent literature (2019–2022) on viral phenotypic impacts, identifying studies where the specific effects of the virus on the fungal host (and specifically virulence and growth rate) could be measured such as after the clearing of a virus from its host (Table 7.1).

Overall, we found that in the published literature it is more common to observe negative effects on the fungal host, both in terms of growth rate and hypovirulence (Fig. 7.2). It is worth noting that these data are likely to be biased toward viruses that have a negative effect on host fitness, because of (1) the strong motivation of researchers to identify candidates for biological control and (2) the range of negative phenotypic effects is broader and more likely to be observed. Moreover, though less and less the case for viral screening with emerging sequencing methods, viruses can be detected via screening of strains for signs of abnormal morphology or hypovirulence (Boland 1992).

Most of the recent studies measured the phenotypic effect of a mycovirus by clearing the virus from an infected strain using either isolation of numerous random asexual spores or hyphal tips. The clearing of strains is also improved when grown in the presence of cycloheximide or ribavirin. These drugs function by inhibition of RNA processing or protein elongation in eukaryotes, which presumably either have a stronger impact on viral components or help dilute out the virus from cells. In this manner virus-free strains can be compared to genetically identical strains with the virus. Another manner of checking phenotype effects of viruses relies on transmitting the virus from a donor strain to a virus-free recipient strain via hyphal anastomosis (Cornejo et al. 2021; Li et al. 2022c). To do this the two fungal strains need to be vegetatively compatible. Strains are paired and following cell

Table 7.1 List of recent studies of the phenotypic effects of mycoviruses in a diversity of fungal taxa. These studies are mostly limited to a clear relationship between a particular virus and its effect on the host could be established, e.g., by clearing the viruses or transferring it to a virus-free host

Host	Virus	Virus family	Phenotypic effect of virus	Method of establishing phenotype	Reference
Alternaria alternata	AalVV1	Totiviridae	Asymptomatic	Protoplast fusion of virus infected and recipient strain	Jamal et al. 2019
Alternaria alternata	AaHV1	Hypoviridae	Reduced growth and hypovirulence	Single spore isolation	Li et al. 2019
Alternaria alternata	AaBbV1	Botybirnaviridae	No effect on growth or morphology	Transfection via viral particles	Shamsi et al. 2019
Alternaria alternata f. sp. mali	AaCV1-QY2	Chrysoviridae	Hypovirulence, lower mycotoxin production, spore germination reduced	Cycloheximide clearing	Li et al. 2022a
Alternaria tenuissima	AaCV1	Chrysoviridae	Reduced growth and sporulation, but increased resistance to antifungals	Protoplasting	Ma et al. 2020
Aspergillus fumigatus	AfuPmV-1	Polymycoviridae	Reduced growth rate, conidiation, and hypovirulence	Single spore isolation	Takahashi-Nakaguchi et al. 2020a
Aspergillus fumigatus	AfuCV41362	Chrysoviridae	Reduced biomass accumulation, lower spore production, reduced stress tolerance, hypovirulence	Single spore isolation	Takahashi-Nakaguchi et al. 2020b
Botryosphaeria dothidea	BdPV2	Partitiviridae	Reduced growth and virulence	Comparisons between naturally infected and uninfected strains	Y. Wang et al. 2022e
Botryosphaeria dothidea	BdBV1	Alphaflexiviridae	Reduced growth and altered morphology, hypovirulence	Single spore isolation	Yang et al. 2021
Botrytis cinerea	BcBoV19	Botourmiaviridae	No effect on growth, virulence, or morphology	Anastomosis	Q. Wang et al. 2022b
Botrytis cinerea	BGDaV1	Genomoviridae	Reduced growth and virulence	Transfection via viral particles	Khalifa and MacDiarmid 2021
Colletotrichum camelliae	CcBV1	Botourmiaviridae	No effect on growth or sporulation	Single spore isolation	Xu et al. 2022b
Colletotrichum fructicola	CfRV1	Polymycoviridae	Reduced growth and virulence	Anastomosis	Fu et al. 2022
Colletotrichum gloeosporioides	CgRV1-Ssa-44.1	Unclassified	Reduced growth and conidiation, hypovirulence	Single spore isolation	Suharto et al. 2022

(continued)

Table 7.1 (continued)

Host	Virus	Virus family	Phenotypic effect of virus	Method of establishing phenotype	Reference
Colletotrichum higginsianum	ChNRV1	Amalgaviridae	No effect on growth rate (except at highest temperatures), conidium width; mild hypervirulence	Cycloheximide clearing with single spore isolation	Olivé and Campo, 2021
Colletotrichum liriopes	ClPV1	Partitiviridae	Reduced sporulation and virulence, also lower growth rate when transferred into another strain	Protoplasting and separately transfection of other strain	Zhu et al. 2021
Cryphonectria naterciae	CnSpV1	Splipalmiviridae	No effect on growth or morphology	Anastomosis	Sato et al. 2022
Cryphonectria naterciae	CnFGV1	Fusagraviridae	Reduced growth and altered morphology, no effect on virulence	Anastomosis	Cornejo et al. 2021
Diaporthe pseudophoenicicola	DpCV1	Chrysoviridae and ourmia-like	Changes to cell wall, hypovirulence, lower resistance to antifungal	Single spore isolation	Xu et al. 2022a
Fusarium equiseti	FePV1	Partitiviridae	Reduced growth and hypovirulence	Single spore isolation	Mahillon et al. 2021
Fusarium graminearum	FgGMTV1	Genomoviridae	Reduced growth and hypovirulence	Transfection via clone	Li et al. 2020
Fusarium graminearum	FgV1-ch	Fusariviridae	Slightly reduced growth rate and asexual sporulation, no effect on virulence.	Anastomosis	L. Zhang et al. 2020b
Fusarium nygamai	HadV1	Polymycoviridae	No effect on growth or morphology	Single spore isolation	Khan et al. 2021a
Fusarium oxysporum	FoMyV1	Mymonaviridae	Reduced growth and conidiation but no effects on virulence	Anastomosis	J. Wang et al. 2022a
Fusarium oxysporum	FoAV1	Alternaviridae	No effect on growth or morphology	Subculture	Wen et al. 2021
Fusarium oxysporum	FoOuLV1	Botourmiaviridae	Hypovirulence, no information on growth rate	Anastomosis	Zhao et al. 2020
Fusarium oxysporum	HadV1	Polymycoviridae	No effect on growth or morphology	Single spore isolation	Sato et al. 2020
Fusarium oxysporum	FodHV2	Hypoviridae	No effect on growth, sporulation, or virulence	Anastomosis	Torres-Trenas et al. 2020

(continued)

Table 7.1 (continued)

Host	Virus	Virus family	Phenotypic effect of virus	Method of establishing phenotype	Reference
Fusarium pseudograminearum	FpgMBV1	Megabirnaviridae	Hypovirulence	Hyphal tip	Xie et al. 2022
Fusarium verticillioides	FvMV1	Mitoviridae	No differences in growth rate, increased conidium production, increased virulence	Comparisons between naturally infected and uninfected strains	Jacquat et al. 2020
Lentinula edodes	LePV1	Partitiviridae	Reduced growth and changes in morphology, pigmentation	Hyphal fragmentation, ribavirin	Sun et al. 2022
Leptosphaeria biglobosa	LbQV-1	Quadriviridae	Changes in pigmentation, increased growth rate, and hypervirulence	Cycloheximide clearing with single spore isolation; transfection with dsRNAs	Shah et al. 2020
Magnaporthe oryzae	MoV2	Totiviridae	No change in growth rate or morphology, pigmentation, but increased growth rate in leaf sheath test (partial evidence of hypervirulence effect)	Single spore isolation	Owashi et al. 2020
Malassezia restricta	MrV40	Totiviridae	No difference in morphology, however larger vacuoles	Serial transfer	Park et al. 2020
Malassezia sympodialis	MsMV1	Totiviridae	Induction of immune response, enhanced skin colonization	Serial transfer at 37 C	Applen Clancey et al. 2020
Melanconiella theae	MtMV1	Mitoviridae	No effect on growth or morphology	Hyphal tips	Shafik et al. 2021
Nigrospora oryzae	NoNRV1	Partitiviridae	Hypovirulence, increased pigmentation, lower thermotolerance	Hyphal tip isolation using ribavirin	X. Wang et al. 2022c
Penicillium crustosum	PcCV1	Chrysoviridae	No effect on growth but reduced resistance to prochloraz antifungal	Conidium isolation using ribavirin	Wang et al. 2019
Penicillium italicum	PiCV1	Chrysoviridae	No differences in growth rate, pigmentation, or virulence	Anastomosis	Zhang et al. 2019

(continued)

Table 7.1 (continued)

Host	Virus	Virus family	Phenotypic effect of virus	Method of establishing phenotype	Reference
Pestalotiopsis theae	PtCV1	Chrysoviridae	Reduced growth rate and altered morphology; hypovirulence-change to endophyte	Transfection via viral particles; anastomosis	Zhou et al. 2021
Pleurotus ostreatus	OSMV-Ch	Tymoviridae	Reduced growth and mushroom yield	Hyphal tip isolation and high temperature	Hu et al. 2022
Pleurotus ostreatus	PoV	Partitiviridae	Reduced growth and sporocarp production, reduced enzyme activity	Hyphal fragmentation	Song et al. 2020
Pseudogymnoascus destructans	PdPV	Partitiviridae	Increased virulence in a pig ear model	Serial transfer with PEG	Thapa et al. 2022
Rhizoctonia solani	RsRhV1	Rhabdoviridae	Hypervirulence	Anastomosis and protoplast regeneration	Li et al. 2022c
Rhizoctonia solani	RsRV5	Partitiviridae	Reduced growth, sclerotium production, and hypovirulence	Protoplasting	Zhang et al. 2021
Rhizoctonia solani	RsHV2	Hypoviridae	No change in growth rate, altered morphology (fewer sclerotia), hypovirulence	Protoplasting	Abdoulaye et al. 2021
Rosellinia necatrix	RnHV2	Hypoviridae	Reduced growth rate and hypovirulence	Anastomosis	Arjona-López et al. 2021
Sclerotinia sclerotiorum	SlaGemV-1	Genomoviridae	Reduced growth, altered reproduction, hypovirulence	Transfection via clone	Feng et al. n. d.
Sclerotinia sclerotiorum	SsEV3	Endornaviridae	Reduced growth and virulence	Removal of other viruses via cycloheximide and ribavirin and protoplasting (original strain was infected with 9 viruses). Univirally infected with SsEV3 was transmitted by anastomosis	Mu et al. 2021
Sclerotinia sclerotiorum	HuSRV1	Unclassified	Reduced growth rate and hypovirulence	Transfection via viral particles	Azhar et al. 2019

(continued)

Table 7.1 (continued)

Host	Virus	Virus family	Phenotypic effect of virus	Method of establishing phenotype	Reference
Trichoderma atroviride	TaMV1-NFCF377	Fusagraviridae	No differences in growth rate, colony morphology, or conidia production; increased production of an antifungal	Single spore and hyphal tip isolation	Chun et al. 2020b
Trichoderma harzianum	ThHV1	Hypoviridae	Reduced sporulation, growth rate, and hypovirulence	Anastomosis	You et al. 2019
Trichoderma harzianum	ThMV1	Unclassified	Greater biomass accumulation and reduced ability to protect plant host from pathogen	Protoplasting	Liu et al. 2019

fusion, strains from the recipient side are isolated and tested for virus. It is important that the donor strain can be distinguished from the recipient, and this is often done through the use of drug markers, such as hygromycin resistance, which can be transformed into the recipient genome. This ensures that the isolated strain has not been the mere product of hyphal invasion into the recipient side by the donor strain. A final method to study viral-induced effects can be done using experimental transfection. Such an approach can be used to move a virus or parts of a virus between strains as well as between species (Eusebio-Cope et al. 2015). One of the confounding factors in understanding viral effects on phenotype is the large number of viruses a single strain may have. If a fungal strain is infected by 5 or more viruses, as is sometimes the case (Khan et al. 2021b; Myers et al. 2020; Y. Wang et al. 2022d), it can be very difficult to assign phenotypic effect to the component viruses. High-throughput production of synthetic viruses would be useful for dissecting these effects using transformation or transduction.

With the hindsight of having detected a mycovirus, it becomes straightforward to look for particular effects by screening cultures for aberrant phenotypes. Despite the biases in the data just explained, it is striking that the majority of them actually do negatively affect the host (Table 7.1, Fig. 7.2). A couple possibilities could explain this prevalence of negative impacts. Firstly, it may be that viruses are transmitted

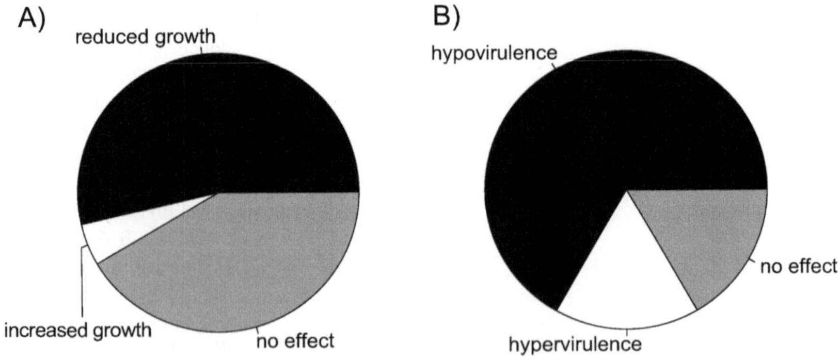

Fig. 7.2 Summary of 53 recent studies on the effects of mycoviruses on fungal host growth (A) and on the effects on the fungal host virulence (B)

horizontally more often than believed, perhaps through endocytosis by one genotype of virions shed by dead fungal cells of another genotype. Alternatively, it could be that from the perspective of growth rate and virulence that the fungal host exists at a local optimum, and it is very difficult for a virus to modify either of these traits in a manner that appears as a net benefit to the host. In other words, since viruses are generally cellular parasites, it is unclear precisely how they might modify molecular pathways within the host organism in order to improve the fitness of both virus and fungus. Nonetheless, for particular virus–fungus combinations, there are instances where this does seem to occur (Jacquat et al. 2020; Owashi et al. 2020; Shah et al. 2020). It is of particular note that these outcomes are context dependent, e.g., host or medium type specific, which does open up the door to dissecting their molecular mechanisms (Filippou et al. 2021). Finally, before turning to the nature of gene expression changes invoked in the host by mycoviruses and the host defense mechanism, we put forward a few examples where context-dependent benefits to the host are observed.

Among the most fascinating mycovirus systems is the infection of panic grass (*Dichanthelium lanuginosum*) with the endophytic fungus *Curvularia protuberata* which itself is infected by an RNA virus CThTV. In a now classical paper, Marquez et al. (2007) showed that the plant requires infection by an endophyte with the virus in order to be able to withstand the high temperature of the soils near geothermal pools at Yellowstone National Park. The mechanism involves a heat stress response and increased osmolyte concentration, however instead of being induced as a stress response to heat, the plants with virus-infected endophytes had a higher constitutive presence of the osmolytes than endophyte-free or endophytes without the virus (Márquez et al. 2007).

A quite different context-dependent phenotype is observed with *Saccharomyces cerevisiae* killer virus system, which is a combination of a L-A helper dsRNA virus, which encodes the capsid and RdRP, and a separate M toxin-encoding dsRNA virus (Schmitt and Breinig 2006). Strains infected with the killer virus system secrete a toxin which kills sensitive cells of *S. cerevisiae* and other yeast species lacking the virus, and only the strains with the killer virus are immune. Immunity appears complicated and dependent on which of the three M encoding viruses is involved (ScV-M1, ScV-M2, and ScV-M28) (Becker and Schmitt 2017; Schmitt and Breinig 2006). Presence of the virus therefore creates an advantage for a virus-infected cell when rare in the environment. Killer viruses are found not only in *S. cerevisiae* but in diverse yeasts, such as *Ustilago maydis* and *Kluyveromyces lactis* (Magliani et al. 1997). There is presumably a strong cost to the production of the toxin as invasion of a non-toxin producing population can only occur under limited conditions of high density and strong spatial structure (Greig and Travisano 2008). Moreover, clones producing the killer virus grown for numerous generations in the lab were shown to lose the ability to both produce and eventually resist the toxin (Buskirk et al. 2020), highlighting both the costs and benefits that virus production and associated resistance entails.

Additional viral-induced phenotypes have been observed which are worth mentioning. Mycoviruses modify the expression and sensitivity of their hosts to antifungals, for example Alternaria alternata chrysovirus 1-N18 (AaCV1-N18) promoted increased resistance of its host to the antifungals difenoconazole or tebuconazole (Ma et al. 2020). Infection by Trichoderma atroviride mycovirus 1 (TaMV1-NFCF377) led to increased production of antifungal metabolites (Chun et al. 2020b) and infection by a totivirus led to increased production of mycotoxin tenuazonic (TeA) acid by *Magnaporthe oryzae* (Ninomiya et al. 2020). Increased sensitivity to antifungals mediated by viral infection is also reported (Niu et al. 2018; Wang et al. 2019). Mycoviruses are also known to reduce the expression of genes involved in programmed cell death triggered by the vegetatively compatibility system (Biella et al. 2002; Li et al. 2008; Shang et al. 2008; Wu et al. 2017). This phenomenon is clearly advantageous to the mycovirus as it will promote the spread of the virus through hyphal fusion of conspecific individuals. Finally, it has been documented that some mycoviruses are able to convert their pathogenic hosts into endophytes (H. Zhang et al. 2020a; Zhou et al. 2021). This may be considered a form of hypovirulence, yet from the perspective of the virus it may lead to promotion of plant growth or increased opportunity for the virus to spread though interactions with other fungal individuals or species in plants.

7.6 Coordinated Virus and Host Response After Infection

The phenotypic changes imparted to the fungal host by the virus presumably occur through the paradigm of altered host gene expression, and this field has been very active describing differences in gene expression using isogenic fungal material with and without viruses (Chun et al. 2020a; Marzano et al. 2018; Wang et al. 2016; Fuyou Deng et al. 2007). For example, an ourmia-like virus of *Sclerotinia sclerotiorum* has only a single ssRNA and one ORF (RdRp), and expression of only RdRp is sufficient for completing the viral cycle (Wang et al. 2020). When more genes are present, the expression of the various viral genes is variable, but can reveal mechanisms by which the host is debilitated by the virus. Dissecting the function of specific viral genes in the fungal host can be accomplished by expression of the viral genes or fragments individually and monitoring phenotypes and host gene expression (Feng et al. n.d.; Gao et al. 2020; Li et al. 2020). For example, the 5 individual ORFs of the polymycovirid AfuPmV-1 M infecting *Aspergillus fumigatus* were individually investigated using expression from a plasmid after transformation (Takahashi-Nakaguchi et al. 2020a). The results show that each of the ORFs has a variable influence and expression level depending on the trait and tissue. In this case, only ORFs 2 and 5 (both with motifs of unknown function) influenced virulence in a mouse model, whereas expression of ORFs 1 (RdRp) & 3 (methyltransferase) influenced the swelling and germination of conidia.

By comparison of isogenic strains with and without a mycovirus, the expression of all host genes can be investigated. Such studies typically find that many genes (hundreds or thousands) exhibit expression differences (Chun et al. 2020a; Gao et al. 2020; Y. Li et al. 2022b), and these data can be used to infer the molecular mechanisms underlying host phenotypes. For example, expression of beta-1,3-glucanase is promoted by infection of *Rhizoctonia solani* with RsPV2 partitivirus (Y. Li et al. 2022b) and may lead to an autolysis of the fungal host associated with hypovirulence toward plants. On the other hand, phenoloxidase and carbohydrate active enzymes were down-regulated in viral-infected *Pleurotus ostreatus* (Song et al. 2020), leading to reduced ability to grow in mushroom production scenarios which may impose a great threat to farming. In general, however, the large number of genes changing expression and their myriad roles highlight that there are few shared mechanisms by which viruses manipulate their host cells.

Despite the diversity of viral and host genes implicated, there are hints that shared mechanisms of fungal host defense exist. Specifically, much emphasis is placed on the expression of genes involved in the RNA silencing (RNAi) pathway, specifically the production of small RNAs by the protein Dicer and the destruction of viral RNA by Argonaute using the small RNAs as a guide (Eusebio-Cope et al. 2015; Li et al. 2013). Gene expression studies of the fungal host show that the RNAi machinery could be up-regulated (Hu et al. 2019) or down-regulated (Aulia et al. 2019; Qu et al. 2021; Yu et al. 2020) during mycoviral infection. Both directions have implications for the outcome of the infection, but from the viral-perspective, reduced expression of RNAi genes should be beneficial for replication. Viral infection also leads to changes in host small RNA expression (Cao et al. 2022a; Marzano et al. 2018), which could relate to the initiation of the RNAi pathway post-infection. Studies of mutants of the DNA methyltransferases (DNMTs) in *Cryphonectria parasitica* suggested their role in defense or tolerance of viral infection as mutants of the DNMTs led to spontaneous loss of the virus despite heavy viral burden (Ko et al. 2021).

Of particular interest is the nature of the viral manipulation of third parties via fungi. A recent example is the upregulation of beta interferon expression in bone marrow-derived macrophages when challenged with virus-infected, relative to virus-cured, *Malassezia sympodialis* (Applen Clancey et al. 2020). These data provide evidence of how a mycovirus can influence third party gene expression which would be indicative of a concomitant induction of immune response. Transcriptome analysis of thermotolerance-promoting virus CThTV infecting the endophyte *Curvularia protuberata* showed increased fungal expression of heat-stress related genes including melanin with virus infection (Morsy et al. 2010), which should promote survival of the fungus at high temperatures, however gene expression in the plant host as modified by the mycovirus has yet to be explored. In contrast, suppression of melanin production was observed with infection of *Magnaporthe oryzae* with chrysovirus 1-D (MoCV1-D) (Higashiura et al. 2019). As melanin is a major virulence factor of many fungi (Nosanchuk and Casadevall 2003), this observation is consistent with a hypovirulence effect of chrysoviruses in many fungi (Higashiura et al. 2019). For the SsHADV-1 infection that converts *S. sclerotiorum* from a necrotrophic pathogen to an endophyte of rapeseed, many genes (3110) of *S. sclerotiorum* are differentially expressed genes (DEGs) during SsHADV-1 infection (Qu et al. 2021). Plant gene expression following inoculation of endophytic strain DT8 with SsHADV-1 showed that expression of genes in the host plant is modified (H. Zhang et al. 2020a), but a direct comparison of plant gene expression with virus infected and uninfected *S. sclerotiorum* would be informative.

7.7 Virus-Mediated Horizontal Gene Transfer

Viruses have the potential to mediate horizontal gene transfer (HGT) between organisms. In bacteria, this process is called transduction and can occur by a number of well-described mechanisms (Chiang et al. 2019). No such mechanisms have been described in fungi, but it is possible that similar ones exist. The most obvious mechanism for gene integration is by integrases, the retroviral enzymes responsible for interweaving viral DNA into cellular DNA. But of the 22 mycoviral families only one, *Pseudoviridae,* encodes integrases (Llorens et al. 2021). Therefore, for most mycoviruses, the integration of genes into their host nuclear genome may be a product of random, erroneous processes. RNA viruses would first require co-option of a reverse-transcriptase to convert to DNA. This viral DNA could then be incorporated into the nuclear genome, perhaps through the co-option of integrases from retrotransposons, or by errors in the repair of double-strand breaks, which has been shown experimentally to be possible (Frank and Wolfe 2009; Moore and Haber 1996; Yu and Gabriel 1999) (Fig. 7.3). Given this, we can make a few predictions. We would expect: (1) DNA viruses or plasmids to be more frequently integrated than RNA since the co-option of a reverse-transcriptase would not be required, (2) variation in the functionality of integrated genes since genes integrate randomly without regard for position of the host's promoter regions or reading frame, and (3) that genes advantageous to the host when carried by a virus are also advantageous when transferred to the nuclear genome and thus would be evolutionarily conserved. Though studies on HGT involving mycoviruses are fairly scant, there have been a few looks into yeast systems that address these predictions.

One of the earliest studies found that 40% of Saccharomycotina genomes tested (n = 26) had copies of viral or plasmid genes integrated into the nuclear genome (Frank and Wolfe 2009). The majority of these were pseudogenic homologs of genes from linear DNA plasmids though some were also from dsRNA viruses and, more rarely, circular DNA plasmids. At least one chromosomal homolog was functional—the *KHS1* gene responsible for the killer phenotype of *S. cerevisiae* strains equipped with this viral gene. Like extrachromosomally-derived killers, strains with the *KHS1* gene secrete a toxin lethal to sensitive cells. Since this study, many more viruses have been characterized which has opened up new possibilities for observing patterns of viral gene integration.

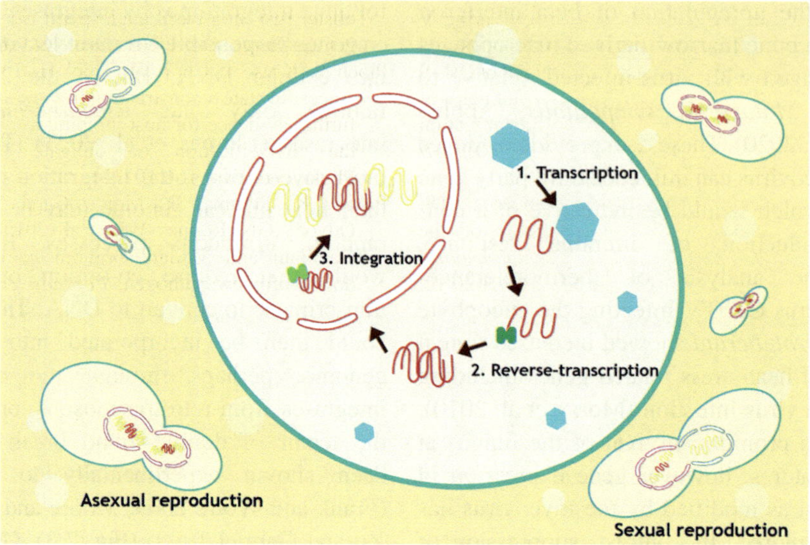

Fig. 7.3 Schematic of a possible mechanism for the integration of viral genetic material into a yeast genome, and subsequent preservation of the viral gene through sexual and asexual reproduction. The virus RNA genome is (1) transcribed and mRNA extruded from the capsid into the cytoplasm where it is (2) reverse-transcribed to DNA through the co-option of a reverse-transcriptase. The DNA molecule enters the nucleus and is (3) integrated into the host genome, perhaps during the repair of a double-stranded break (as predicted by Frank and Wolfe 2009) or by the co-option of an integrase of a retrotransposon. The newly integrated viral gene could be maintained in the population through mitotic division of the cell (asexual reproduction/budding), or by sexual reproduction resulting in variation in the presence of the viral gene in the offspring

Recently, Fredericks et al. identified a new dsRNA virus-associated killer toxin in *Saccharomyces paradoxus,* related to the canonical "killer" in *Saccharomyces cerevisiae*, and found chromosomal homologs to this toxin sporadically throughout the Saccharomycotina, which they showed to be biologically active in killing competitors (2021a). These results implicate dsRNAs as agents of multiple independent HGT events that were, notably, between species. Further, gene duplication has occurred in multiple species with evidence of this gene family evolving under purifying selection.

Virally-derived toxin-encoding genes have clearly made their way into genomes on multiple occasions throughout the Saccharomycotina. In a different approach, Satwika et al. fully sequenced an active plasmid in *Debaryomyces hansenii* and identified multiple homologs of plasmid ORFs of unknown function in the nuclear genome (2012). This result suggests that with more complete characterization of plasmids and viruses, we are likely to find the sources of additional chromosomal elements of fungi. It is interesting to consider how much of fungal genomes may well be from non-fungal sources. And further, how frequently these random, aberrant processes must occur such that eventually the HGT of a viral gene is both functional and does not interfere with the transcription of another critical gene.

Based on these few studies in yeasts, our predictions mostly hold up. But, whether HGT events have occurred in hyphal fungi remains to be tested. Indeed, whether HGT of viral genes is truly more common in yeasts than other fungi or whether this trend is a result of biased sampling is unclear. However, we might expect HGT to be both more common and easier to detect in yeasts. As long as a HGT event is not lethal, it should be possible to maintain in a yeast population since every cell division results in vertical transmission of the virus, whether sexual or asexual (Fig. 7.3). In multicellular fungi the HGT event would have to occur in conidia or a cell that undergoes sexual reproduction in to proliferate vertically. Without vertical transmission, it is unlikely that individual nuclei containing genes of viral origin would mitotically reproduce enough to be widely detected, unless they imbued a very strong fitness advantage. In yeasts, we already know of viral genes that provide strong fitness advantages, and thus searching for these genes in particular has enabled studies of HGT. It is clear how the serendipitous transfer of such a valuable gene as one encoding a secreted toxin could be evolutionarily maintained. Considering that some of these yeasts are typically the dominant microbe of their environment, particularly in artificial environments such as in fermentation production, individuals with these genes would rapidly overtake the populations as individuals susceptible to the toxin would be eliminated. While there are some examples of beneficial viruses in hyphal fungi, there are not such well-understood mechanisms of mutualism as those in the yeast "killers." Characterization of the function of more mycoviral genes may provide more insight and motivate studies of HGT in more diverse fungi.

Whether mycoviruses are mediating genetic exchange between species has yet to be thoroughly investigated, but because mycoviruses may be limited in cross-species transmission it is expected to be less common. Interestingly, a pseudogene homolog of a *Penicillium* virus was found in the nuclear genome of *Vanderwaltozyma polyspora*, possibly suggesting past cross-species viral transmission (Frank and Wolfe 2009). In concept, rare events of cross-species virus transmission could facilitate HGT resulting in functional genes new to the receiving species. This has almost certainly occurred at some point, despite the miniscule probabilities involved, given the vastness of time.

7.8 Applications to Agriculture, Wildlife and Human Health

An increased recognition of the threat that fungal pathogens pose to agriculture, human and wildlife health has motivated interest in alternatives to chemical interventions, which tend to have damaging side effects and be neither practical nor sustainable. Biological control, in which the antagonistic relationship between a pathogen and its own natural parasites is leveraged to minimize the harm done by the primary pathogen, is one alternative typically considered to have a lesser negative impact on the ecosystem. Due to their ability to induce hypovirulence, the prospect of using mycoviruses for biological control has received a lot of interest, particularly in regard to fungal pathogens of agriculturally important crops but also to control fungal diseases of wildlife.

The best studied system involves the hypovirulence-inducing viruses in *Cryphonectria parasitica*. The fungus arrived in North America in the early 1900s and is solely responsible for the blight that effectively eradicated chestnut trees in North America, forever altering the landscape. A few decades later the pathogen arrived in Europe but has caused considerably less damage, in part due to the greater resistance of the European chestnut species and in part due to success of natural biological control. Extensive reviews of the application of hypoviruses for biological control of chestnut blight in both North America and Europe have been published, and the reader is pointed to these for in-depth analysis (Heiniger and Rigling 1994; Milgroom and Cortesi 2004).

In North America, attempts at biological control with hypoviruses have largely failed due to limitations of transmission in nature. In most regions of North America, *C. parasitica* has high levels of sexual reproduction, which restricts hypovirus transmission because these viruses do not transmit in sexual spores, and sexual reproduction maintains diversity in vegetative compatibility loci resulting in reduced horizontal transmission of the viruses. This system arguably poses substantial challenges relative to those in which the viruses, for instance, readily transmit via sexual spores and whose host populations have low diversity in vegetative compatibility loci. Biological control may be more successful for virus/host pairs in which the virus is more efficiently transmitted vertically. However, this creates a paradox, because if a virus transmits faithfully with its host, it will be selected to minimize impacts to host reproduction. An alternative may be to introduce viruses to non-native hosts in the lab to induce hypovirulence, and then apply it to the field (García-Pedrajas et al. 2019). Virus–

host pairs in which the virus spreads more efficiently horizontally should also be more successful for biocontrol. One interesting approach is to genetically engineer "super donors" through knock-outs of *vic* genes to create fungal strains capable of horizontally transmitting hypovirulence-inducing mycoviruses to genotypes with various *vic* alleles (Zhang and Nuss 2016). Nevertheless, the lack of an extracellular route of transmission generally in mycoviruses is a major limitation to their effective and efficient utilization for biological control.

There is one example showing exceptional promise as a biological control agent: *Sclerotinia sclerotiorum* hypovirulence-associated DNA virus 1 (SsHADV-1) is a small circular DNA virus that, unlike other known hypovirulence-inducing viruses, can infect the host when applied extracellularly (Yu et al. 2010). Its natural host, *S. sclerotiorum*, causes white mold on many agriculturally important crops and is difficult to manage. Purified particles of SsHADV-1, when inoculated on plants, reduces the likelihood of disease, minimizes disease severity, and improves crop yield (H. Zhang et al. 2020a). As research progresses, perhaps particularly with DNA mycoviruses, examples of such crop-protective virus application may be more common.

It remains unclear and largely unexplored whether a similar biocontrol strategy could be attempted to control fungal pathogens of humans or wildlife. Clinical isolates of various human pathogens have been shown to be infected by mycoviruses, but the impacts of the viruses are unknown (Kinsella et al. 2022). One promising avenue bypassing the need for extracellular transmission of mycoviruses to the pathogenic strain involves using mycovirally-derived killer toxins therapeutically. Numerous studies have shown the inhibitory capacity of various killer yeasts on pathogenic fungi, including clinical and drug-resistant isolates (Buzzini and Martini 2001; Fredericks et al. 2021b; Giovati et al. 2018; Hodgson et al. 1995; Middelbeek et al. 1980; Schmitt et al. 1997; Walker et al. 1995). Killer toxins may prove to be excellent therapeutic agents against pathogenic yeasts, such as *Candida* sp., because they are nontoxic to human cells and could potentially be applied locally, as opposed to orally-administered fungicidal drugs, while remaining active at low pH, which is required for local treatment of vaginal candidiasis (Fredericks et al. 2021b). Overall therapeutic applications of mycoviruses may be possible and should be explored given the ever-increasing threat of fungal pathogens, but research focus is needed.

7.9 Conclusions and Future Directions

Here we overviewed many of the major areas in mycoviral research with emphasis on the effects of the viruses on their fungal hosts. Viruses come in several flavors, and consideration of the complexity of viral taxonomy requires appreciation that there are multiple independent origins of viruses. A major theme in virus research is that a little can go a long way. Viral fragments with single ORFs are sufficient for self-replication and are sufficient for invoking major phenotypic effects on their hosts. Despite the major biases in reporting of viral phenotypes, we find that actually the majority of the studies in the literature find that viruses negatively impact growth rate and virulence of their hosts, much as might be expected for a pathogen. Nonetheless the variety of different impacts on host phenotypes provides a glimpse into the coevolutionary process of virus and host as well as the implications of these viral effects on fungal biology. Mutualistic benefits are not uncommon, yet it is very likely that the benefits are always context-dependent viruses come with costs. A better, unbiased assessment of virus impacts on fungal phenotypes and how often mutualisms occur is needed. Given the incredible unexplored diversity of mycoviruses, it is likely that emerging systems for their study will arise. On the other hand, continued emphasis on the few model systems, such as Cryphonectria hypovirus 1 (CHV1) and Rosellinia necatrix megabirnavirus 1 (RnMBV1) are needed (Chiba et al. 2009; Eusebio-Cope et al. 2015), as the underlying mechanisms of viral manipulation of host and fungal defense against viruses are poorly understood. There is a clear association of the RNAi pathway and the presence of killer viruses in yeasts (Drinnenberg et al. 2011), and RNAi is involved in viral silencing (Nuss 2011), however the RNAi model lacks a component by which fungi could acquire immunity toward mycoviruses, a phenomenon which is a major open area in mycoviral research.

There is a great promise for using mycoviruses to manipulate fungi in ways that benefit society. As discussed, mycoviruses could be commandeered to help as biocontrol agents. There are numerous other applications which could involve use of virus infections. Specifically, since they can change susceptibility to stressors, such as osmotic, temperature, and antifungal agents, they could be transfected into fungi in agricultural or clinical settings. As mycoviruses are involved in horizontal transfer of fungal genes, and they might be co-opted for use in fungal transformation systems. Fungi are hidden workhorses for natural and derived products and are becoming ever more of interest for their potential applications in the biofuel, fiber and textile, supplement, and food industries. How mycoviruses impact or may potentially enhance fungal industry applications is a largely unexplored area.

References

Abdoulaye AH, Hai D, Tang Q, Jiang D, Fu Y, Cheng J, Lin Y, Li B, Kotta-Loizou I, Xie J (2021) Two distant helicases in one mycovirus: evidence of horizontal gene transfer between mycoviruses, coronaviruses and other nidoviruses. Virus Evol 7:veab043. https://doi.org/10.1093/ve/veab043

Al Rwahnih M, Daubert S, Urbez-Torres JR, Cordero F, Rowhani A (2011) Deep sequencing evidence from single grapevine plants reveals a virome dominated by mycoviruses. Arch Virol 156:397–403. https://doi.org/10.1007/s00705-010-0869-8

Anagnostakis SL (1982) Biological control of chestnut blight. Science 215:466–471. https://doi.org/10.1126/science.215.4532.466

Anagnostakis SL (1988) Cryphonectria Parasitica, cause of chestnut blight. In: Sidhu GS (ed) Advances in plant pathology, genetics of plant pathogenic fungi. Academic Press, pp 123–136. https://doi.org/10.1016/B978-0-12-033706-4.50011-6

Andika IB, Wei S, Cao C, Salaipeth L, Kondo H, Sun L (2017) Phytopathogenic fungus hosts a plant virus: a naturally occurring cross-kingdom viral infection. Proc Natl Acad Sci 114:12267–12272. https://doi.org/10.1073/pnas.1714916114

Applen Clancey S, Ruchti F, Leibund Gut-Landmann S, Heitman J, Ianiri G (2020) A novel mycovirus evokes transcriptional rewiring in the fungus Malassezia and stimulates Beta interferon production in macrophages. mBio 11:e01534-20. https://doi.org/10.1128/mBio.01534-20

Arjona-Lopez JM, Telengech P, Jamal A, Hisano S, Kondo H, Yelin MD, Arjona-Girona I, Kanematsu S, Lopez-Herrera CJ, Suzuki N (2018) Novel, diverse RNA viruses from mediterranean isolates of the phytopathogenic fungus, Rosellinia necatrix: insights into evolutionary biology of fungal viruses. Environ Microbiol 20:1464–1483. https://doi.org/10.1111/1462-2920.14065

Arjona-López JM, Telengech P, Suzuki N, López-Herrera CJ (2021) A moderate level of hypovirulence conferred by a hypovirus in the avocado white root rot fungus, Rosellinia necatrix. Fungal Biol 125:69–76. https://doi.org/10.1016/j.funbio.2020.10.007

Aulia A, Andika IB, Kondo H, Hillman BI, Suzuki N (2019) A symptomless hypovirus, CHV4, facilitates stable infection of the chestnut blight fungus by a coinfecting reovirus likely through suppression of antiviral RNA silencing. Virology 533:99–107. https://doi.org/10.1016/j.virol.2019.05.004

Azhar A, Mu F, Huang H, Cheng J, Fu Y, Hamid MR, Jiang D, Xie J (2019) A novel RNA virus related to Sobemoviruses confers Hypovirulence on the Phytopathogenic fungus Sclerotinia sclerotiorum. Viruses 11:759. https://doi.org/10.3390/v11080759

Becker B, Schmitt MJ (2017) Yeast killer toxin K28: biology and unique strategy of host cell intoxication and killing. Toxins 9:333. https://doi.org/10.3390/toxins9100333

Bevan EA, Herring AJ, Mitchell DJ (1973) Preliminary characterization of two species of ds RNA in yeast and their relationship to the "killer" character. Nature 245: 81–86. https://doi.org/10.1038/245081b0

Biella S, Smith ML, Aist JR, Cortesi P, Milgroom MG (2002) Programmed cell death correlates with virus transmission in a filamentous fungus. Proc R Soc Lond B Biol Sci 269:2269–2276. https://doi.org/10.1098/rspb.2002.2148

Boland GJ (1992) Hypovirulence and double-stranded RNA in Sclerotinia sclerotiorum. Can J Plant Pathol 14:10–17. https://doi.org/10.1080/07060669209500900

Bonfante P, Desirò A (2017) Who lives in a fungus? The diversity, origins and functions of fungal endobacteria living in Mucoromycota. ISME J 11:1727–1735. https://doi.org/10.1038/ismej.2017.21

Bouneb M, Turchetti T, Nannelli R, Roversi PF, Paoli F, Danti R, Simoni S (2016) Occurrence and transmission of mycovirus Cryphonectria hypovirus 1 from dejecta of Thyreophagus corticalis (Acari, Acaridae). Fungal Biol 120:351–357. https://doi.org/10.1016/j.funbio.2015.11.004

Buskirk SW, Rokes AB, Lang GI (2020) Adaptive evolution of nontransitive fitness in yeast. eLife 9:e62238. https://doi.org/10.7554/eLife.62238

Buzzini P, Martini A (2001) Discrimination between Candida albicans and other pathogenic species of the genus Candida by their differential sensitivities to toxins of a panel of killer yeasts. J Clin Microbiol 39:3362–3364. https://doi.org/10.1128/JCM.39.9.3362-3364.2001

Cao CTH, Derbyshire MC, Regmi R, Li H, Jones MGK, Wylie SJ (2022a) Small RNA analyses of a Ceratobasidium isolate infected with three endornaviruses. Viruses 14:2276. https://doi.org/10.3390/v14102276

Cao X, Liu J, Pang J, Kondo H, Chi S, Zhang J, Sun L, Andika IB (2022b) Common but nonpersistent acquisitions of plant viruses by plant-associated fungi. Viruses 14:2279. https://doi.org/10.3390/v14102279

Chen B, Choi GH, Nuss DL (1994) Attenuation of fungal virulence by synthetic infectious hypovirus transcripts. Science 264:1762–1764. https://doi.org/10.1126/science.8209256

Chiang YN, Penadés JR, Chen J (2019) Genetic transduction by phages and chromosomal islands: the new and noncanonical. PLoS Pathog 15:e1007878. https://doi.org/10.1371/journal.ppat.1007878

Chiba S, Salaipeth L, Lin Y-H, Sasaki A, Kanematsu S, Suzuki N (2009) A novel bipartite double-stranded RNA Mycovirus from the white root rot fungus Rosellinia necatrix: molecular and biological characterization, taxonomic considerations, and potential for biological control. J Virol 83:12801–12812. https://doi.org/10.1128/JVI.01830-09

Chu Y-M, Lim W-S, Yea S-J, Cho J-D, Lee Y-W, Kim K-H (2004) Complexity of dsRNA Mycovirus isolated from Fusarium graminearum. Virus Genes 28:135–143. https://doi.org/10.1023/B:VIRU.0000012270.67302.35

Chun J, Ko Y-H, Kim D-H (2020a) Transcriptome analysis of Cryphonectria parasitica infected with Cryphonectria hypovirus 1 (CHV1) reveals distinct genes related to fungal metabolites, virulence, antiviral RNA-silencing, and their regulation. Front. Microbiol 11:1711

Chun J, Na B, Kim D-H (2020b) Characterization of a novel dsRNA mycovirus of Trichoderma atroviride NFCF377 reveals a member of "Fusagraviridae" with changes in antifungal activity of the host fungus. J Microbiol 58:1046–1053. https://doi.org/10.1007/s12275-020-0380-1

Commer B, Shaw BD (2021) Current views on endocytosis in filamentous fungi. Mycology 12:1–9. https://doi.org/10.1080/21501203.2020.1741471

Cornejo C, Hisano S, Bragança H, Suzuki N, Rigling D (2021) A new double-stranded RNA mycovirus in Cryphonectria naterciae is able to cross the species barrier and is deleterious to a new host. J Fungi 7:861. https://doi.org/10.3390/jof7100861

Cortesi P, McCulloch CE, Song H, Lin H, Milgroom MG (2001) Genetic control of horizontal virus transmission in the chestnut blight fungus, *Cryphonectria parasitica*. Genetics 159:107–118. https://doi.org/10.1093/genetics/159.1.107

Coutts RHA, Livieratos IC (2003) A rapid method for sequencing the 5′- and 3′-termini of double-stranded RNA viral templates using RLM-RACE. J Phytopathol 151:525–527. https://doi.org/10.1046/j.1439-0434.2003.00755.x

Dalzoto PR, Glienke-Blanco C, Kava-Cordeiro V, Ribeiro JZ, Kitajima EW, Azevedo JL (2006) Horizontal transfer and hypovirulence associated with double-stranded RNA in Beauveria bassiana. Mycol Res 110:1475–1481. https://doi.org/10.1016/j.mycres.2006.08.009

Deng F, Allen T, Hillman B, Nuss D (2007) *Comparative analysis of alterations in host phenotype and transcript accumulation following hypovirus and mycoreovirus infections of the chestnut blight fungus Cryphonectria parasitica* [WWW document]. https://doi.org/10.1128/EC.00166-07

Drinnenberg IA, Fink GR, Bartel DP (2011) Compatibility with killer explains the rise of RNAi-deficient fungi. Science 333:1592. https://doi.org/10.1126/science.1209575

Du Z, Tang Y, Zhang S, She X, Lan G, Varsani A, He Z (2014) Identification and molecular characterization of a single-stranded circular DNA virus with similarities to Sclerotinia sclerotiorum hypovirulence-associated DNA virus 1. Arch Virol 159:1527–1531. https://doi.org/10.1007/s00705-013-1890-5

Eusebio-Cope A, Sun L, Tanaka T, Chiba S, Kasahara S, Suzuki N (2015) The chestnut blight fungus for studies on virus/host and virus/virus interactions: from a natural to a model host. Virology 477:164–175. https://doi.org/10.1016/j.virol.2014.09.024

Feldman TS, Morsy MR, Roossinck MJ (2012) Are communities of microbial symbionts more diverse than communities of macrobial hosts? Fungal Biol 116:465–477. https://doi.org/10.1016/j.funbio.2012.01.005

Feng C, Feng J, Wang Z, Pedersen C, Wang X, Saleem H, Domier L, Marzano S-YL (n.d.) Identification of the viral determinant of hypovirulence and host range in sclerotiniaceae of a genomovirus reconstructed from the plant metagenome. J. Virol. 95:e00264-21. https://doi.org/10.1128/JVI.00264-21

Filippou C, Diss RM, Daudu JO, Coutts RHA, Kotta-Loizou I (2021) The polymycovirus-mediated growth enhancement of the entomopathogenic fungus Beauveria bassiana is dependent on carbon and nitrogen metabolism. Front. Microbiol. 12:606366

Frank AC, Wolfe KH (2009) Evolutionary capture of viral and plasmid DNA by yeast nuclear chromosomes. Eukaryot Cell 8:1521–1531. https://doi.org/10.1128/EC.00110-09

Fredericks LR, Lee MD, Crabtree AM, Boyer JM, Kizer EA, Taggart NT, Roslund CR, Hunter SS, Kennedy CB, Willmore CG, Tebbe NM, Harris JS, Brocke SN, Rowley PA (2021a) The species-specific acquisition and diversification of a K1-like family of killer toxins in budding yeasts of the Saccharomycotina. PLoS Genet 17:e1009341. https://doi.org/10.1371/journal.pgen.1009341

Fredericks LR, Lee MD, Eckert HR, Li S, Shipley MA, Roslund CR, Boikov DA, Kizer EA, Sobel JD, Rowley PA (2021b) Vaginal isolates of Candida glabrata are uniquely susceptible to ionophoric killer toxins produced by Saccharomyces cerevisiae. Antimicrob Agents Chemother 65:e02450-20. https://doi.org/10.1128/AAC.02450-20

Fu M, Zhang H, Yin M, Han Z, Bai Q, Peng Y, Shafik K, Zhai L, Hong N, Xu W, Wang G, Kotta-Loizou I (2022) A novel heptasegmented positive-sense single-stranded RNA virus from the phytopathogenic fungus Colletotrichum fructicola. J Virol 96:e00318-22. https://doi.org/10.1128/jvi.00318-22

Gao Z, Wu J, Jiang D, Xie J, Cheng J, Lin Y (2020) ORF I of Mycovirus SsNSRV-1 is associated with debilitating symptoms of Sclerotinia sclerotiorum. Viruses 12:456. https://doi.org/10.3390/v12040456

García-Pedrajas MD, Cañizares MC, Sarmiento-Villamil JL, Jacquat AG, Dambolena JS (2019) Mycoviruses in biological control: from basic research to field implementation. Phytopathology 109:1828–1839. https://doi.org/10.1094/PHYTO-05-19-0166-RVW

Ghabrial SA (1998) Origin, adaptation and evolutionary pathways of fungal viruses. Virus Genes 16:119–131. https://doi.org/10.1023/A:1007966229595

Ghabrial SA, Castón JR, Jiang D, Nibert ML, Suzuki N (2015) 50-plus years of fungal viruses. Virology 60th Anniversary Issue 479–480:356–368. https://doi.org/10.1016/j.virol.2015.02.034

Gilbert KB, Holcomb EE, Allscheid RL, Carrington JC (2019) Hiding in plain sight: new virus genomes discovered via a systematic analysis of fungal public transcriptomes. PLoS One 14:e0219207. https://doi.org/10.1371/journal.pone.0219207

Giovati L, Santinoli C, Ferrari E, Ciociola T, Martin E, Bandi C, Ricci I, Epis S, Conti S (2018) Candidacidal activity of a novel killer toxin from Wickerhamomyces anomalus against fluconazole-susceptible and -resistant strains. Toxins 10:68. https://doi.org/10.3390/toxins10020068

Göker M, Scheuner C, Klenk H-P, Stielow JB, Menzel W (2011) Codivergence of mycoviruses with their hosts. PLoS One 6:e22252. https://doi.org/10.1371/journal.pone.0022252

Gorbalenya AE, Krupovic M, Mushegian A, Kropinski AM, Siddell SG, Varsani A, Adams MJ, Davison AJ, Dutilh BE, Harrach B, Harrison RL, Junglen S, King AMQ, Knowles NJ, Lefkowitz EJ, Nibert ML, Rubino L, Sabanadzovic S, Sanfaçon H, Simmonds P, Walker PJ, Zerbini FM, Kuhn JH, International Committee on Taxonomy of Viruses Executive Committee (2020) The new scope of virus taxonomy: partitioning the virosphere into 15 hierarchical ranks. Nat Microbiol 5:668–674. https://doi.org/10.1038/s41564-020-0709-x

Greig D, Travisano M (2008) Density-dependent effects on allelopathic interactions in yeast. Evolution 62:521–527. https://doi.org/10.1111/j.1558-5646.2007.00292.x

Heiniger U, Rigling D (1994) Biological control of chestnut blight in Europe. Annu Rev Phytopathol 32:581–599

Herrero N, Dueñas E, Quesada-Moraga E, Zabalgogeazcoa I (2012) Prevalence and diversity of viruses in the entomopathogenic fungus Beauveria bassiana. Appl Environ Microbiol 78:8523–8530. https://doi.org/10.1128/AEM.01954-12

Higashiura T, Katoh Y, Urayama S-I, Hayashi O, Aihara M, Fukuhara T, Fuji S-I, Kobayashi T, Hase S, Arie T, Teraoka T, Komatsu K, Moriyama H (2019) Magnaporthe oryzae chrysovirus 1 strain D confers growth inhibition to the host fungus and exhibits multiform viral structural proteins. Virology 535:241–254. https://doi.org/10.1016/j.virol.2019.07.014

Hillman BI, Suzuki N (2004) Viruses of the chestnut blight fungus, Cryphonectria parasitica. In: Maramorosch K, Shatkin AJ (eds) Advances in virus research. Elsevier/Academic Press, Amsterdam, pp 423–472. https://doi.org/10.1016/S0065-3527(04)63007-7

Hodgson VJ, Button D, Walker GM (1995) Anti-Candida activity of a novel killer toxin from the yeast Williopsis mrakii. Microbiol Read Engl 141(Pt 8):2003–2012. https://doi.org/10.1099/13500872-141-8-2003

Hollings M (1962) Viruses associated with a die-back disease of cultivated mushroom. Nature 196:962–965

Hu W, Luo H, Yang Y, Wang Q, Hong N, Wang G, Wang A, Wang L (2019) Comprehensive analysis of full genome sequence and Bd-mil RNA/target mRNAs to discover the mechanism of hypovirulence in Botryosphaeria dothidea strains on pear infection with BdCV1 and BdPV1. IMA Fungus 10:3. https://doi.org/10.1186/s43008-019-0008-4

Hu H-J, Wang J-R, Cheng X-H, Liu Y, Zhang X-Y (2022) Preliminary studies on the effects of oyster mushroom spherical virus China strain on the mycelial growth and fruiting body yield of the edible mushroom Pleurotus ostreatus. Biology 11:574. https://doi.org/10.3390/biology11040574

Ihrmark K, Johannesson H, Stenström E, Stenlid J (2002) Transmission of double-stranded RNA in Heterobasidion annosum. Fungal Genet Biol 36:147–154. https://doi.org/10.1016/S1087-1845(02)00011-7

Jacquat AG, Theumer MG, Cañizares MC, Debat HJ, Iglesias J, García Pedrajas MD, Dambolena JS (2020) A survey of mycoviral infection in Fusarium spp. isolated from maize and sorghum in Argentina identifies the first mycovirus from Fusarium verticillioides. Viruses 12:1161. https://doi.org/10.3390/v12101161

Jamal A, Sato Y, Shahi S, Shamsi W, Kondo H, Suzuki N (2019) Novel victorivirus from a Pakistani isolate of Alternaria alternata lacking a typical translational stop/restart sequence signature. Viruses 11:577. https://doi.org/10.3390/v11060577

Janda M, Ahlquist P (1993) RNA-dependent replication, transcription, and persistence of brome mosaic virus RNA replicons in S. cerevisiae. Cell 72:961–970. https://doi.org/10.1016/0092-8674(93)90584-D

Kanematsu S, Sasaki A, Onoue M, Oikawa Y, Ito T (2010) Extending the fungal host range of a Partitivirus and a Mycoreovirus from Rosellinia necatrix by inoculation of protoplasts with virus particles. Phytopathology 100:922–930. https://doi.org/10.1094/PHYTO-100-9-0922

Khalifa ME, Mac Diarmid RM (2021) A mechanically transmitted DNA Mycovirus is targeted by the defence machinery of its host, Botrytis cinerea. Viruses 13: 1315. https://doi.org/10.3390/v13071315

Khan HA, Sato Y, Kondo H, Jamal A, Bhatti MF, Suzuki N (2021a) A second capsidless hadakavirus strain with 10 positive-sense single-stranded RNA genomic segments from Fusarium nygamai. Arch Virol 166: 2711–2722. https://doi.org/10.1007/s00705-021-05176-x

Khan HA, Shamsi W, Jamal A, Javaied M, Sadiq M, Fatma T, Ahmed A, Arshad M, Waseem M, Babar S, Dogar MM, Virk N, Janjua HA, Kondo H, Suzuki N, Bhatti MF (2021b) Assessment of mycoviral diversity in Pakistani fungal isolates revealed infection by 11 novel viruses of a single strain of Fusarium mangiferae isolate SP1. J Gen Virol 102:001690. https://doi.org/10.1099/jgv.0.001690

Kinsella CM, Deijs M, Gittelbauer HM, van der Hoek L, van Dijk K (2022) Human clinical isolates of pathogenic fungi are host to diverse mycoviruses. Microbiol Spectr 10:e01610-22. https://doi.org/10.1128/spectrum.01610-22

Ko Y-H, So K-K, Chun J, Kim D-H (2021) Distinct roles of two DNA methyltransferases from Cryphonectria parasitica in fungal virulence, responses to Hypovirus infection, and viral clearance. mBio 12:e02890-20. https://doi.org/10.1128/mBio.02890-20

Kondo H, Botella L, Suzuki N (2022) Mycovirus diversity and evolution revealed/inferred from recent studies. Annu Rev Phytopathol 60:307–336. https://doi.org/10.1146/annurev-phyto-021621-122122

Kotta-Loizou I, Coutts RHA (2017a) Studies on the Virome of the Entomopathogenic fungus Beauveria bassiana reveal novel dsRNA elements and mild Hypervirulence. PLoS Pathog 13:e1006183. https://doi.org/10.1371/journal.ppat.1006183

Kotta-Loizou I, Coutts RHA (2017b) Mycoviruses in aspergilli: a comprehensive review. Front. Microbiol. 8:1699

Lee K-M, Yu J, Son M, Lee Y-W, Kim K-H (2011) Transmission of Fusarium boothii Mycovirus via protoplast fusion causes Hypovirulence in other Phytopathogenic fungi. PLoS One 6:e21629. https://doi.org/10.1371/journal.pone.0021629

Lerer V, Shlezinger N (2022) Inseparable companions: fungal viruses as regulators of fungal fitness and host adaptation. Front. Cell. Infect. Microbiol. 12:1556

Li H, Fu Y, Jiang D, Li G, Ghabrial SA, Yi X (2008) Down-regulation of Sclerotinia sclerotiorum gene expression in response to infection with Sclerotinia sclerotiorum debilitation-associated RNA virus. Virus Res 135:95–106. https://doi.org/10.1016/j.virusres.2008.02.011

Li Y, Lu J, Han Y, Fan X, Ding S-W (2013) RNA interference functions as an antiviral immunity mechanism in mammals. Science 342:231–234. https://doi.org/10.1126/science.1241911

Li H, Bian R, Liu Q, Yang L, Pang T, Salaipeth L, Andika IB, Kondo H, Sun L (2019) Identification of a novel Hypovirulence-inducing Hypovirus from Alternaria alternata. Front. Microbiol. 10:1076

Li P, Wang S, Zhang L, Qiu D, Zhou X, Guo L (2020) A tripartite ssDNA mycovirus from a plant pathogenic fungus is infectious as cloned DNA and purified virions. Sci Adv 6:eaay9634. https://doi.org/10.1126/sciadv.aay9634

Li B, Cao Y, Ji Z, Zhang J, Meng X, Dai P, Hu T, Wang S, Cao K, Wang Y (2022a) Coinfection of two Mycoviruses confers Hypovirulence and reduces the production of Mycotoxin Alternariol in Alternaria alternata f sp. mali. Front Microbiol 13:910712. https://doi.org/10.3389/fmicb.2022.910712

Li Y, Li S, Liang Z, Cai Q, Zhou T, Zhao C, Wu X (2022b) RNA-seq analysis of Rhizoctonia solani AG-4HGI strain BJ-1H infected by a new viral strain of Rhizoctonia solani Partitivirus 2 reveals a potential mechanism for Hypovirulence. Phytopathology 112:1373–1385. https://doi.org/10.1094/PHYTO-08-21-0349-R

Li Y, Lyu R, Hai D, Jia J, Jiang D, Fu Y, Cheng J, Lin Y, Xie J (2022c) Two novel Rhabdoviruses related to Hypervirulence in a Phytopathogenic fungus. J Virol 96:e00012–e00022. https://doi.org/10.1128/jvi.00012-22

Liu X, Gorovsky MA (1993) Mapping the 5′ and 3′ ends of Tetrahymena thermophila mRNAs using RNA ligase mediated amplification of cDNA ends (RLM-RACE). Nucleic Acids Res 21:4954–4960. https://doi.org/10.1093/nar/21.21.4954

Liu Y-C, Linder-Basso D, Hillman BI, Kaneko S, Milgroom MG (2003) Evidence for interspecies transmission of viruses in natural populations of filamentous fungi in the genus Cryphonectria. Mol Ecol 12: 1619–1628. https://doi.org/10.1046/j.1365-294X.2003.01847.x

Liu H, Fu Y, Li B, Yu X, Xie J, Cheng J, Ghabrial SA, Li G, Yi X, Jiang D (2011) Widespread horizontal gene transfer from circular single-stranded DNA viruses to eukaryotic genomes. BMC Evol Biol 11:276. https://doi.org/10.1186/1471-2148-11-276

Liu S, Xie J, Cheng J, Li B, Chen T, Fu Y, Li G, Wang M, Jin H, Wan H, Jiang D (2016) Fungal DNA virus

infects a mycophagous insect and utilizes it as a transmission vector. Proc Natl Acad Sci 113:12803–12808. https://doi.org/10.1073/pnas.1608013113
Liu C, Li M, Redda ET, Mei J, Zhang J, Wu B, Jiang X (2019) A novel double-stranded RNA mycovirus isolated from Trichoderma harzianum. Virol J 16: 113. https://doi.org/10.1186/s12985-019-1213-x
Llorens C, Soriano B, Krupovic M (2021) ICTV virus taxonomy profile: pseudoviridae. J Gen Virol 102: 001563. https://doi.org/10.1099/jgv.0.001563
Ma G, Zhang X, Hua H, Zhou T, Wu X (2020) Molecular and biological characterization of a novel strain of Alternaria alternata chrysovirus 1 identified from the pathogen Alternaria tenuissima causing watermelon leaf blight. Virus Res 280:197904. https://doi.org/10.1016/j.virusres.2020.197904
Magliani W, Conti S, Gerloni M, Bertolotti D, Polonelli L (1997) Yeast killer systems. Clin Microbiol Rev 10: 369–400
Mahillon M, Decroës A, Caulier S, Tiendrebeogo A, Legrève A, Bragard C (2021) Genomic and biological characterization of a novel partitivirus infecting Fusarium equiseti. Virus Res 297:198386. https://doi.org/10.1016/j.virusres.2021.198386
Márquez LM, Redman RS, Rodriguez RJ, Roossinck MJ (2007) A virus in a fungus in a plant: three-way Symbiosis required for thermal tolerance. Science 315:513–515. https://doi.org/10.1126/science.1136237
Marzano S-YL, Domier LL (2016) Novel mycoviruses discovered from metatranscriptomics survey of soybean phyllosphere phytobiomes. Virus Res 213:332–342. https://doi.org/10.1016/j.virusres.2015.11.002
Marzano S-YL, Nelson BD, Ajayi-Oyetunde O, Bradley CA, Hughes TJ, Hartman GL, Eastburn DM, Domier LL (2016) Identification of diverse mycoviruses through metatranscriptomics characterization of the viromes of five major fungal plant pathogens. J Virol 90:6846–6863. https://doi.org/10.1128/JVI.00357-16
Marzano S-YL, Neupane A, Domier L (2018) Transcriptional and small RNA responses of the white Mold fungus Sclerotinia sclerotiorum to infection by a virulence-attenuating hypovirus. Viruses 10:713. https://doi.org/10.3390/v10120713
Middelbeek EJ, Hermans JM, Stumm C, Muytjens HL (1980) High incidence of sensitivity to yeast killer toxins among Candida and Torulopsis isolates of human origin. Antimicrob Agents Chemother 17: 350–354. https://doi.org/10.1128/AAC.17.3.350
Milgroom M, Cortesi P (2004) Biological control of chestnut blight with hypovirulence: a critical analysis. Annu Rev Phytopathol 42:311–338
Moore JK, Haber JE (1996) Capture of retrotransposon DNA at the sites of chromosomal double-strand breaks. Nature 383:644–646. https://doi.org/10.1038/383644a0
Morris TJ, Dodds JA (1979) Isolation and analysis of double-stranded RNA from virus-infected plant and fungal tissue. Phytopathology 69:854. https://doi.org/10.1094/Phyto-69-854
Morsy MR, Oswald J, He J, Tang Y, Roossinck MJ (2010) Teasing apart a three-way symbiosis: transcriptome analyses of Curvularia protuberata in response to viral infection and heat stress. Biochem Biophys Res Commun 401:225–230. https://doi.org/10.1016/j.bbrc.2010.09.034
Mu F, Li B, Cheng S, Jia J, Jiang D, Fu Y, Cheng J, Lin Y, Chen T, Xie J (2021) Nine viruses from eight lineages exhibiting new evolutionary modes that co-infect a hypovirulent phytopathogenic fungus. PLoS Pathog 17:e1009823. https://doi.org/10.1371/journal.ppat.1009823
Myers JM, James TY (2022) Mycoviruses. Curr Biol 32: R150–R155. https://doi.org/10.1016/j.cub.2022.01.049
Myers JM, Bonds AE, Clemons RA, Thapa NA, Simmons DR, Carter-House D, Ortanez J, Liu P, Miralles-Durán A, Desirò A, Longcore JE, Bonito G, Stajich JE, Spatafora JW, Chang Y, Corrochano LM, Gryganskyi A, Grigoriev IV, James TY (2020) Survey of early-diverging lineages of fungi reveals abundant and diverse mycoviruses. mBio 11:e02027-20. https://doi.org/10.1128/mBio.02027-20
Nerva L, Ciuffo M, Vallino M, Margaria P, Varese GC, Gnavi G, Turina M (2016) Multiple approaches for the detection and characterization of viral and plasmid symbionts from a collection of marine fungi. Virus Res 219:22–38. https://doi.org/10.1016/j.virusres.2015.10.028
Nerva L, Varese GC, Falk BW, Turina M (2017) Mycoviruses of an endophytic fungus can replicate in plant cells: evolutionary implications. Sci Rep 7:1908. https://doi.org/10.1038/s41598-017-02017-3
Ning S, Kang Q, Liu H, Lu Y, Sui L, Xu W, Shi W, Li Q, Zhang Z (2022) Interspecific spread of dsRNA mycoviruses in entomogenous fungi Beauveria spp. Virus Res 322:198933. https://doi.org/10.1016/j.virusres.2022.198933
Ninomiya A, Urayama S, Suo R, Itoi S, Fuji S, Moriyama H, Hagiwara D (2020) Mycovirus-induced Tenuazonic acid production in a Rice blast fungus Magnaporthe oryzae. Front. Microbiol. 11:1641
Niu Y, Yuan Y, Mao J, Yang Z, Cao Q, Zhang T, Wang S, Liu D (2018) Characterization of two novel mycoviruses from Penicillium digitatum and the related fungicide resistance analysis. Sci Rep 8:5513. https://doi.org/10.1038/s41598-018-23807-3
Nosanchuk JD, Casadevall A (2003) The contribution of melanin to microbial pathogenesis. Cell Microbiol 5: 203–223. https://doi.org/10.1046/j.1462-5814.2003.00268.x
Nuss DL (2005) Hypovirulence: Mycoviruses at the fungal–plant interface. Nat Rev Microbiol 3:632–642. https://doi.org/10.1038/nrmicro1206
Nuss DL (2011) Mycoviruses, RNA silencing, and viral RNA recombination. Adv Virus Res 80:25–48. https://doi.org/10.1016/B978-0-12-385987-7.00002-6
Olivé M, Campo S (2021) The dsRNA mycovirus ChNRV1 causes mild hypervirulence in the fungal

phytopathogen Colletotrichum higginsianum. Arch Microbiol 203:241–249. https://doi.org/10.1007/s00203-020-02030-7

Owashi Y, Aihara M, Moriyama H, Arie T, Teraoka T, Komatsu K (2020) Population structure of double-stranded RNA Mycoviruses that infect the Rice blast fungus Magnaporthe oryzae in Japan. Front. Microbiol. 11:593784

Panavas T, Nagy PD (2003) Yeast as a model host to study replication and recombination of defective interfering RNA of tomato bushy stunt virus. Virology 314:315–325. https://doi.org/10.1016/S0042-6822(03)00436-7

Park M, Cho Y-J, Kim D, Yang C-S, Lee SM, Dawson TL, Nakamizo S, Kabashima K, Lee YW, Jung WH (2020) A novel virus alters gene expression and vacuolar morphology in Malassezia cells and induces a TLR3-mediated inflammatory immune response. mBio 11: e01521-20. https://doi.org/10.1128/mBio.01521-20

Pearson MN, Beever RE, Boine B, Arthur K (2009) Mycoviruses of filamentous fungi and their relevance to plant pathology. Mol Plant Pathol 10:115–128. https://doi.org/10.1111/j.1364-3703.2008.00503.x

Petrzik K, Sarkisova T, Starý J, Koloniuk I, Hrabáková L, Kubešová O (2016) Molecular characterization of a new monopartite dsRNA mycovirus from mycorrhizal Thelephora terrestris (Ehrh.) and its detection in soil oribatid mites (Acari: Oribatida). Virology 489:12–19. https://doi.org/10.1016/j.virol.2015.11.009

Prospero S, Botella L, Santini A, Robin C (2021) Biological control of emerging forest diseases: how can we move from dreams to reality? For Ecol Manag 496:119377. https://doi.org/10.1016/j.foreco.2021.119377

Qu Z, Fu Y, Lin Y, Zhao Z, Zhang X, Cheng J, Xie J, Chen T, Li B, Jiang D (2021) Transcriptional responses of Sclerotinia sclerotiorum to the infection by SsHADV-1. J Fungi 7:493. https://doi.org/10.3390/jof7070493

Sato Y, Shamsi W, Jamal A, Bhatti MF, Kondo H, Suzuki N (2020) Hadaka virus 1: a Capsidless eleven-segmented positive-sense single-stranded RNA virus from a Phytopathogenic fungus, Fusarium oxysporum. mBio 11:e00450-20. https://doi.org/10.1128/mBio.00450-20

Sato Y, Shahi S, Telengech P, Hisano S, Cornejo C, Rigling D, Kondo H, Suzuki N (2022) A new tetra-segmented splipalmivirus with divided RdRP domains from Cryphonectria naterciae, a fungus found on chestnut and cork oak trees in Europe. Virus Res 307: 198606. https://doi.org/10.1016/j.virusres.2021.198606

Satwika D, Klassen R, Meinhardt F (2012) Repeated capture of a cytoplasmic linear plasmid by the host nucleus in Debaryomyces hansenii. Yeast 29:145–154. https://doi.org/10.1002/yea.2893

Schmitt MJ, Breinig F (2006) Yeast viral killer toxins: lethality and self-protection. Nat Rev Microbiol 4: 212–221. https://doi.org/10.1038/nrmicro1347

Schmitt MJ, Poravou O, Trenz K, Rehfeldt K (1997) Unique double-stranded RNAs responsible for the anti-Candida activity of the yeast Hanseniaspora uvarum. J Virol 71:8852–8855. https://doi.org/10.1128/jvi.71.11.8852-8855.1997

Shafik K, Umer M, You H, Aboushedida H, Wang Z, Ni D, Xu W (2021) Characterization of a novel Mitovirus infecting Melanconiella theae isolated from tea plants. Front. Microbiol. 12:757556

Shah UA, Kotta-Loizou I, Fitt BDL, Coutts RHA (2020) Mycovirus-induced Hypervirulence of Leptosphaeria biglobosa enhances systemic acquired resistance to Leptosphaeria maculans in Brassica napus. Mol Plant-Microbe Interact® 33:98–107. https://doi.org/10.1094/MPMI-09-19-0254-R

Shamsi W, Sato Y, Jamal A, Shahi S, Kondo H, Suzuki N, Bhatti MF (2019) Molecular and biological characterization of a novel botybirnavirus identified from a Pakistani isolate of Alternaria alternata. Virus Res 263:119–128. https://doi.org/10.1016/j.virusres.2019.01.006

Shang J, Wu X, Lan X, Fan Y, Dong H, Deng Y, Nuss DL, Chen B (2008) Large-scale expressed sequence tag analysis for the chestnut blight fungus Cryphonectria parasitica. Fungal Genet Biol 45:319–327. https://doi.org/10.1016/j.fgb.2007.11.002

Sinden JW, Hauser E (1950) Report on two new mushroom diseases. Mushroom Sci 1:96–100

Smit WA, Wingfield BD, Wingfield MJ (1996) Reduction of laccase activity and other Hypovirulence-associated traits in dsRNA-containing strains of Diaporthe ambigua. Phytopathology 86:1311–1316

Son M, Yu J, Kim K-H (2015) Five questions about mycoviruses. PLOS Pathog 11:e1005172. https://doi.org/10.1371/journal.ppat.1005172

Song H-Y, Kim N, Kim D-H, Kim J-M (2020) The PoV mycovirus affects extracellular enzyme expression and fruiting body yield in the oyster mushroom, Pleurotus ostreatus. Sci Rep 10:1094. https://doi.org/10.1038/s41598-020-58016-4

Suharto AR, Jirakkakul J, Eusebio-Cope A, Salaipeth L (2022) Hypovirulence of Colletotrichum gloeosporioides associated with dsRNA Mycovirus isolated from a mango orchard in Thailand. Viruses 14:1921. https://doi.org/10.3390/v14091921

Sun Y, Guo M, Wang J, Bian Y, Xu Z (2022) Curing two predominant viruses occurring in Lentinula edodes by chemotherapy and mycelial fragmentation methods. J Virol Methods 300:114370. https://doi.org/10.1016/j.jviromet.2021.114370

Sutela S, Vainio EJ (2020) Virus population structure in the ectomycorrhizal fungi Lactarius rufus and L. tabidus at two forest sites in southern Finland. Virus Res 285:197993. https://doi.org/10.1016/j.virusres.2020.197993

Sutela S, Poimala A, Vainio EJ (2019) Viruses of fungi and oomycetes in the soil environment. FEMS Microbiol Ecol 95:fiz 119. https://doi.org/10.1093/femsec/fiz119

Suzaki K, Ikeda K, Sasaki A, Kanematsu S, Matsumoto N, Yoshida K (2005) Horizontal transmission and host-virulence attenuation of totivirus in violet root rot fungus Helicobasidium mompa. J Gen Plant Pathol 71:161–168. https://doi.org/10.1007/s10327-005-0181-8

Takahashi-Nakaguchi A, Shishido E, Yahara M, Urayama S, Ninomiya A, Chiba Y, Sakai K, Hagiwara D, Chibana H, Moriyama H, Gonoi T (2020a) Phenotypic and molecular biological analysis of Polymycovirus Afu PmV-1M from Aspergillus fumigatus: reduced fungal virulence in a mouse infection model. Front. Microbiol. 11:607795

Takahashi-Nakaguchi A, Shishido E, Yahara M, Urayama S, Sakai K, Chibana H, Kamei K, Moriyama H, Gonoi T (2020b) Analysis of an intrinsic Mycovirus associated with reduced virulence of the human pathogenic fungus aspergillus fumigatus. Front. Microbiol. 10:3045

Thapa V, Roossinck MJ (2019) Determinants of coinfection in the Mycoviruses. Front. Cell. Infect. Microbiol. 9:169

Thapa V, Keller NP, Roossinck MJ (2022) Evaluation of virus-free and wild-type isolates of pseudogymnoascus destructans using a porcine ear model. mSphere 7: e01022-21. https://doi.org/10.1128/msphere.01022-21

Torres-Trenas A, Cañizares MC, García-Pedrajas MD, Pérez-Artés E (2020) Molecular and biological characterization of the first Hypovirus identified in Fusarium oxysporum. Front. Microbiol. 10:3131

Tran TT, Li H, Nguyen DQ, Jones MGK, Wylie SJ (2019) Co-infection with three Mycoviruses stimulates growth of a Monilinia fructicola isolate on nutrient medium, but does not induce Hypervirulence in a natural host. Viruses 11:89. https://doi.org/10.3390/v11010089

Vainio EJ, Müller MM, Korhonen K, Piri T, Hantula J (2015) Viruses accumulate in aging infection centers of a fungal forest pathogen. ISME J 9:497–507. https://doi.org/10.1038/ismej.2014.145

Vainio EJ, Pennanen T, Rajala T, Hantula J (2017) Occurrence of similar mycoviruses in pathogenic, saprotrophic and mycorrhizal fungi inhabiting the same forest stand. FEMS Microbiol Ecol 93:fix 003. https://doi.org/10.1093/femsec/fix003

Van Alfen NK, Jaynes RA, Anagnostakis SL, Day PR (1975) Chestnut blight: biological control by transmissible Hypovirulence in Endothia parasitica. Science 189:890–891. https://doi.org/10.1126/science.189.4206.890

van Diepeningen AD, Debets AJM, Hoekstra RF (1998) Intra- and interspecies virus transfer in Aspergilli via protoplast fusion. Fungal Genet Biol 25:171–180. https://doi.org/10.1006/fgbi.1998.1096

van Diepeningen A, Debets A, Slakhorst S, Fekete C, Hornok L, Hoekstra R (2000) Interspecies virus transfer via protoplast fusions between Fusarium poae and black Aspergillus strains. Fungal Genet Rep 47:99–100. https://doi.org/10.4148/1941-4765.1216

Varga J, Vágvölgyi C, Tóth B (2005) Recent advances in mycovirus research. Acta Microbiol Immunol Hung 50:77–94. https://doi.org/10.1556/amicr.50.2003.1.8

Walker GM, Mcleod AH, Hodgson VJ (1995) Interactions between killer yeasts and pathogenic fungi. FEMS Microbiol Lett 127:213–222. https://doi.org/10.1111/j.1574-6968.1995.tb07476.x

Walker PJ, Siddell SG, Lefkowitz EJ, Mushegian AR, Adriaenssens EM, Alfenas-Zerbini P, Dempsey DM, Dutilh BE, García ML, Curtis Hendrickson R, Junglen S, Krupovic M, Kuhn JH, Lambert AJ, Łobocka M, Oksanen HM, Orton RJ, Robertson DL, Rubino L, Sabanadzovic S, Simmonds P, Smith DB, Suzuki N, Van Doorslaer K, Vandamme A-M, Varsani A, Zerbini FM (2022) Recent changes to virus taxonomy ratified by the International Committee on Taxonomy of Viruses (2022). Arch Virol 167: 2429–2440. https://doi.org/10.1007/s00705-022-05516-5

Wang S, Zhang J, Li P, Qiu D, Guo L (2016) Transcriptome-based discovery of Fusarium graminearum stress responses to FgHV1 infection. Int J Mol Sci 17:1922. https://doi.org/10.3390/ijms17111922

Wang S, Yang Z, Zhang T, Li N, Cao Q, Li G, Yuan Y, Liu D (2019) Molecular characterization of a Chrysovirus isolated from the citrus pathogen Penicillium crustosum and related fungicide resistance analysis. Front. Cell. Infect. Microbiol. 9:156

Wang Q, Mu F, Xie J, Cheng J, Fu Y, Jiang D (2020) A single ssRNA segment encoding RdRp is sufficient for replication, infection, and transmission of Ourmia-like virus in fungi. Front Microbiol 11:379. https://doi.org/10.3389/fmicb.2020.00379

Wang J, Li C, Song P, Qiu R, Song R, Li X, Ni Y, Zhao H, Liu H, Li S (2022a) Molecular and biological characterization of the first Mymonavirus identified in Fusarium oxysporum. Front. Microbiol. 13:1302

Wang Q, Zou Q, Dai Z, Hong N, Wang G, Wang L (2022b) Four novel Mycoviruses from the Hypovirulent Botrytis cinerea SZ-2-3y isolate from Paris polyphylla: molecular characterisation and mitoviral sequence transboundary entry into plants. Viruses 14:151. https://doi.org/10.3390/v14010151

Wang X, Lai J, Hu H, Yang J, Zang K, Zhao F, Zeng G, Liao Q, Gu Z, Du Z (2022c) Infection of Nigrospora nonsegmented RNA virus 1 has important biological impacts on a fungal host. Viruses 14:795. https://doi.org/10.3390/v14040795

Wang Y, Xu Z, Hai D, Huang H, Cheng J, Fu Y, Lin Y, Jiang D, Xie J (2022d) Mycoviromic analysis unveils complex virus composition in a Hypovirulent strain of Sclerotinia sclerotiorum. J Fungi 8:649. https://doi.org/10.3390/jof8070649

Wang Y, Zhao H, Cao J, Yin X, Guo Y, Guo L, Wu H, Zhang M (2022e) Characterization of a novel mycovirus from the Phytopathogenic fungus Botryosphaeria dothidea. Viruses 14:331. https://doi.org/10.3390/v14020331

Webb RJ, Roberts AA, Wylie S, Kosch T, Toledo LF, Merces M, Skerratt LF, Berger L (2022) Non-detection of mycoviruses in amphibian chytrid fungus (Batrachochytrium dendrobatidis) from Australia.

Fungal Biol 126:75–81. https://doi.org/10.1016/j.funbio.2021.10.004

Wen C, Wan X, Zhang Y, Du H, Wei C, Zhong R, Zhang H, Shi Y, Xie J, Fu Y, Zhao Y (2021) Molecular characterization of the first Alternavirus identified in Fusarium oxysporum. Viruses 13:2026. https://doi.org/10.3390/v13102026

Williamson KE, Fuhrmann JJ, Wommack KE, Radosevich M (2017) Viruses in soil ecosystems: an unknown quantity within an unexplored territory. Annu Rev Virol 4:201–219. https://doi.org/10.1146/annurev-virology-101416-041639

Wu S, Cheng J, Fu Y, Chen T, Jiang D, Ghabrial SA, Xie J (2017) Virus-mediated suppression of host non-self recognition facilitates horizontal transmission of heterologous viruses. PLoS Pathog 13:e1006234. https://doi.org/10.1371/journal.ppat.1006234

Xie Y, Wang Z, Li K, Liu D, Jia Y, Gao F, Dai J, Zhang S, Zhang X, Li H (2022) A megabirnavirus alleviates the pathogenicity of Fusarium pseudograminearum to wheat. Phytopathology 112:1175–1184. https://doi.org/10.1094/PHYTO-03-21-0126-R

Xu G, Zhang X, Liang X, Chen D, Xie C, Kang Z, Zheng L (2022a) A novel hexa-segmented dsRNA mycovirus confers hypovirulence in the phytopathogenic fungus Diaporthe pseudophoenicicola. Environ Microbiol 24: 4274–4284. https://doi.org/10.1111/1462-2920.15963

Xu X, Hai D, Li J, Huang F, Wang Y (2022b) Molecular characterization of a novel penoulivirus from the phytopathogenic fungus Colletotrichum camelliae. Arch Virol 167:641–644. https://doi.org/10.1007/s00705-021-05334-1

Yang M, Xu W, Zhou X, Yang Z, Wang Y, Xiao F, Guo Y, Hong N, Wang G (2021) Discovery and characterization of a novel bipartite Botrexvirus from the Phytopathogenic fungus Botryosphaeria dothidea. Front. Microbiol. 12:696125

You J, Zhou K, Liu X, Wu M, Yang L, Zhang J, Chen W, Li G (2019) Defective RNA of a novel Mycovirus with high transmissibility detrimental to biocontrol properties of Trichoderma spp. Microorganisms 7: 507. https://doi.org/10.3390/microorganisms7110507

Yu X, Gabriel A (1999) Patching broken chromosomes with extranuclear cellular DNA. Mol Cell 4:873–881. https://doi.org/10.1016/S1097-2765(00)80397-4

Yu X, Li B, Fu Y, Jiang D, Ghabrial SA, Li G, Peng Y, Xie J, Cheng J, Huang J, Yi X (2010) A geminivirus-related DNA mycovirus that confers hypovirulence to a plant pathogenic fungus. Proc Natl Acad Sci 107: 8387–8392. https://doi.org/10.1073/pnas.0913535107

Yu X, Li B, Fu Y, Xie J, Cheng J, Ghabrial SA, Li G, Yi X, Jiang D (2013) Extracellular transmission of a DNA mycovirus and its use as a natural fungicide. Proc Natl Acad Sci 110:1452–1457. https://doi.org/10.1073/pnas.1213755110

Yu J, Park JY, Heo J-I, Kim K-H (2020) The ORF2 protein of Fusarium graminearum virus 1 suppresses the transcription of FgDICER2 and FgAGO1 to limit host antiviral defences. Mol Plant Pathol 21:230–243. https://doi.org/10.1111/mpp.12895

Zhang D-X, Nuss DL (2016) Engineering super mycovirus donor strains of chestnut blight fungus by systematic disruption of multilocus vic genes. Proc Natl Acad Sci 113:2062–2067. https://doi.org/10.1073/pnas.1522219113

Zhang T, Li N, Yuan Y, Cao Q, Chen Y, Tan B, Li G, Liu D (2019) Blue-white colony selection of virus-infected isogenic recipients based on a Chrysovirus isolated from Penicillium italicum. Virol Sin 34:688–700. https://doi.org/10.1007/s12250-019-00150-z

Zhang H, Xie J, Fu Y, Cheng J, Qu Z, Zhao Z, Cheng S, Chen T, Li B, Wang Q, Liu X, Tian B, Collinge DB, Jiang D (2020a) A 2-kb Mycovirus converts a pathogenic fungus into a beneficial endophyte for Brassica protection and yield enhancement. Mol Plant, Special Issue on Plant-Microbe Interactions 13:1420–1433. https://doi.org/10.1016/j.molp.2020.08.016

Zhang L, Chen X, Bhattacharjee P, Shi Y, Guo L, Wang S (2020b) Molecular characterization of a novel strain of Fusarium graminearum virus 1 infecting Fusarium graminearum. Viruses 12:357. https://doi.org/10.3390/v12030357

Zhang M, He Z, Huang X, Shu C, Zhou E (2021) Genome organizations and functional analyses of a novel Gammapartitivirus from Rhizoctonia solani AG-1 IA strain D122. Viruses 13:2254. https://doi.org/10.3390/v13112254

Zhao L, Rosario K, Breitbart M, Duffy S (2019) Chapter three-eukaryotic circular rep-encoding single-stranded DNA (CRESS DNA) viruses: ubiquitous viruses with small genomes and a diverse host range. In: Kielian M, Mettenleiter TC, Roossinck MJ (eds) Advances in virus research. Academic Press, pp 71–133. https://doi.org/10.1016/bs.aivir.2018.10.001

Zhao Y, Zhang Y, Wan X, She Y, Li M, Xi H, Xie J, Wen C (2020) A novel Ourmia-like Mycovirus confers Hypovirulence-associated traits on fusarium oxysporum. Front. Microbiol. 11:569869

Zhao L, Lavington E, Duffy S (2021) Truly ubiquitous CRESS DNA viruses scattered across the eukaryotic tree of life. J Evol Biol 34:1901–1916. https://doi.org/10.1111/jeb.13927

Zhou L, Li X, Kotta-Loizou I, Dong K, Li S, Ni D, Hong N, Wang G, Xu W (2021) A mycovirus modulates the endophytic and pathogenic traits of a plant associated fungus. ISME J 15:1893–1906. https://doi.org/10.1038/s41396-021-00892-3

Zhu JZ, Guo J, Hu Z, Zhang XT, Li XG, Zhong J (2021) A novel Partitivirus that confer Hypovirulence to the plant pathogenic fungus Colletotrichum liriopes. Front. Microbiol. 12:653809

Bacterial Endosymbionts of Mucoromycota Fungi: Diversity and Function of their Interactions

8

Jessie K. Uehling, Alessandra Salvioli, Kevin R. Amses, Laila P. Partida-Martínez, Gregory Bonito, and Paola Bonfante

Abstract

Endosymbiotic bacteria are commonly inhabitants of environmental isolates of Mucoromycota fungi. Described bacterial endosymbionts of Mucoromycota are related to *Burkholderia* and *Mycoplasma* lineages. Bacterial endosymbionts of Mucoromycota tend to be obligately dependent on their fungal host for nutrition, but bacterial endosymbionts are not always needed for normal vegetative growth by their fungal hosts. When present in the fungal cytosol, bacterial endosymbionts alter fungal physiology including gene expression, protein synthesis, metabolism, and sporulation. Many bacterial endosymbionts have steamlined genomes due to reductive evolution, and the categories of functional gene loss differ by fungal host lineage. Much remains to be learned about bacterial endosymbionts of fungi regarding their ubiquity, transmission, and fungal host interactions.

Keywords

Mycoavidus, Mycetohabitans, Candidatus Glomeribacter, Phylogenomics, Co-evolution, Fungal–bacterial interactions

8.1 Introduction

Mucoromycota is a phylum of fungi that inhabit diverse terrestrial environments. These species are largely microscopic and ecologically diverse. Many are plant, microbial, and soil-associated saprotrophs, some are pathogens of plants or animals, and others are mutualistic or commensal plant rhizosphere inhabitants (Spatafora et al. 2016). Notorious Mucoromycota fungi include *Rhizopus* species that regularly colonize decomposing plant-based food products and opportunistically infect animals, as well as arbuscular mycorrhizal fungi (AMF) that are associated with nearly ¾ of the terrestrial plant species. Mucoromycota hyphae are coenocytic and have few or no cross walls. They reproduce asexually via chlamydospore and sporangiospore production. The plesiomorphy uniting many fungi in this group is sexual reproduction via gametangial conjugation and the production of

J. K. Uehling (✉) · K. R. Amses
Department of Botany and Plant Pathology, Oregon State University, Corvallis, OR, USA
e-mail: Jessie.Uehling@OregonState.edu

A. Salvioli · P. Bonfante
Department of Life Sciences and Systems Biology, University of Torino, Torino, Italy

L. P. Partida-Martínez
Department of Genetic Engineering, Centro de Investigación y de Estudios Avanzados del Instituto Politécnico Nacional (Cinvestav) - Unidad Irapuato, Irapuato, Mexico

G. Bonito
Department of Plant Soil and Microbial Sciences, Michigan State University, East Lansing, MI, USA

© The Author(s), under exclusive license to Springer Nature Switzerland AG 2023

S. Pöggeler, T. James (eds.), *Evolution of Fungi and Fungal-Like Organisms*, The Mycota 14,
https://doi.org/10.1007/978-3-031-29199-9_8

zygospores, although in several lineages sexual reproduction is poorly understood (Spatafora et al. 2016).

Mucoromycota and Zoopagomycota are two monophyletic clades that are basal to the Dikarya and that were formerly classified as Zygomycota (Spatafora et al. 2016). Historically, Zygomycota classification was based on morphological, chemical, and life history traits (Moreau 1952). However, phylogenetic and phylogenomic reconstructions demonstrate the paraphyly of the former Zygomycota, and that Zoopagomycota and Mucoromycota form monophyletic clades basal to Dikarya (James et al. 2006; Liu et al. 2009; Chang et al. 2015a). Although other phylum-level classifications based on ribosomal DNA have been proposed for Fungi (Tedersoo et al. 2018), we use Mucoromycota as defined by Spatafora et al. (2016) and present insights into the endosymbionts of its three monophyletic subphyla the Glomeromycotina, Mucoromycotina, and Mortierellomycotina (Spatafora et al. 2016).

Associations between Mucoromycota fungi and their bacterial endosymbionts share many commonalities and a few key differences that reflect the evolution of their interactions. Mechanistic, functional, and symbiotic interaction insights have been gleaned through comparative genomics and physiological assays using genetically identical fungal isolates with and without endosymbionts (Desirò et al. 2018; Uehling et al. 2017; Salvioli et al. 2016; Partida-Martinez et al. 2007; Büttner et al. 2021). Across all three subphyla, it is clear that the endosymbionts are predominantly related to *Burkholderia* and *Mycoplasma* bacterial lineages, although novel genomic and computational efforts will undoubtedly elucidate additional endohyphal bacterial and viral diversity (Robinson et al. 2021; Myers et al. 2020).

The relative abundance and function of bacterial endosymbionts and their fungal hosts is context dependent (Desirò et al. 2018; Salvioli et al. 2008). Bacterial endosymbionts can influence fungal growth, transcription, protein abundance, metabolism, and sporulation (Uehling et al. 2017; Salvioli et al. 2016; Partida-Martinez et al. 2007; Mondo et al. 2017; Li et al. 2017; Salvioli di Fossalunga et al. 2017; Dearth et al. 2018; Partida-Martinez and Hertweck 2005). However, the directionality and magnitude of those shifts differ between symbionts and host systems. For instance, while clearing *Rhizopus microsporus* (Mucoromycotina) of *Mycetohabitans* endosymbionts led to reduced zygospore production, clearing *Linnemannia sugarandi* (Mortierellomycotina) of *Mycoavidus* increases zygospore production (Mondo et al. 2017; Takashima et al. 2020). These intimate fungal–endosymbiont interactions coupled with the observation that several Mucoromycota fungi host closely related bacterial endosymbionts have led to the early invasion hypothesis. This hypothesis posits that select lineages of fungi and bacterial endosymbionts have co-evolved, in some cases for over 400 million years (Uehling et al. 2017; Mondo et al. 2012; Bonfante and Desirò 2017). One genomic feature that supports this hypothesis is that some bacterial endosymbiont genomes are extraordinarily streamlined. This gene loss is thought to reflect host adaptation over long periods of time and has resulted in decreased endosymbiont ability to live outside their hosts (Uehling et al. 2017; Lackner et al. 2011a; Ghignone et al. 2012; Fujimura et al. 2014; Moebius et al. 2014). Investigations into the role of ecology and transmission mode on endobacterial genome evolution and function as they relate to the hyphal environment are ongoing and are an exciting research frontier in mycology.

8.2 History of the Field

While the recent rise of next generation genome sequencing has enabled much greater appreciation for endosymbiont ubiquity and diversity, the first observations of endosymbiotic bacteria within Mucoromycota fungi were made over 50 years ago with electron microscopy. Pioneering observations were made by Barbara Mosse who demonstrated that plant biomass increases result from interactions with arbuscular mycorrhizal fungi. Through these transmission electron microscopy studies, she detailed how the fungus, which now bears her name (*Funneliformis mosseae*), not only increases plant biomass but also produces spores containing bacteria-like organisms (BLOs) (Mosse 1970; Mosse 1962). Her results were unprecedented as DNA sequence-based confirmation did not yet exist and comparable examples were very limited. Examples of bacterial endosymbionts were scant in fungi, but included the well cited, gram-negative, rod-shaped bacteria living inside host-derived vesicles of *Amoeba discoides* (Jeon 1972). Further use of electron microscopy rapidly expanded the diversity of fungi hypothesized to be harboring apparently non-pathogenic BLOs.

From 1975 to the 1990s, observations of BLOs were made by many research groups in several AMF species, and other fungal species associated with photosynthetic symbionts including, *Jimgerdemannia flammicorona* = *Endogone flammicorona,* and *Geosiphon pyriforme* (Scannerini and Bonfante 1975; Scannerini and Bonfante 1991; MacDonald and Chandler 1981; Sward 1981; Schüßler et al. 1994). Later the technical development of freeze-substitution allowed higher resolution imaging of bacteria inside *Gigaspora margarita* vacuoles (Bonfante et al. 1994). These detailed morphological observations identified two main types of BLOs: rod-shaped BLOs, which were enclosed inside a fungal vacuole, and irregularly coccoid BLOs directly embedded in the fungal cytoplasm. Thus, decades before the wide application of the polymerase chain reaction (PCR) and the use of ribosomal DNA (rDNA) researchers had already detected through electron microscopy two classes of endobacteria that colonize Mucoromycota cells. In parallel, bacterial symbionts have been documented in the Ascomycota and Basidiomycota (Hoffman and Arnold 2010; Bertaux et al. 2003), but much remains to be understood about how these bacteria impact their fungal hosts. Here we focus on bacterial endosymbiont diversity, ubiquity, and functioning in Mucoromycota fungi.

8.3 Mucoromycota Associated Bacterial Diversity

Researchers who study bacterial endosymbionts of fungi have been limited by challenges in imaging and cultivating bacterial endosymbionts in monoculture in lab settings with several notable exceptions. Images such as transmission electron micrographs (Fig. 8.1) or fluorescent images are essential to unequivocally demonstrate the intracellular nature of endosymbionts and differentiate endosymbionts from free-living bacterial associates. The observation some endosymbionts are uncultivable has led to the categorization of obligate (uncultivable) and the so-called facultative (cultivable) endosymbionts in the literature (Uehling et al. 2017; Mondo et al. 2012; Bonfante and Desirò 2017; Mondo et al. 2016; Ohshima et al. 2016). However, some obligate endobacteria have been cultured on custom media supplemented to complement functional gene loss (auxotrophy) identified through genome analysis (Ohshima et al. 2016; Glaeser et al. 2015) prompting a growing consensus this dichotomy may be artificial and endobacterial genome evolution in response to fungi is a plastic process. Therefore, the term obligate endobacteria may be better used to refer to whether the bacteria are nutritionally dependent on the fungal host. Often these relationships are facultative for the fungus, which may maintain its specific features, for example mycorrhization capacity irrespective of the endosymbiont presence (Bonfante and Desirò 2017; Deveau et al. 2018).

The endobacteria of Mucoromycota fall into two taxonomic groups closely related to *Burkholderia* and *Mycoplasma. Burkholderia*-Related Endosymbionts (BREs) are present in all three fungal subphyla. The best studied example is the endosymbiont *Mycetohabitans rhizoxinica*, which facilitates plant pathogenicity in *Rhizopus microsporus a*nd protection from micropredators (Partida-Martinez and Hertweck 2005; Richter et al. 2022) (Figs. 8.2 and 8.3), and was recently observed in Mortierellomycotina. Other well-studied examples of BRE include *Candidatus* Glomeribacter gigasporarum reported from some Glomeromycotina (Mondo et al. 2012; Bianciotto et al. 2003; Desirò et al. 2014), and *Mycoavidus cysteinexigens* in some Mortierellomycotina (Uehling et al. 2017; Ohshima et al. 2016; Sato et al. 2010) (Figs. 8.2 and 8.3). *Mycoavidus*

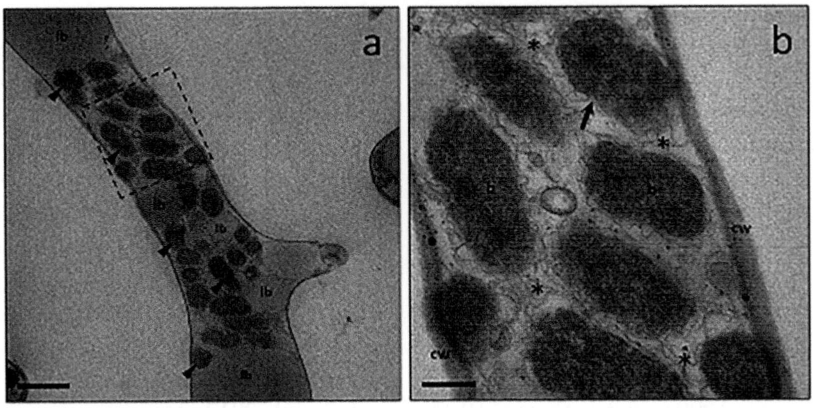

Fig. 8.1 Intracellular confirmation of *Mycoavidus* endosymbionts in *Linnemannia elongata.* Figure reproduced with permission from Uehling et al. 2017, doi 10.1111/1462-2920.13669 Transmission electron microscopy of wild type (a, b) *Linnemannia (Mortierella) elongata* NVP64. (**a**). A group of clustered rod-shaped *Mycoavidus* cells (arrowhead) within the fungal mycelium in proximity of several, usually large, lipid bodies (lb). (**b**). Magnification of square from panel (a): *Mycoavidus* cells (NVP64) (b) have a jagged cell wall and are surrounded by a tangled complex of membranes (asterisk). Endobacteria are engaged in cell division, as suggested by the central constriction (arrow). Image by Alessandro Desirò

diversity falls into 3 clades (Ohshima et al. 2016) called BRE clades A, B, and C with *M. cysteinexigens* AG77 in Clade A (Figs. 8.2 and 8.3). Mollicutes-Related Endosymbionts (MREs) are a second class of obligate endobacteria and are also associated with members of all three subphyla of Mucoromycota (Desirò et al. 2014; Naumann et al. 2010; Desirò et al. 2013; Toomer et al. 2015) (Figs. 8.2 and 8.3). Novel species and genera were erected to house this taxon, including *Candidatus* Moeniiplasma glomeromycotorum which is hosted by several Glomeromycotina fungi (Toomer et al. 2015; Naito et al. 2017). Novel related taxa that reside within Mortierellomycotina fungi are thought to be distinctive from *Candidatus* Moeniiplasma glomeromycotorum based on genome rDNA and sequence divergence but have yet to be formally classified (Fig. 8.3). Genomes of BRE and MRE bacteria are often reduced, and some are GC skewed compared to their free-living relatives (Fig. 8.4).

Striking patterns of similarity and difference between BRE and MRE have emerged. These include varied degrees of fungal host dependence; variable intraspecific genome reduction; loss of motility; abridged primary metabolism; and impacts on fungal host physiology. Fungal isolates harboring endosymbionts can be cleared without impacting host viability in most cases through antibiotic culture passaging (Desirò et al. 2018; Uehling et al. 2017; Partida-Martinez and Hertweck 2005; Okrasińska et al. 2021). A second similarity is that endobacterial populations can persist in individual fungal isolates, a phenomenon hypothesized to reflect their potential for vertical transmission in nature (Uehling et al. 2017; Partida-Martinez et al. 2007). Many fungal endobacteria adapt to fungal hosts resources, resulting in endosymbiont genome streamlining (Desirò et al. 2018; Uehling et al. 2017; Lastovetsky et al. 2016) (Fig. 8.4, Table 8.1). Differences between Mucoromycota-endobacteria pairs include physiological or niche differentiation in their fungal hosts, the categories of functional endosymbiont genes lost during genome streamlining, and the diversity of the endosymbiont populations with a single hypha. For example, while most Mucoromycota BRE populations are thought to be genetically uniform, diverse MRE populations can co-reside within hyphae or spores of mycorrhizal AMF and Endogonales (Naumann et al. 2010; Chang et al. 2019).

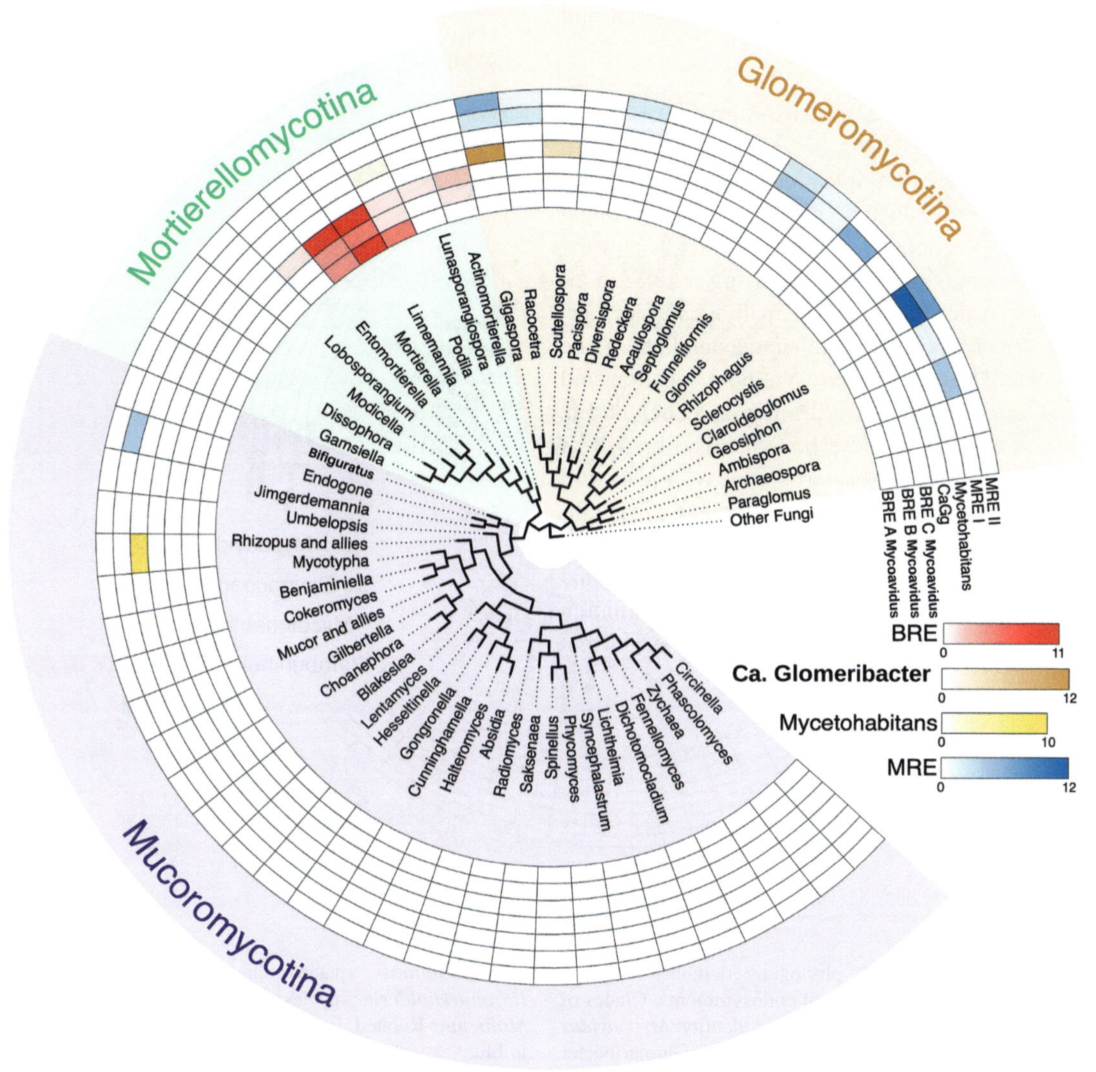

Fig. 8.2 Endobacterial diversity across the Mucoromycota. The interior cladogram represents current phylogenetic diversity and relationships between genera of the Mucoromycota colored by subphylum on the interior. The exterior heat maps displays endosymbionts bacterial identity and abundance by genus. *Mycoavidus* (BRE) are colored in red, *Candidatus* Glomeribacter gigasporarum (CaGg) are colored in brown, *Mycetohabitans* species are colored in yellow, and *Mollicutes*-Related Endosymbionts (MREs) are colored in blue

8.4 Mucoromycota Lineage Specific Insights

8.4.1 Glomeromycotina Fungi

Glomeromycotina are cosmopolitan root symbionts of plants, widely known as arbuscular mycorrhizal fungi. Over 72% of land plants host arbuscular mycorrhizal fungi, which are known to increase plant growth rates and stress tolerance (Chang et al. 2015a; Brundrett and Tedersoo 2018; Genre et al. 2020). AMF form intracellular arbuscules within root cortical cells which serve as an exchange site for mineral nutrients and reduced carbon (Smith and Read 2010). Plant hosts of arbuscular mycorrhizal fungi are taxonomically diverse, including non-vascular spore-bearing and vascular seed-bearing plants

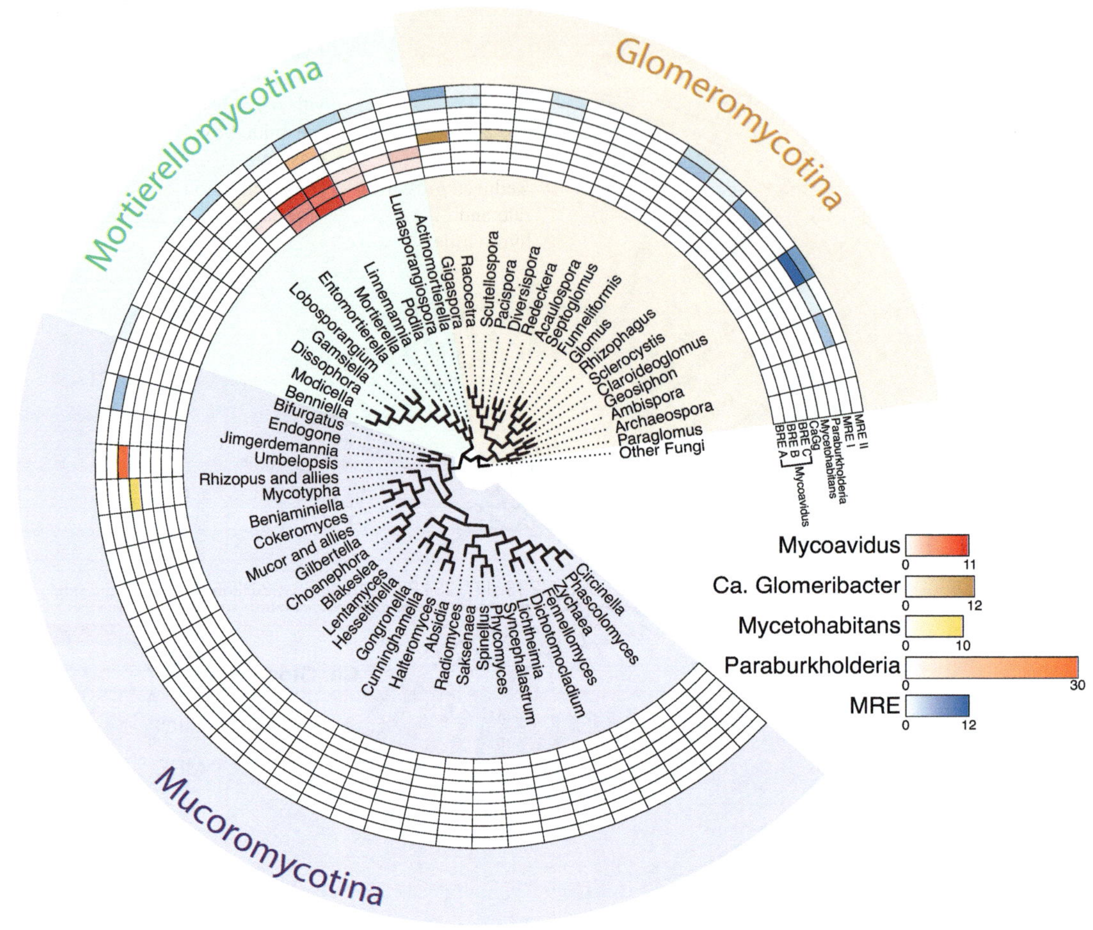

Fig. 8.3 Bacterial 16S phylogeny demonstrating the diversity and identity of fungal endosymbionts. Clades of bacteria are colored by phylogenetic identity. *Mycoavidus* (BRE) are colored in red, *Candidatus* Glomeribacter gigasporarum (CaGg) are colored in brown, *Mycetohabitans* species are colored in yellow, *Paraburkholderia* species are colored in orange, and *Mollicutes*-Related Endosymbionts (MREs) are colored in blue

(Brundrett and Tedersoo 2018; Genre et al. 2020; Smith and Read 2010). While the majority of Glomeromycotina fungi form symbioses with plants, one notable exception is *Geosiphon pyriforme*. This species hosts the photosynthetic cyanobacterium *Nostoc punctiforme* as a primary symbiont but individuals can simultaneously be infected with MRE (Schüßler et al. 1996). Glomeromycotina are one of the most ancient terrestrial fungal lineages, dating back to over 400 million years ago in the Ordovician where they were believed to have facilitated primary plant colonization across terrestrial Earth (Taylor et al. 1995; Remy et al. 1994; Redecker et al. 2000). Glomeromycotina species host both obligate BRE and MRE species of endobacteria (Toomer et al. 2015; Naito et al. 2015).

8.4.1.1 Glomeromycotina Endosymbiotic Bacteria

Since the first microscopic observations carried out in the 1970s revealed the presence of endobacteria in many isolates of arbuscular mycorrhizal fungi (Scannerini and Bonfante 1991), modern molecular tools including 16S rDNA sequencing have identified several endobacteria related to *Burkholderia* (Bianciotto et al. 1996) (Figs. 8.2,8.3). This research began

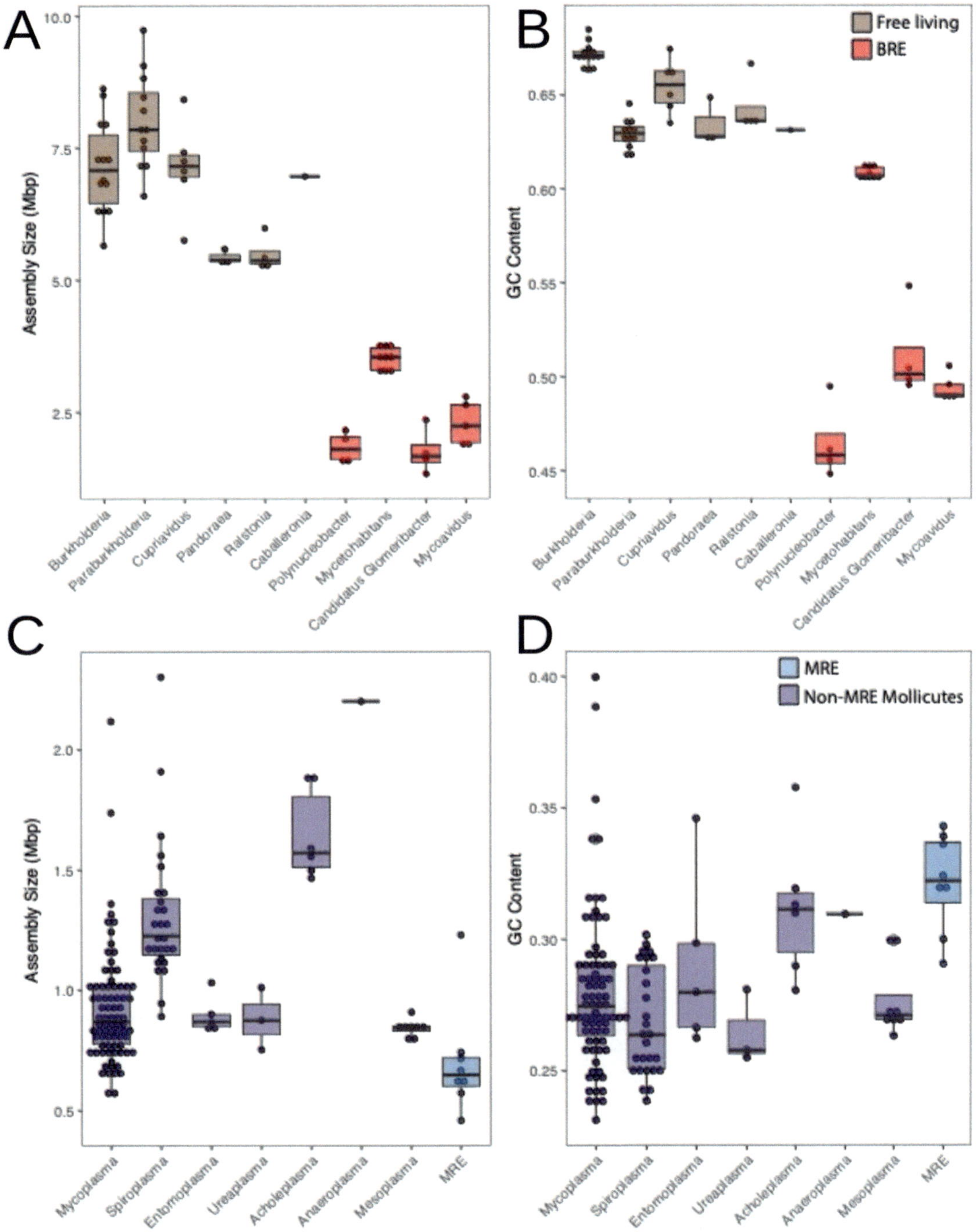

Fig. 8.4 Genome size and GC content of bacterial endosymbionts of fungi. Figure compares genome sizes and GC contents among genomes of *Burkholderia*-Related Endosymbionts (BREs) and *Mycoplasma* Related Endosymbionts (MREs) compared to their free-living relatives. Genome assembly size in megabases (Mbp) is reported by taxa panels **A** and **C.** The percentage of DNA base pairs that are either Guanine (G) or Cysteine (C) are reported by taxa panels **B** and **D.** *All* BRE are colored in red, while all MRE are colored in blue

Table 8.1 Bacterial endosymbiont genomic resources. Table lists taxonomic identity and genome data including size, assembly status, and literature citations for both endosymbionts and fungal hosts

BRE							
Taxon	Isolate	Accession	Length	Contigs	Fungal host	Fungal host accession	Paper ctiations
Mycoavidus cysteinexigens	AG77	JAPELI000000000	2.64 Mb	1	*Linnemannia elongata* AG77	PRJNA196039	Uehling et al. 2017
Mycoavidus cysteinexigens	B1-EB	AP018150 / PRJDB5716	2.8 Mb	1	*Linnemannia elongata* FMR23–6	–	Sharmin et al. 2018
Mycoavidus cysteinexigens	B2-EB	AP021872	1.88 Mb	1	*Entomortierella parvispora* E1425	PRJDB10536	Guo et al. 2020
Mycoavidus cysteinexigens	HK-1	PRJNA733818	2.24 Mb	1	*Podila verticillata* NRRL6337	GCA_000739165.1	Büttner et al. 2021
Mycoavidus cysteinexigens	SF98555	CP102085/ ASM2458273v1	2.2 Mb	1	*Podila verticillata*	–	Büttner et al. 2021
Glomeribacter gigasporarum	BEG1	PRJNA276133/ ASM168402v1	2.35 Mb	851	*Racocetra castanea* BEG1	–	Mondo et al. 2016
Glomeribacter gigasporarum	BEG34	PRJEA69997	1.72 Mb	128	*Gigaspora margarita* BEG34	PRJNA575165	Ghignone et al. 2012
Glomeribacter gigasporarum	JA201A	PRJNA276133/ ASM168415v1	1.62 Mb	1175	*Gigaspora margarita* JA201A	–	Mondo et al. 2016
Glomeribacter gigasporarum	IN211	PRJNA276133/ ASM168417v1	1.34 Mb	737	*Cetraspora pellucida* IN211	PRJEB45340; SAMEA8911299	Mondo et al., 2016
Mycetohabitans rhizoxinica	HKI454 or B1	PRJEA51915	3.75	3	*Rhizopus microsporus* ATCC 62417	PRJEB7410	Lackner et al. 2011a, 2011b; Horn et al. 2015
Mycetohabitans sp.	B2	PRJNA734142	3.79	211	*Rhizopus microsporus* ATCC 20577	–	Niehs et al., 2022
Mycetohabitans sp.	B6	PRJNA734142	3.71	328	*Rhizopus microsporus* ATCC 52811	–	Niehs et al. 2022
Mycetohabitans endofungorum	HKI456 or B5	PRJNA734142, PRJNA370785	3.29	143	*Rhizopus microsporus* CBS 112285	–	Estrada de los Santos et al. 2018, Niehs et al. 2022
Mycetohabitans sp.	B4 or B13	PRJNA734142, PRJEB18879	3.5	153	*Rhizopus microsporus* ATCC 52813	PRJNA430271, PRJNA205957	Niehs et al. 2022
Mycetohabitans sp.	B7 or B14	PRJNA734142, PRJEB18862	3.59	167	*Rhizopus microsporus* ATCC 52814	PRJNA330886	Niehs et al. 2022
Mycetohabitans sp.	B3	PRJNA734142	3.54	226	*Rhizopus microsporus* CBS 111563	–	Niehs et al. 2022

Mycetohabitans sp.	B8	PRJNA734142	3.28	78	*Rhizopus microsporus* CBS 308.87	–	Niehs et al. 2022
MRE							
Taxon	Isolate	Accession	Length	Contigs	Fungal host	Fungal host accession	Paper citations
Candidatus Moeniiplasma glomeromycotorum	DhMRE	PRJEB8356	0.7 Mb	3	*Dentiscutata heterogama*	GCA_910591775.1 (not sure it is the same strain)	Torres-Cortes et al. 2015
Candidatus Moeniiplasma glomeromycotorum	MRE-RV	JQIB00000000	1.2 Mb	99	*Racocetra verrucosa* VA103A	–	Naito et al. 2015
Candidatus Moeniiplasma glomeromycotorum	MRE-CE	JPXH00000000	0.66 Mb	67	*Claroideoglomus etunicatum* CA-OT135	–	Naito et al. 2015
Candidatus Moeniiplasma glomeromycotorum	MRE-RC	JPXG00000000	0.74 Mb	34	*Rhizophagus clarus NB112A*	–	Naito et al. 2015
Candidatus Moeniiplasma glomeromycotorum	Several MREs	PRJNA670049	0.5–1.5 Mb	5–54	Rhizophagus irregularis	PRJNA670049	Chang et al. 2019
Candidatus Moeniiplasma glomeromycotorum	DeMRE (I-1,I-2, III)	PRJNA407498	0,57–0,63 Mb	9	Diversispora epigaea IT104	PRJNA431769	Sun et al. 2019

Table 8.2 Model bacterial endosymbiont research systems. Table indicates fungal identity and taxonomic information, as well as ecology and endosymbiont interaction data

Mucoromycotina			Mortierellomycotina	Glomeromycotina	
Mucorales	Endogonales		Moriterellales	AM fungi	Geosiphonales
Rhizopus, Mucor, Umbelopsis	*Densospora*	*Jimgerdemannia, Endogone*	*Mortierella, Actinomortierella, Linnemannia, Podila, Lobosporangium, Modicella, Gamsiella, Dissophora, Benniella, Entomortierella, Lunasporangiospora*	*Diversispora, Dentiscutata, Gigaspora, Glomus, Ambispora, Scutellospora, Rhizophagus, Racocetra, Racocetra, Claroideoglomus*	*Geosiphon*
Plant pathogens, soil saprotrophs, opportunistic animal pathogens	Endophytes of bryophytes, lycophytes, ferns, Trifolium	Ectomycorrhizal symbionts, saprotrophs	Saprotrophs, endophytes, plant associates	Obligate mycorrhizal biotrophs with 80% of land plants	Symbiotic with cyanobacteria
BRE	MRE	MRE	MRE, BRE	MRE, BRE	MRE

on *Gigaspora margarita* BEG 34 and was extended to other members of the Gigasporaceae. The endobacteria were characterized as *Candidatus* Glomeribacter gigasporarum (Desirò et al. 2014) and observed in the cytoplasm occurring singularly or in groups, surrounded by a vacuole-like membrane (Bianciotto et al. 1996). Current phylogenetic analyses indicate that *Candidatus* Glomeribacter gigasporarum endobacteria form a sister lineage to *Mycoavidus,* a clade of endosymbionts that colonize Mortierellomycotina fungi, discussed further below. In contrast with *Mycoavidus*, attempts to isolate and cultivate *Candidatus* Glomeribacter gigasporarum in monoculture have been thus far unsuccessful. These bacteria are presumed to be primarily transmitted vertically through spores to progeny (Mondo et al. 2016; Bianciotto et al. 2003; Jargeat et al. 2004). However, patterns of host-switching owing to a lack of co-divergence have also been reported (Mondo et al. 2012).

To obtain a *Candidatus* Glomeribacter gigasporarum-free fungal isolate, commonly used antibiotic-based clearing methods have been attempted and have thus far failed (Jargeat et al. 2004). Instead, consecutive monosporic generations of *Gigaspora margarita* BEG 34 have produced a fungal cured line, devoid of the endobacteria as confirmed by molecular and microscopic methods (Lumini et al. 2007) (Table 8.2). This technique enabled functional studies on the fungal–endobacterial association via comparisons of fungal behavior in an isogenic background with and without endosymbionts and has become a standard protocol for studying interactions between AMF and endobacteria. We discuss these patterns below in the functional data section.

Spores and hyphae of Glomeromycotina are known to host MRE (Mosse 1970; Mosse 1962) (Fig. 8.2). Glomeromycotina hosts of MRE include *Ambispora appendicula* Att1235–2, *Ambispora fennica* Att200–26, *Geosiphon pyriformis* AS402, *Gigaspora margarita* MAFF520054, *Glomus versiforme* Att475–45 and Att1226–4, *Glomus etunicatum* Att239–4, MUCL47650, MUCL49411, and CA-OT126–3-2, *Scutellospora gilmorei* Att590–16, *Glomus caledonium* Att263–23, and *Glomus mosseae* Att109–20 and Bol17 (Naumann et al. 2010). Using the 16S rDNA sequence as a phylogenetic marker, the close relationship between endobacteria with Mollicutes was demonstrated, despite the presence of an electron dense cell wall-like layer (Naumann et al. 2010). More recently, this lineage has been placed within Mycoplasmataceae (Naito et al. 2015; Torres-Cortes et al. 2015). MREs are presumed to be transmitted vertically under laboratory conditions (Naito et al. 2015). However, there is a high level of genetic recombination in MRE genomes analyzed so far, indicating potential roles for horizontal transmission and recombination within bacterial populations, or even hosts (Toomer et al. 2015). Recently named *Candidatus* Moeniiplasma glomeromycotorum

(Naito et al. 2015), these cytoplasm residing endobacteria belong to the MRE, and so far have not been cultured outside the fungal host.

8.4.1.2 Glomeromycotina endosymbiont Diversity and Ubiquity

The occurrence of *Candidatus* Glomeribacter gigasporarum (BRE) within AMF is limited to Gigasporaceae, where they have been investigated in *Gigaspora* and *Scutellospora* (Table 8.2) (Mondo et al. 2012; Bianciotto et al. 2003). There are many different isolates of *Candidatus* Glomeribacter gigasporarum (BRE) and although their rDNA sequences are conserved, each fungal host harbors a genetically uniform *Candidatus* Glomeribacter population (Mondo et al. 2012; Desirò et al. 2014) (Fig. 8.2, Table 8.2). The association between arbuscular mycorrhizal fungi and *Candidatus* Glomeribacter gigasporarum is inferred to be at least 400 million years old, predating the diversification of Gigasporaceae (VanKuren et al. 2013). The frequency of *Candidatus* Glomeribacter gigasporarum observations among populations varies across the family and within closely related fungal genotypes, and has been observed in ~15% (18 out of 115) fungal isolates surveyed (Mondo et al. 2012). For example, *Candidatus* Glomeribacter is not present within *G. rosea* BEG9 (Bianciotto et al. 2003) but it has been detected from other isolates of the same species (Toomer et al. 2015), suggesting that endobacterial populations may fluctuate depending on environmental conditions, their physical location within the fungus, or fungal development.

In contrast, *Candidatus* Moeniiplasma glomeromycotorum (MRE) occur frequently across Glomeromycotina (Fig. 8.2). Naumann and colleagues surveyed phylogenetically diverse AMF and detected MRE in members of Archaeosporales, Glomerales, and Diversisporales, as well as in *G. pyriforme* (Naumann et al. 2010). MRE hosts include *Rhizophagus clarus* (= *Glomus clarum*), *Rhizophagus intraradices* (= *G. intraradices*), and *Glomus* DAOM197198 (= *R. irregularis*). In natural populations, the incidence of MRE in AMF was reported to reach 88%, much higher than the 2% reported for BRE across the same populations of AMF (Lastovetsky et al. 2018). Unlike BRE, multiple divergent MRE sequences can be amplified from a single AMF individual (Naumann et al. 2010; Desiro et al. 2013; Toomer et al. 2015).

Some arbuscular mycorrhizal fungi contain both BRE and MRE (Table 8.2), as is the case for *G. margarita.* While the two coexisting endobacteria persist within the same fungal line, it was determined that *Candidatus* Glomeribacter gigasporarum maintained a higher relative abundance within the fungal host. From a phylogenetic point of view, the widespread MRE distribution across AMF suggests that the invasion by the common ancestor occurred before the radiation of Glomeromycotina, at least 400 million years ago (Bonfante and Desirò 2017; Naumann et al. 2010).

8.4.1.3 Glomeromycotina and Endosymbiotic Genetic Resources

The first arbuscular mycorrhizal fungal genome was published in 2013 from the model species *Rhizophagus irregularis* DAOM197198 (Tisserant et al. 2013). Spores which contain thousands of nuclei but no endobacteria offered strictly fungal genomic DNA for the assembly. Arbuscular mycorrhizal fungi have some of the largest genomes in the fungal kingdom, spanning from ~140 Mbp of in *R. irregularis* to ~770 Mbp in *G. margarita* BEG34 (Venice et al. 2020a). The total gene number in these fungi is higher than in many other fungi and can be attributed to transposable elements (Venice et al. 2020a). Because arbuscular mycorrhizal fungi are coenocytic, they have offered opportunities to address novel biological questions such as the diversity of nuclear genomic variants within an individual. It has been established that intra-isolate polymorphisms mirror that of homokaryotic or heterokaryotic isolates behaving as functional diploids (Ropars et al. 2016).

The first genome sequencing of endobacteria associated with AMF preceded the genome description of its fungal host: *Candidatus* Glomeribacter gigasporarum from *G. margarita*

BEG34 was obtained in 2012 using a combination of Sanger and 454 sequencing technologies (Ghignone et al. 2012). The size of the genome and its functional annotation (1.7–1.9 Mb, with 1736 coding sequences) are consistent with the unculturable status of these endobacteria (Table 8.1). For example, *Ca.* Glomeribacter gigasporarum displays limited sugar catabolism and lacks the ability to synthesize amino acids such as arginine, isoleucine, leucine, methionine, phenylalanine, tryptophan, histidine, and valine (Ghignone et al. 2012). At the same time, its genome encodes many amino-acid permeases and transporters, indicating that *Ca.* Glomeribacter gigasporarum could take up such compounds from its fungal host as an energy source (Ghignone et al. 2012). Similar to *Mycoavidus, Ca.* Glomeribacter retains essential genes involved in DNA replication and repair, but has lost individual glycolytic genes such as phosphofructokinase (Uehling et al. 2017; Ghignone et al. 2012; Guo et al. 2020). Interestingly, *Ca.* Glomeribacter contains the full operon for vitamin B12 synthesis, genes for antibiotic resistance and toxin-antitoxin systems, which may contribute to the ecological fitness of the endosymbiont plus host, or holobiont (Salvioli di Fossalunga et al. 2017; Ghignone et al. 2012). While phylogenetic analyses place *Ca.* Glomeribacter in the Burkholderiaceae, a hierarchical clustering of KEGG metabolic pathways grouped it together with insect endosymbionts, suggesting that convergent evolution could have shaped an adaptation to the intracellular lifestyle (Ghignone et al. 2012). Events of horizontal gene transfer have been postulated in the *Ca.* Glomeribacter genome. Genes encoding for hybrid non-ribosomal peptide synthases-polyketide synthases (NRPS-PKS) have been analyzed, with one cluster showing a close relationship with sequences belonging to *Myxococcus (*Venice et al. 2020b*)*.

Single isolate MRE have been genome sequenced from *Dentiscutata heterogama* and three metagenome assemblies of MRE from *Racocetra verrucosa, Claroideoglomus etunicatum,* and *Rhizophagus clarus* are available (Naito et al. 2015; Torres-Cortes et al. 2015) (Table 8.1). In contrast to BRE, MRE show even further reduced capacities (Fig. 8.4). The severely reduced genomes of these MRE populations (0.7–1.3 Mb) support their direct dependence on their fungal hosts. For example, gene loss has been observed in glycolysis, the tricarboxylic acid cycle, oxidative phosphorylation, and DNA replication pathways. As only a few putative nutrient transporters were annotated in MRE genomes, alternative ATP-producing pathways are also lacking such as the arginine dihydrolase pathway. As such, the mechanism of energy production in MRE still remains unclear (Naito et al. 2015; Torres-Cortes et al. 2015). A metabolic hierarchical clustering applied to MRE places them with obligate insect endosymbionts with very reduced metabolic capacities (Torres-Cortes et al. 2015). Sun and colleagues sequenced and assembled the genomes of the AMF *Diversispora epigaea* and its MRE (Sun et al. 2019), and found two distinct MRE populations that were approximately 88% similar on a genomic scale. Interestingly, the MRE genome assemblies obtained from *D. epigaea* contained several coding sequences suggested to have been obtained by horizontal gene transfer (Naito et al. 2015; Torres-Cortes et al. 2015). These genes are involved in fungal metabolism, such as chitin deacetylase, glycosyltransferase family, chitin synthase, and malic enzyme and were expressed in different conditions, suggesting that they could have a functional importance for the fungal-bacterial holobiont (Naito et al. 2015; Torres-Cortes et al. 2015). In the same study, the authors noticed that MRE cannot synthesize purine and pyrimidine de novo, and instead imported these from their fungal host. After observing a significant expansion of a nucleoside salvage pathway family in *D. epigaea*, the authors suggested that the host competes with the MRE for these resources and that a pathogen defense system exists in AMF including *G. margarita (Venice et al.* 2020a*)*.

The generation of genomic resources for Glomeromycotina endobacteria has enabled analyses of the evolution and function in these microbes. Genomic commonalities of heritable endobacteria include molecular evolution rate acceleration, genome contraction and accumulation of deleterious mutations in functional genes

(Fig. 8.4, Table 8.1) (Bennett and Moran 2013). Both *Ca.* Glomeribacter and *Ca.* Moeniiplasma appear to evolve faster than their free-living relatives. However, the calculated rate of mutation accumulation for *Ca.* Glomeribacter is comparable to that of free-living bacteria and much lower than the one estimated for insect endobacteria. This observation led researchers to hypothesize that the evolutionary rate acceleration in *Ca.* Glomeribacter relies on the long-term maintenance of a large clonal population together with a low occurrence of recombination and host switching (Mondo et al. 2016). In contrast, *Ca.* Moeniiplasma are characterized by very high genome plasticity, intrahost genetic diversity, and population-level recombination (Naumann et al. 2010; Toomer et al. 2015; Naito et al. 2017; Naito et al. 2015). They possess active recombination machinery and display an ultrarapid mutation rate, posited to be one of the fastest rates of evolution observed in bacteria (Naito and Pawlowska 2016). *Ca.* Moeniiplasma are widely presumed to be vertically transmitted in arbuscular mycorrhizal fungi, although recent studies suggest they may have originated from a host-switching event that involved an animal-infecting MRE ancestor (Naito et al. 2015), rendering the origin of the fungal-*Ca.* Moeniiplasma symbiosis enigmatic still.

8.4.1.4 Glomeromycotina and Endosymbiotic Functional Data

Genomic and empirical data indicate that while Glomeromycotina endobacteria are dependent on their fungal hosts for nutrition, endobacteria are non-essential to AMF fungi. However, *Ca.* Glomeribacter increases fungal hosts' fitness in some settings. These insights are made possible by deriving cleared fungal lines with no bacteria in the same genetic background as wildtype. In experiments carried out with both fungal lines, *Gigaspora margarita* BEG34 cured of its endosymbionts germinated slower and was slower to grow and colonize plant host roots at the time of the curing experiment, but revealed adaptive capacities along the years and eventually became similar to the WT line. As arbuscular mycorrhizal fungi are obligate biotrophs, in natural conditions the reduced ability of the cleared lines to grow during the pre-symbiotic stage of its life cycle might lead to a reduced rate of root colonization, and thus to a lower chance of success as compared with the wildtype isolates. In addition, fungal fatty acids are less abundant in the cured spores (Salvioli et al. 2010). Comparative transcriptomic analyses of *G. margarita* demonstrate that mitochondrial functions and pathways (ATP synthesis, oxidative phosphorylation, reactive oxygen species metabolism) are enhanced by endobacterial presence (Salvioli et al. 2016). Combining the transcriptomic, proteomic, metabolomic, and cellular biology approaches has provided insights into endobacterial effect of bioenergetic capacity of its fungal host (Salvioli et al. 2016; Dearth et al. 2018; Vannini et al. 2016). Wild type isolates elicit ATP production and respiration associated with a greater ROS detoxification potential (Salvioli et al. 2016; Dearth et al. 2018; Vannini et al. 2016).

Plant host health is also impacted via interactions with fungi and their endobacteria. For example, in *Trifolium* plants, roots colonized with the wildtype arbuscular mycorrhizal fungal isolates containing endobacteria showed a decrease in protein carbonylation, which is a hallmark of oxidative damage (Vannini et al. 2016). Similar findings were reported in *Lotus japonicus* roots, where transcriptomics and proteomics revealed a diverse activation of many pathways in roots colonized with either wild type or cleared arbuscular mycorrhizal fungi (Venice et al. 2020a). These studies demonstrated that *Ca.* Glomeribacter presence affects central arbuscular mycorrhiza functioning, including antioxidant metabolism, respiration, lipid biosynthesis, and the activity of the phosphate transporter, which is the iconic marker of mycorrhizal functionality in plants (Venice et al. 2021).

Outside of the model system *G. margarita* BEG34, understanding of functional effects of endobacteria on their Glomeromycotina hosts is still in its infancy. A recent study revealed C flux within a quadruplet symbiosis that included *Allium* roots and *Gigaspora margarita*

MAFF520054 that hosts both BRE and MRE (Kuga et al. 2021). Using C13 isotope labeling, C flow was followed from the plant to fungi to BRE and MRE and showed that MRE represent a higher cost to their fungal host in terms of carbon compared to BRE (Kuga et al. 2021). These observations corroborate the hypothesis that MRE is more opportunistic than BRE. All attempts to clear arbuscular mycorrhizal fungi of MRE to date have been unsuccessful, and thus the interaction of MRE and their hosts cannot yet be identified as mutualistic or parasitic.

8.4.2 Mucoromycotina Fungi

Mucoromycotina fungi have several lifestyles. Some are saprotrophic and live in soils or eutrophic environments such as animal wastes or plant refuse, while others live as endophytes or ectomycorrhizal symbionts. A few species cause pathogenesis in plants (Gnanesh et al. 2021; Cruz-Lachica et al. 2017; Gho et al. 1978) and animals, including humans (Desirò et al. 2018; Hibbett et al. 2007; Jeong et al. 2019). This subphylum is the most speciose within Mucoromycota, containing 354 described species distributed across the Mucorales, Endogonales, and Umbelopsidales, as well as many other species that remain to be described.

Among the orders, Endogonales include taxa that can form mycorrhizal associations with diverse plants, including with members of the Pinaceae, Fagaceae, liverworts and hornworts (Desiro et al. 2013; Warcup 1990; Walker 1985; Field et al. 2015; Yamamoto et al. 2017). Many species of Endogonales form small hypogeous sporocarps, containing zygospores; thus, they account for rare cases of sporocarp production within Mucoromycota (Smith et al. 2013). The Endogonales are comprised of two family-level clades (Desirò et al. 2017); Endogonaceae which contains *Endogone* and *Jimgerdemannia*, and Densosporaceae, containing *Densospora* and *Sphaerocreas*. While members of the Endogonaceae tend to reproduce by producing sequestrate (closed) sporocarps, Densosporaceae produce tiny sporocarps which are quite cryptic in nature (Desirò et al. 2017; Hirose et al. 2014). Fungi in both families remain quite under sampled, but they are represented by considerable phylogenetic diversity known from environmental plant tissue samples, including arbuscule-forming fine root endophytes (Desirò et al. 2017; Walker et al. 2018). *Bifiguratus* and *Calcarisporiella* are the most basal Mucoromycotina (Torres-Cruz et al. 2017; Hirose et al. 2012).

In contrast to the Endogonales, fungi in the Mucorales are saprotrophic and associated with decaying plant materials. Some Mucoralean fungi including species in *Rhizopus*, *Lichtheimia*, and *Mucor*, cause opportunistic human fungal infections, collectively called Mucoromycosis. Fungi in the Umbelopsidales are saprotrophic, soil and plant-associated fungi (Spatafora et al. 2016; Meyer and Walter 2003) and are well-known to produce lipids including polyunsaturated fatty acids and house endosymbiotic bacteria (Papanikolaou and Aggelis 2019).

8.4.2.1 Mucoromycotina endosymbiont Diversity and Ubiquity

Despite the great diversity of Mucoromycotina fungi, only a few genera have been reported to be associated with BRE or MRE (Fig. 8.2, Table 8.2). These include *Rhizopus*, *Mucor*, and *Umbelopsis* that harbor BRE (Partida-Martinez and Hertweck 2005; Okrasińska et al. 2021; Lackner et al. 2009), and *Endogone* and *Jimgerdemannia* that associate with MRE (Desirò et al. 2015a; Chang et al. 2015b) (Fig. 8.2, Table 8.2).

Discovery of BRE symbionts in *Rhizopus* began with the search for the biosynthetic genes of the phytotoxin rhizoxin (Partida-Martinez and Hertweck 2005). Rhizoxin is the causative agent of rice seedling blight (Gho et al. 1978; Iwasaki et al. 1984) and a potent antimitotic, antifungal macrolide. This toxin is produced by several isolates of *Rhizopus microsporus* obtained from diverse habitats including rice seedlings, soils, necrotic human tissues, ground nuts, and fermented foods such as tempeh and sufu. The biosynthetic genes found in all rhizoxin-producing *Rhizopus* isolates have high homology with Type I PKS bacterial genes, suggesting the presence of bacterial symbionts within

Table 8.3 Secondary metabolites produced by bacterial endosymbionts of fungi. Table describes secondary metabolite diversity by endosymbiont including biosynthetic gene clusters, functional insights, and literature citations

Secondary metabolite	Produced by	Biosynthetic gene cluster	Function	References
Mucoromyocotina (*Rhizopus microsporus* mainly)				
Rhizoxin	*Mycetohabitans* B1-B8, B9 *P. fluorescens* pf-5	Rhi,AM411073, (hybrid PKS-NRPS)	1. Causal agent of rice seedling blight, 2. Antimitotic, binds to N (AA 100th) in beta tubulin and inhibits mitosis, 3. Defense against nematodes and other micropredators of fungi	(Partida-Martinez and Hertweck 2005, 2007; Brendel et al. 2007; Scherlach et al. 2006, 2012; Richter et al. 2022)
Rhizonin	*Mycetohabitans endofungorum* B5		Mycotoxin, capable of producing lethal hepatic failure	(Partida-Martinez et al. 2007)
Galactofuranose lipopolysaccharide	Experiments done in *M. rhizoxinica* B1	BGC 29 genes, 22 LPS	LPS contains an O-antigen that mimics fungi and allows colonization of *R. microsporus*	(Leone et al. 2010)
T2SS Chitinase	Experiments done in *M. rhizoxinica* B1		T2SS helps delivers chitinase to allow bacterial penetration into fungal hyphae	(Moebius et al. 2014)
Exopolysaccharide	Experiments done in *M. rhizoxinica* B1 (megaplasmid)	WcaA, WcaB, RBRH04235	Exopolysaccharide contributes to aggregation of symbiotic bacteria	(Uzum et al. 2015)
Holrhizin A	All sequenced *Mycetohabitans* (B1-B8)	HolA, RBRH01792 (6 modules, NRPS)	Reduces surface tension, regulates biofilm formation, and contributes to fungal colonization	(Niehs et al. 2018b)
Heptarhizin	*Mycetohabitans* (Pacific clade, B1, B2, and B6)	HepA (7 modules, NRPS)	Metabolite so far only known from Mycetohabitans strains from the Pacific clade	(Niehs et al. 2018a)
Endopyrroles	*Mycetohabitans* clades A and B (Pacific), C and D (Eurasian)	EpyA-E, MN081797 (8 modules, NRPS)	Produced under salt stress only in symbiosis	(Niehs et al. 2019)
Lasso peptides, burhizin-23, mycetohabin-16, mycetohabin-15	Experiments done in *M. rhizoxinica* B1.	Mycetohabin-16-like BGC also in Mycoavidus sp. from Mortierella elongata AG77 (Glomeribacter sp. 10,166,415)	Produced in symbiosis, both in Mycetohabitans and Mycoavidus spp.	(Cheng and Hua 2020)
TAL effector	*Mycetohabitans* sp. B13 (B4)	Btl19–13	Effector of T3SS, affects host transcription and one enhances tolerance to	(Carter et al. 2020)

(continued)

Table 8.3 (continued)

Secondary metabolite	Produced by	Biosynthetic gene cluster	Function	References
			cell membrane stress in the fungus	
Arw effector, TAL effector (3 bats)	All sequenced *Mycetohabitans* (B1-B8) Bat1 (RBRH_01844), Bat2 (RBRH_01776) and Bat3 (RBRH_01777)	Single, double, and triple mutants	Arw and Bat1–3 effector proteins of the T3SS. Arw had no effect on sporulation.	(Richter et al. 2020)
Necroxime A, B, C, and D	*Mycetohabitans* sp. B8 and BRE from P. *verticillata* NRRL 6337 (*Ca.* Mycoavidus necroximicus)	Nec BGC MN734804	Cytotoxic benzolactones that help defend host fungi from fungivorous nematodes	(Büttner et al. 2021; Niehs et al. 2020)
Habitasporins A, B, and C	*Mycetohabitans* spp. (B1 to B8)	habA, habB, accession numbers: MZ520316, MZ52031, respectively.	Needed for effective sporulation	(Niehs et al. 2022)
Glomeromycotina				
Compounds were not identified, only putative BGCs in both fungal and bacterial partners	NR-PKS with bacterial signature			(Venice et al. 2020a)

R. microsporus (Tables 8.2 and 8.3). This hypothesis was confirmed with 16S rRNA gene sequencing, and the presence of living bacterial cells inside the fungal cytoplasm was confirmed with Syto9 staining and microscopy. Phylogenetic analyses of the 16S sequences from *Rhizopus* associated, rhizoxin-producing BRE revealed these bacteria comprise a new clade within the genus *Burkholderia*, warranting creation of a new genus *Mycetohabitans (*Partida-Martinez and Hertweck 2007*)*. The type strains were designated as *M. rhizoxinica* and *M. endofungorum* (= *Burkholderia rhizoxinica* and *B. endofungorum)* (Estrada-de Los Santos et al. 2018) (Table 8.1). Through curing and reintroduction of BRE symbionts from the fungus employing antibiotics and by independently culturing *Mycetohabitans* endosymbionts, it was demonstrated rhizoxin was directly produced by the symbionts (Partida-Martinez and Hertweck 2005). This pioneering work demonstrated that endosymbiont derived toxins in plant pathogenic fungi alter ecological functions. The *Rhizopus-Mycetohabitans* symbiosis currently serves as a model system for understanding fungal–bacterial interactions due to the cultivability of bacterial symbionts; ability to derive bacterial free fungal lineages with antibiotic clearing; and widespread occurrence of naturally *Rhizopus* isolates with and without endobacteria.

Rhizopus microsporus and related Mucorales isolates are often found in the environment, and some can opportunistically infect mammals including humans (Spatafora et al. 2016; Petrikkos et al. 2014; Frey-Klett et al. 2011). Some clinical *R. microsporus* isolates have been shown to form stable associations with BREs including *Mycetohabitans rhizoxinica* (=*Burkholderia rhizoxinica*) (Ibrahim et al. 2008; Itabangi et al. 2022) and *Ralstonia pickettii* (Itabangi et al. 2022). Both *R. microsporus* and free-living *Burkholderia* species are pathogens with unique virulence mechanisms (Walsh et al. 2012; Petrikkos et al. 2012; Wiersinga et al. 2006). These observations have given rise to the hypothesis that BRE may directly or indirectly increase the virulence of their pathogen fungal hosts. Given the demonstrated antimitotic role of rhizoxin in plant pathogenic *R. microsporus* (Partida-Martinez and Hertweck 2005; Partida-Martinez and Hertweck 2007) efforts have been made to study the presence and absence of *rhi* biosynthetic gene cluster in BRE of clinical *R. microporous* isolates. This work concluded that because the Rhi genes are not universally present in bacterial endosymbionts of fungi and because rhizoxin has a non-appreciable effect on mammalian phagocytes at biologically relevant ratios that endosymbionts do not influence the outcomes of *R. microsporus* infections in the human body (Ibrahim et al. 2008). In conflicting results, it was shown that the presence of a second BRE, *Ralstonia pickettii* from *Rhizopus microsporus* shaped fungal transcription and in some cases the outcome of phagocytotic assays and fungal cell wall composition (Itabangi et al. 2022). Further, it has been hypothesized that environmental interactions with phagocytic soil microbes including cellular slime molds, amoebae, and nematodes favor acquisition of virulence traits including secondary metabolite diversity in bacterial endosymbionts of fungi (Richter et al. 2022; Itabangi et al. 2022). Consistent with this, it was recently shown that rhizoxin derivatives prevent fungal killing by fungivorous amoeba (Richter et al. 2022). In addition, it was recently shown that a cancer patient succumbed to *Mycetohabitans rhizoxinica* bacteremia, and that genomic DNA from both *M. rhizoxinica* and *R. microsporus* was present in her body (Yang et al. 2022) further supporting a role for environmentally acquired endosymbionts in the virulence of their fungal hosts. Many open research questions remain including whether the putative added virulence of BREs in Mucoromycotina human fungal pathogens is direct or indirect, how functionally similar BRE of *R. microsporus* and other human fungal pathogen are, what pathogenicity mechanisms are at play, and how frequently isolates transit between the environment and clinics.

Efforts to quantify endosymbionts in Mucoromycotina isolates demonstrate they are not present in every isolate. Screening a collection of 22 *Rhizopus microsporus* resulted in detection of *Mycetohabitans* symbionts in 8 of 22 (36%) of isolates (Partida-Martinez and Hertweck 2007; Lackner et al. 2011b; Partida-Martinez 2013). Another investigation found by screening 33 *R. microsporus,* 31 *R. arrhizus*, and 6 *R. delemar* isolates from the CBS-KNAW Fungal Biodiversity Center reference collection, that 11% of all isolates and 21% of *R. microsporus* housed *Mycetohabitans* species (Dolatabadi et al. 2016). In *Umbelopsis* spp., it was observed that 23/40 (57%) of isolates contained *Paraburkholderia* symbionts (Okrasińska et al. 2021) with homology to free-living bacteria based on molecular sequencing. In the same study 1 of 15 (6%) isolates of *Mucor,* harbored *Paraburkholderia* symbionts (Okrasińska et al. 2021). In contrast, studies investigating MRE abundance demonstrate in approximately 45% of the Endogonales fruiting bodies contained endobacteria (Desirò et al. 2015b).

More recently, metagenomes of MRE and their hosts were generated by sequencing *Endogone* and *Jimgerdemannia* fruiting bodies (Table 8.2) (Chang et al. 2019). Three of the four metagenomes sequenced had MRE within them. MRE contigs were assembled for *J. flammicorona* AD002 (0.9 Mb), *J. lactiflua* OSC162217 (1.8 Mb), and *Endogone* sp. FLAS F-59071 (2.9 Mb). Given that MRE genomes are predicted to be ~1 Mb, it was hypothesized that some of the sampled sporulating bodies contained multiple MRE phylotypes, which were further confirmed through phylogenetics of extracted 16S rDNA assemblies from these metagenomes (Chang et al. 2019).

8.4.2.2 Mucoromycotina and endosymbiont Genetic Resources

There are rapidly growing genomic resources for Mucoromycotina fungi, with over 186 publicly available genomes on the National Center for Biotechnology's website (176 Mucoromycetes; 5 Endogonomycetes; and 5 Umbelopsidomycetes). For BRE symbionts, the full genome of *Mycetohabitans rhizoxinica* HKI-454 (B1) was first assembled and published in 2011 (Lackner et al. 2011a; Lackner et al. 2011b; Lackner et al. 2011c). To date, seven other draft genomes of *Mycetohabitans* spp. associated with *R. microsporus*, including the type strain *M. endofungorum* HKI-456 (B5), are publicly available (Table 8.1). A remarkable feature observed from all sequenced *Mycetohabitans* symbionts to date is that, despite their rather reduced size, they possess a great number of biosynthetic gene clusters (BGCs) as described in (Table 8.3) when compared to free-living *Burkholderia* spp. (Büttner et al. 2021; Moebius et al. 2014; Richter et al. 2022; Venice et al. 2020b; Partida-Martinez and Hertweck 2007; Mullins and Mahenthiralingam 2021; Niehs et al. 2022; Brendel et al. 2007; Scherlach et al. 2006; Leone et al. 2010; Uzum et al. 2015; Carter et al. 2020).

For MRE symbionts, there are currently six available genomes or metagenomes available. (Table 8.1). Contigs of MRE associated with *Endogone* and *Jimgerdemannia* are also publicly available (Table 8.1).

8.4.2.3 Mucoromycotina endosymbiont Functional Data

The *Rhizopus microsporus-Mycetohabitans* symbiosis is a model system for functional studies of fungal–endosymbiont interactions due to genetic tractability of both organisms, and the establishment of Koch's postulate to test functional hypotheses leveraging cultivability and genetic transformation of endosymbionts (Partida-Martinez and Hertweck 2005; Partida-Martinez and Hertweck 2007). In *R. microsporus,* the *Mycetohabitans* symbionts increase the metabolic capabilities of their fungal hosts through the production of the potent toxins rhizoxin, rhizonin, and several other secondary metabolites that increase fungal pathogenesis (Partida-Martinez and Hertweck 2005; Partida-Martinez and Hertweck 2007) (Table 8.3). Moreover, some *Mycetohabitans* can influence asexual sporulation (Partida-Martinez et al. 2007) and the rate of sexual zygospore production (Mondo et al. 2017). In both cases, bacterial symbionts are vertically transmitted through the sporangiospores and zygospores, respectively. Further studies in *R. microsporus* demonstrated *Mycetohabitans* contain Burkholderia TAL-like (Btl) proteins that act as effectors on fungal host DNA and may influence fungal transcription (Carter et al. 2020).

A recent study demonstrated that *R. delemar* isolates derived from agricultural soils can also harbor *Mycetohabitans* species. Interestingly, a cleared line of *R. delemar* showed no differential growth rates compared to their wild type counterparts, and neither the symbiotic or cured lines produced the secondary metabolite rhizoxin nor had the *rhi* gene cluster. Further, cleared *R. delemar* isolates did not differ in their asexual sporulation rates (Cabrera-Rangel et al. 2022). Much remains to be learned about the diversity of *Mycetohabitans* and how distribution of functional endosymbiont genes impacts evolution of these bacterial fungal symbioses. Collectively, these studies suggest that fungi, as most plants and animals, are indeed holobionts that depend on microbiota (bacteria and/or viruses) for their fitness and survival (Partida-Martínez 2017).

8.4.3 Mortierellomycotina Fungi

The subphylum Mortierellomycotina contains one single family, Mortierellaceae, with 13 monophyletic genera (Vandepol et al. 2020). Like their relatives in Mucoromycotina and Glomeromycotina, Mortierellomycotina fungi occur on every continent. However, we currently lack clear understanding about the niches these fungi occupy and their ecological roles in the environment. Mortierellomycotina have been isolated from plant-associated soils, living roots, dead organic substrates, and insect tissues. Most taxa in this group are considered to be saprotrophs (Vandepol et al. 2020; Macias et al. 2019; Gams 1977; Dixon-Stewart 1932) and many are chitinolytic (Uehling et al. 2017), cycling nitrogen in soils via chitin degradation (Liao 2021; Liao et al. 2019). Several key Mortierellaceae species are utilized in applied and industrial mycology for their polyunsaturated fatty acid biosynthesis (Mamani et al. 2019; Goyzueta-Mamani et al. 2021). Fungi in the Mortierellomycotina associate with diverse species of bacteria, including *Paraburkholderia*,

Paenibacillus, and *Pseudomonas,* the Enterobacteriaceae, and with the obligate MRE and BRE (Desirò et al. 2018; Uehling et al. 2017; Ohshima et al. 2016; Sato et al. 2010; Okrasińska et al. 2021) such as *Candidatus* Mycoplasma, *Mycoavidus,* and *Mycetohabitans* species. Genera of Mortierellomycotina that host bacterial endosymbionts include *Mortierella, Actinomortierella, Linnemannia, Podila, Benniella, Modicella, Lobosporangium, Dissophora,* and *Gamsiella* (Desirò et al. 2018; Uehling et al. 2017; Büttner et al. 2021; Robinson et al. 2021; Okrasińska et al. 2021; Takashima et al. 2018a; Telagathoti et al. 2021).

8.4.3.1 Discovery of Mortierellomycotina Endosymbionts

Mortierellomycotina fungi host both BRE and MRE (Fig. 8.2, Table 8.2). Detection of BRE within Mortierellomycotina was first described by Sato and colleagues within an isolate of *Linnemannia elongata* (=*Mortierella elongata*) from Japan (Table 8.2). Originally reported to be *Ca.* Glomeribacter, these endosymbionts are now described as *Mycoavidus cysteinexigens* after analyzing their genome sequences and physiology (Fujimura et al. 2014; Ohshima et al. 2016). Later the genomes of both *Linnemannia elongata* AG77 and *Mycoavidus cysteinexigens AG77* were sequenced from an isolate obtained from Duke Forest in the United States (Uehling et al. 2017). *Mycoavidus* has since been documented throughout the Mortierellaceae including in *Mortierella* (*M. alpina, M. ambigua, M. chienii, M. oedorhiza, M. polycephala, M. sossauensis, M. sugadairana*); *Linnemannia* (*L. exigua, L. gamsii*); *Podila* (*P. horticola, P. humilis, P. minutissima, P. verticillata*); *Entomortierella* (*E. lignicola, E. parvispora*); *Actinomortierella* (*A. formicae*); and *Lunasporangiospora* (*L. microzygospora, L. rishikesha, L. rostafinski, L. selenospora) (Büttner et al.* 2021*; Okrasińska et al.* 2021*; Telagathoti et al.* 2021*; Takashima et al.* 2018b*; Herlambang et al.* 2022*)* (Fig. 8.2).

MRE have also been reported in Mortierellomycotina since 2018 (Desirò et al. 2018). Desirò and colleagues showed that MRE bacteria are present in *Lobosporangium transversale, Linnemannia elongata, L. gamsii, Mortierella capitata, M. thaxteri, Podila humilis, P. minutissima, Lunasporangiospora selenospora,* and *L. rostafinskii* (Desirò et al. 2018) (Fig. 8.2, Table 8.2). Given their presence in Glomeromycotina (Naumann et al. 2010), Mucoromycotina (Desirò et al. 2018), MRE are hypothesized to be widely distributed amongst Mucoromycota fungi and commensal or opportunistic symbionts (Kuga et al. 2021).

8.4.3.2 Diversity and Ubiquity of Endosymbionts in Mortierellomycotina

The intraspecific abundance of BRE within Mortierellomycotina fungi including *Linnemannia* varies by study, but all studies show that within any given species, many naturally occurring isolates do not host endobacteria. Utilizing 16S rDNA genes as a proxy for endosymbiotic presence, Takashima et al. (2018b) found that 22% (53/238) of environmental Mortierellaceae isolates, and 37% (22/59) of *Mortierella* isolates harbored BREs, validated with fluorescent and electron transmission microscopy (Takashima et al. 2018b). Imaging such as these are necessary to distinguish the intrahyphal endosymbionts from free-living hyphal associates, which are common. Telagathoti and colleagues later determined that environmental Mortierellaceae isolates from Europe including *Mortierella* sensu stricto, *Dissophora, Entomortierella, Gamsiella, Linnemannia,* and *Podila* frequently have a BRE bacterial signal; they report that 30% of Mortierellaceae isolates were BRE 16S positive as determined through PCR (Okrasińska et al. 2021). Finally, Okrasinská and peers reported that approximately 20% of Mucoromycotina and Mortierellomycotina isolates screened harbored BRE, which were validated with fluorescent imaging (Okrasińska et al. 2021). In all these studies bacteria were screened from isolates obtained from the environment on media lacking antibiotics to not bias results. Taken together, these results indicate that endosymbionts are present in ~20–30% of environmental Mortierellaceae isolates. Future research is needed to better understand the frequency of BRE endosymbionts, why isolates naturally differ in endosymbiont status, and how the endosymbiont presence impacts fungal functioning in different environments and selective pressures.

In contrast to the BRE, MRE have been detected within approximately 3% of Mortierellaceae isolates (Desirò et al. 2018). While they found that MRE ubiquity was relatively low, they showed that MRE may persist in isolates for decades in the case of *Lobosporangium*. Desirò and colleagues went on to demonstrate the relationship between culture temperature, culture age, and MRE abundance within the fungal colony (Desirò et al. 2018). Interestingly, hosts grew best at lower temperatures, and slower at high temperatures, while MRE cells were more abundant in cultures at higher temperatures (Desirò et al. 2018).

8.4.3.3 Mortierellomycotina and Endosymbiont Genetic Resources

There are rapidly growing genomic resources for Mortierellomycotina fungi (Spatafora et al. 2016; Uehling et al. 2017; Vandepol et al. 2020) and for their endosymbionts, with over 30 host genomes currently publicly available on the NCBI GenBank. There are currently 5 published genome sequences for *Mycoavidus* species available (Table 8.1). These include *Mycoavidus cysteinexigens* from *Linnemannia elongata*, *Entomortierella parvispora*, and *Podila verticillata* (Table 8.1). *Ca.* Mycetohabitans necroximicus was also detected from *P. verticillata* (Uehling et al. 2017; Büttner et al. 2021; Guo et al. 2020; Sharmin et al. 2018). These genomes range in length from 1.9 to 2.8Gb (Fig. 8.4, Table 8.1), and although they share nucleotide and protein homology, each also contains unique genes and substantial rearrangements (Guo et al. 2020). Comparisons between these endosymbiont genomes show reductive evolution and specialization of metabolic capacities including decreased abilities to produce amino acids and complete glycolysis (Büttner et al. 2021; Guo et al. 2020; Herlambang et al. 2022).

BRE *Mycoavidus* genomes in the Mortierellomycotina are similar, with ~81–84% of genes shared with high homology between individuals (Uehling et al. 2017; Guo et al. 2020). While each of the 5 *Mycoavidus* genomes from Mortierellomycotina hosts shares many similar gene sequences, the relative proportion of unique content varies by isolate from 8 to 12% (Guo et al. 2020). There are several well supported clades of *Mycoavidus* BRE that will likely lead to novel genera, and are noted by BRE clades A-C (Takashima et al. 2018b) (Figs. 8.2 and 8.3). Conversely, single copy genes constitute 64–80% of the *M. cysteinexigens* in B1-EB and B2-EB, respectively (Guo et al. 2020). Last, all *Mycoavidus* genomes sequenced to date have similar numbers of rRNAs, tRNAs, and ~ 50% GC content (Uehling et al. 2017; Guo et al. 2020; Sharmin et al. 2018). Future research evaluating the driving factors for these similarities and differences given the isolation of each endosymbiotic lineage will yield valuable insights into evolution of these organisms.

BRE endosymbionts in these fungi have reduced genomes, to their closest free-living Burkholderiales relatives in the genera *Burkholderia, Paraburkholderia, Ralstonia,* and *Cupriavidus* which have 5–12 Mb genomes and are divided into circular chromosomes and secondary replicons such as plasmids (Partida-Martinez and Hertweck 2007; Carter et al. 2020; Szabó et al. 2022) (Fig. 8.4). In contrast, *Mycetohabitans* BRE have genomes ranging from 3.3 to 3.8 Mb (Lackner et al. 2011b, c) and *Mycoavidus* BRE have genomes ranging from 1.9 to 2.8 Mb and have approximately 1600–2300 genes (Uehling et al. 2017; Fujimura et al. 2014; Guo et al. 2020; Sharmin et al. 2018) (Fig. 8.4, Table 8.1). The most reduced genomes are observed in *Ca.* Glomeribacter genomes which range from 1.3 to 2.3 Mb genes (Ghignone et al. 2012; Jargeat et al. 2004) (Fig. 8.4, Table 8.1). However, many of these genomes are incomplete and future work on entire endosymbiont bacterial chromosomes will be enlightening.

The extraordinary gene loss within endosymbionts of fungi is thought to reflect selective pressures imparted by the fungal host (Uehling et al. 2017; Guo et al. 2020). For example, *Mycoavidus cysteinexigens* AG77 has lost individual glycolytic genes and nearly entire pathways such as Entner–Doudoroff, while maintaining transporters and catalytic genes for host-derived lipids and fatty acids (Uehling et al. 2017) and these resources come from their fungal hosts. In contrast to what has been observed in insect endosymbionts, *Mycoavidus* genomes retain genes in DNA repair pathways, which may contribute to purging of deleterious mutations (Guo et al. 2020). While the debate about whether fungal endobacteria are undergoing stable evolutionary trajectories continues (Guo et al. 2020), it remains clear that a hallmark of fungal endosymbiosis is reduced gene content compared to related free-living bacteria. Given that many bacterial endosymbionts of fungi are thought to be transmitted vertically, evolutionary theory also predicts that molecular evolution will be accelerated in these smaller populations which experience bottlenecks during transmission (Ohta 1972). Genomic data for bacterial endosymbionts of fungi is still scarce relative to other symbiotic systems, yet preliminary data suggests that genomes of bacterial endosymbionts are streamlined reflecting host adaptation resulting from unique intracellular selective pressures (Uehling et al. 2017; Mondo et al. 2016; Pawlowska et al. 2018) (Fig. 8.4, Table 8.1).

8.4.3.4 Mortierellomycotina and Endosymbiont Functional Data

Functional interaction data from comparative assays of cleared and wild type fungi of the same background genotype have shown that bacterial endosymbionts impact fungal physiology, transcription, metabolism, and growth variably. Both BRE and MRE presence impact fungal growth rates (Desirò et al. 2018; Uehling et al. 2017) metabolism of fatty acids and sugars (Li et al. 2017), volatile production (Uehling et al. 2017; Guo et al. 2020). BRE, but not MRE, decrease their hosts ability to reproduce sexually (Takashima et al. 2020). Research on the benefits for fungi on hosting bacterial endosymbionts points toward holo-organismal utilization of bacterial secondary metabolites for modulating interactions between fungi and other organisms. Early work in the Mucoromycotina demonstrated the role of the hybrid PKA-NRPS derived rhizoxin in endosymbiont–plant interactions (Partida-Martinez and Hertweck 2007). Recent work by Buttner et al. and Niehs et al. showed that both *Mycetohabitans* and *Ca.* Mycoavidus necroximicus share the *nec* genes, a hybrid PKS-NRPS biosynthetic gene cluster which encodes cytotoxic benzolactones named necroximes (Niehs et al. 2022) (Table 8.3). It was shown that necroximes enable their Mortierellomycotina host fungi to evade fungiverous grazing of nematodes (Büttner et al. 2021; Partida-Martinez and Hertweck 2007). This hypothesis is supported by the general depletion of Mucoromycota genomes in secondary metabolic gene clusters compared to other clades of filamentous fungi and raises the hypothesis that specialized metabolisms in fungi may be derived from metabolisms of bacterial endosymbionts (Koczyk et al. 2021; Wurlitzer et al. 2021; Tabima et al. 2020). These findings also suggest that the horizontal transfer of biosynthetic gene clusters, such as *nec,* among different lineages of endofungal BRE symbionts is likely.

8.4.3.5 Bacterial and Viral Endosymbionts in Mucoromycota

Viruses which reside inside of fungal hyphae are covered in detail in Chap. 8 by Myers et al. and elsewhere (Kondo et al. 2022). Here, we present Mucoromycota specific insights into mycoviruses that co-reside with endosymbiotic bacteria in fungal hyphae and spores. Mycoviruses are estimated to be present in approximately 27% of Mucoromycota isolates, and appear to be most abundant in the Glomeromycotina subphylum (67%), followed by Mortierellomycotina (34%) and Mucoromycotina (19%) (Myers 2020). Co-occurrence of bacterial and viral symbionts within the phylum Mucoromycota has been reported in four cases, including *Geosiphon pyriformis*, *Gigaspora margarita* BEG34, and *Rhizopus microsporus* ATCC52814. *Geosiphon pyriformis* is associated with *Nostoc punctiforme* and a Tymo-like virus (Myers 2020). *G. margarita* BEG34, which harbors *Ca.* Glomeribacter gigasporarum (Bianciotto et al. 2003), possesses a diverse viral community formed by four mitoviruses, one Ourmia-like virus, one Giardia-like virus, and two sequences related to *Fusarium graminearum* mycoviruses (Turina et al. 2018). *Rhizopus microsporus* ATCC52814 associates with *Mycetohabitans* sp. B7 and two narnaviruses called RmNV-20S and RmNV-23S (Espino-Vázquez et al. 2020). *Benniella erionea* GBAus27b contains MRE bacteria which carry a belonging to the Inoviridae (Roux et al. 2019), adding even more complexity to life within fungal hyphae. From a functional standpoint, it has been demonstrated that narnaviruses and Mycetohabitans endobacteria co-residing in *R. microsporus* cytoplasm are required for successful sexual reproduction (Espino-Vázquez et al. 2020) and each microbe alone reduces or abolishes vegetative sporulation (Partida-Martinez et al. 2007). These observations underscore the complexity of Mucoromycota fungal holobionts whose ecological and evolutionary trajectories are influenced by both bacterial and viral symbionts.

8.4.3.6 Future Directions and Open Questions

Intriguing research questions for all subphyla include how endobacteria influence their fungal hosts' interactions with other organisms and the environment, including impacting virulence of opportunistic human fungal pathogens. We currently lack a fundamental understanding of the spatial dynamics of endosymbiont location and relative abundance within fungal cells, or the factors that affect these patterns. The branch of mycology studying bacterial endosymbiosis will continue to benefit from development of visualization technologies, application of novel genome editing tools, and development of computational genomic pipelines. Generating robust population genomic data sets will allow analysis of current dogmas surrounding partner specificity, co-evolution, and endosymbiont transmission dynamics within and between fungal hosts. Last, novel genome sequences of hosts and their endosymbionts will provide an opportunity to evaluate functional and evolutionary similarities with other eukaryotic-endosymbiont partners, which will in turn illuminate fundamental rules for intracellular life.

Acknowledgments The authors would like to acknowledge the following funding sources for supporting this work: Support was provided by NSF 2202410, NSF 2030338 to JKU, DOESFA LANLF59T, NSF 1737898 to GB, and Conacyt FOINS-2015-01-006 to LPPM.

References

Bennett GM, Moran NA (2013) Small, smaller, smallest: the origins and evolution of ancient dual symbioses in a phloem-feeding insect. Genome Biol Evol 5(9): 1675–1688. https://doi.org/10.1093/gbe/evt118

Bertaux J, Schmid M, Prevost-Boure NC, Churin JL, Hartmann A, Garbaye J, Frey-Klett P (2003) In situ identification of intracellular bacteria related to Paenibacillus spp. in the mycelium of the ectomycorrhizal fungus Laccaria bicolor S238N. Appl Environ Microbiol 69(7):4243–4248. https://doi.org/10.1128/aem.69.7.4243-4248.2003

Bianciotto V, Bandi C, Minerdi DANIELA, Sironi M, Volker H, Titchy A, Bonfante P (1996) An obligately endosymbiotic mycorrhizal fungus itself harbors obligately intracellular bacteria. Appl Environ Microbiol 62(8):3005–3010

Bianciotto V, Lumini E, Bonfante P, Vandamme P (2003) 'Candidatus Glomeribacter gigasporarum'gen. Nov., sp. nov., an endosymbiont of arbuscular mycorrhizal fungi. Int J Syst Evol Microbiol 53(1):121–124

Bonfante P, Desirò A (2017) Who lives in a fungus? The diversity, origins and functions of fungal endobacteria living in Mucoromycota. ISME J 11(8):1727–1735

Bonfante P, Balestrini R, Mend Gen K (1994) Storage and secretion processes in the spore of Gigaspora margarita Becker &Hall as revealed by high-pressure freezing and freeze substitution. New Phytol 128(1):93–101

Brendel N, Partida-Martinez LP, Scherlach K, Hertweck C (2007) A cryptic PKS–NRPS gene locus in the plant commensal Pseudomonas fluorescens pf-5 codes for the biosynthesis of an antimitotic rhizoxin complex. Org Biomol Chem 5(14):2211–2213

Brundrett MC, Tedersoo L (2018) Evolutionary history of mycorrhizal symbioses and global host plant diversity. New Phytol 220(4):1108–1115

Büttner H, Niehs SP, Vandelannoote K, Cseresnyés Z, Dose B, Richter I, Gerst R, Figge MT, Stinear TP,

Pidot SJ (2021) Bacterial endosymbionts protect beneficial soil fungus from nematode attack. Proc Natl Acad Sci 118(37):e2110669118

Cabrera-Rangel JF, Mendoza-Servın JV, Cordova-Lopez-G, Alcalde-Vazquez R, Garcıa-Estrada RS, Winkler R, Partida-Martınez LP (2022) Symbiotic and toxinogenic Rhizopus spp. isolated from soils of different papaya producing regions in Mexico

Carter ME, Carpenter SC, Dubrow ZE, Sabol MR, Rinaldi FC, Lastovetsky OA, Mondo SJ, Pawlowska TE, Bogdanove AJ (2020) A TAL effector-like protein of an endofungal bacterium increases the stress tolerance and alters the transcriptome of the host. Proc Natl Acad Sci 117(29):17122–17129

Chang Y, Wang S, Sekimoto S, Aerts AL, Choi C, Clum A, LaButti KM, Lindquist EA, Yee Ngan C, Ohm RA (2015a) Phylogenomic analyses indicate that early fungi evolved digesting cell walls of algal ancestors of land plants. Genome Biol Evol 7(6): 1590–1601

Chang Y, Wang S, Sekimoto S, Aerts AL, Choi C, Clum A, LaButti KM, Lindquist EA, Ngan CY, Ohm RA (2015b) Phylogenomic analyses indicate that early fungi evolved digesting cell walls of algal ancestors of land plants. Genome Biol Evol 7(6):1590–1601

Chang Y, Desirò A, Na H, Sandor L, Lipzen A, Clum A, Barry K, Grigoriev IV, Martin FM, Stajich JE (2019) Phylogenomics of Endogonaceae and evolution of mycorrhizas within Mucoromycota. New Phytol 222(1):511–525

Cheng C, Hua ZC (2020) Lasso peptides: heterologous production and potential medical application. Front Bioeng Biotechnol 8:571165. https://doi.org/10.3389/fbioe.2020.571165

Cruz-Lachica I, Márquez-Zequera I, García-Estrada RS, Carrillo-Fasio JA, León-Félix J, Allende-Molar R (2017) Identification of mucoralean fungi causing soft rot in papaya (Carica papaya L.) fruit in Mexico. Revista mexicana de fitopatología 35(3):397–417

Dearth SP, Castro HF, Venice F, Tague ED, Novero M, Bonfante P, Campagna SR (2018) Metabolome changes are induced in the arbuscular mycorrhizal fungus Gigaspora margarita by germination and by its bacterial endosymbiont. Mycorrhiza 28(5):421–433

Desirò A, Naumann M, Epis S, Novero M, Bandi C, Genre A, Bonfante P (2013) Mollicutes-related endobacteria thrive inside liverwort-associated arbuscular mycorrhizal fungi. Environ Microbiol 15(3):822–836. https://doi.org/10.1111/j.1462-2920.2012.02833.x

Desirò A, Salvioli A, Ngonkeu EL, Mondo SJ, Epis S, Faccio A, Kaech A, Pawlowska TE, Bonfante P (2014) Detection of a novel intracellular microbiome hosted in arbuscular mycorrhizal fungi. ISME J 8(2):257–270

Desirò A, Faccio A, Kaech A, Bidartondo MI, Bonfante P (2015a) Endogone, one of the oldest plant-associated fungi, host unique Mollicutes-related endobacteria. New Phytol 205(4):1464–1472

Desirò A, Salvioli A, Bonfante P (2015b) Investigating the Endobacteria which thrive in arbuscular mycorrhizal fungi. Methods Mol Biol (Clifton, NJ) 1399:29–53

Desirò A, Rimington WR, Jacob A, Pol NV, Smith ME, Trappe JM, Bidartondo MI, Bonito G (2017) Multigene phylogeny of Endogonales, an early diverging lineage offungi associated with plants. IMA Fungus 8(2):245–257

Desirò A, Hao Z, Liber JA, Benucci GMN, Lowry D, Roberson R, Bonito G (2018) Mycoplasma-related endobacteria within Mortierellomycotina fungi: diversity, distribution and functional insights into their lifestyle. ISME J 12(7):1743

Deveau A, Bonito G, Uehling J, Paoletti M, Becker M, Bindschedler S, Hacquard S, Hervé V, Labbé J, Lastovetsky O (2018) Bacterial-fungal interactions: ecology, mechanisms and challenges. FEMS Microbiol Rev 42(3):335–352

Dixon-Stewart D (1932) Species of Mortierella isolated from soil. Trans Br Mycol Soc 17(3):208–220

Dolatabadi S, Scherlach K, Figge M, Hertweck C, Dijksterhuis J, Menken SB, de Hoog GS (2016) Food preparation with mucoralean fungi: a potential biosafety issue? Fungal Biol 120(3):393–401

Espino-Vázquez AN, Bermúdez-Barrientos JR, Cabrera-Rangel J, Córdova-López G, Cardoso-Martínez F, Martínez-Vázquez A, Camarena-Pozos DA, Mondo SJ, Pawlowska TE, Abreu-Goodger C (2020) Narnaviruses: novel players in fungal–bacterial symbioses. ISME J 14(7):1743–1754

Estrada-de Los Santos P, Palmer M, Chávez-Ramírez B, Beukes C, Steenkamp ET, Briscoe L, Khan N, Maluk M, Lafos M, Humm E (2018) Whole genome analyses suggests that Burkholderia sensu lato contains two additional novel genera (Mycetohabitans gen. Nov., and Trinickia gen. Nov.): implications for the evolution of diazotrophy and nodulation in the Burkholderiaceae. Genes 9(8):389

Field KJ, Rimington WR, Bidartondo MI, Allinson KE, Beerling DJ, Cameron DD, Duckett JG, Leake JR, Pressel S (2015) First evidence of mutualism between ancient plant lineages (H aplomitriopsida liverworts) and M ucoromycotina fungi and its response to simulated P alaeozoic changes in atmospheric CO 2. New Phytol 205(2):743–756

Frey-Klett P, Burlinson P, Deveau A, Barret M, Tarkka M, Sarniguet A (2011) Bacterial-fungal interactions: hyphens between agricultural, clinical, environmental, and food microbiologists. Microbiol Mol Biol Rev 75(4):583–609. https://doi.org/10.1128/MMBR.00020-11

Fujimura R, Nishimura A, Ohshima S, Sato Y, Nishizawa T, Oshima K, Hattori M, Narisawa K, Ohta H (2014) Draft genome sequence of the Betaproteobacterial endosymbiont associated with the fungus Mortierella elongata FMR23-6. Genome Announc 2(6):e01272-14. https://doi.org/10.1128/genomeA.01272-14

Gams W (1977) A key to the species of Mortierella. Persoonia 9(3):381–391

Genre A, Lanfranco L, Perotto S, Bonfante P (2020) Unique and common traits in mycorrhizal symbioses. Nat Rev Microbiol 18(11):649–660

Ghignone S, Salvioli A, Anca I, Lumini E, Ortu G, Petiti L, Cruveiller S, Bianciotto V, Piffanelli P, Lanfranco L, Bonfante P (2012) The genome of the obligate endobacterium of an AM fungus reveals an interphylum network of nutritional interactions. ISME J 6(1):136–145. https://doi.org/10.1038/ismej.2011.110

Gho N, Yaoita T, Aoyagi K, Sato Z (1978) Studies on the control of Rhizopus in the nursery cases of rice seedlings, 5: influence of a phytotoxic substance produced by Rhizopus in growth of rice. In: Proceedings of the Association for Plant Protection of Hokuriku

Glaeser SP, Imani J, Alabid I, Guo H, Kumar N, Kämpfer P, Hardt M, Blom J, Goesmann A, Rothballer M (2015) Non-pathogenic rhizobium radiobacter F4 deploys plant beneficial activity independent of its host Piriformospora indica. ISME J 10(4):871–884

Gnanesh B, Tejaswi A, Arunakumar G, Supriya M, Manojkumar H, Tewary P (2021) Molecular phylogeny, identification and pathogenicity of Rhizopus oryzae associated with root rot of mulberry in India. J Appl Microbiol 131(1):360–374

Goyzueta-Mamani L, de Carvalho J, Magalhães A Jr, Soccol C (2021) Production of arachidonic acid by Mortierella alpina using wastes from potato chips industry. J Appl Microbiol 130(5):1592–1601

Guo Y, Takashima Y, Sato Y, Narisawa K, Ohta H, Nishizawa T (2020) Mycoavidus sp. strain B2-EB: comparative genomics reveals minimal genomic features required by a cultivable Burkholderiaceae-related endofungal bacterium. Appl Environ Microbiol 86(18):e01018–e01020

Herlambang A, Guo Y, Takashima Y, Narisawa K, Ohta H, Nishizawa T (2022) Whole-genome sequence of Entomortierella parvispora E1425, a Mucoromycotan fungus associated with Burkholderiaceae-related endosymbiotic bacteria. Microbiol Resour Announc 11(1):e01101–e01121

Hibbett DS, Binder M, Bischoff JF, Blackwell M, Cannon PF, Eriksson OE, Huhndorf S, James T, Kirk PM, Lücking R (2007) A higher-level phylogenetic classification of the fungi. Mycol Res 111(5):509–547

Hirose D, Degawa Y, Inaba S, Tokumasu S (2012) The anamorphic genus Calcarisporiella is a new member of the Mucoromycotina. Mycoscience 53(4):256–260

Hirose D, Degawa Y, Yamamoto K, Yamada A (2014) Sphaerocreas pubescens is a member of the Mucoromycotina closely related to fungi associated with liverworts and hornworts. Mycoscience 55(3): 221–226

Hoffman MT, Arnold AE (2010) Diverse bacteria inhabit living hyphae of phylogenetically diverse fungal endophytes. Appl Environ Microbiol 76(12): 4063–4075. https://doi.org/10.1128/AEM.02928-09

Horn F, Üzüm Z, Möbius N, Guthke R, Linde J, Hertweck C (2015) Draft genome sequences of symbiotic and nonsymbiotic Rhizopus microsporus strains CBS 344.29 and ATCC 62417. Genome Announc 3(1): e01370–e01314. https://doi.org/10.1128/genomeA.01370-14

Ibrahim AS, Gebremariam T, Liu M, Chamilos G, Kontoyiannis DP, Mink R, Kwon-Chung KJ, Fu Y, Skory CD, Edwards JE Jr (2008) Bacterial endosymbiosis is widely present among zygomycetes but does not contribute to the pathogenesis of mucormycosis. J Infect Dis 198(7):1083–1090

Itabangi H, Sephton-Clark PC, Tamayo DP, Zhou X, Starling GP, Mahamoud Z, Insua I, Probert M, Correia J, Moynihan PJ (2022) A bacterial endosymbiont of the fungus Rhizopus microsporus drives phagocyte evasion and opportunistic virulence. Current biology 32(5):1115–1130. e1116

Iwasaki S, Kobayashi H, Furukawa J, Namikoshi M, Okuda S, Sato Z, Matsuda I, Noda T (1984) Studies on macrocyclic lactone antibiotics. vii structure of a phytotoxin. J Antibiot 37(4):354–362

James TY, Kauff F, Schoch CL, Matheny PB, Hofstetter V, Cox CJ, Celio G, Gueidan C, Fraker E, Miadlikowska J, Lumbsch HT, Rauhut A, Reeb V, Arnold AE, Amtoft A, Stajich JE, Hosaka K, Sung GH, Johnson D, O'Rourke B, Crockett M, Binder M, Curtis JM, Slot JC, Wang Z, Wilson AW, Schussler A, Longcore JE, O'Donnell K, Mozley-Standridge S, Porter D, Letcher PM, Powell MJ, Taylor JW, White MM, Griffith GW, Davies DR, Humber RA, Morton JB, Sugiyama J, Rossman AY, Rogers JD, Pfister DH, Hewitt D, Hansen K, Hambleton S, Shoemaker RA, Kohlmeyer J, Volkmann-Kohlmeyer B, Spotts RA, Serdani M, Crous PW, Hughes KW, Matsuura K, Langer E, Langer G, Untereiner WA, Lucking R, Budel B, Geiser DM, Aptroot A, Diederich P, Schmitt I, Schultz M, Yahr R, Hibbett DS, Lutzoni F, McLaughlin DJ, Spatafora JW, Vilgalys R (2006) Reconstructing the early evolution of fungi using a six-gene phylogeny. Nature 443(7113):818–822. https://doi.org/10.1038/nature05110

Jargeat P, Cosseau C, Ola'h B, Jauneau A, Bonfante P, Batut J, Becard G (2004) Isolation, free-living capacities, and genome structure of "Candidatus Glomeribacter gigasporarum," the endocellular bacterium of the mycorrhizal fungus Gigaspora margarita. J Bacteriol 186(20):6876–6884

Jeon KW (1972) Development of cellular dependence on infective organisms: micrurgical studies in amoebas. Science 176(4039):1122–1123

Jeong W, Keighley C, Wolfe R, Lee WL, Slavin M, Kong D, Chen S-A (2019) The epidemiology and clinical manifestations of mucormycosis: a systematic review and meta-analysis of case reports. Clin Microbiol Infect 25(1):26–34

Koczyk G, Pawłowska J, Muszewska A (2021) Terpenoid biosynthesis dominates among secondary metabolite clusters in mucoromycotina genomes. J Fungi 7(4):285

Kondo H, Botella L, Suzuki N (2022) Mycovirus diversity and evolution revealed/inferred from recent studies. Annu Rev Phytopathol 60:307–336

Kuga Y, Wu T-D, Sakamoto N, Katsuyama C, Yurimoto H (2021) Allocation of carbon from an arbuscular mycorrhizal fungus, Gigaspora margarita, to its gram-negative and positive Endobacteria revealed by high-resolution secondary ion mass spectrometry. Microorganisms 9(12):2597

Lackner G, Hertweck C (2011) Impact of endofungal bacteria on infection biology, food safety, and drug development. PLoS Pathog 7(6):e1002096

Lackner G, Partida-Martinez LP, Hertweck C (2009) Endofungal bacteria as producers of mycotoxins. Trends Microbiol 17(12):570–576. https://doi.org/10.1016/j.tim.2009.09.003

Lackner G, Moebius N, Hertweck C (2011a) Endofungal bacterium controls its host by an hrp type III secretion system. ISME J 5(2):252–261. https://doi.org/10.1038/ismej.2010.126

Lackner G, Moebius N, Partida-Martinez LP, Boland S, Hertweck C (2011b) Evolution of an endofungal lifestyle: deductions from the Burkholderia rhizoxinica genome. BMC Genomics 12(1):1–13

Lackner G, Moebius N, Partida-Martinez L, Hertweck C (2011c) Complete genome sequence of Burkholderia rhizoxinica, an endosymbiont of Rhizopus microsporus. J Bacteriol 193(3):783–784

Lastovetsky OA, Gaspar ML, Mondo SJ, LaButti KM, Sandor L, Grigoriev IV, Henry SA, Pawlowska TE (2016) Lipid metabolic changes in an early divergent fungus govern the establishment of a mutualistic symbiosis with endobacteria. Proc Natl Acad Sci 113(52): 201615148

Lastovetsky OA, Ahn E, Mondo SJ, Toomer KH, Zhang A, Johnson LM, Pawlowska TE (2018) Distribution and population structure of endobacteria in arbuscular mycorrhizal fungi at North Atlantic dunes. ISME J 12(12):3001–3013

Leone MR, Lackner G, Silipo A, Lanzetta R, Molinaro A, Hertweck C (2010) An unusual galactofuranose lipopolysaccharide that ensures the intracellular survival of toxin-producing bacteria in their fungal host. Angew Chem Int Ed 49(41):7476–7480

Li Z, Yao Q, Dearth SP, Entler MR, Castro Gonzalez HF, Uehling JK, Vilgalys RJ, Hurst GB, Campagna SR, Labbé JL (2017) Integrated proteomics and metabolomics suggests symbiotic metabolism and multimodal regulation in a fungal-endobacterial system. Environ Microbiol 19(3):1041–1053

Liao H-L (2021) The plant-growth-promoting fungus, Mortierella elongata: its biology, ecological distribution, and growth-promoting activities: SS679/SL466, 3/2021. EDIS 2021(2)

Liao H-L, Bonito G, Rojas JA, Hameed K, Wu S, Schadt CW, Labbé J, Tuskan GA, Martin F, Grigoriev IV (2019) Fungal endophytes of Populus trichocarpa alter host phenotype, gene expression, and rhizobiome composition. Mol Plant-Microbe Interact 32(7): 853–864

Liu Y, Steenkamp ET, Brinkmann H, Forget L, Philippe H, Lang BF (2009) Phylogenomic analyses predict sistergroup relationship of nucleariids and fungi and paraphyly of zygomycetes with significant support. BMC Evol Biol 9(1):1–11

Lumini E, Bianciotto V, Jargeat P, Novero M, Salvioli A, Faccio A, Becard G, Bonfante P (2007) Presymbiotic growth and sporal morphology are affected in the arbuscular mycorrhizal fungus Gigaspora margarita cured of its endobacteria. Cell Microbiol 9(7): 1716–1729. https://doi.org/10.1111/j.1462-5822.2007.00907.x

MacDonald R, Chandler MR (1981) Bacterium-like organelles in the vesicular-arbuscular mycorrhizal fungus Glomus caledonius. New Phytol 89(2):241–246

Macias AM, Marek PE, Morrissey EM, Brewer MS, Short DP, Stauder CM, Wickert KL, Berger MC, Metheny AM, Stajich JE (2019) Diversity and function of fungi associated with the fungivorous millipede, Brachycybe lecontii. Fungal Ecol 41:187–197

Mamani LDG, Magalhães AI Jr, Ruan Z, de Carvalho JC, Soccol CR (2019) Industrial production, patent landscape, and market trends of arachidonic acid-rich oil of Mortierella alpina. Biotechnol Res Innov 3(1):103–119

Meyer W, Walter G (2003) Delimitation of Umbelopsis (Mucorales, Umbelopsidaceae fam. Nov.) based on ITS sequence and RFLP data. Mycol Res 107(3): 339–350

Moebius N, Üzüm Z, Dijksterhuis J, Lackner G, Hertweck C (2014) Active invasion of bacteria into living fungal cells. elife 3:e03007

Mondo SJ, Toomer KH, Morton JB, Lekberg Y, Pawlowska TE (2012) Evolutionary stability in a 400-million-year-old heritable facultative mutualism. Evolution 66(8):2564–2576

Mondo SJ, Salvioli A, Bonfante P, Morton JB, Pawlowska TE (2016) Nondegenerative evolution in ancient heritable bacterial endosymbionts of fungi. Mol Biol Evol 33(9):msw086

Mondo SJ, Lastovetsky OA, Gaspar ML, Schwardt NH, Barber CC, Riley R, Sun H, Grigoriev IV, Pawlowska TE (2017) Bacterial endosymbionts influence host sexuality and reveal reproductive genes of early divergent fungi. Nat Commun 8(1):1843

Moreau F (1952) Les champignons, physiologie, morphologie développement ét systematique

Mosse B (1962) The establishment of vesicular-arbuscular mycorrhiza under aseptic conditions. Microbiology 27(3):509–520

Mosse B (1970) Honey-coloured, sessile Endogone spores: II. Changes in fine structure during spore development. Arch Mikrobiol 74(2):129–145

Mullins AJ, Mahenthiralingam E (2021) The hidden genomic diversity, specialized metabolite capacity, and revised taxonomy of Burkholderia sensu lato. Front Microbiol 12:726847

Myers J (2020) The diversity of Mycoviruses in early-diverging fungi, and their evolutionary implications (Doctoral dissertation)
Myers J, Bonds A, Clemons R, Thapa N, Simmons D, Carter-House D, Ortanez J, Liu P, Miralles-Durán A, Desirò A (2020) Survey of early-diverging lineages of fungi reveals abundant and diverse mycoviruses. MBio 11(5):e02027–e02020
Naito M, Pawlowska TE (2016) Defying Muller's ratchet: ancient heritable endobacteria escape extinction through retention of recombination and genome plasticity. MBio 7(3):e02057–e02015
Naito M, Morton JB, Pawlowska TE (2015) Minimal genomes of mycoplasma-related endobacteria are plastic and contain host-derived genes for sustained life within Glomeromycota. Proc Natl Acad Sci U S A 112(25):7791–7796. https://doi.org/10.1073/pnas.1501676112
Naito M, Desirò A, González JB, Tao G, Morton JB, Bonfante P, Pawlowska TE (2017) 'Candidatus Moeniiplasma glomeromycotorum', an endobacterium of arbuscular mycorrhizal fungi. Int J Syst Evol Microbiol 67(5):1177–1184
Naumann M, Schussler A, Bonfante P (2010) The obligate endobacteria of arbuscular mycorrhizal fungi are ancient heritable components related to the Mollicutes. ISME J 4(7):862–871. https://doi.org/10.1038/ismej.2010.21
Niehs SP, Scherlach K, Hertweck C (2018a) Genomics-driven discovery of a linear lipopeptide promoting host colonization by endofungal bacteria. Org Biomol Chem 16(37):8345–8352. https://doi.org/10.1039/C8OB01515E
Niehs SP, Dose B, Scherlach K, Roth M, Hertweck C (2018b) Genomics-driven discovery of a symbiont-specific cyclopeptide from bacteria residing in the rice seedling blight fungus. Chembiochem 19 (20):2167–2172. https://doi.org/10.1002/cbic.201800400
Niehs SP, Dose B, Scherlach K, Pidot SJ, Stinear TP, Hertweck C (2019) Genome mining reveals endopyrroles from a nonribosomal peptide assembly line triggered in fungal–bacterial symbiosis. ACS Chem Biol 14(8):1811–1818. https://doi.org/10.1021/acschembio.9b00406
Niehs SP, Dose B, Richter S, Pidot SJ, Dahse HM, Stinear TP, Hertweck C (2020) Mining symbionts of a spider-transmitted fungus illuminates uncharted biosynthetic pathways to cytotoxic benzolactones. Angew Chem Int Ed 59(20):7766–7771. https://doi.org/10.1002/anie.201916007
Niehs SP, Scherlach K, Dose B, Uzum Z, Stinear TP, Pidot SJ, Hertweck C (2022) A highly conserved gene locus in endofungal bacteria codes for the biosynthesis of symbiosis-specific cyclopeptides. PNAS Nexus 1(4): pgac152
Ohshima S, Sato Y, Fujimura R, Takashima Y, Hamada M, Nishizawa T, Narisawa K, Ohta H (2016) Mycoavidus cysteinexigens gen. Nov., sp. nov., an endohyphal bacterium isolated from a soil isolate of the fungus Mortierella elongata. Int J Syst Evol Microbiol 66(5):2052–2057
Ohta T (1972) Population size and rate of evolution. J Mol Evol 1(4):305–314
Okrasińska A, Bokus A, Duk K, Gęsiorska A, Sokołowska B, Miłobędzka A, Wrzosek M, Pawłowska J (2021) New endohyphal relationships between mucoromycota and burkholderiaceae representatives. Appl Environ Microbiol 87(7): e02707–e02720
Papanikolaou S, Aggelis G (2019) Sources of microbial oils with emphasis to Mortierella (Umbelopsis) isabellina fungus. World J Microbiol Biotechnol 35(4):1–19
Partida-Martinez LP (2013) A model for bacterial-fungal interactions. LAP LAMBERT Academic Publishing, London
Partida-Martinez LP, Hertweck C (2005) Pathogenic fungus harbours endosymbiotic bacteria for toxin production. Nature 437(7060):884–888. https://doi.org/10.1038/nature03997
Partida-Martinez LP, Hertweck C (2007) A gene cluster encoding rhizoxin biosynthesis in "Burkholderia rhizoxina", the bacterial endosymbiont of the fungus Rhizopus microsporus. Chembiochem 8(1):41–45. https://doi.org/10.1002/cbic.200600393
Partida-Martinez LP, Monajembashi S, Greulich KO, Hertweck C (2007) Endosymbiont-dependent host reproduction maintains bacterial-fungal mutualism. Curr Biol 17(9):773–777. https://doi.org/10.1016/j.cub.2007.03.039. Epub 2007 Apr 5. PMID: 17412585
Partida-Martínez LP (2017) The fungal holobiont: evidence from early diverging fungi. Environ Microbiol 19(8):2919–2923. https://doi.org/10.1111/1462-2920.13731. Epub 2017 Apr 3. PMID: 28295957
Pawlowska TE, Gaspar ML, Lastovetsky OA, Mondo SJ, Real-Ramirez I, Shakya E, Bonfante P (2018) Biology of fungi and their bacterial endosymbionts. Annu Rev Phytopathol 56(1):289–309
Petrikkos G, Skiada A, Lortholary O, Roilides E, Walsh TJ, Kontoyiannis DP (2012) Epidemiology and clinical manifestations of mucormycosis. Clin Infect Dis 54 (suppl_1):S23–S34
Petrikkos G, Skiada A, Drogari-Apiranthitou M (2014) Epidemiology of mucormycosis in Europe. Clin Microbiol Infect 20:67–73
Redecker D, Kodner R, Graham LE (2000) Glomalean fungi from the Ordovician. Science 289(5486): 1920–1921
Remy W, Taylor TN, Hass H, Kerp H (1994) Four hundred-million-year-old vesicular arbuscular mycorrhizae. Proc Natl Acad Sci 91(25):11841–11843
Richter I, Uzum Z, Stanley CE, Moebius N, Stinear TP, Pidot SJ, Ferling I, Hillmann F, Hertweck C (2020) Secreted TAL effectors protect symbiotic bacteria from entrapment within fungal hyphae. bioRxiv:2020–03. https://doi.org/10.1101/2020.03.28.013177

Richter I, Radosa S, Cseresnyés Z, Ferling I, Büttner H, Niehs SP, Gerst R, Scherlach K, Figge MT, Hillmann F (2022) Toxin-producing endosymbionts shield pathogenic fungus against micropredators. MBio 13(5): e01440–e01422

Robinson AJ, House GL, Morales DP, Kelliher JM, Gallegos-Graves LV, LeBrun ES, Davenport KW, Palmieri F, Lohberger A, Bregnard D (2021) Widespread bacterial diversity within the bacteriome of fungi. Commun Biol 4(1):1–13

Ropars J, Toro KS, Noel J, Pelin A, Charron P, Farinelli L, Marton T, Krüger M, Fuchs J, Brachmann A (2016) Evidence for the sexual origin of heterokaryosis in arbuscular mycorrhizal fungi. Nat Microbiol 1(6):1–9

Roux S, Krupovic M, Daly RA, Borges AL, Nayfach S, Schulz F, Sharrar A, Matheus Carnevali PB, Cheng J-F, Ivanova NN (2019) Cryptic inoviruses revealed as pervasive in bacteria and archaea across Earth's biomes. Nat Microbiol 4(11):1895–1906

Salvioli di Fossalunga A, Lipuma J, Venice F, Dupont L, Bonfante P (2017) The endobacterium of an arbuscular mycorrhizal fungus modulates the expression of its toxin–antitoxin systems during the life cycle of its host. ISME J 11(10):2394–2398

Salvioli A, Lumini E, Anca IA, Bianciotto V, Bonfante P (2008) Simultaneous detection and quantification of the unculturable microbe Candidatus Glomeribacter gigasporarum inside its fungal host Gigaspora margarita. New Phytol 180(1):248–257

Salvioli A, Chiapello M, Fontaine J, Hadj-Sahraoui AL, Grandmougin-Ferjani A, Lanfranco L, Bonfante P (2010) Endobacteria affect the metabolic profile of their host Gigaspora margarita, an arbuscular mycorrhizal fungus. Environ Microbiol 12(8):2083–2095

Salvioli A, Ghignone S, Novero M, Navazio L, Bagnaresi P, Bonfante P (2016) Symbiosis with an endobacterium increases the fitness of a mycorrhizal fungus, raising its bioenergetic potential. ISME J 10(1):130–144

Sato Y, Narisawa K, Tsuruta K, Umezu M, Nishizawa T, Tanaka K, Yamaguchi K, Komatsuzaki M, Ohta H (2010) Detection of Betaproteobacteria inside the mycelium of the fungus Mortierella elongata. Microbes Environ 25(4):321–324. https://doi.org/10.1264/jsme2.ME10134

Scannerini S, Bonfante P 1975 Fontana a ultrastructural model for the host-symbiont interaction in the endotrophic mycorrhizae of Ornithogalum umbellatum L. In: Endomycorrhizas; Proceedings of a Symposium

Scannerini S, Bonfante P (1991) Bacteria and bacteria like objects in endomycorrhizal fungi (Glomaceae). Symbiosis as Source of Evolutionary Innovation: Speciation and Morphogenesis:273–287

Scherlach K, Partida-Martinez LP, Dahse H-M, Hertweck C (2006) Antimitotic Rhizoxin derivatives from a cultured bacterial endosymbiont of the Rice pathogenic fungus Rhizopus m icrosporus. J Am Chem Soc 128(35):11529–11536

Scherlach K, Busch B, Lackner G, Paszkowski U, Hertweck C (2012) Symbiotic cooperation in the biosynthesis of a phytotoxin. Angew Chem Int Ed 51 (38):9615–9618. https://doi.org/10.1002/anie.201204540

Schüßler A, Mollenhauer D, Schnepf E, Kluge M (1994) Geosiphon pyriforme, an endosymbiotic association of fungus and cyanobacteria: the spore structure resembles that of arbuscular mycorrhizal (AM) fungi. Botanica Acta 107(1):36–45

Schüßler A, Bonfante P, Schnepf E et al (1996) Characterization of the Geosiphon pyriforme symbiosome by affinity techniques: confocal laser scanning microscopy (CLSM) and electron microscopy. Protoplasma 190:53–67. https://doi.org/10.1007/BF01281194

Sharmin D, Guo Y, Nishizawa T, Ohshima S, Sato Y, Takashima Y, Narisawa K, Ohta H (2018) Comparative genomic insights into endofungal lifestyles of two bacterial endosymbionts, Mycoavidus cysteinexigens and Burkholderia rhizoxinica. Microbe Environ 33(1): 17138

Smith SE, Read DJ (2010) Mycorrhizal symbiosis. Academic press

Smith ME, Gryganskyi A, Bonito G, Nouhra E, Moreno-Arroyo B, Benny G (2013) Phylogenetic analysis of the genus Modicella reveals an independent evolutionary origin of sporocarp-forming fungi in the Mortierellales. Fungal Genet Biol 61:61–68

Spatafora JW, Chang Y, Benny GL, Lazarus K, Smith ME, Berbee ML, Bonito G, Corradi N, Grigoriev I, Gryganskyi A (2016) A phylum-level phylogenetic classification of zygomycete fungi based on genome-scale data. Mycologia 108(5):1028–1046

Sun X, Chen W, Ivanov S, MacLean AM, Wight H, Ramaraj T, Mudge J, Harrison MJ, Fei Z (2019) Genome and evolution of the arbuscular mycorrhizal fungus Diversispora epigaea (formerly glomus versiforme) and its bacterial endosymbionts. New Phytol 221(3):1556–1573

Sward R (1981) The structure of the spores of Gigaspora margaritaI The dormant spore. New Phytol 87(4): 761–768

Szabó G, Schulz F, Manzano-Marín A, Toenshoff ER, Horn M (2022) Evolutionarily recent dual obligatory symbiosis among adelgids indicates a transition between fungus-and insect-associated lifestyles. ISME J 16(1):247–256

Tabima JF, Trautman IA, Chang Y, Wang Y, Mondo S, Kuo A, Salamov A, Grigoriev IV, Stajich JE, Spatafora JW (2020) Phylogenomic analyses of non-Dikarya fungi supports horizontal gene transfer driving diversification of secondary metabolism in the amphibian gastrointestinal symbiont, Basidiobolus. G3: genes, genomes, Genetics 10(9):3417–3433

Takashima Y, Degawa Y, Ohta H, Narisawa K (2018a) Mortierella sugadairana, a new homothallic species related to the firstly described heterothallic species in the genus. Mycoscience 59(3):200–205

Takashima Y, Seto K, Degawa Y, Guo Y, Nishizawa T, Ohta H, Narisawa K (2018b) Prevalence and intrafamily phylogenetic divergence of Burkholderiaceae-related endobacteria associated with species of Mortierella. Microbe Environ 33(4):18081

Takashima Y, Degawa Y, Nishizawa T, Ohta H, Narisawa K (2020) Aposymbiosis of a Burkholderiaceae-related endobacterium impacts on sexual reproduction of its fungal host. Microbes Environ 35(2):ME19167

Taylor TN, Remy W, Hass H, Kerp H (1995) Fossil arbuscular mycorrhizae from the early Devonian. Mycologia 87(4):560–573

Tedersoo L, Sánchez-Ramírez S, Kõljalg U, Bahram M, Döring M, Schigel D, May T, Ryberg M, Abarenkov K (2018) High-level classification of the fungi and a tool for evolutionary ecological analyses. Fungal Divers 90(1):135–159

Telagathoti A, Probst M, Peintner U (2021) Habitat, snow-cover and soil pH, affect the distribution and diversity of mortierellaceae species and their associations to bacteria. Front Microbiol 12:1817

Tisserant E, Malbreil M, Kuo A, Kohler A, Symeonidi A, Balestrini R, Charron P, Duensing N, Freidit Frey N, Gianinazzi-Pearson V, Gilbert LB, Handa Y, Herr JR, Hijri M, Koul R, Kawaguchi M, Krajinski F, Lammers PJ, Masclaux FG, Murat C, Morin E, Ndikumana S, Pagni M, Petitpierre D, Requena N, Rosikiewicz P, Riley R, Saito K, San Clemente H, Shapiro H, van Tuinen D, Becard G, Bonfante P, Paszkowski U, Shachar-Hill YY, Tuskan GA, Young JP, Sanders IR, Henrissat B, Rensing SA, Grigoriev IV, Corradi N, Roux C, Martin F (2013) Genome of an arbuscular mycorrhizal fungus provides insight into the oldest plant symbiosis. Proc Natl Acad Sci U S A 110(50): 20117–20122. https://doi.org/10.1073/pnas.1313452110

Toomer KH, Chen X, Naito M, Mondo SJ, den Bakker HC, VanKuren NW, Lekberg Y, Morton JB, Pawlowska TE (2015) Molecular evolution patterns reveal life history features of mycoplasma-related endobacteria associated with arbuscular mycorrhizal fungi. Mol Ecol 24(13):3485–3500

Torres-Cortes G, Ghignone S, Bonfante P, Schussler A (2015) Mosaic genome of endobacteria in arbuscular mycorrhizal fungi: Transkingdom gene transfer in an ancient mycoplasma-fungus association. Proc Natl Acad Sci U S A 112(25):7785–7790. https://doi.org/10.1073/pnas.1501540112

Torres-Cruz TJ, Billingsley Tobias TL, Almatruk M, Hesse CN, Kuske CR, Desirò A, Benucci GMN, Bonito G, Stajich JE, Dunlap C (2017) Bifiguratus adelaidae, gen. Et sp. nov., a new member of Mucoromycotina in endophytic and soil-dwelling habitats. Mycologia 109(3):363–378

Turina M, Ghignone S, Astolfi N, Silvestri A, Bonfante P, Lanfranco L (2018) The virome of the arbuscular mycorrhizal fungus Gigaspora margarita reveals the first report of DNA fragments corresponding to replicating non-retroviral RNA viruses in fungi. Environ Microbiol 20(6):2012–2025

Uehling J, Gryganskyi A, Hameed K, Tschaplinski T, Misztal P, Wu S, Desirò A, Vande Pol N, Du Z, Zienkiewicz A (2017) Comparative genomics of Mortierella elongata and its bacterial endosymbiont Mycoavidus cysteinexigens. Environ Microbiol 19(8):2964–2983

Uzum Z, Silipo A, Lackner G, De Felice A, Molinaro A, Hertweck C (2015) Structure, genetics and function of an exopolysaccharide produced by a bacterium living within fungal hyphae. Chembiochem 16(3):387–392

Vandepol N, Liber J, Desirò A, Na H, Kennedy M, Barry K, Grigoriev IV, Miller AN, O'Donnell K, Stajich JE (2020) Resolving the Mortierellaceae phylogeny through synthesis of multi-gene phylogenetics and phylogenomics. Fungal Divers 104(1):267–289

VanKuren NW, den Bakker HC, Morton JB, Pawlowska TE (2013) Ribosomal RNA gene diversity, effective population size, and evolutionary longevity in asexual Glomeromycota. Evolution: international journal of organic. Evolution 67(1):207–224

Vannini C, Carpentieri A, Salvioli A, Novero M, Marsoni M, Testa L, Pinto MC, Amoresano A, Ortolani F, Bracale M (2016) An interdomain network: the endobacterium of a mycorrhizal fungus promotes antioxidative responses in both fungal and plant hosts. New Phytol 211(1):265–275

Venice F, Ghignone S, Salvioli di Fossalunga A, Amselem J, Novero M, Xianan X, Sędzielewska Toro K, Morin E, Lipzen A, Grigoriev IV (2020a) At the nexus of three kingdoms: the genome of the mycorrhizal fungus Gigaspora margarita provides insights into plant, endobacterial and fungal interactions. Environ Microbiol 22(1):122–141

Venice F, Desirò A, Silva G, Salvioli A, Bonfante P (2020b) The mosaic architecture of NRPS-PKS in the arbuscular mycorrhizal fungus gigaspora margarita shows a domain with bacterial signature. Front Microbiol 11:581313

Venice F, Chialva M, Domingo G, Novero M, Carpentieri A, Salvioli di Fossalunga A, Ghignone S, Amoresano A, Vannini C, Lanfranco L (2021) Symbiotic responses of Lotus japonicus to two isogenic lines of a mycorrhizal fungus differing in the presence/absence of an endobacterium. Plant J 108(6): 1547–1564

Walker C (1985) Endogone lactiflua forming ectomycorrhizas with Pinus contorta. Trans Br Mycol Soc 84(2):353–355

Walker C, Gollotte A, Redecker D (2018) A new genus, Planticonsortium (Mucoromycotina), and new combination (P. tenue), for the fine root endophyte, Glomus tenue (basionym Rhizophagus tenuis). Mycorrhiza 28(3):213–219

Walsh TJ, Gamaletsou MN, McGinnis MR, Hayden RT, Kontoyiannis DP (2012) Early clinical and laboratory diagnosis of invasive pulmonary, extrapulmonary, and disseminated mucormycosis (zygomycosis). Clin Infect Dis 54(suppl_1):S55–S60

Warcup J (1990) Taxonomy, culture and mycorrhizal associations of some zygosporic Endogonaceae. Mycol Res 94(2):173–178

Wiersinga WJ, Van der Poll T, White NJ, Day NP, Peacock SJ (2006) Melioidosis: insights into the pathogenicity of Burkholderia pseudomallei. Nat Rev Microbiol 4(4):272–282

Wurlitzer JM, Stanišić A, Wasmuth I, Jungmann S, Fischer D, Kries H, Gressler M (2021) Bacterial-like nonribosomal peptide synthetases produce cyclopeptides in the zygomycetous fungus Mortierella alpina. Appl Environ Microbiol 87(3):e02051–e02020

Yamamoto K, Endo N, Degawa Y, Fukuda M, Yamada A (2017) First detection of Endogone ectomycorrhizas in natural oak forests. Mycorrhiza 27(3):295–301

Yang S, Anikst V, Adamson PC (2022) Endofungal Mycetohabitans rhizoxinica Bacteremia associated with Rhizopus microsporus respiratory tract infection. Emerg Infect Dis 28(10):2091

Fungi and their Environmental Micropredators

9

Silvia Radosa, Nauman Saeed, and Falk Hillmann

Abstract

Predator–prey interactions are obvious drivers in the evolution among higher animals, but until today, there is little known about such relationships in the microverse. Like any other microbes, fungi can be prone to predation by bacteria, amoebae, or nematodes. While the boundaries between parasitism, biotrophy, and predation appear to be less sharp for bacteria, highly sophisticated mechanisms to feed on are obvious for unicellular amoebae and nematodes. Even phagocytic amoebae are not restricted to the ingestion and intracellular killing of single yeast cells or conidia but they can also attack and invade entire hyphae of filamentous fungi. Nematodes, in turn, can specifically open hyphae via injection-needle like stylets. In this chapter, we provide a selected overview of examples from recent years on how fungal preys exploit a variety of strategies to either escape or survive attacks by these specialized environmental predators. At least some of these factors have been shown to fulfil a dual function, serving in predator defence but also against innate immune cells when colonizing higher animals. Whether predatory defence has thus acted as a selection pressure towards virulence determinants in fungi is a matter of ongoing research.

Keywords

Amoeba · Predation · Phagocytosis · Virulence

S. Radosa
Junior Research Group Evolution of Microbial Interactions, Leibniz Institute for Natural Product Research and Infection Biology – Hans Knöll Institute (HKI), Jena, Germany

N. Saeed
Junior Research Group Evolution of Microbial Interactions, Leibniz Institute for Natural Product Research and Infection Biology – Hans Knöll Institute (HKI), Jena, Germany

Biochemistry/Biotechnology, Faculty of Engineering, Wismar University of Applied Sciences Technology, Business and Design, Wismar, Germany

F. Hillmann (✉)
Biochemistry/Biotechnology, Faculty of Engineering, Wismar University of Applied Sciences Technology, Business and Design, Wismar, Germany
e-mail: falk.hillmann@hs-wismar.de

9.1 Introduction

What is long known for higher eukaryotes becomes increasingly clear also on the microbial scale: Predator–prey relationships are abundant in natural environments and can impose selection pressure on both partners. More and more examples show that not only bacteria but also filamentous fungi and yeasts are exposed to micropredators. While insects and even vertebrates feed mainly on fungal fruiting bodies, this chapter will focus on micropredators, unseen

© The Author(s), under exclusive license to Springer Nature Switzerland AG 2023

S. Pöggeler, T. James (eds.), *Evolution of Fungi and Fungal-Like Organisms*, The Mycota 14,
https://doi.org/10.1007/978-3-031-29199-9_9

bacteria, nematodes, and especially the ubiquitous amoebae that have adapted highly specific tools to find, engulf, or open fungal cells. While predatory interactions between fungi and eukaryotic micropredators were described half a century ago, laboratory- controlled bipartite interactions studies aiming to reveal the mechanistic cellular, biochemical, and molecular details were only initiated in the last 20 years. Apart from unique killing mechanisms on the predator side, these studies have also provided accumulating evidence that coevolution between the two partners has triggered fungal defence mechanisms that not only allow predator escape, but may also be exploited against immune cells and thus favour the emergence of traits considered to be virulence determinants in higher animals.

9.2 Fungal Micropredators

There are numerous examples for highly evolved mutualistic interactions between fungi and other taxonomic groups. The most famous are the highly complex symbioses in lichens or the formation of fungal roots called mycorrhizae and are known for several decades. Without any doubt, these cross-kingdom partnerships have driven the evolution of the fungal kingdom, but frequently occurring antagonistic interactions have also shaped the fungal tree. The two most simple antagonisms between two fungal species occur when one partner simply feeds on or acquires nutrients from other fungi in close proximity (Fig. 9.1). Such a lifestyle is considered necrotrophic when nutrients are derived from a dead fungal host cell. Necrotrophic mycoparasites are less specific to fungal prey and therefore exhibit a rather broad host range. Members of the *Trichoderma, Escovopsis, Tolypocladium,* and *Clonostachys* genera are the most common examples (Karlsson et al. 2017). When in contrast, a living organism is directly attacked, a fungal lifestyle becomes biotrophic and can be considered mycopredatory or mycoparasitic (Barnett and Binder 1973). Since biotrophic mycoparasites usually coevolve with their hosts, they generally exhibit a narrow host range and thus, rather few organisms have been studied in detail. *Ampelomyces quisqualis*, a natural fungal opponent of Erysiphales fungi, the causative agents of powdery mildew disease, is one of them (Németh et al. 2019). Such mycoparasites detect the existence of prospective hosts, move towards, and attach to them, followed by the frequent formation of coils around the prey to produce appressoria-like infectious structures that aid host penetration (Chet et al. 1981; Elad et al. 1983; Lu et al. 2004). Furthermore, numerous cell-wall disintegrating enzymes and antifungal metabolites are released throughout necrotrophic or biotrophic mycoparasitism to digest and destroy fungal hosts and utilize nutrients (Lu et al. 2004; Almeida et al. 2007; Druzhinina et al. 2012). In addition to fungal predation, the best-known examples of mycoparasites have the ability to facilitate plant growth through rhizosphere competence (Harman et al. 2004; Chatterton and Punja 2010; Dubey et al. 2014). Therefore, their use as an active ingredient in biocontrol formulations is not surprising. A prominent example of such a biocontrol agent (BCA) is *Coniothyrium minitans*, a fungal parasite whose conidia are used as a bio-fungicide against the white mould causing fungus *Sclerotinia sclerotiorum* (de Vrije et al. 2001).

A wide range of bacteria are also fungivorous and have the capability to feed on fungal cells (Boer et al. 2005; Fritsche et al. 2006). Like mycoparasites, bacteria exhibit three phenomenal mechanisms of acquiring nutrients, namely exo- and endocellular biotrophy and exocellular necrotrophy (Fig. 9.1). In exocellular biotrophic interactions, bacteria coexist with fungal hyphae or may colonize their surfaces to assimilate the nutrients released by active fungal cells. The synthesis of antibacterial compounds by fungal cells can either be tolerated or suppressed by biotrophs, and they may be able to modify fungal metabolism to facilitate the release of nutrients (Leveau and Preston 2008). However, endocellular mycoparasitic bacteria reside in the living fungal host and actively absorb the nutrients from the cytoplasm, such as *Burkholderia mallei* and *Burkholderia pseudomallei* (Levy et al. 2003).

Parasitic and predatory interactions	Organisms	Description	Micropredators
Necrotrophic mycoparasitism	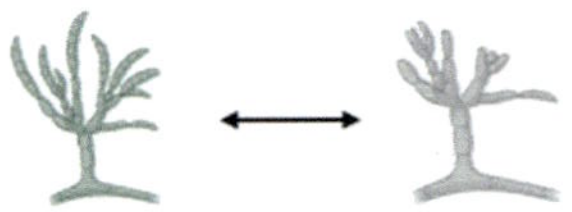	Nutrients are released from dead fungal host cell. Less specific to fungal prey, broad host range	*Trichoderma, Escovospsis, Tolypocladium, Clonostachys* (Karlsson et al. 2017)
Biotrophic mycoparasitism	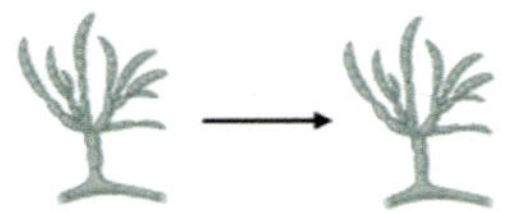	Living fungal cell is directly attacked by another fungus. Mycoparasites usually coevolve with their host, therefor exhibit a narrow host range.	*Ampelomyces quisqualis* (Nemeth et al. 2019) *Coniothyrium minitans* (de Vrije et al. 2001)
Exocellular biotrophy	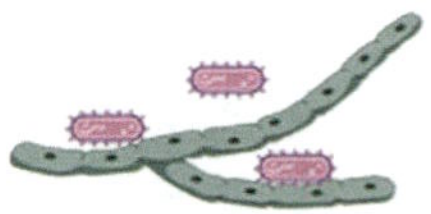	Bacteria coexist with or on the surface of fungi and utilise the nutrients released by the fungus. Often able to modify fungal metabolism.	*Pseudomonas aeruginosa* (Davis-Hanna et al. 2008; Mowat et al. 2010; Rella et al. 2012)
Endocellular biotrophy	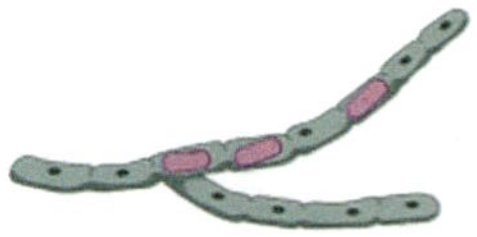	Bacteria reside in the living fungal host and actively absorb nutrients from cytoplasm.	*Burkholderia mallei, B. pseudomallei* (Levy et al. 2003)
Exocellular necrotrophy	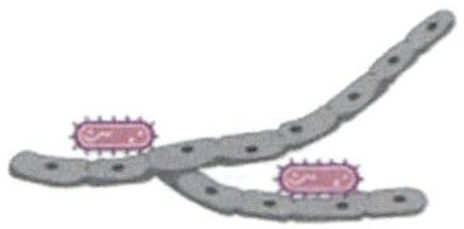	Bacteria release antifungal metabolites to kill their fungal prey and feed on them.	*Pseudomonas tolaasii* (Rainy et al. 1993; jolivet et al. 1999; Lo Cantore et al. 2006) *janthinobacterium agaricidamnosum* (Graupner et al. 2012)
Nematode predation	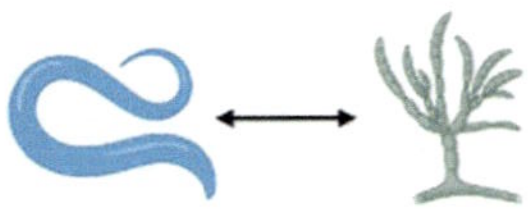	Mycophagous nematodes feed on fungi whereas, nematophagous fungi feed on nematodes using trapping systems.	*Aphelenchus avenae,* Bursaphelenchus okinawaensis, Caenorhabditid elegans (Zhang et al. 2020; Tayyrov et al. 2018; Fischer and Requena 2022)
Amoeba predation	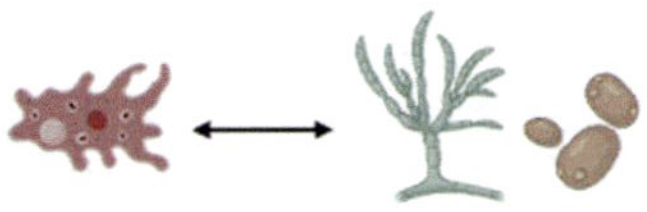	Fungivorous amoeba feed predominantly on the fungal cells via phagocytosis or ruphocytosis.	*Acanthamoeba castellanii* (Goncalves et al. 2019; Steenbergen et al. 2004), *Allovahlkampfia spelaean, Vermamoeba vermiformis* (Albuquerque et al. 2019) *Protostelium aurantium* (Hillmann et al. 2018; Radosa et al. 2019; Ferling et al. 2020)

Fig. 9.1 Overview table of major micropredatory interactions among the fungal kingdom

In contrast, exocellular necrotrophic bacteria release antifungal metabolites to kill their fungal prey and feed on them (Leveau and Preston 2008). Actinobacteria, β-proteobacteria, and myxobacteria are only a few of the diverse bacterial groups that have been shown to necrotize fungal hosts (Boer et al. 2005). These mycophagous bacteria produce a variety of distinct hydrolytic enzymes, toxins, and antibiotics. Nevertheless, most of the bacterial–fungal interaction studies have identified low molecular weight bacterial toxins as the primary causative agents of fungal necrosis. *Pseudomonas tolaasii*'s toxin tolaasin is one of the prominent examples of such predation factors causing brown blotch disease in the button mushroom *Agaricus bisporus* (Rainey et al. 1993; Jolivet et al. 1999; Lo Cantore et al. 2006). Tolaasin's ability to disrupt the cell membrane by pore formation or biosurfactant activity has been extensively investigated by Jourdan and others (Scherlach et al. 2013; Jourdan et al. 2003). A similar mode of action has been shown for Jagaricin from *Janthinobacterium agaricidamnosum*, another membrane active toxin against a major causative agent in soft rot disease of mushrooms like *Agaricus bisporus* (Graupner et al. 2012). In addition to toxins, necrotizing bacteria may subsequently secrete an array of hydrolytic enzymes such as chitinases, lipases, glucanases, and proteases to disintegrate cell components and absorb fungal nutrients (Lorito et al. 1994; Fogliano et al. 2002; Woo et al. 2002; Someya et al. 2007). Moreover, biotrophic mycophagous bacteria are anticipated to modify three key aspects of fungal physiology, i.e., growth, membrane permeability, and metabolism (nutrient efflux) to feed on the fungal host (Leveau and Preston 2008). For example, *Pseudomonas aeruginosa* quorum sensing molecules (QSM) 3-oxo-C12 homoserine lactone (HSL) and 2-heptyl-3,4-dihydroxyquinoline have the ability to inhibit filamentation in *Candida albicans* (Davis-Hanna et al. 2008), biofilm formation in *Aspergillus fumigatus* (Mowat et al. 2010), and growth in *Cryptococcus neoformans* (Rella et al. 2012), respectively. This might help the bacteria to benefit from the fungal host.

Single cell eukaryotes from the kingdom Amoebozoa are dominant environmental micropredator that feeds on a variety of microorganisms, including fungi (Chakraborty et al. 1983; Old and Darbyshire 1978; Magnet et al. 2015). However, only few of them such as *Acanthamoeba castellanii*, *Allovahlkampfia spelaea*, *Protostelium aurantium*, and *Vermamoeba vermiformis* have been well documented as true fungal predators (Albuquerque et al. 2019; Tosetti et al. 2014; Hillmann et al. 2015; Radosa et al. 2019; Radosa et al. 2021).

Interestingly, these fungivorous amoebae also display distinct predatory tendencies based on the species and environment they live in. For example, *P. aurantium*, a widespread inhabitant of plant leaves, can feed on many members of *Candida* clade but not on any of those of the *Saccharomyces* clade (Radosa et al. 2019). Consequently, their role is crucial in determining the composition of the microbial communities in the environment (German et al. 2013; Siddiqui and Khan 2012). When considering that fungi and their amoeba predators may have interacted for millions of years it appears obvious that predatory selection pressure could have boosted the ramification of the fungal evolutionary tree.

The amoeba *P. aurantium* utilizes phagocytosis and ruphocytosis mechanisms to feed on the opportunistic human pathogenic yeast *Candida parapsilosis* and the filamentous fungus *A. fumigatus*, respectively. Following phagocytosis, the yeast cells of *C. parapsilosis* encounter oxidative stress and are lyzed in acidified phagolysosomes of *P. aurantium*. A similar mechanism of intracellular killing was observed for swollen conidia of *A. fumigatus*. However, before swelling, dormant conidia of *Aspergillus* and a variety of other fungi are covered by protective coats of amorphous polymers and hydrophobin proteins. These melanins and hydrophobin layers can mask the pathogen (or prey) associated molecular patterns (PAMPs), thereby protecting the dormant conidia from predation by *P. aurantium* and other amoebae (Ferling et al. 2020; Geib et al. 2016; Radosa et al. 2019; Radosa et al. 2021).

Germlings and hyphae have either none or drastically reduced melanin layers, allowing amoebae like *P. aurantium* to perforate the cell membrane of *A. fumigatus*. Germlings and hyphae are further invaded by the amoebae and the complete cytoplasmic fluid is resorbed within minutes. This mechanism was coined ruphocytosis (Fig. 9.2)., based on the Greek word *ρουφω,* "to suck", or "slurp" (Radosa et al. 2019).

The first observation of fungivorous amoeba dates back to the original discovery of a predation on *Cryptococcus neoformans* (Castellani 1930). *A. castellanii* can phagocytose a wide variety of fungal species, but feeding in laboratory conditions depends on environmental factors, more precisely on balance of divalent cations Mg^{2+} and Ca^{2+} (Fu and Casadevall 2018). The phospholipid-dependent expression of polysaccharide capsule in the yeast *C. neoformans* and temperature-mediated filamentous growth of dimorphic fungi also protect them from intracellular killing by *A. castellanii* (Steenbergen et al. 2004; Chrisman et al. 2011; Guimaraes et al. 2016). To further analyse the impact of fungal predation, Gonçalves and colleagues infected the wax moth *Galleria mellonella* with *C. neoformans*, *C. albicans*, *Paracoccidioides brasiliensis*, *Sporothrix brasiliensis*, *Histoplasma capsulatum*, and *Saccharomyces cerevisiae* cells recovered from amoeba and in all cases the larvae were killed quicker than controls (Gonçalves et al. 2019). Hence, predation studies with fungivorous amoebae such as *A. castellanii* and *P. aurantium* support the hypothesis that environmental interactions trigger the evolution of pathogenic traits or resistance to immune cells.

9.3 Dual-Use of Fungal Predatory Defence Mechanisms as Virulence Factors

Environmentally acquired human pathogenic fungi naturally occur as saprophytes involved in the processing of decayed organic matter, nutrient, and carbon recycling, intensely shaping the microbiome they are inhabiting. In this hostile milieu, they are constantly exposed to changing conditions such as temperature, light, oxygen, salinity, or humidity, as well as to competition for nutrients with other microorganisms and more importantly, they often represent a central dietary resource for soil protists, nematodes, or arthropods (Bahram and Netherway 2021).

In this respect, fungi have evolved sophisticated strategies to enhance their competitiveness for nutrient acquisition and to protect themselves against eukaryotic predators (Künzler 2018). In turn, predators continuously evolved new strategies to secure food from their preys. This process of an evolutionary arms race has widely been described for fungi and nematodes. The latter even includes fungivorous nematodes as well as nematophagous fungi (Zhang et al. 2020; Bleuler-Martínez et al. 2011; Tayyrov

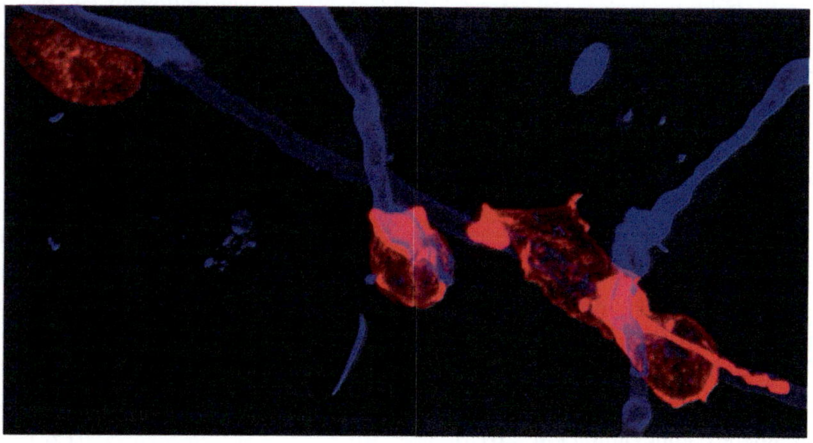

Fig. 9.2 Ruphocytic interactions of *Protostelium aurantium* with the filamentous fungus *Aspergillus fumigatus*. The amoebae attack and invade hyphal filaments. F-actin filaments of *P. aurantium* and chitin of the fungal cell wall are stained in red and blue, respectively

et al. 2018; Fischer and Requena 2022; Yang et al. 2020; Youssar et al. 2019). Both antagonists can contribute significantly to the immune status of surrounding plants and gain a lot of interest due to their possible role in biocontrol applications mainly in agriculture and forestry (Zhang et al. 2020).

For several human pathogenic fungi it has been demonstrated that some of these predator defence factors fulfil a so-called *dual-use* in the sense that they can be equally relevant for the survival in the natural environment as well as within the human host (Novohradská et al. 2017). This appears convincing since an infection of humans or mammals with a fungal conidium or an environmental yeast like *C. neoformans* seems to be only accidental and any pathogenic outcome depends almost exclusively on the host susceptibility. Most of the environmentally acquired pathogenic fungi like *A. fumigatus* or *C. neoformans* have defined ecological niches as decomposers of organic material in natural environments, such as in compost heaps or decaying wood and thus, lack the specific need for a human or animal host (Casadevall and Pirofski 2007). This idea has already been extensively studied in bacteria like *Legionella* spp. or *Mycobacterium* spp., classic example of intracellular pathogens, which can employ amoebae as reservoirs and exploit similar mechanisms during replication and survival in human phagocytes (Molmeret et al. 2005).

9.3.1 Melanins

The chemically diverse and irregularly structured polymers are among the most prominent examples of dual-use factors contributing to the defence against phagocytic immune cells and amoebae. An important role for melanins in protecting fungi against a variety of abiotic stresses is known for a long time and includes UV light protection, metal binding, energy harvesting, thermoregulation, protection against oxidative stress, but also inhibitory activities against microbial lytic enzymes and defensins, thereby enhancing the fungal survival in the environment (Heinekamp et al. 2013).

In *A. fumigatus*, green pigment dihydroxynaphthalene (DHN) melanin regulates host proinflammatory cytokine response by physically masking fungal PAMPs from immune recognition (Chai et al. 2010). During the germination of fungal conidia, the melanin layer is gradually degraded, exposing fungal PAMPs and allowing recognition by pathogen recognition receptors (PRRs). A pigment-less mutant of *A. fumigatus*, Δ*pksP*, was taken up at significantly higher rates by human macrophages than a respective wild type strain (Luther et al. 2007). Interestingly, similar outcomes have resulted from the interaction of *A. fumigatus* with amoeboid soil predator *Dictyostelium discoideum* (Hillmann et al. 2015). Here, fungal spores were efficiently phagocytosed by *D. discoideum*, but ingestion was higher when conidia were devoid of the DHN-melanin. What is more, the same study proved the co-occurrence of *A. fumigatus* and *Dictyostelids* sharing the same micro-habitats indicating that such predatory–prey interactions could be abundant in nature. The biosynthesis of DHN-melanin is, in fact, restricted to the fruiting bodies, suggesting its possible protective role against environmental predators.

Further, by dissecting the endocytic pathway using confocal microscopy, it was found that DHN-melanin highly reduced acidification of phagolysosomes of murine-derived alveolar macrophages, human monocyte-derived macrophages, human neutrophils, and epithelial cells (Thywißen et al. 2011; Amin et al. 2014). In *D. discoideum*, phagocytosis of fungal conidia was followed by the formation of the nascent phagosome, but here the phagolysosome was rapidly, but only transiently acidified, then neutralized, before the final exocytosis of the fungal spore. However, melanized conidia displayed an extended retention time within the phagocytic cells. As a repercussion, the swelling of the fungal spore induced enhanced phagolysosomal damage. At the same time, the membrane repair machinery of the amoeba was inefficiently recruited to the damage sites, thereby interfering with the canonical autophagy pathway of the amoeba (Ferling et al. 2020).

The cinnamon-brown conidia of the closely related *Aspergillus terreus* are known to be covered by non-ribosomally synthetized Asp-melanin. *A. terreus* lacks the highly conserved naphthopyrone synthase, and as a consequence, it is not able to inhibit phagolysosomal acidification in macrophages (Slesiona et al. 2012). Nevertheless, Asp-melanin accounts for resistance against UV light and even hampers the phagocytosis by amoeba *D. discoideum*, thus likely contributing to the survival of this fungus in its natural environment (Geib et al. 2016).

As a sign of convergent evolution, a comparable protective mechanism has also been engaged by the basidiomycetous yeast *C. neoformans*. Like in *A. fumigatus*, melanin of *C. neoformans* provides protection from ultraviolet light and scavenges ROS generated by phagocytes (Liu et al. 2021). *C. neoformans* has long been studied as a typical example of an environmentally acquired fungal pathogen whose virulence determinants for humans have evolved as a consequence of selection pressure imposed by amoeboid predators (Derengowski Lda et al. 2013). Striking parallels in phagocytic processing by amoeba can be found when comparing to fungus–macrophage interactions (Gaylord et al. 2020). Melanin and polysaccharide capsules (mentioned later) are major virulence factors of *C. neoformans*. When phagocytosed by the amoeba *Acanthamoeba castellanii*, both melanized and non-melanized strains survived and replicated inside of the host cell which has led to the death of amoeba cells. However, a difference in survival was observed when cells devoid of melanin also lacked a capsule. These cells were rapidly killed (Steenbergen et al. 2001). Hence, melanin significantly contributes to the resistance against amoeba predation, but the capsule seems to play an even more important role in this interaction.

9.3.2 Capsule

The characteristic polysaccharide capsule of *C. neoformans* consists of two large repeating polymers, glucuronoxylomannan (GXM) and glucuronoxylomannogalactan (GXMGal). Although the capsule itself is weakly immunogenic, it masks the underlying cell wall PAMPs, such as β-glucans, α-mannans, and chitin, from the recognition by the innate immune system (Gaylord et al. 2020).

Incubation of *C. neoformans* with *A. castellanii* elicited an approximately four-fold increase in the cryptococcal capsular volume compared to yeast cells grown in the absence of amoeba (Chrisman et al. 2011). Similar capsular enlargement has been observed when yeast cells were confronted with macrophages. Environmental strains of *C. neoformans* have in general smaller capsule, in contrast with the situation found during infection whereby cells with very large capsules are commonly found in mammalian tissue (Zaragoza and Casadevall 2004). While in human cells, the capsular enlargement arises from high CO_2 and iron deprivation, in amoeba, this effect occurs in response to phospholipids of amoeba origin. *C. neoformans* can cause enzymatic damage to the cell membrane of amoebae and macrophages thus releasing membrane phospholipids. These are subsequently cleaved by fungal extracellular phospholipase B, releasing their polar heads that are in turn sensed by *C. neoformans*, triggering capsule enlargement and the formation of giant cells (Chrisman et al. 2011).

9.3.3 Phospholipase

Phospholipase activity is another important virulence factor and a factor of dual-use in *Cryptococcus* survival. Phospholipase B-deficient mutant, Δ*plb1*, was found to be dramatically impaired in virulence in both the mouse inhalational model and the rabbit meningitis model. Furthermore, Δ*plb1* strains exhibited a growth defect in a J774 macrophage-like cell line (Cox et al. 2001).

This mutant was also unable to growth in the presence of the amoeba *A. castellanii,* suggesting a significance of this virulence factor in environment as well as during infection (Steenbergen et al. 2001).

In addition, phospholipase B seems to be implicated in the process of non-lytic exocytosis, also known as non-lytic phagosomal extrusion or vomocytosis, as mutants deficient in Plb1 secretion are impaired in non-lytic exocytosis (NLE) (Gaylord et al. 2020). Identical mechanisms of WASH-mediated constitutive exocytosis and non-lytic exocytosis are employed by the fungus to escape both amoeba and human macrophages (Watkins et al. 2018; Ma et al. 2006).

9.3.4 Mannoprotein Coat

In a first line of fungal defence stands the hiding from this recognition, and fungi have, in this respect, covered their PAMPs under the thick layers of cell wall composites, such as aforementioned capsule in *C. neoformans*. In the human commensal *Candida albicans,* the functional role of the capsule is at least partially fulfilled by the presence of a thick surface coating with mannoproteins (Erwig and Gow 2016).

Over the past decade, it has clearly been demonstrated that the mannoprotein coat is essential for *C. albicans* to escape the recognition of innate immune cells (McKenzie et al. 2010; Bain et al. 2014). The uptake and phagocytosis by macrophages were significantly reduced in mutants defected in phosphomannan biosynthesis and *O*- and *N*-linked mannosylation.

However, only lately, the role of this protective coat has been proven to be effective also in the context of the fungivorous amoeboid predator *Protostelium aurantium* (Radosa et al. 2019). Treatment with sub-inhibitory concentrations of caspofungin or mannosidase turned the previously unrecognizable *C. albicans* to a readily ingested food source.

In accordance with this, two mannose-binding proteins on the surface of *A. castellanii* have been identified to mediate the recognition of the fungal prey (Gonçalves et al. 2019). Besides, the interaction between *A. castellanii* and several fungi was inhibited by the addition of mannose, while no inhibition was observed with N-acetylgalactosamine (GalNac) and galactose.

9.3.5 Reactive Oxygen Species

Apart from recognition-shielding function, melanin and capsule structures on the surface of many fungi were thought to be antioxidants against host reactive oxygen and nitrogen species (ROS, RNS) (Zaragoza et al. 2008; Liu et al. 2021). However, the idea that melanins would serve a direct protective role against ROS was recently challenged for DHN-melanin of *A. fumigatus*, as conidia of mutants lacking DHN-melanin biosynthesis were as sensitive to ROS as melanized wild-type conidia (Keizer et al. 2022). Reactive oxygen species, such as superoxide ($O2^{\bullet -}$) and hydrogen peroxide (H_2O_2) are by-products of normal aerobic metabolism, and together with more reactive singlet oxygen($^1O_2$), peroxide radical ($HO^{\bullet}_2$), peroxide ion (HO^-_2), hydroxyl radical ($HO^{\bullet}$), and nitric oxide ($NO^{\bullet}$) cause DNA damage, protein inactivation, and lipid peroxidation via oxidation of iron sulphur centre and cysteine residues in virtually any cell molecule (Aguirre et al. 2006; Gessler et al. 2007). Cells, in turn, evolved a number of mechanisms to reduce the intracellular ROS levels, like superoxide dismutases, catalases, peroxidases, glutathione peroxidases, or peroxiredoxins. Activation of this antioxidant response is not only an important virulence trait, but also many fungi produce and handle ROS as a signal transduction mechanism in multicellular development. In turn, innate immune cells may exploit ROS to activate the fungal apoptosis-like pathway, thereby killing ingested conidia of the human pathogen *A. fumigatus* (Shlezinger et al. 2017). Besides immune cells, the generation of ROS as a key component of cell autonomous defence against various pathogens has also been observed for environmental predators, such as insects (Bergin et al. 2005), nematodes (McCallum and Garsin 2016), or amoebae (Dunn et al. 2018).

The insect immune response possesses a multitude of structural and functional similarities to the innate immune response of mammals. Haemocytes of the greater wax moth *G. mellonella* were found to be capable of phagocytosing *C. albicans* cells in a manner similar to human neutrophils via the production of superoxide (Bergin et al. 2005). The study further identified several proteins being homologous to the NADPH oxidase complex of human neutrophils.

The innate immune response of a nematode *Caenorhabditis elegans* is also accompanied by an increase of reactive oxygen species. The genome of *C. elegans* encodes for two NOX family enzymes, however ROS production has been functionally characterized only for the BLI-3 enzyme. BLI-3 generated ROS play an important defensive role against both epidermal and intestinal fungal infection (McCallum and Garsin 2016).

The amoeba *D. discoideum* encodes for three NADPH oxidase catalytic subunits, i.e. noxA to noxC, with NoxA and NoxB being even homologous of the mammalian gp91phox subunit (Lardy et al. 2005). In co-incubation experiments with *A. fumigatus,* ROS production coincides with neutralization of *A. fumigatus*-containing phagolysosomes, as previously shown for dendritic cells, having a proposed function in DHN-melanin degradation (Ferling et al. 2020).

In another example, the fungivorous amoeba *P. aurantium* directly targets cell redox homeostasis upon predation on the yeast *Candida parapsilosis* (Radosa et al. 2021). Yeast cells were exposed to host ROS shortly after uptake and resided within an acidic compartment, a process widely conserved in phagocytic cells of immune system.

These studies suggest that soil fungi have been constantly exposed to the ROS generated by their environmental predators and conversely, must have evolved strategies to counter this biotic stress. Further functional genomics of *C. parapsilosis* confronted with *P. aurantium* revealed the highly expressed *PRX1* gene in *C. parapsilosis*, encoding a thioredoxin-linked peroxidase accountable for cell redox homeostasis and response to oxidative stress (Radosa et al. 2021). A *C. parapsilosis PRX1* mutant was found to be not only impaired in the growth when subjected to organic hydroperoxide but also displays lower survival rate when confronted with amoeba as well as primary macrophages.

Finally, however, the production of reactive oxygen species by fungi as a direct response to environmental predators has not been clearly observed yet. Even more, it seems plausible that host-derived ROS act rather as a signal molecule in phagosome maturation (Ferling et al. 2020).

9.3.6 Nutritional Immunity

Every living organism requires a trace amount of metals like copper, iron, zinc, and manganese for its cellular functions. In human pathogenic fungi, maintenance of metal homeostasis has been extensively studied as a possible virulence factor, since many host cells, as a form of defence mechanism, actively sequestrate the amount of these micronutrient for the pathogen availability—a process collectively called nutritional immunity (Malavia et al. 2017).

The mechanisms how fungi deal with the lack of trace metals seems to be highly adapted to the environment in which they have evolved. Environmentally acquired fungi like *A. fumigatus* have developed more generic siderophore systems to acquire the iron, unlike human-adapted fungi like *C. albicans*, which relies on mechanisms to obtain iron from host-specific sources like heme or ferritin (Siscar-Lewin et al. 2022).

Amoeba cells are able to modulate zinc and copper homeostasis in an effort to attenuate microbial pathogenicity. This process has been described for bacterial cells, in fact, the same strategy has been now observed in fungi (German et al. 2013; Hao et al. 2016).

The acidic environment of amoebal phagolysosomes moreover increases the protease activity and enhances the toxicity of trace metals such as Cu(I) and Zn(II), and respectively decreases their bioavailability. Transcriptional profiling of *C. neoformans* during ingestion by *A. castellanii* and macrophages revealed a common set of upregulated genes involved in nutrient uptake (Derengowski Lda et al. 2013). An experimental study of Ribeiro et al. provided further evidence that amoebae can reduce the zinc availability for *Cryptococcus gattii* in order to reduce fungal proliferation (Ribeiro et al. 2017). In parallel, the P1-type copper exporter CRP1 of *C. parapsilosis* was found to be the highest upregulated gene in response to amoeba predation, whereas an orthologue to CTR1 copper importer was the second most downregulated gene (Radosa et al. 2021). Additionally, sensitivity of the *C. parapsilosis* deletion mutant *Δ/Δcrp1* towards high copper concentrations was only apparent when cells were exposed to high Cu at acidic pH, suggesting that intoxication of fungal cells inside of the phagosomal compartments represents an evolutionary antimicrobial activity evolved by phagocytic cells.

9.3.7 Production of Secondary Metabolites and Peptides

Production of secondary metabolites and peptides (either ribosomally or non-ribosomally synthesized) acts as a direct killing factor and main chemical defence of fungi.

These bioactive compounds are involved in numerous inter-species interactions such as communication, chemical defence, endosymbiotic relationship, or pathogenicity (Boysen et al. 2021).

Interestingly, it has been proposed that the effector molecules directed against microbial competitors are secreted by fungal hyphae or mycelium, whereas molecules against metazoan predators are usually stored intracellularly within the fungal hyphae or in the fruiting bodies and are taken up during predation (Künzler 2018). Gliotoxin, aflatoxin B1, cyclosporin A, lovastatin, or penicillin are all autonomously secreted examples of fungal defence effectors, naturally meant to target either fungi, bacteria, or insects living in close proximity and competing for the same source of nutrients. Several secondary metabolites have been already identified to act effectively against environmental predators, possessing either amoebicidal or insecticidal activity (Boysen et al. 2021). For example, the model mushroom *Coprinopsis cinerea* produces a mycelium-localized toxic compound upon challenge with the fungivorous nematode *Aphelenchus avenae* (Plaza et al. 2016). Moreover, the functional analysis identified a specific induction of several genes encoding previously characterized nematotoxic lectins with high toxicity towards the bacterivorous nematode *C. elegans*, and several putative antibacterial proteins (Plaza et al. 2016).

One of the best known fungal secondary metabolite is aflatoxin, produced mainly by filamentous fungus *Aspergillus flavus*, contaminating grain, seeds, spices, edible nuts, and even milk. Aflatoxin is extremely hepatocarcinogenic in humans, further causing immune suppression, acute aflatoxicosis, and underdevelopment in small children (Wild and Gong 2009). However, not all strains of *A. flavus* produce aflatoxin, and it has been shown that selection for aflatoxin production is driven by interaction with insects (Drott et al. 2017). Aflatoxin greatly reduced the fitness of *Drosophila melanogaster* and the mortality of its larvae increased when infected with toxicogenic fungal strains, in comparison to non-toxicogenic strains. Interestingly, toxicogenic strains showed even slightly higher fitness during the larvae infection but not in the absence of larvae. Moreover, the addition of external aflatoxin greatly increased fungal fitness but only in the presence of *D. melanogaster* larvae. These data suggest that aflatoxin production is selected for through an interaction with insects and the cost of aflatoxin production in the absence of susceptible insects will favour non-toxigenic isolates (Drott et al. 2017).

The dual-use of cyclosporin A produced by *Tolypocladium inflatum* has been shown, as it blocks a crucial step in calcium dependent signal transduction in T-cells via inhibition of calcineurin, resulting in immunosuppression, but at the same time it can also suppress insect defence cells (Bushley et al. 2013) and poses even antifungal properties (Cruz et al. 2000).

Gliotoxin is a non-ribosomal peptide derived toxin of several fungal genera such as *Aspergillus, Penicillium, Trichoderma,* and *Leptosphaeria*, whose production is furthermore tightly regulated by a plethora of exogenous biotic and abiotic factors (Boysen et al. 2021). Together with the spore-borne toxin trypacidin and meroterpenoid fumagilin, these secondary metabolites are not only cytotoxic to humans but play parallel roles in natural environment against amoeboid predators such as *D. discoideum* and *Entamoeba histolytica*, or against eukaryotic parasites such as *Trypanosoma* and *Plasmodium* (Hillmann et al. 2015; Mattern et al. 2015; Arico-Muendel et al. 2009).

Spore diffusates of non-germinated *A. fumigatus* conidia, of both clinical and environmental isolates, inhibited certain functions of neutrophil phagocytes as well as the growth of the amoeba *Naegleria gruberi* (Hobson 2000), suggesting that dormant conidia might have to defend against phagocytic predators.

In this sense, it is worth mentioning that co-cultivation experiments of fungi with other species often lead to the activation of silent gene clusters and thus, to the discovery of new metabolites with antimicrobial properties (Rutledge and Challis 2015). For example, grazing by larval, *D. melanogaster* induced the expression of several putative resistance genes in *Aspergillus nidulans*, including the secondary metabolite master regulator LaeA (Caballero Ortiz et al. 2013). *A. nidulans*, in fact, relies on secondary metabolite production to avoid predation by the fungivorous springtail *Folsomia candida*. Moreover, arthropod-induced production of secondary metabolites further accounted for a fungal phenotype that showed enhanced resistance to fungivory (Döll et al. 2013).

9.3.8 Filamentous Growth

A key component of fungal virulence is admittedly phenotypical switching from yeast to hyphae and filamentous growth. Despite the large number of fungal species that have the ability to grow filamentously, only a few of them switch to hyphae upon infection. These include *C. albicans*, *Candida dubliniensis*, *Malassezia* spp., or *Trichophyton rubrum* (Brand 2012). In the case of *C. albicans*, the combination of body temperature and serum is a key activator of hyphae formation, suggesting the adaptation of this fungus to commensal lifestyle. In fact, it seems very reasonable that the mechanism of hyphae formation in many fungi serve other purposes, such as substrate adherence, translocation between environments, nutrient acquisition, but induced filamentation or even the formation of biofilms may again serve a dual use to defend against natural predators.

Several dimorphic fungi, such as *Histoplasma capsulatum, Blastomyces dermatitidis,* and *Sporothrix schenckii* switch to a hyphae formation upon interaction with the amoeba *A. castellanii* occurring even under the condition where the yeast form in preferred (Steenbergen et al. 2004). *C. albicans* forms filaments when ingested by nematode *C. elegans* (Breger et al. 2007). Hyphae of the filamentous fungus *A. fumigatus* are directly targeted by fungivorous amoeba *P. aurantium* via "ruphocytosis", a strategy of amoeba to perforate the tip of the hyphae, invade it, and feed on the cytoplasmic content (Radosa et al. 2019).

An incubation of *D. discoideum* with non-filamentous *C. albicans* mutants defected in the hyphal growth, such as Δ*cph1*/Δ*efg1* or Δ*hgc1* mutant, resulted in reduced resistance towards amoeba predation. Besides, co-incubation with the amoeba at 22 °C induced the hyphae formation in *C. albicans* wild type even without the presence of any known hyphae-inducing substance in the medium in the contact-independent manner (Koller et al. 2016).

9.3.9 Dual-Use Virulence Factors of Phytopathogenic and Entomopathogenic Fungi

The number of studies concerning tripartite interaction of soil microorganism as an evolutionary driver underlying the acquisition of virulence traits and establishment of endosymbiosis between fungi and bacteria is recently increasing. The phytopathogenic mucoromycete *Rhizopus microsporus* was found to harbour different bacterial endosymbionts that are necessary for full fungal virulence and serve the fungal host in fending off protozoan and metazoan predators (Itabangi et al. 2022; Richter et al. 2022).

The endosymbiotic beta-proteobacterium *Mycetohabitans rhizoxinica*, residing within the fungal hyphae, produces the extremely potent toxin rhizoxin. At least one ecological role of this phytotoxin seems to be the inhibition of microtubule polymerization in fungal micropredators, such as the amoeba *P. aurantium* and nematodes like *Aphelenchus avenae*, and thereby protecting the host fungus from these two fungivores (Richter et al. 2022).

This strategy to defend against environmental micropredators seems to be a common trait for Mucormycota fungi, since all toxic compounds identified in Mucormycota fungi are produced by their bacterial endosymbionts. On the other hand, the important plant-growth promoting fungus *Mortierella verticillata* forms an endosymbiotic relationship with the toxin-producing bacterium *Mycoavidus necroximicus*, which protects the fungus, and thereby also the plant, from nematode attacks (Büttner et al. 2021).

A so far unknown compound is also produced by endosymbiotic bacterium *Ralstonia pickettii* of *R. microsporus*. Conditioned media from endosymbiont-containing fungal cultures inhibited antiphagocytic activity of macrophages as well as of *D. discoideum* amoeba. Besides, the presence of the endosymbiotic bacterium was necessary for the virulence of *R. microsporus* in the zebrafish model (Itabangi et al. 2022).

Last but not least, even the virulence of entomopathogenic fungi *Metarhizium anisopliae* and *Beauveria bassiana* could be explained as a consequence of fungal adaptation to avoid amoeba grazing (Bidochka et al. 2010). As mentioned earlier, phagocytic haemocytes, circulating in the hemolymph of insect, are analogous to mammalian macrophages. Interestingly, the authors further proposed the idea, that the ability of fungi to avoid amoeboid predators seems to be restricted to zoopathogenic fungi, including the invertebrate pathogens, potentially due to the absence of phagocytic cells that would reinforce such an adaptation.

Hence, studies on fungal interactions with non-mammalian phagocytes shed the light on the origin and evolution of fungal virulence determinants. Predator-induced resistance to fungivory therefore significantly contributes to the understanding of the evolutionary distinct strategies of fungal pathogenicity.

9.4 Experimental Efforts on Evolution of Fungal Virulence

The major fungal lineages are very ancient and have emerged approximately 1.5–two million years ago, long before the appearance of innate immunity, which is estimated to have emerged 300–400 million years ago (Hedges et al. 2004). From over 3–five million fungal species, approximately only 500 are able to cause disease in humans (Siscar-Lewin et al. 2022). Thus, to establish the infection or commensal stage within the human host, fungi must have already been endowed with the ability to counteract with innate immune system. These properties they have either acquired earlier during the common evolution in the environment or lately by within the host selection.

A commonly accepted idea for the evolution of microbial virulence, known as continuum hypothesis, proposes that pathogens should always evolve towards avirulence or reduced virulence, especially if there is a higher possibility for vertical transmission. However, when there is a trade-off between virulence and vertical transmission, selection may favour horizontal transmission and higher virulence. In other words, virulence factors are maintained to support the exploitation, proliferation, and transmission of the parasite (Alizon et al. 2009). This seems to be typical for fungal plant pathogens. Less virulent fungal strains displayed a lower rate for vertical transmission, probably due to their slow growth, thereby providing enough time for more chemical defences to be mobilized by the plant before fungal hyphae reached the seed. *Atkinsonella hypoxylon* is a fungal pathogen of the grass *Danthonia compressa*. Fully virulent strains reduce fitness of the grass by 50% or more by castrating the inflorescence, despite reducing its own change for vertical transmission by seeds. Less virulent strains of *A. hypoxylon* are causing less damage to the plant inflorescence but interestingly, those have only rarely been found in the nature (Kover and Clay 1998). Since the 90 s, the "trade-off" hypothesis has been significantly challenged, suggesting that the trade-offs exist, but may not be as simple as it was usually considered. Consequently, at least two alternative hypotheses have been proposed to explain the evolution of microbial virulence: the "short-sighted" and the "coincidental" hypothesis.

The coincidental evolution hypothesis states that factors responsible for virulence were not generated for virulence per se, but rather to increase the fitness of the pathogen in a non-parasitic context (Levin and Svanborg Edén 1990; Adiba et al. 2010). This idea was strongly supported by a series of microevolution experiments, in which continuous passage of an avirulent or low virulent strain of a certain pathogen through an environmental predator, would subsequently led to the generation of such a trait, that would increase the resistance of fungus in either murine infection or enhance the survival when confronted with innate immune cells.

The avirulent strain of *Histoplasma capsulatum*, CIB 1980, was passaged with and without amoeba *A. castellanii* and then used to infect mice. While no colony-forming units were recovered from mice infected with non-passaged strain, mice infected with amoeba-passaged strain suffered from persistent fungal infection. Exposure of *H. capsulatum* to *A. castellanii,* moreover resulted in an increase in hyphal cells (Steenbergen et al. 2004). Similarly, yeast cells of *H. capsulatum* are under normal condition, not able to kill the larvae of *G. mellonella*. However, the same cells recovered from an *A. castellanii* interaction, killed 100% of larvae within 6 days. A similar enhanced virulence phenotype against *Galleria* larvae after co-culture with *A. castellanii* was observed for the dimorphic fungi *Sporothrix brasiliensis* and *Paracoccidioides brasiliensis* as well as for *C. neoformans, C. albicans,* and *S. cerevisiae* (Gonçalves et al. 2019).

An encapsulated strain of *C. neoformans* increased its virulence by the formation of a larger capsule and faster time to melanization when continuously incubated with *D. discoideum* (Steenbergen et al. 2003). Even here, the mice infected with *C. neoformans* grown with live *D. discoideum* had significantly shorter survival time than mice infected with *C. neoformans* grown either alone or in the presence of dead amoeba. Furthermore, this study also proved the concept that even an avirulent strain can turn out to be virulent in a susceptible host, as the incubation of acapsular *C. neoformans* with amoeba impaired in phagocytosis resulted in infection.

Interestingly, exposing the environmental isolates of *C. neoformans* to another amoeba *A. castellanii* over the prolonged period resulted in complex pleiotropic changes such as pseudohyphal growth, larger capsule thickness, increased urease activity, enhanced macrophage toxicity but reduced melanin production. However, this in turn was accompanied by less virulence in mice because the strain elicited a strong antifungal immune response (Fu et al. 2021). Variations in phenotypic and genetic changes between evolved strains suggested that the microevolution happens frequently and rapidly when exposed to amoeba.

In another study, previous interaction of *C. neoformans* with *A. castellanii* generated fungal cells more efficient in killing the wax moth *G. mellonella* and more resistant to oxygen- and nitrogen-derived molecular species. Here, increased urease activity and amoeba-induced changes in cell wall architecture accounted proposedly for the enhanced survival in this invertebrate model (Rizzo et al. 2017).

The thermodimorphic fungus *Paracoccidioides* spp., endemic to Latin America, is a causative agent of a neglected tropical disease called paracoccidioidomycosis despite the fact for its life cycle does not need to infect humans or other animals. Cultivation of soil samples revealed the presence of numerous amoeba species, such as *Allovahlkampfia spelaea*, *Vermamoeba vermiformis,* and *Acanthamoeba* sp. All of these efficiently ingested and killed yeast cells of *P. brasiliensis*. Later, a sequential co-cultivation of *Paracoccidioides* with *A. castellanii* selected for the induction of virulence transcripts and accumulation of cell wall alpha-glucans, polysaccharides that mask the recognition of fungal PAMPs. Survived fungal cells were further confronted with amoeba and J774 murine macrophages and used to infect BALB/c mice and *G. mellonella.* When compared to non-passaged cells, the proportion of dead amoeba increased and the number of fungal colony-forming units was significantly higher in both amoeba and murine macrophages. Additionally, evolved strains were also able to kill both animal models significantly faster (Albuquerque et al. 2019).

Noticeably, in the centre of these experiments stands an amoeboid predator, giving the evidence that amoeboid predators act as an important mediator for the maintenance of fungal virulence. The idea that environmentally acquired pathogens evolved their virulence factors upon interaction with soil amoeba in the so-called environmental school of virulence, has been studied over decades. However, only recently it has been summarized as "Amoeboid predator-animal virulence" hypothesis (Casadevall et al. 2019). In summary, this hypothesis states that constant amoeboid predation on fungal species during evolution selected for the virulence traits that supported the virulence in certain animal hosts. Given the countless number of potential soil predators, this hypothesis can also be universally extended to other microorganisms and their soil predators. The whole phenomenon seems to follow a similar pattern and seems to have occurred many times during the evolution: first, amoeba "feast" on the microorganism for its own nutrient acquisition; prey microorganisms adapt to the host cell and became more "fit" in terms of selecting for and maintenance of virulence factors; the fitness adaptation to the evolutionary more complex hosts results in the "fist" for the survival within the host milieu and consequently the host damage as "feat" outcome and returning the pathogen back to the environment; comprehensively named as "feast-fit-fist-feat" or "4F" hypothesis (da Silva Ferreira et al. 2021).

References

Adiba S, Nizak C, van Baalen M, Denamur E, Depaulis F (2010) From grazing resistance to pathogenesis: the coincidental evolution of virulence factors. PLoS One 5(8):e11882. https://doi.org/10.1371/journal.pone.0011882

Aguirre J, Hansberg W, Navarro R (2006) Fungal responses to reactive oxygen species. Med Mycol 44: S101–S107. https://doi.org/10.1080/13693780600900080

Albuquerque P, Nicola AM, Magnabosco DAG, Derengowski LS, Crisóstomo LS, Xavier LCG, Frazão SO, Guilhelmelli F, de Oliveira MA, Dias JN, Hurtado FA, Teixeira MM, Guimarães AJ, Paes HC, Bagagli E, Felipe MSS, Casadevall A, Silva-Pereira I (2019) A hidden battle in the dirt: soil amoebae interactions with Paracoccidioides spp. PLoS Negl Trop Dis 13(10): e0007742. https://doi.org/10.1371/journal.pntd.0007742

Alizon S, Hurford A, Mideo N, Baalen M (2009) Virulence evolution and the trade-off hypothesis: history, current state of affairs and the future. J Evol Biol 22: 245–259. https://doi.org/10.1111/j.1420-9101.2008.01658.x

Almeida F, Cerqueira FM, Silva RN, Ulhoa CJ, Lima AL (2007) Mycoparasitism studies of Trichodermaharzianum strains against Rhizoctonia solani: evaluation of coiling and hydrolytic enzyme production. Biotechnol Lett 29:1189–1193

Amin S, Thywissen A, Heinekamp T, Saluz HP, Brakhage AA (2014) Melanin dependent survival of Apergillus fumigatus conidia in lung epithelial cells. Int J Med Microbiol 304(5–6):626–636. https://doi.org/10.1016/j.ijmm.2014.04.009

Arico-Muendel C, Centrella PA, Contonio BD, Morgan BA, O'Donovan G, Paradise CL, Skinner SR, Sluboski B, Svendsen JL, White KF, Debnath A, Gut J, Wilson N, McKerrow JH, DeRisi JL, Rosenthal PJ, Chiang PK (2009) Antiparasitic activities of novel, orally available fumagillin analogs. Bioorg Med Chem Lett 19(17):5128–5131. https://doi.org/10.1016/j.bmcl.2009.07.029

Bahram M, Netherway T (2021) Fungi as mediators linking organisms and ecosystems. FEMS Microbiol

Rev 46(2):fuab058. https://doi.org/10.1093/femsre/fuab058
Bain JM, Louw J, Lewis LE, Okai B, Walls CA, Ballou ER, Walker LA, Reid D, Munro CA, Brown AJ, Brown GD, Gow NA, Erwig LP (2014) Candida albicans hypha formation and mannan masking of β-glucan inhibit macrophage phagosome maturation. mBio 5(6):e01874. https://doi.org/10.1128/mBio.01874-14
Barnett H, Binder F (1973) The fungal host-parasite relationship. Annu Rev Phytopathol 11(1):273–292
Bergin D, Reeves EP, Renwick J, Wientjes FB, Kavanagh K (2005) Superoxide production in Galleria mellonella hemocytes: identification of proteins homologous to the NADPH oxidase complex of human neutrophils. Infect Immun 73(7):4161–4170. https://doi.org/10.1128/iai.73.7.4161-4170.2005
Bidochka MJ, Clark DC, Lewis MW, Keyhani NO (2010) Could insect phagocytic avoidance by entomogenous fungi have evolved via selection against soil amoeboid predators? Microbiol (Reading) 156(Pt 7):2164–2171. https://doi.org/10.1099/mic.0.038216-0
Bleuler-Martínez S, Butschi A, Garbani M, Wälti MA, Wohlschlager T, Potthoff E, Sabotiĉ J, Pohleven J, Lüthy P, Hengartner MO, Aebi M, Künzler M (2011) A lectin-mediated resistance of higher fungi against predators and parasites. Mol Ecol 20(14):3056–3070. https://doi.org/10.1111/j.1365-294X.2011.05093.x
Boer W, Folman LB, Summerbell RC, Boddy L (2005) Living in a fungal world: impact of fungi on soil bacterial niche development. FEMS Microbiol Rev 29(4):795–811
Boysen JM, Saeed N, Hillmann F (2021) Natural products in the predatory defence of the filamentous fungal pathogen Aspergillus fumigatus. Beilstein J Org Chem 17:1814–1827. https://doi.org/10.3762/bjoc.17.124
Brand A (2012) Hyphal growth in human fungal pathogens and its role in virulence. Int J Microbiol 2012:517529. https://doi.org/10.1155/2012/517529
Breger J, Fuchs BB, Aperis G, Moy TI, Ausubel FM, Mylonakis E (2007) Antifungal chemical compounds identified using a C. elegans pathogenicity assay. PLoS Pathog 3(2):e18. https://doi.org/10.1371/journal.ppat.0030018
Bushley KE, Raja R, Jaiswal P, Cumbie JS, Nonogaki M, Boyd AE, Owensby CA, Knaus BJ, Elser J, Miller D, Di Y, McPhail KL, Spatafora JW (2013) The genome of tolypocladium inflatum: evolution, organization, and expression of the cyclosporin biosynthetic gene cluster. PLoS Genet 9(6):e1003496. https://doi.org/10.1371/journal.pgen.1003496
Büttner H, Niehs SP, Vandelannoote K, Cseresnyés Z, Dose B, Richter I, Gerst R, Figge MT, Stinear TP, Pidot SJ, Hertweck C (2021) Bacterial endosymbionts protect beneficial soil fungus from nematode attack. Proc Natl Acad Sci U S A 118(37):e2110669118. https://doi.org/10.1073/pnas.2110669118
Caballero Ortiz S, Trienens M, Rohlfs M (2013) Induced fungal resistance to insect grazing: reciprocal fitness consequences and fungal gene expression in the drosophila-aspergillus model system. PLoS One 8(8): e74951. https://doi.org/10.1371/journal.pone.0074951
Casadevall A, Pirofski L-A (2007) Accidental virulence, cryptic pathogenesis, Martians, lost hosts, and the pathogenicity of environmental microbes. Eukaryot Cell 6(12):2169–2174. https://doi.org/10.1128/EC.00308-07
Casadevall A, Fu MS, Guimaraes AJ, Albuquerque P (2019) The 'amoeboid predator-fungal animal virulence' hypothesis. J Fungi 5(1):10
Castellani A (1930) An amoeba found in cultures of a yeast preliminary note-third note-an amoeba growing in cultures of a yeast second note-fourth note. J Trop Med London 33:160–160
Chai LYA, Netea MG, Sugui J, Vonk AG, van de Sande WWJ, Warris A, Kwon-Chung KJ, Jan Kullberg B (2010) Aspergillus fumigatus conidial melanin modulates host cytokine response. Immunobiology 215(11):915–920. https://doi.org/10.1016/j.imbio.2009.10.002
Chakraborty S, Old K, Warcup J (1983) Amoebae from a take-all suppressive soil which feed on Gaeumannomyces graminis tritici and other soil fungi. Soil Biol Biochem 15(1):17–24
Chatterton S, Punja ZK (2010) Factors influencing colonization of cucumber roots by Clonostachys rosea f. catenulata, a biological disease control agent. Biocontrol Sci Tech 20(1):37–55. https://doi.org/10.1080/09583150903350253
Chet I, Harman GE, Baker R (1981) Trichoderma hamatum: its hyphal interactions with Rhizoctonia solani and Pythium spp. Microb Ecol 7(1):29–38. https://doi.org/10.1007/BF02010476
Chrisman CJ, Albuquerque P, Guimaraes AJ, Nieves E, Casadevall A (2011) Phospholipids trigger Cryptococcus neoformans capsular enlargement during interactions with amoebae and macrophages. PLoS Pathog 7(5):e1002047. https://doi.org/10.1371/journal.ppat.1002047
Cox GM, McDade HC, Chen SCA, Tucker SC, Gottfredsson M, Wright LC, Sorrell TC, Leidich SD, Casadevall A, Ghannoum MA, Perfect JR (2001) Extracellular phospholipase activity is a virulence factor for Cryptococcus neoformans. Mol Microbiol 39(1):166–175. https://doi.org/10.1046/j.1365-2958.2001.02236.x
Cruz MC, Del Poeta M, Wang P, Wenger R, Zenke G, Quesniaux VF, Movva NR, Perfect JR, Cardenas ME, Heitman J (2000) Immunosuppressive and nonimmunosuppressive cyclosporine analogs are toxic to the opportunistic fungal pathogen Cryptococcus neoformans via cyclophilin-dependent inhibition

of calcineurin. Antimicrob Agents Chemother 44(1): 143–149. https://doi.org/10.1128/aac.44.1.143-149.2000

da Silva Ferreira M, de Souza GD, Medeiros EG, Peralta JM, Guimarães AJ (2021) "Feast-fit-fist-feat": overview of free-living amoeba interactions with fungi and virulence as a foundation for success in battle. Curr Trop Med Rep 8(1):18–31. https://doi.org/10.1007/s40475-020-00220-3

Davis-Hanna A, Piispanen AE, Stateva LI, Hogan DA (2008) Farnesol and dodecanol effects on the Candida albicans Ras1-cAMP signalling pathway and the regulation of morphogenesis. Mol Microbiol 67(1):47–62. https://doi.org/10.1111/j.1365-2958.2007.06013.x

de Vrije T, Antoine N, Buitelaar RM, Bruckner S, Dissevelt M, Durand A, Gerlagh M, Jones EE, Lüth P, Oostra J, Ravensberg WJ, Renaud R, Rinzema A, Weber FJ, Whipps JM (2001) The fungal biocontrol agent Coniothyrium minitans: production by solid-state fermentation, application and marketing. Appl Microbiol Biotechnol 56(1–2):58–68. https://doi.org/10.1007/s002530100678

Derengowski Lda S, Paes HC, Albuquerque P, Tavares AH, Fernandes L, Silva-Pereira I, Casadevall A (2013) The transcriptional response of Cryptococcus neoformans to ingestion by Acanthamoeba castellanii and macrophages provides insights into the evolutionary adaptation to the mammalian host. Eukaryot Cell 12(5):761–774. https://doi.org/10.1128/ec.00073-13

Döll K, Chatterjee S, Scheu S, Karlovsky P, Rohlfs M (2013) Fungal metabolic plasticity and sexual development mediate induced resistance to arthropod fungivory. Proc R Soc B Biol Sci 280(1771): 20131219. https://doi.org/10.1098/rspb.2013.1219

Drott MT, Lazzaro BP, Brown DL, Carbone I, Milgroom MG (2017) Balancing selection for aflatoxin in Aspergillus flavus is maintained through interference competition with, and fungivory by insects. Proc R Soc B Biol Sci 284(1869):20172408. https://doi.org/10.1098/rspb.2017.2408

Druzhinina IS, Shelest E, Kubicek CP (2012) Novel traits of Trichoderma predicted through the analysis of its secretome. FEMS Microbiol Lett 337(1):1–9

Dubey MK, Jensen DF, Karlsson M (2014) Hydrophobins are required for conidial hydrophobicity and plant root colonization in the fungal biocontrol agent Clonostachys rosea. BMC Microbiol 14(1):18. https://doi.org/10.1186/1471-2180-14-18

Dunn JD, Bosmani C, Barisch C, Raykov L, Lefrançois LH, Cardenal-Muñoz E, López-Jiménez AT, Soldati T (2018) Eat prey, live: Dictyostelium discoideum as a model for cell-autonomous Defenses. Front Immunol 8:1906. https://doi.org/10.3389/fimmu.2017.01906

Elad Y, Barak R, Chet I (1983) Possible role of lectins in mycoparasitism. J Bacteriol 154(3):1431–1435

Erwig LP, Gow NA (2016) Interactions of fungal pathogens with phagocytes. Nat Rev Microbiol 14(3): 163–176. https://doi.org/10.1038/nrmicro.2015.21

Ferling I, Dunn JD, Ferling A, Soldati T, Hillmann F (2020) Conidial melanin of the human-pathogenic fungus Aspergillus fumigatus disrupts cell autonomous defenses in amoebae. mBio 11:3. https://doi.org/10.1128/mBio.00862-20

Fischer R, Requena N (2022) Small-secreted proteins as virulence factors in nematode-trapping fungi. Trends Microbiol 30(7):615–617. https://doi.org/10.1016/j.tim.2022.03.005

Fogliano V, Ballio A, Gallo M, Woo S, Scala F, Lorito M (2002) Pseudomonas lipodepsipeptides and fungal cell wall-degrading enzymes act synergistically in biological control. Mol Plant-Microbe Interact 15(4): 323–333

Fritsche K, Leveau JH, Gerards S, Ogawa S, de Boer W, van Veen JA (2006) Collimonas fungivorans and bacterial mycophagy. IOBC WPRS BULLETIN 29(2):27

Fu MS, Casadevall A (2018) Divalent metal cations potentiate the predatory capacity of amoeba for Cryptococcus neoformans. Appl Environ Microbiol 84:3. https://doi.org/10.1128/aem.01717-17

Fu MS, Liporagi-Lopes LC, Dos Santos SRJ, Tenor JL, Perfect JR, Cuomo CA, Casadevall A (2021) Amoeba predation of Cryptococcus neoformans results in pleiotropic changes to traits associated with virulence. mBio 12(2):e00567-21. https://doi.org/10.1128/mBio.00567-21

Gaylord EA, Choy HL, Doering TL (2020) Dangerous liaisons: interactions of Cryptococcus neoformans with host phagocytes. Pathogens 9(11):891. https://doi.org/10.3390/pathogens9110891

Geib E, Gressler M, Viediernikova I, Hillmann F, Jacobsen ID, Nietzsche S, Hertweck C, Brock M (2016) A non-canonical melanin biosynthesis pathway protects Aspergillus terreus conidia from environmental stress. Cell Chem Biol 23(5):587–597. https://doi.org/10.1016/j.chembiol.2016.03.014

German N, Doyscher D, Rensing C (2013) Bacterial killing in macrophages and amoeba: do they all use a brass dagger? Future Microbiol 8(10):1257–1264. https://doi.org/10.2217/fmb.13.100

Gessler NN, Aver'yanov AA, Belozerskaya TA (2007) Reactive oxygen species in regulation of fungal development. Biochem Mosc 72(10):1091–1109. https://doi.org/10.1134/S0006297907100070

Gonçalves DS, Ferreira MS, Gomes KX, Rodríguez-de La Noval C, Liedke SC, da Costa GCV, Albuquerque P, Cortines JR, Saramago Peralta RH, Peralta JM, Casadevall A, Guimarães AJ (2019) Unravelling the interactions of the environmental host Acanthamoeba castellanii with fungi through the recognition by mannose-binding proteins. Cell Microbiol 21(10): e13066. https://doi.org/10.1111/cmi.13066

Graupner K, Scherlach K, Bretschneider T, Lackner G, Roth M, Gross H, Hertweck C (2012) Imaging mass spectrometry and genome mining reveal highly antifungal virulence factor of mushroom soft rot pathogen. Angew Chem Int Ed Engl 51(52):13173–13177. https://doi.org/10.1002/anie.201206658

Guimaraes AJ, Gomes KX, Cortines JR, Peralta JM, Peralta RH (2016) Acanthamoeba spp. as a universal host for pathogenic microorganisms: one bridge from environment to host virulence. Microbiol Res 193:30–38. https://doi.org/10.1016/j.micres.2016.08.001

Hao X, Lüthje F, Rønn R, German NA, Li X, Huang F, Kisaka J, Huffman D, Alwathnani HA, Zhu YG, Rensing C (2016) A role for copper in protozoan grazing - two billion years selecting for bacterial copper resistance. Mol Microbiol 102(4):628–641. https://doi.org/10.1111/mmi.13483

Harman GE, Howell CR, Viterbo A, Chet I, Lorito M (2004) Trichoderma species — opportunistic, avirulent plant symbionts. Nat Rev Microbiol 2(1):43–56. https://doi.org/10.1038/nrmicro797

Hedges SB, Blair JE, Venturi ML, Shoe JL (2004) A molecular timescale of eukaryote evolution and the rise of complex multicellular life. BMC Evol Biol 4(1):2. https://doi.org/10.1186/1471-2148-4-2

Heinekamp T, Thywissen A, Macheleidt J, Keller S, Valiante V, Brakhage A (2013) Aspergillus fumigatus melanins: interference with the host endocytosis pathway and impact on virulence. Front Microbiol 3:440. https://doi.org/10.3389/fmicb.2012.00440

Hillmann F, Novohradská S, Mattern DJ, Forberger T, Heinekamp T, Westermann M, Winckler T, Brakhage AA (2015) Virulence determinants of the human pathogenic fungus aspergillus fumigatus protect against soil amoeba predation. Environ Microbiol 17(8): 2858–2869. https://doi.org/10.1111/1462-2920.12808

Hobson RP (2000) The effects of diffusates from the spores of aspergillus fumigatus and A. terreus on human neutrophils, Naegleria gruberi and Acanthamoeba castellanii. Med Mycol 38(2): 133–141. https://doi.org/10.1080/mmy.38.2.133.141

Itabangi H, Sephton-Clark PCS, Tamayo DP, Zhou X, Starling GP, Mahamoud Z, Insua I, Probert M, Correia J, Moynihan PJ, Gebremariam T, Gu Y, Ibrahim AS, Brown GD, King JS, Ballou ER, Voelz K (2022) A bacterial endosymbiont of the fungus Rhizopus microsporus drives phagocyte evasion and opportunistic virulence. Curr Biol 32(5):1115–1130. e1116. https://doi.org/10.1016/j.cub.2022.01.028

Jolivet S, Arpin N, Olivier J, Wichers H (1999) Biochemical and physiological aspects of brown blotch disease. FEMS Microbiol Rev 23:591–614

Jourdan F, Lazzaroni S, Méndez BL, Lo Cantore P, de Julio M, Amodeo P, Iacobellis NS, Evidente A, Motta A (2003) A left-handed α-helix containing both L-and D-amino acids: the solution structure of the antimicrobial lipodepsipeptide tolaasin. Proteins: Struct Funct Bioinform 52(4):534–543

Karlsson M, Atanasova L, Jensen DF, Zeilinger S (2017) Necrotrophic mycoparasites and their genomes. Microbiol Spectr 5(2):5.2.08. https://doi.org/10.1128/microbiolspec.FUNK-0016-2016

Keizer EM, Valdes ID, McCann BL, Bignell EM, Wösten HAB, de Cock H (2022) The protective role of 1,8-dihydroxynaphthalene-melanin on conidia of the opportunistic human pathogen Aspergillus fumigatus revisited: no role in protection against hydrogen peroxide and superoxides. mSphere 7(1):e0087421. https://doi.org/10.1128/msphere.00874-21

Koller B, Schramm C, Siebert S, Triebel J, Deland E, Pfefferkorn AM, Rickerts V, Thewes S (2016) Dictyostelium discoideum as a novel host system to study the interaction between phagocytes and yeasts. Front Microbiol 7:1665. https://doi.org/10.3389/fmicb.2016.01665

Kover PX, Clay K (1998) Trade-off between virulence and vertical transmission and the maintenance of a virulent plant pathogen. Am Nat 152(2):165–175. https://doi.org/10.1086/286159

Künzler M (2018) How fungi defend themselves against microbial competitors and animal predators. PLoS Pathog 14(9):e1007184–e1007184. https://doi.org/10.1371/journal.ppat.1007184

Lardy B, Bof M, Aubry L, Paclet MH, Morel F, Satre M, Klein G (2005) NADPH oxidase homologs are required for normal cell differentiation and morphogenesis in Dictyostelium discoideum. Biochim Biophys Acta (BBA) - Mol Cell Res 1744(2): 199–212. https://doi.org/10.1016/j.bbamcr.2005.02.004

Leveau JH, Preston GM (2008) Bacterial mycophagy: definition and diagnosis of a unique bacterial–fungal interaction. New Phytol 177(4):859–876

Levin BR, Svanborg Edén C (1990) Selection and evolution of virulence in bacteria: an ecumenical excursion and modest suggestion. Parasitology 100(Suppl): S103–S115. https://doi.org/10.1017/s0031182000073054

Levy A, Chang BJ, Abbott LK, Kuo J, Harnett G, Inglis TJ (2003) Invasion of spores of the arbuscular mycorrhizal fungus Gigaspora decipiens by Burkholderia spp. Appl Environ Microbiol 69(10):6250–6256

Liu S, Youngchim S, Zamith-Miranda D, Nosanchuk JD (2021) Fungal melanin and the mammalian immune system. J fungi (Basel, Switzerland) 7(4):264. https://doi.org/10.3390/jof7040264

Lo Cantore P, Lazzaroni S, Coraiola M, Serra MD, Cafarchia C, Evidente A, Iacobellis NS (2006) Biological characterization of white line-inducing principle (WLIP) produced by pseudomonas reactans NCPPB1311. Mol Plant-Microbe Interact 19(10): 1113–1120

Lorito M, Peterbauer C, Hayes CK, Harman GE (1994) Synergistic interaction between fungal cell wall degrading enzymes and different antifungal compounds enhances inhibition of spore germination. Microbiology 140(3):623–629

Lu Z, Tombolini R, Woo S, Zeilinger S, Lorito M, Jansson JK (2004) In vivo study of Trichoderma-pathogen-plant interactions, using constitutive and inducible green fluorescent protein reporter systems. Appl Environ Microbiol 70(5):3073–3081

Luther K, Torosantucci A, Brakhage AA, Heesemann J, Ebel F (2007) Phagocytosis of Aspergillus fumigatus conidia by murine macrophages involves recognition by the dectin-1 beta-glucan receptor and toll-like receptor 2. Cell Microbiol 9(2):368–381. https://doi.org/10.1111/j.1462-5822.2006.00796.x

Ma H, Croudace JE, Lammas DA, May RC (2006) Expulsion of live pathogenic yeast by macrophages. Curr Biol 16(21):2156–2160. https://doi.org/10.1016/j.cub.2006.09.032

Magnet A, Peralta R, Gomes T, Izquierdo F, Fernandez-Vadillo C, Galvan A, Pozuelo M, Pelaz C, Fenoy S, Del Águila C (2015) Vectorial role of Acanthamoeba in legionella propagation in water for human use. Sci Total Environ 505:889–895

Malavia D, Crawford A, Wilson D (2017) Chapter Three - nutritional immunity and fungal pathogenesis: the struggle for micronutrients at the host–pathogen interface. In: Poole RK (ed) Advances in microbial physiology, vol 70. Academic Press, pp 85–103. https://doi.org/10.1016/bs.ampbs.2017.01.006

Mattern DJ, Schoeler H, Weber J, Novohradská S, Kraibooj K, Dahse HM, Hillmann F, Valiante V, Figge MT, Brakhage AA (2015) Identification of the antiphagocytic trypacidin gene cluster in the human-pathogenic fungus Aspergillus fumigatus. Appl Microbiol Biotechnol 99(23):10151–10161. https://doi.org/10.1007/s00253-015-6898-1

McCallum KC, Garsin DA (2016) The role of reactive oxygen species in modulating the Caenorhabditis elegans immune response. PLoS Pathog 12(11): e1005923. https://doi.org/10.1371/journal.ppat.1005923

McKenzie CG, Koser U, Lewis LE, Bain JM, Mora-Montes HM, Barker RN, Gow NA, Erwig LP (2010) Contribution of Candida albicans cell wall components to recognition by and escape from murine macrophages. Infect Immun 78(4):1650–1658. https://doi.org/10.1128/iai.00001-10

Molmeret M, Horn M, Wagner M, Santic M, Abu Kwaik Y (2005) Amoebae as training grounds for intracellular bacterial pathogens. Appl Environ Microbiol 71(1): 20–28. https://doi.org/10.1128/aem.71.1.20-28.2005

Mowat E, Rajendran R, Williams C, McCulloch E, Jones B, Lang S, Ramage G (2010) Pseudomonas aeruginosa and their small diffusible extracellular molecules inhibit Aspergillus fumigatus biofilm formation. FEMS Microbiol Lett 313(2):96–102. https://doi.org/10.1111/j.1574-6968.2010.02130.x

Németh MZ, Pintye A, Horváth ÁN, Vági P, Kovács GM, Gorfer M, Kiss L (2019) Green fluorescent protein transformation sheds more light on a widespread Mycoparasitic interaction. Phytopathology 109(8): 1404–1416. https://doi.org/10.1094/phyto-01-19-0013-r

Novohradská S, Ferling I, Hillmann F (2017) Exploring virulence determinants of filamentous fungal pathogens through interactions with soil amoebae. Front Cell Infect Microbiol 7:497. https://doi.org/10.3389/fcimb.2017.00497

Old K, Darbyshire J (1978) Soil fungi as food for giant amoebae. Soil Biol Biochem 10(2):93–100

Plaza DF, Schmieder SS, Lipzen A, Lindquist E, Künzler M (2016) Identification of a novel Nematotoxic protein by challenging the model mushroom Coprinopsis cinerea with a Fungivorous nematode. G3 Genes|Genomes|Genet 6(1):87–98. https://doi.org/10.1534/g3.115.023069

Radosa S, Ferling I, Sprague JL, Westermann M, Hillmann F (2019) The different morphologies of yeast and filamentous fungi trigger distinct killing and feeding mechanisms in a fungivorous amoeba. Environ Microbiol 21(5):1809–1820

Radosa S, Sprague JL, Lau SH, Tóth R, Linde J, Krüger T, Sprenger M, Kasper L, Westermann M, Kniemeyer O, Hube B, Brakhage AA, Gácser A, Hillmann F (2021) The fungivorous amoeba Protostelium aurantium targets redox homeostasis and cell wall integrity during intracellular killing of Candida parapsilosis. Cell Microbiol 23(11):e13389. https://doi.org/10.1111/cmi.13389

Rainey PB, Brodey CL, Johnstone K (1993) Identification of a gene cluster encoding three high-molecular-weight proteins, which is required for synthesis of tolaasin by the mushroom pathogen Pseudomonas tolaasii. Mol Microbiol 8(4):643–652

Rella A, Yang MW, Gruber J, Montagna MT, Luberto C, Zhang YM, Del Poeta M (2012) Pseudomonas aeruginosa inhibits the growth of Cryptococcus species. Mycopathologia 173(5–6):451–461. https://doi.org/10.1007/s11046-011-9494-7

Ribeiro NS, dos Santos FM, Garcia AWA, Ferrareze PAG, Fabres LF, Schrank A, Kmetzsch L, Rott MB, Vainstein MH, Staats CC (2017) Modulation of zinc homeostasis in Acanthamoeba castellanii as a possible antifungal strategy against Cryptococcus gattii. Front Microbiol 8:1626. https://doi.org/10.3389/fmicb.2017.01626

Richter I, Radosa S, Cseresnyés Z, Ferling I, Büttner H, Niehs SP, Gerst R, Scherlach K, Figge MT, Hillmann F (2022) Toxin-producing endosymbionts shield pathogenic fungus against micropredators. mBio 13(5): e01440-22

Rizzo J, Albuquerque PC, Wolf JM, Nascimento R, Pereira MD, Nosanchuk JD, Rodrigues ML (2017) Analysis of multiple components involved in the interaction between Cryptococcus neoformans and Acanthamoeba castellanii. Fungal Biol 121(6–7): 602–614. https://doi.org/10.1016/j.funbio.2017.04.002

Rutledge PJ, Challis GL (2015) Discovery of microbial natural products by activation of silent biosynthetic gene clusters. Nat Rev Microbiol 13(8):509–523. https://doi.org/10.1038/nrmicro3496

Scherlach K, Lackner G, Graupner K, Pidot S, Bretschneider T, Hertweck C (2013) Biosynthesis and mass spectrometric imaging of tolaasin, the virulence factor of brown blotch mushroom disease.

Chembiochem 14(18):2439–2443. https://doi.org/10.1002/cbic.201300553

Shlezinger N, Irmer H, Dhingra S, Beattie SR, Cramer RA, Braus GH, Sharon A, Hohl TM (2017) Sterilizing immunity in the lung relies on targeting fungal apoptosis-like programmed cell death. Science 357(6355):1037–1041. https://doi.org/10.1126/science.aan0365

Siddiqui R, Khan NA (2012) Biology and pathogenesis of Acanthamoeba. Parasit Vectors 5:6. https://doi.org/10.1186/1756-3305-5-6

Siscar-Lewin S, Hube B, Brunke S (2022) Emergence and evolution of virulence in human pathogenic fungi. Trends Microbiol 30(7):693–704. https://doi.org/10.1016/j.tim.2021.12.013

Slesiona S, Gressler M, Mihlan M, Zaehle C, Schaller M, Barz D, Hube B, Jacobsen ID, Brock M (2012) Persistence versus escape: Aspergillus terreus and Aspergillus fumigatus employ different strategies during interactions with macrophages. PLoS One 7(2): e31223. https://doi.org/10.1371/journal.pone.0031223

Someya N, Tsuchiya K, Yoshida T, Noguchi MT, Akutsu K, Sawada H (2007) Co-inoculation of an antibiotic-producing bacterium and a lytic enzyme-producing bacterium for the biocontrol of tomato wilt caused by fusarium oxysporum f. sp. lycopersici. Biocontrol Sci 12(1):1–6

Steenbergen JN, Shuman HA, Casadevall A (2001) <em>Cryptococcus neoformans</em> interactions with amoebae suggest an explanation for its virulence and intracellular pathogenic strategy in macrophages. Proc Natl Acad Sci 98(26):15245–15250. https://doi.org/10.1073/pnas.261418798

Steenbergen JN, Nosanchuk JD, Malliaris SD, Casadevall A (2003) Cryptococcus neoformans virulence is enhanced after growth in the genetically malleable host Dictyostelium discoideum. Infect Immun 71(9): 4862–4872. https://doi.org/10.1128/iai.71.9.4862-4872.2003

Steenbergen JN, Nosanchuk JD, Malliaris SD, Casadevall A (2004) Interaction of Blastomyces dermatitidis, Sporothrix schenckii, and Histoplasma capsulatum with Acanthamoeba castellanii. Infect Immun 72(6): 3478–3488. https://doi.org/10.1128/iai.72.6.3478-3488.2004

Tayyrov A, Schmieder SS, Bleuler-Martinez S, Plaza DF, Künzler M (2018) Toxicity of potential fungal Defense proteins towards the Fungivorous nematodes Aphelenchus avenae and Bursaphelenchus okinawaensis. Appl Environ Microbiol 84(23): e02051-18. https://doi.org/10.1128/aem.02051-18

Thywißen A, Heinekamp T, Dahse HM, Schmaler-Ripcke J, Nietzsche S, Zipfel PF, Brakhage AA (2011) Conidial Dihydroxynaphthalene melanin of the human pathogenic fungus Aspergillus fumigatus interferes with the host endocytosis pathway. Front Microbiol 2:96. https://doi.org/10.3389/fmicb.2011.00096

Tosetti N, Croxatto A, Greub G (2014) Amoebae as a tool to isolate new bacterial species, to discover new virulence factors and to study the host-pathogen interactions. Microb Pathog 77:125–130. https://doi.org/10.1016/j.micpath.2014.07.009

Watkins RA, Andrews A, Wynn C, Barisch C, King JS, Johnston SA (2018) Cryptococcus neoformans escape from Dictyostelium amoeba by both WASH-mediated constitutive exocytosis and Vomocytosis. Front Cell Infect Microbiol 8:108. https://doi.org/10.3389/fcimb.2018.00108

Wild CP, Gong YY (2009) Mycotoxins and human disease: a largely ignored global health issue. Carcinogenesis 31(1):71–82. https://doi.org/10.1093/carcin/bgp264

Woo S, Fogliano V, Scala F, Lorito M (2002) Synergism between fungal enzymes and bacterial antibiotics may enhance biocontrol. Antonie Van Leeuwenhoek 81(1): 353–356

Yang CT, Vidal-Diez de Ulzurrun G, Gonçalves AP, Lin HC, Chang CW, Huang TY, Chen SA, Lai CK, Tsai IJ, Schroeder FC, Stajich JE, Hsueh YP (2020) Natural diversity in the predatory behavior facilitates the establishment of a robust model strain for nematode-trapping fungi. Proc Natl Acad Sci U S A 117(12): 6762–6770. https://doi.org/10.1073/pnas.1919726117

Youssar L, Wernet V, Hensel N, Yu X, Hildebrand HG, Schreckenberger B, Kriegler M, Hetzer B, Frankino P, Dillin A, Fischer R (2019) Intercellular communication is required for trap formation in the nematode-trapping fungus Duddingtonia flagrans. PLoS Genet 15(3): e1008029. https://doi.org/10.1371/journal.pgen.1008029

Zaragoza O, Casadevall A (2004) Experimental modulation of capsule size in Cryptococcus neoformans. Biol Proced Online 6:10–15. https://doi.org/10.1251/bpo68

Zaragoza O, Chrisman CJ, Castelli MV, Frases S, Cuenca-Estrella M, Rodríguez-Tudela JL, Casadevall A (2008) Capsule enlargement in Cryptococcus neoformans confers resistance to oxidative stress suggesting a mechanism for intracellular survival. Cell Microbiol 10(10):2043–2057. https://doi.org/10.1111/j.1462-5822.2008.01186.x

Zhang Y, Li S, Li H, Wang R, Zhang KQ, Xu J (2020) Fungi-nematode interactions: diversity, ecology, and biocontrol prospects in agriculture. J Fungi (Basel) 6(4):206. https://doi.org/10.3390/jof6040206

Global Fungal Diversity Estimated from High-Throughput Sequencing 10

Petr Baldrian, Petr Kohout, and Tomáš Větrovský

Abstract

Fungi are key players in vital ecosystem services, spanning carbon cycling, decomposition, and varied plant symbioses. Due to their cryptic lifestyle, it was difficult to assess their diversity until the advent of methods of high-throughput sequencing. Based on the papers utilizing high-throughput sequencing approaches to study fungi in natural habitats using the nuclear ribosomal internal transcribed spacer 2 (ITS2) contained in the public open database GlobalFungi (https://globalfungi.com), the current estimate of global fungal diversity is 6.3 million species, considering 97% sequence similarity as a species-level threshold. Of the observed fungi, most belong to Ascomycota and Basidiomycota: 57% and 37% of taxa, respectively. Soil and litter represent the habitats with the highest alpha diversity of fungi followed by air, plant shoots, plant roots, and deadwood. Based on the high-throughput sequencing data, the highest proportion of unknown fungal species is associated with samples of lichen and plant tissues. Climate was identified as the key driver of fungal biogeography. In contrast to plants and most other taxa, fungal diversity in tropics appears to be lower than at high latitudes. Despite limitations, the use of high-throughput sequencing is an important tool for the assessment of diversity, biogeography, and ecology of fungi.

Keywords

Fungal diversity · High-throughput sequencing · Metabarcoding · Biogeography · Fungal ecology

10.1 Introduction

Fungi represent one of the most diverse groups of organisms in the world (Purvis and Hector 2000; Hawksworth 2001). However, the cryptic lifestyle of most fungal species makes estimates of global biodiversity challenging. Given their critical roles in many ecosystem processes (Peay et al. 2016), understanding the extent of global fungal biodiversity and the factors that determine it can provide crucial information about the stability and functioning of ecosystems. In the past, mycology has mostly relied on morphological characteristics, but early studies of amplified molecular markers derived from high-throughput sequencing (HTS) (Buée et al. 2009; Amend et al. 2010; Baldrian et al. 2012) have suggested that traditional approaches greatly underestimate the diversity of fungi. The ability to recognize and

P. Baldrian (✉) · P. Kohout · T. Větrovský
Laboratory of Environmental Microbiology, Institute of Microbiology of the Czech Academy of Sciences, Praha, Czech Republic
e-mail: baldrian@biomed.cas.cz

© The Author(s), under exclusive license to Springer Nature Switzerland AG 2023

S. Pöggeler, T. James (eds.), *Evolution of Fungi and Fungal-Like Organisms*, The Mycota 14,
https://doi.org/10.1007/978-3-031-29199-9_10

classify fungi based on appropriate molecular markers obtained by DNA sequencing has opened up the opportunity to explore the composition and diversity of fungal communities (Koljalg et al. 2013), and HTS-based metabarcoding has demonstrated its scientific predictive power for studying global fungal diversity and biogeography (Tedersoo et al. 2014; Davison et al. 2015) as well as for elucidating regional drivers of fungal community diversity and composition (Talbot et al. 2014; Tedersoo et al. 2020b; Odriozola et al. 2021). Moreover, mycological efforts supported by HTS are able to address a wide range of questions regarding ecological and functional aspects of fungal communities (Baldrian et al. 2012; Peay et al. 2016; Nilsson et al. 2019a). Metabarcoding of fungal communities has become a well-established method readily available to the broader scientific community with a solid methodological backing (Lindahl et al. 2013; Větrovský and Baldrian 2013; Anslan et al. 2018; Nilsson et al. 2019a). As a result, several hundred papers have been published to date using high-throughput sequencing to analyse fungal community composition (Větrovský et al. 2020).

With the exception of a few large-scale attempts (Tedersoo et al. 2014; Davison et al. 2015), the power of HTS to determine the global diversity of fungi by assembling data from multiple studies is only now beginning to be exploited. This cumulative approach allows for more precise estimates of global fungal diversity (Větrovský et al. 2019; Větrovský et al. 2020; Baldrian et al. 2022).

10.2 Global Database of Fungal Metabarcoding Data

An appreciation of the number of studies that use metabarcoding to characterize fungal communities in terrestrial and aquatic communities has led to the creation of a community resource: a database of results from all available studies that use fungal internal transcribed spacer (ITS) as a metabarcoding marker—the GlobalFungi database (Větrovský et al. 2020). This curated database (https://globalfungi.com) is a public repository of metabarcoding data that continuously catalogues the fungal records generated by high-throughput sequencing studies (Fig. 10.1). To achieve the goal of making published data findable, accessible, interoperable, and reusable (FAIR), the web interface allows users to search the entire database or within a selected set of samples for individual sequences, fungal species hypotheses—groups of sequences that share similarity approximately corresponding to a species level created using the data of the UNITE database (Koljalg et al. 2013), species, or genera; obtain a visual representation of their distribution in the environment; and access and download sequence data and metadata (Fig. 10.1). In addition, the user interface allows authors to submit data from studies not yet included, thereby contributing to building a resource for the community of researchers in fungal systematics, biogeography, and ecology (Větrovský et al. 2020).

The GlobalFungi database includes high-throughput studies of fungal community composition published since 2009. All studies that fulfilled the following criteria are included: (1) the samples represent natural habitats that have not been experimentally treated in a laboratory or greenhouse, (2) the exact geographic location of each sample is available, (3) the primers used for amplicon sequencing were designed to target the entire fungal community (rather than subsets of it), (4) internal transcribed spacer regions (ITS1, ITS2, or both) were the subject of amplification, and (5) the sequencing data were publicly available. In its third release, the database covered up to 36,684 samples of fungal communities from 367 studies that contained over 1.1 billion sequence records from more than 200 million unique variants (Fig. 10.2).

10.3 Global Fungal Diversity Estimate

It was long in the past when scientists realized that fungi are extremely diverse. Since then, there have been repeated to estimate the total number of

Fig. 10.1 User interface of the GlobalFungi database (https://globalfungi.com) provides a quick yet complex access to the data in the GlobalFungi Database. It is possible to search for individual fungal taxa—genera, species, or Species Hypotheses (Taxon search) or to search for sequences and taxa by sequence similarity (Sequence search). The list of studies can be accessed through the Studies button and regional mycoflora can be retrieved using Geosearch. There is also the option to submit new data using the Submit your study option

fungal species on Earth. The estimates were originally obtained by quantifying of host-specific fungi and multiplying them with putative number of fungal specialists per host (i.e. plant or insect) species (Hawksworth and Lücking 2017; Hyde et al. 2020). The high-throughput sequencing opened in the recent years another opportunity to assess the global fungal richness, and with increasing numbers of studies, the opportunities offered by this approach increase.

As GlobalFungi is currently the largest collection of fungal ITS marker data, it can serve as a data source for estimating global fungal diversity. Based on 97% sequence similarity clustering of full-length fungal ITS2 sequences, we recently identified more than a million fungal nonsingleton

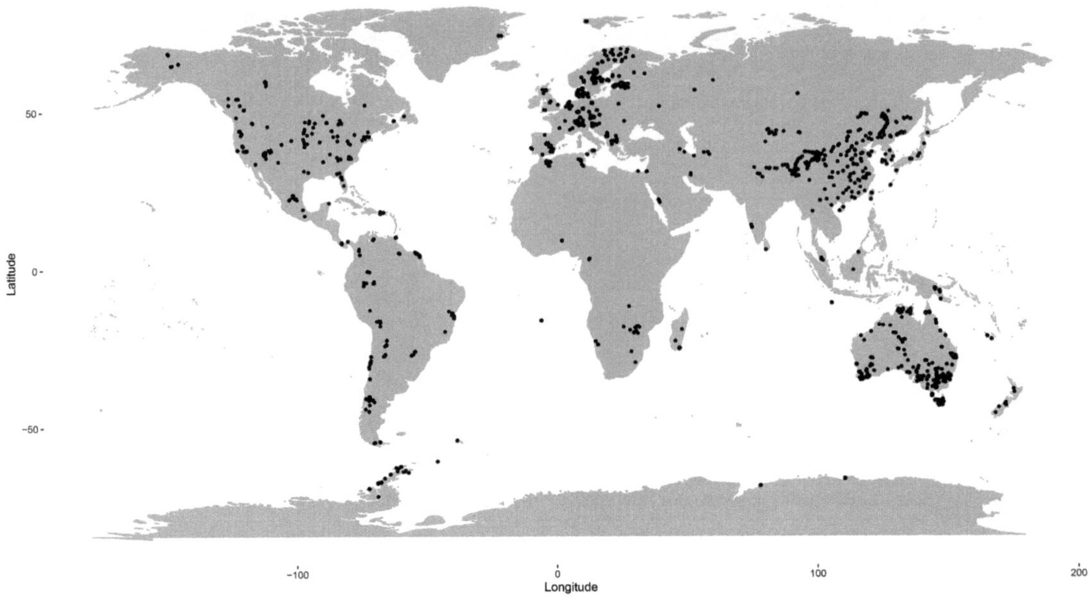

Fig. 10.2 Location of 36,684 samples covered in GlobalFungi database, release 3 (https://globalfungi.com)

Operational Transcription Units (OTUs) that were already reported and are contained in the database. The total global diversity of fungi was estimated to be approximately 6.3 million OTUs based on the Chao1 estimate calculated from the entire dataset (Baldrian et al. 2022).

Although this estimate is significantly higher than some previously proposed estimates of global fungal diversity, e.g., 2.2 or 3.8 million species (Hawksworth and Lücking 2017; Hyde et al. 2020), there are several reasons why this might be a rather conservative value. First, the threshold for judging a sequence as fungal was set at a conservative e-value of e-50 after using BLAST. Second, although some of the global doubletons and OTUs with higher sequence counts are likely to be undetected chimaeras or otherwise compromised data, many of the global singletons that were excluded from the analysis as potential technical sequencing errors or chimeric sequences are likely to represent true fungal species. Third, although some fungal species harbour several different copies of ITS2 with less than 97% similarity (Stockinger et al. 2010; Lindner et al. 2013; Větrovský et al. 2016), and some OTUs may represent rRNA pseudogenes (Glass et al. 2013), identical or very similar ITS sequences shared between different species are known from many species-rich genera of Pezizomycotina, including *Cladosporium*, *Penicillium*, *Fusarium,* and *Aspergillus* (Schoch et al. 2012). It should also be noted that there are several geographical gaps in knowledge where our information about fungal communities is limited. The high-throughput sequencing work published to date shows considerable geographic variation with one sample per 4000 km2 reported from Europe and only one sample per 200,000 km^2 in Africa. Habitat sampling has thus far been largely dominated by studies focusing on soils, particularly forest soils (Větrovský et al. 2020). It is very likely that more representative sampling covering a wider range of environmental conditions would lead to higher estimates of overall diversity. In addition, there is for sure additional fungal diversity associated with other habitats: marine environments, animal-associated microbiomes, or engineered systems. As has been shown for bacteria, these habitats contain a high proportion of specific microbial taxa (Thompson et al. 2017), and the same pattern should be true in fungi. It should be noted that metabarcoding studies on fungal communities in marine habitats are even fewer than studies on bacteria. Additionally, surveys of animal-host-associated mycobiomes are much less frequent than studies on host-associated bacteria (Peay et al. 2016; Nilsson et al. 2019a; Seibold et al. 2019). Finally, many fungal taxa, such as the phylum Glomeromycota, show limited recovery when general fungal primers are used (Kohout et al. 2014).

Metabarcoding has previously been used to demonstrate the existence of so-far undescribed but widespread groups of fungi and provide their putative taxonomic placement (Nilsson et al. 2016; Tedersoo et al. 2020b). The present dataset also lends itself to such an endeavour. The majority of OTUs in our data belonged to Ascomycota (613,755 OTUs; 57%), followed by Basidiomycota (395,877; 37%), Glomeromycota (12,780; 1%), Mucoromycota (8263; 0.8%), Mortierellomycota (4621; 0.4%), Rozellomycota (3027; 0.3%), and fungal organisms with an unclear phylum-level classification (39,246 OTUs; 3.6%, Fig. 10.3). The most common families are listed in Table 10.1 (Baldrian et al. 2022).

The largest proportion of the terrestrial habitats was represented by soil samples—approximately 60% of all samples. Together with other samples from the rhizosphere or litter, these topsoil samples accounted for almost 70% of the total. Aboveground plant biomass (shoots) and plant roots were also frequently targeted by HTS (14% and 6%, respectively) as was deadwood (9% of samples). When examining the diversity across all samples, the Chao1 estimate increased with sampling depth across the dataset, but Chao1 estimates at fixed sampling depths (5000 sequences) were higher for soil than for deadwood, roots, and shoots (Baldrian et al. 2022) (Fig. 10.4).

Accordingly, when comparing samples of different habitat types after random subsampling to 5000 sequences, the highest OTU richness was found in soil (mean ± SE: 366 ± 4), followed by litter, with deadwood showing the lowest value (82 ± 2). At this sampling depth, Chao1 estimates were highest for soil and lowest for shoots, roots, and deadwood (Table 10.2) (Baldrian et al. 2022). Although no significant correlation was found between the volume of material sampled and alpha diversity (Větrovský et al. 2019), we can assume that larger samples or samples that are composite and represent larger volume of material capture higher diversity of potential fungal habitats and thus more fungal species (Smith and Peay 2014).

The comparison of clustered sequences used in a study with the UNITE reference corpus (Baldrian et al. 2022) allowed us to estimate the relative proportion of taxa and sequences that belong to previously undescribed fungal species. Across all samples, 78% of sequences (representing 52% of OTUs) mapped to previously characterized taxa, showing a similarity of greater than 97% to any UNITE 8.2 species hypothesis. The relative proportion of sequences of known species ranged from 75% in soil samples to almost 97% in air samples. The lichen habitat appears to be the most promising for finding new fungal species, as 69% of recorded OTUs were not mapped to any characterized taxon; this share was significantly higher than in any other habitat. This observation is consistent with previous reports documenting a surprisingly high number of unrecognized species among lichenized fungi (Lücking et al. 2014). In the litter, the proportion of new taxa averaged a moderate 35%, perhaps reflecting the fact that this has been investigated by mycologists for several centuries. It should be noted that fungi associated with animal hosts were not included in the analysis due to limited data availability. These and other undersampled habitats may still harbour an unknown proportion of total fungal diversity (Wu et al. 2019) and definitely worth further exploration.

10.4 Global Patterns and Determinants of Fungal Diversity

In the context of ongoing global change, it is critical to determine how climate and other environmental factors affect the diversity and distribution of fungal communities (Hillebrand 2004). Fungal symbionts may benefit plants by mitigating abiotic stressors associated with climate change such as heat and drought

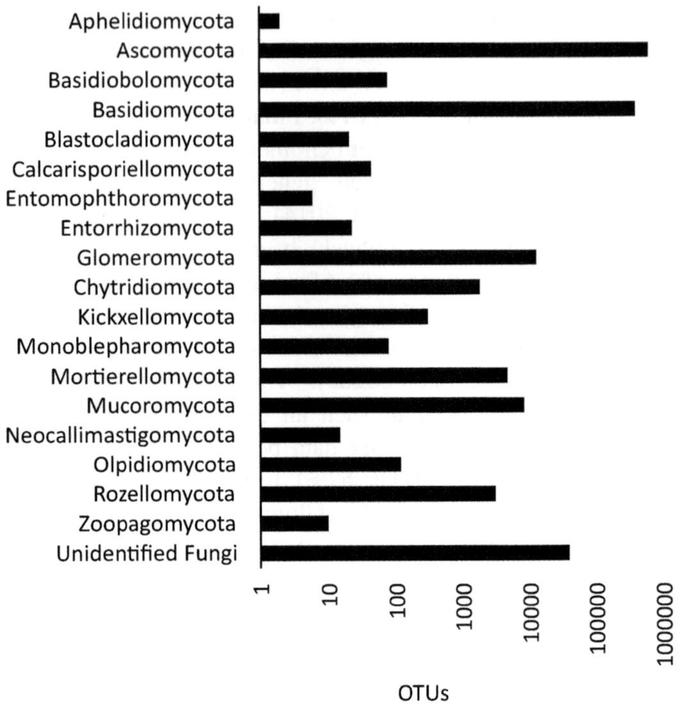

Fig. 10.3 Taxonomic breakdown of the best hits of nonsingleton OTUs present in the GlobalFungi database (Baldrian et al. 2022). The taxonomy follows (Tedersoo et al. 2018)

(Kivlin et al. 2013), but the distribution of these symbionts may be controlled by mechanisms other than climate (Kreft and Jetz 2007; Bardgett and van der Putten 2014). Because the geographic distribution and environmental preferences of almost all fungi remain unknown (Tedersoo et al. 2014), it is difficult to assess their current status and future threats to their existence (Jetz et al. 2012; Joppa et al. 2016). Given the importance of plant–fungal interactions, the ability to predict shifts in fungal distribution could help to understand or predict ecosystem-level changes. Recently, fungal community composition has been found to be influenced by climatic

Table 10.1 Families with the most predicted fungal species-level taxa (OTUs) based on the GlobalFungi database (https://globalfungi.com), release 3. The numbers indicate predicted OTU counts in each family and the numbers of sequences of fungi in each family in the database

Family	OTUs	Sequences (thousands)
Cortinariaceae (Basidiomycota)	67,940	4265
Inocybaceae (Basidiomycota)	65,952	5358
Hydnangiaceae (Basidiomycota)	37,583	1166
Aspergillaceae (Ascomycota)	36,708	3717
Nectriaceae (Ascomycota)	34,350	5480
Herpotrichiellaceae (Ascomycota)	31,699	5733
Bolbitiaceae (Basidiomycota)	27,700	933
Tricholomataceae (Basidiomycota)	26,127	2665
Pezizaceae (Ascomycota)	24,886	1545
Agaricaceae (Basidiomycota)	24,004	1000
Chaetomiaceae (Ascomycota)	23,959	4705
Thelephoraceae (Basidiomycota)	16,774	1468
Helotiaceae (Ascomycota)	15,615	5265
Pleosporaceae (Ascomycota)	15,358	4603

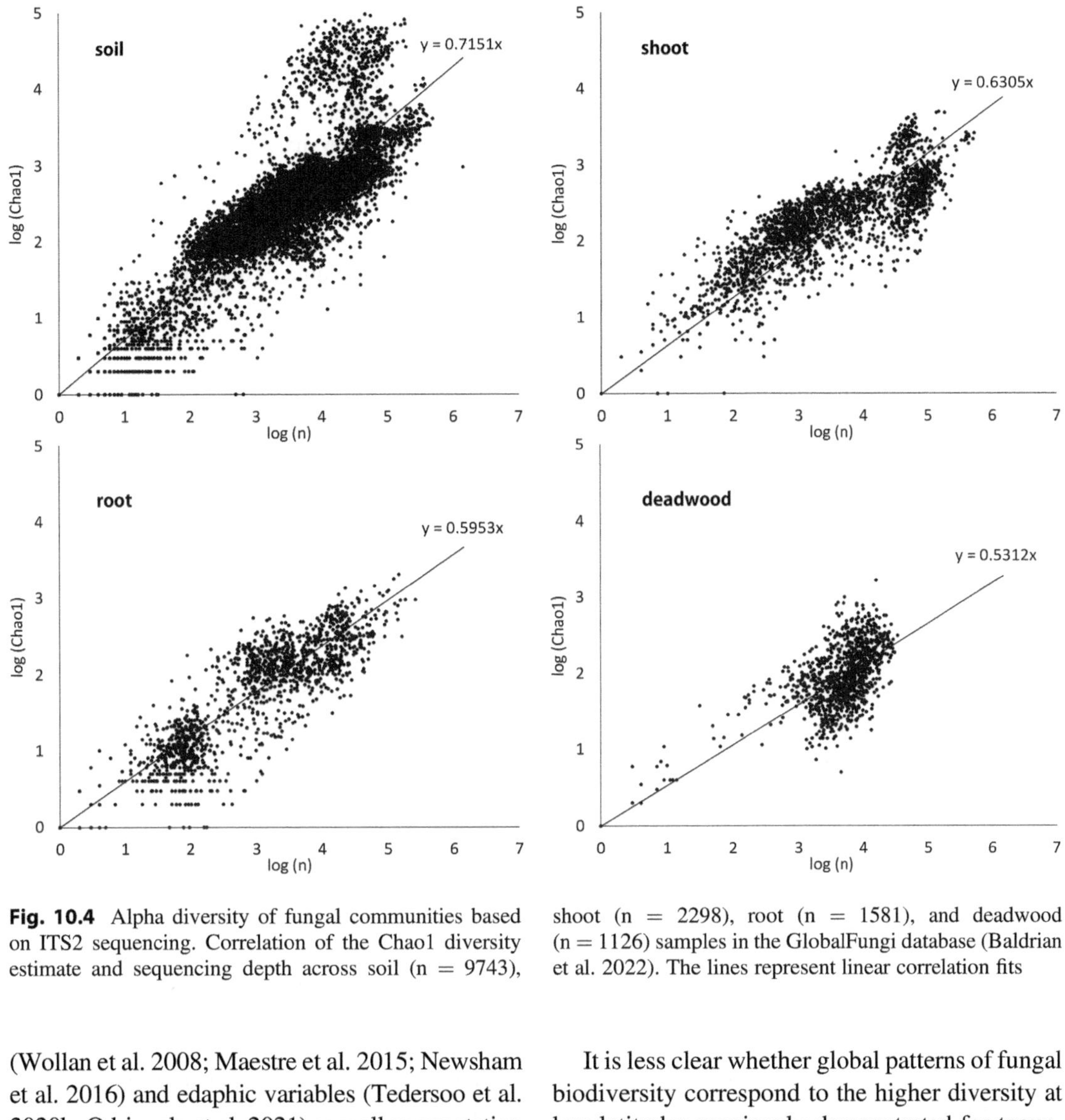

Fig. 10.4 Alpha diversity of fungal communities based on ITS2 sequencing. Correlation of the Chao1 diversity estimate and sequencing depth across soil (n = 9743), shoot (n = 2298), root (n = 1581), and deadwood (n = 1126) samples in the GlobalFungi database (Baldrian et al. 2022). The lines represent linear correlation fits

(Wollan et al. 2008; Maestre et al. 2015; Newsham et al. 2016) and edaphic variables (Tedersoo et al. 2020b; Odriozola et al. 2021) as well as vegetation elements (Crowther et al. 2016).

It is less clear whether global patterns of fungal biodiversity correspond to the higher diversity at low latitudes previously demonstrated for terrestrial macroorganisms and bacteria (Amundson

Table 10.2 Mean fungal species-level taxon (OTU) richness and Chao1 estimates in samples from various habitats. The values were calculated after subsampling each sample to 5000 sequences, and the data are based on the GlobalFungi database (https://globalfungi.com), release 3

Habitat	n	Chao1	OTU Richness
Soil	4113	1219 ± 48	366 ± 4
Litter	178	569 ± 18	332 ± 9
Air	36	392 ± 24	202 ± 10
Lichen	84	276 ± 16	134 ± 9
Shoot	933	228 ± 6	107 ± 3
Root	416	215 ± 6	135 ± 4
Deadwood	630	140 ± 5	82 ± 2

et al. 2015; Thompson et al. 2017). The distribution of plant and animal species also exhibits strong biogeographic partitioning (Kreft and Jetz 2010). Although some biogeographic patterns have been observed for bacteria, other data indicate that dominant microbes are widely distributed and thus can occur in all regions of suitable environments (Hanson et al. 2012; Delgado-Baquerizo et al. 2018).

In a previous study (Větrovský et al. 2019), data from 3084 soil samples in 36 independent studies covering a wide range of climatic conditions were evaluated in terms of the diversity of these communities. Both species richness and the Chao1 index across the different sets consistently revealed relatively low fungal diversity in the tropics. In contrast, areas of higher latitude showed considerable variation in fungal diversity with the most diverse areas concentrated near the high latitudes (Fig. 10.5). Based on these results, there is no convincing support for high fungal diversity in the tropics, which contrasts dramatically with well-known patterns for plants, arthropods, vertebrates (Amundson et al. 2015), and some bacteria (Thompson et al. 2017). Moreover, fungal diversity showed a moderate inverse relationship with temperature, confirming higher diversity at higher latitudes with colder and more seasonal climates (Větrovský et al. 2019). Although these results will need to be further refined and re-evaluated as more data become available, particularly from currently understudied areas, the present comparisons strongly suggest surprisingly low fungal diversity in the tropics.

Fungi can be categorized into ecological guilds reflecting their nutrition or other features (Põlme et al. 2020). When comparing the distribution of fungal species-level taxa belonging to different ecological guilds, ectomycorrhizal fungi have a lower upper temperature limit than that of fungi from other guilds. The temperature valence, defined as a range of temperatures where 90% of all observations of a particular taxon were observed, averaged 4.2 °C in ectomycorrhizal fungi compared with those in other fungal guilds (6.7–9.6 °C). The precipitation valence (330 mm) of this group was also narrower than those of all other guilds (430 to 860 mm), except for endophytes and wood-associated saprotrophs, which was due to the relatively low abundance of ectomycorrhizal fungi in samples with low precipitation. The narrower climatic niche of ectomycorrhizal fungi suggests that these mutualistic plant symbionts may need to respond to climate change-related problems by altering their phenology, range of distribution, or physiology. If they fail to do so, they are likely to become extinct under future climatic conditions. Plant pathogens, in contrast, appear to have much broader climatic niches (Fig. 10.6) (Větrovský et al. 2019).

10.5 Limitations of Molecular Barcoding of Fungi

Although it is tempting to consider HTS as the best solution to explore fungal diversity and biogeography, molecular barcoding and metabarcoding approaches have several limitations that prevent their use as the sole tool for diversity studies. The first important problem is the definition of species. Species are biological entities whose existence is subject to assessment by experimental tools, although these tools may be difficult to apply to all species. Metabarcoding relies on the mathematical construction of taxa. These constructions are an approximation of reality and typically build on an average barcoding gap (sequence divergence) that delineates DNA sequences into molecular taxa or OTUs. The barcoding gap varies significantly among fungi, such that some biological species appear to be multiple OTUs, whereas others may be clustered into a single OTU (Schoch et al. 2012; Liu et al. 2015; Větrovský et al. 2016). Another important issue is the scale of observations. The distribution of fungi can be extremely heterogeneous in space, and samples collected over a small area may share only a few dominant taxa (Štursova et al. 2016). Metabarcoding is also prone to errors from technical sample processing as well as platform-specific sequencing issues, which adds more sources of bias (Nilsson et al. 2019a). One such potential bias is nucleic acid extraction, and

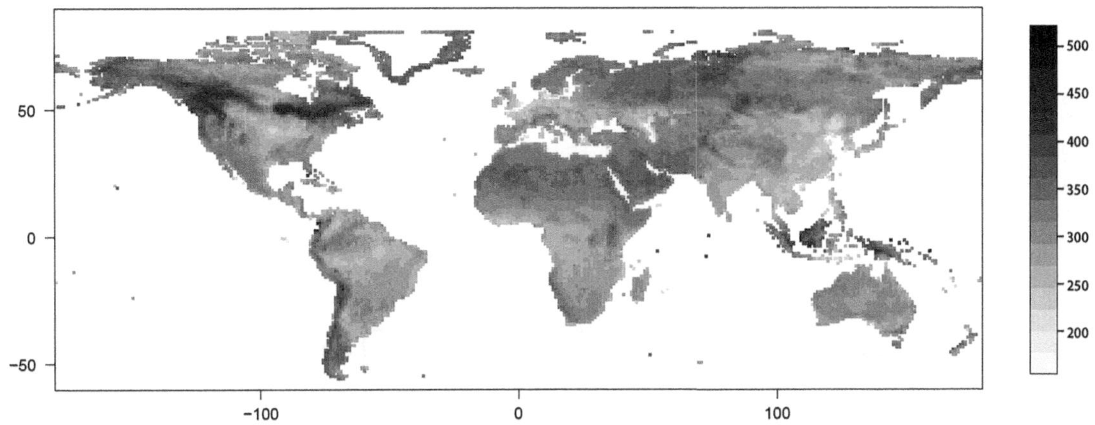

Fig. 10.5 Inferred patterns of fungal species diversity (OTU richness) predicted by the best-subset GLM (Větrovský et al. 2019)

another is the formation of chimeric sequences during marker amplification. The limited ability of bioinformatics tools to recognize chimaeras results in an unknown number of OTUs representing artificial taxa, whereas an unknown number of authentic OTUs are incorrectly removed as potential chimaeras (Schloss et al. 2011). For alpha diversity studies, cross-contamination of samples is a major and largely undescribed problem. There are also many important aspects related to the choice of molecular marker (Schoch et al. 2012). The ITS2 region has the advantage of high taxonomic resolution and limited length variability; moreover, a battery of well-tested PCR primers with varying degrees of specificity is available (Ihrmark et al. 2012; Schoch et al. 2012; Tedersoo et al. 2015). Despite the reasonable coverage of some ITS2 primers, all fungus-specific primer pairs used thus far in HTS studies, which were designed to be generally applicable to all fungi, miss at least a few of the 467 most abundant fungal taxa in GlobalFungi (Větrovský et al. 2019). It should be noted that existing ITS primers have been designed based on databases of available sequences, and the extent to which they work well for previously undiscovered and possibly genetically divergent fungal groups remains largely unknown.

Due to the fact that the number of copies of rDNA (and thus ITS) per genome varies among fungi (Lofgren et al. 2019), single-copy markers were considered as an alternative metabarcoding marker. While some of them were proven successful for analysis of closely taxonomically related taxa and appeared to possibly better estimate species-level diversity, they typically miss a large proportion of taxa across the fungal tree of life and show highly variable barcoding gap (Větrovský et al. 2016).

10.6 Conclusions

Using a large number of metabarcoding datasets from the GlobalFungi database, the conservative estimate of global fungal species richness was 6.3 million species (Baldrian et al. 2022). The global distribution of fungal diversity is shaped by multiple environmental factors, but climate appears to be the major factor affecting the most common fungal species. This finding underscores how profound the impacts of ongoing climate change can be on ecosystem functioning and food security. For example, beneficial ectomycorrhizal fungi exhibit narrow climate niches compared with those of plant pathogens. Thus, the impact of climate on fungal distribution could seriously impair plant productivity. These results open the way for further such research and identify climate as a major determinant of biogeographic patterns

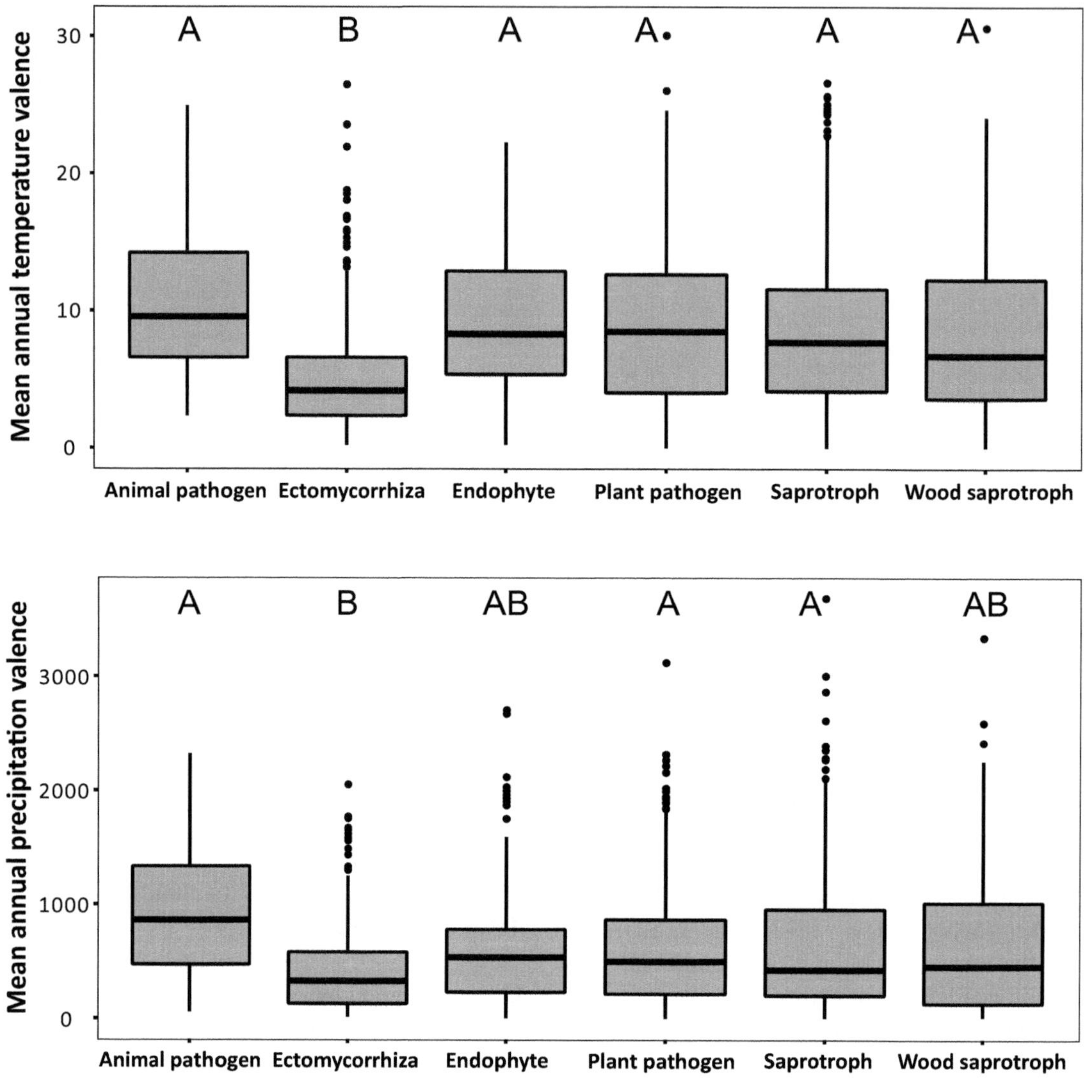

Fig. 10.6 Climatic determinants of ecological guilds of fungi. (**a**) The first and ninth deciles of sample mean annual temperature and (**b**) annual precipitation for SHs belonging to selected ecological guilds with occurrence in >10 samples (Větrovský et al. 2019)

in fungi, which appears to differ dramatically from those known for other eukaryotes and bacteria. It is clear that the estimation of diversity from HTS data needs to be viewed in light of the many potential sources of bias mentioned above. However, attempts to estimate extant fungal diversity without considering the HTS perspective appear unrealistic. Mycology as a whole stands to gain from the successful implementation of HTS, and classical mycology is entering the age of big data; there is considerable benefit for the future of fungal ecology and biogeography arising from this paradigm change.

Acknowledgements This work was supported by the Czech Science Foundation (21-17749S).

References

Amend AS, Seifert KA, Samson R, Bruns TD (2010) Indoor fungal composition is geographically patterned and more diverse in temperate zones than in the tropics. Proc Natl Acad Sci U S A 107:13748–13753

Amundson R, Berhe AA, Hopmans JW, Olson C, Sztein AE, Sparks DL (2015) Soil and human security in the 21st century. Science 348:1261071

Anslan S, Nilsson RH, Wurzbacher C, Baldrian P, Tedersoo L, Bahram M (2018) Great differences in performance and outcome of high-throughput sequencing data analysis platforms for fungal metabarcoding. MycoKeys 39:29–40

Baldrian P, Kolařík M, Štursová M et al (2012) Active and total microbial communities in forest soil are largely different and highly stratified during decomposition. ISME J 6:248–258

Baldrian P, Větrovský T, Lepinay C, Kohout P (2022) High-throughput sequencing view on the magnitude of global fungal diversity. Fungal Divers 114:539–547

Bardgett RD, van der Putten WH (2014) Belowground biodiversity and ecosystem functioning. Nature 515: 505–511

Buéé M, Reich M, Murat C, Morin E, Nilsson RH, Uroz S, Martin F (2009) 454 pyrosequencing analyses of forest soils reveal an unexpectedly high fungal diversity. New Phytol 184:449–456

Crowther TW, Todd-Brown KEO, Rowe CW et al (2016) Quantifying global soil carbon losses in response to warming. Nature 540:104–108

Davison J, Moora M, Opik M et al (2015) Global assessment of arbuscular mycorrhizal fungus diversity reveals very low endemism. Science 349:970–973

Delgado-Baquerizo M, Oliverio AM, Brewer TE, Benavent-González A, Eldridge DJ, Bardgett RD, Maestre FT, Singh BK, Fierer N (2018) A global atlas of the dominant bacteria found in soil. Science 359:320–325

Glass DJ, Takebayashi N, Olson LE, Taylor DL (2013) Evaluation of the authenticity of a highly novel environmental sequence from boreal forest soil using ribosomal RNA secondary structure modeling. Mol Phylogenet Evol 67:234–245

Hanson CA, Fuhrman JA, Horner-Devine MC, Martiny JBH (2012) Beyond biogeographic patterns: processes shaping the microbial landscape. Nat Rev Microbiol 10:497–506

Hawksworth DL (2001) The magnitude of fungal diversity: the 1.5 million species estimate revisited. Mycol Res 105:1422–1432

Hawksworth DL, Lücking R (2017) Fungal diversity revisited: 2.2 to 3.8 million species. Microbiol Spectr 5:4

Hillebrand H (2004) On the generality of the latitudinal diversity gradient. Am Nat 163:192–211

Hyde KD, Jeewon R, Chen Y-J et al (2020) The numbers of fungi: is the descriptive curve flattening? Fungal Divers 103:219–271

Ihrmark K, Bodeker ITM, Cruz-Martinez K et al (2012) New primers to amplify the fungal ITS2 region - evaluation by 454-sequencing of artificial and natural communities. FEMS Microbiol Ecol 82:666–677

Jetz W, McPherson JM, Guralnick RP (2012) Integrating biodiversity distribution knowledge: toward a global map of life. Trends Ecol Evol 27:151–159

Joppa LN, O'Connor B, Visconti P et al (2016) Filling in biodiversity threat gaps. Science 352:416–418

Kivlin SN, Emery SM, Rudgers JA (2013) Fungal symbionts alter plant responses to global change. Am J Bot 100:1445–1457

Kohout P, Sudova R, Janouskova M, Ctvrtlikova M, Hejda M, Pankova H, Slavikova R, Stajerova K, Vosatka M, Sykorova Z (2014) Comparison of commonly used primer sets for evaluating arbuscular mycorrhizal fungal communities: is there a universal solution? Soil Biol Biochem 68:482–493

Koljalg U, Nilsson RH, Abarenkov K et al (2013) Towards a unified paradigm for sequence-based identification of fungi. Mol Ecol 22:5271–5277

Kreft H, Jetz W (2007) Global patterns and determinants of vascular plant diversity. Proc Natl Acad Sci U S A 104:5925–5930

Kreft H, Jetz W (2010) A framework for delineating biogeographical regions based on species distributions. J Biogeogr 37:2029–2053

Lindahl BD, Nilsson RH, Tedersoo L et al (2013) Fungal community analysis by high-throughput sequencing of amplified markers - a user's guide. New Phytol 199: 288–299

Lindner DL, Carlsen T, Nilsson RH, Davey M, Schumacher T, Kauserud H (2013) Employing 454 amplicon pyrosequencing to reveal intragenomic divergence in the internal transcribed spacer rDNA region in fungi. Ecol Evol 3:1751–1764

Liu XZ, Wang QM, Goker M et al (2015) Towards an integrated phylogenetic classification of the Tremellomycetes. Stud Mycol 81:85–147

Lofgren LA, Uehling JK, Branco S, Bruns TD, Martin F, Kennedy PG (2019) Genome-based estimates of fungal rDNA copy number variation across phylogenetic scales and ecological lifestyles. Mol Ecol 28:721–730

Lücking R, Dal-Forno M, Sikaroodi M, Gillevet PM, Bungartz F, Moncada B, Yynez-Ayabaca A, Chaves JL, Coca LF, Lawrey JD (2014) A single macrolichen constitutes hundreds of unrecognized species. Proc Natl Acad Sci U S A 111:11091–11096

Maestre FT, Delgado-Baquerizo M, Jeffries TC et al (2015) Increasing aridity reduces soil microbial diversity and abundance in global drylands. Proc Natl Acad Sci U S A 112:15684–15689

Newsham KK, Hopkins DW, Carvalhais LC, Fretwell PT, Rushton SP, O'Donnell AG, Dennis PG (2016) Relationship between soil fungal diversity and temperature in the maritime Antarctic. Nat Clim Change 6:182–186

Nilsson RH, Wurzbacher C, Bahram M et al (2016) Top 50 most wanted fungi. MycoKeys 12:29–40

Nilsson RH, Anslan S, Bahram M, Wurzbacher C, Baldrian P, Tedersoo L (2019a) Mycobiome diversity: high-throughput sequencing and identification of fungi. Nat Rev Microbiol 17:95–109

Nilsson RH, Larsson KH, Taylor AFS et al (2019b) The UNITE database for molecular identification of fungi: handling dark taxa and parallel taxonomic classifications. Nucleic Acids Res 47:D259–D264

Odriozola I, Navrátilová D, Tláskalová P, Klinerová T, Červenková Z, Kohout P, Větrovský T, Čížková P, Starý M, Baldrian P (2021) Predictors of soil fungal biomass and community composition in temperate mountainous forests in Central Europe. Soil Biol Biochem 161:108366

Peay KG, Kennedy PG, Talbot JM (2016) Dimensions of biodiversity in the earth mycobiome. Nat Rev Microbiol 14:434–447

Põlme S, Abarenkov K, Henrik Nilsson R et al (2020) Fungal traits: a user-friendly traits database of fungi and fungus-like stramenopiles. Fungal Divers 105:1–16

Purvis A, Hector A (2000) Getting the measure of biodiversity. Nature 405:212–219

Schloss PD, Gevers D, Westcott SL (2011) Reducing the effects of PCR amplification and sequencing Artifacts on 16S rRNA-based studies. PLoS One 6:14

Schoch CL, Seifert KA, Huhndorf S et al (2012) Nuclear ribosomal internal transcribed spacer (ITS) region as a universal DNA barcode marker for fungi. Proc Natl Acad Sci U S A 109:6241–6246

Seibold S, Müller J, Baldrian P, Cadotte MW, Štursová M, Biedermann PHW, Krah F-S, Bässler C (2019) Fungi associated with beetles dispersing from dead wood – Let's take the beetle bus! Fungal Ecol 39: 100–108

Smith DP, Peay KG (2014) Sequence depth, not PCR replication, improves ecological inference from next generation DNA sequencing. PLoS One 9:e90234

Stockinger H, Kruger M, Schussler A (2010) DNA barcoding of arbuscular mycorrhizal fungi. New Phytol 187:461–474

Štursova M, Bárta J, Šantručková H, Baldrian P (2016) Small-scale spatial heterogeneity of ecosystem properties, microbial community composition and microbial activities in a temperate mountain forest soil. FEMS Microbiol Ecol 92:fiw185

Talbot JM, Bruns TD, Taylor JW et al (2014) Endemism and functional convergence across the north American soil mycobiome. Proc Natl Acad Sci 111: 6341–6346

Tedersoo L, Bahram M, Polme S et al (2014) Global diversity and geography of soil fungi. Science 346: 1256688

Tedersoo L, Anslan S, Bahram M et al (2015) Shotgun metagenomes and multiple primer pair-barcode combinations of amplicons reveal biases in metabarcoding analyses of fungi. MycoKeys 10:1–43

Tedersoo L, Sánchez-Ramírez S, Kõljalg U, Bahram M, Döring M, Schigel D, May T, Ryberg M, Abarenkov K (2018) High-level classification of the fungi and a tool for evolutionary ecological analyses. Fungal Divers 90:135–159

Tedersoo L, Anslan S, Bahram M, Kõljalg U, Abarenkov K (2020a) Identifying the 'unidentified' fungi: a global-scale long-read third-generation sequencing approach. Fungal Divers 103:273–293

Tedersoo L, Anslan S, Bahram M et al (2020b) Regional-scale in-depth analysis of soil fungal diversity reveals strong pH and plant species effects in Northern Europe. Front Microbiol 11:1953

Thompson LR, Sanders JG, McDonald D et al (2017) A communal catalogue reveals Earth's multiscale microbial diversity. Nature 551:457–463

Větrovský T, Baldrian P (2013) Analysis of soil fungal communities by amplicon pyrosequencing: current approaches to data analysis and the introduction of the pipeline SEED. Biol Fertil Soils 49:1027–1037

Větrovský T, Kolařík M, Žifčáková L, Zelenka T, Baldrian P (2016) The *rpb2* gene represents a viable alternative molecular marker for the analysis of environmental fungal communities. Mol Ecol Resour 16:388–401

Větrovský T, Kohout P, Kopecký M et al (2019) A meta-analysis of global fungal distribution reveals climate-driven patterns. Nat Commun 10:5142

Větrovský T, Morais D, Kohout P et al (2020) GlobalFungi, a global database of fungal occurrences from high-throughput-sequencing metabarcoding studies. Sci Data 7:228

Wollan AK, Bakkestuen V, Kauserud H, Gulden G, Halvorsen R (2008) Modelling and predicting fungal distribution patterns using herbarium data. J Biogeogr 35:2298–2310

Wu B, Hussain M, Zhang W, Stadler M, Liu X, Xiang M (2019) Current insights into fungal species diversity and perspective on naming the environmental DNA sequences of fungi. Mycology 10:127–140

Part IV

Evolution of Metabolism and Development

Activation of Secondary Metabolite Production in Fungi

11

Lukas M. Zehner, Mario K. C. Krespach, Maria C. Stroe, Maira Rosin, Volker Schroeckh, and Axel A. Brakhage

Abstract

Fungi can be found in virtually every habitat on earth. Over time, they evolved strategies to survive even the most challenging conditions, leading to their enormous diversity. For their survival in the habitat they have acquired the ability to produce a multitude of secondary metabolites, also named natural products. These unusual low-molecular-weight compounds exhibit a variety of effects, ranging from antimicrobial activity to protective compounds and information molecules for neighboring microorganisms. Through these effects, our recent data substantiate the hypothesis that secondary metabolites shape the composition of microbial consortia (microbiomes) by reducing or promoting the growth of certain microorganisms or changing their metabolic activity. Genetic analyses indicate that fungi have the potential to produce far more secondary metabolites than have been identified yet. Their encoding biosynthesis gene clusters remain silent under laboratory conditions and the ecological context is required for their activation. Therefore, we have initiated research on the activation of such silent gene clusters by microbial communication. Since then, many studies attempted to mimic naturally inducing conditions, e.g., by varying culture conditions, genetic manipulation of the producing organism, or co-culturing of multiple microorganisms. Elucidating the ecological roles of these compounds and the underlying triggers leading to their production is of major importance for our understanding of how microbial communities are shaped and how complex interactive networks can be formed with their profound influence on human health and the environment. Further, understanding the ecological trigger regulating the production of secondary metabolites will allow us to shed light on the pool of yet unidentified compounds in search for potential new antibiotics and other useful compounds.

L. M. Zehner · M. K. C. Krespach · M. Rosin ·
A. A. Brakhage (✉)
Department of Molecular and Applied Microbiology, Leibniz Institute for Natural Product Research and Infection Biology (Leibniz-HKI), Jena, Germany

Institute for Microbiology, Friedrich Schiller University Jena, Jena, Germany
e-mail: lukas.zehner@leibniz-hki.de; maira.rosin@leibniz-hki.de; axel.brakhage@leibniz-hki.de; axel.brakhage@hki-jena.de

M. C. Stroe
Department of Molecular and Applied Microbiology, Leibniz Institute for Natural Product Research and Infection Biology (Leibniz-HKI), Jena, Germany

Present Address: Department of Microbiology, Institute of Applied Biosciences, Karlsruhe Institute of Technology (KIT) – South Campus, Karlsruhe, Germany
e-mail: maria.stroe@kit.edu

V. Schroeckh
Department of Molecular and Applied Microbiology, Leibniz Institute for Natural Product Research and Infection Biology (Leibniz-HKI), Jena, Germany

© The Author(s), under exclusive license to Springer Nature Switzerland AG 2023

S. Pöggeler, T. James (eds.), *Evolution of Fungi and Fungal-Like Organisms*, The Mycota 14,
https://doi.org/10.1007/978-3-031-29199-9_11

Keywords

Secondary metabolites · Fungi · Silent BGCs · Microbial interaction · Cross-kingdom communication · Microbiome · Chemical signaling · Infection · Mycotoxins · Antibiotics

11.1 Introduction

Secondary metabolites (SMs), also called natural products or specialized metabolites, are compounds that are mostly not essential for the general survival of the producing organism, but instead provide benefits under specific conditions (Brakhage 1998). Filamentous fungi are renowned producers of a plethora of SMs. These compounds can function as messenger molecules that enable communication, including cross-kingdom communication between microorganisms, provide accessibility to limited nutrients and resources, or act as chemical weapons to fight competitors (Fig. 11.1) (Netzker et al. 2015; Khalid and Keller 2021). These diverse roles suggest that they are important factors in microbial communication and interaction with the biotic and abiotic environment. SMs, especially mycotoxins, often play important roles in fungal infections. Some support the infection process, *e.g.*, by influencing the host immune system, while others directly damage the host (Park et al. 2015; Broom 2015; Scharf et al. 2014). The targets of these compounds are diverse. Many SMs have found useful applications, *e.g.*, in the food or pharmaceutical industry (Narsing Rao et al. 2017; McCormack and Perry 2005; Kardos and Demain 2011). While some SMs have rather broad effects, influencing various targets in the vicinity of the producer, others are highly specialized, only required and applied in specific situations or against certain opponents.

Genes encoding the biosynthesis enzymes of these compounds are often located in close proximity to each other, forming so-called biosynthetic gene clusters (BGCs) (Brakhage 2013). As the compounds are not constantly needed and their production requires energy, it is beneficial for the producers to synthesize SMs only when the respective conditions are met. Throughout the evolution of fungi, a variety of mechanisms arose that regulate the production of SMs in response to biotic and abiotic signals or triggers. These include the presence of compounds indicative of competing organisms, low availability of nutrients, entering and leaving specific phases of the microbial life cycle, and infection of a host but also changes in temperature or pH (Fig. 11.1). As a consequence, SMs are often produced under rather specific environmental or growth conditions and are therefore frequently overlooked in laboratory set-ups. Identifying conditions that lead to SM formation as well as understanding the underlying genetic and molecular mechanisms are therefore of high value, especially for the medical and pharmaceutical industry. Such knowledge would also provide the opportunity to discover hitherto unknown compounds with putative new targets and modes of action. This is of special importance in antibiotics research, as the number of multiresistant microorganisms is growing, continuously reducing the effectiveness of the existing antibiotic compounds and thus enhancing the risk of patients succumbing to infections (Mendelson et al. 2022). Furthermore, studying the cross-talk of microorganisms at the molecular level will advance our efforts to better our understanding of complex ecosystems. Discovering the mechanisms that shape microbial communities and the processes that change their compositions will be a valuable asset when predicting the consequences of disruptive environmental processes like climate change and environmental pollution (Coban et al. 2022).

In this chapter, we summarize ecological triggers that lead to the induction of SM production in fungi, starting with fungal interactions with bacteria and other fungi in various ecosystems, *via* abiotic and environmental influences on the induction of fungal BGCs, and fungus–host interactions in infection and ending with interactions of fungi with other eukaryotes.

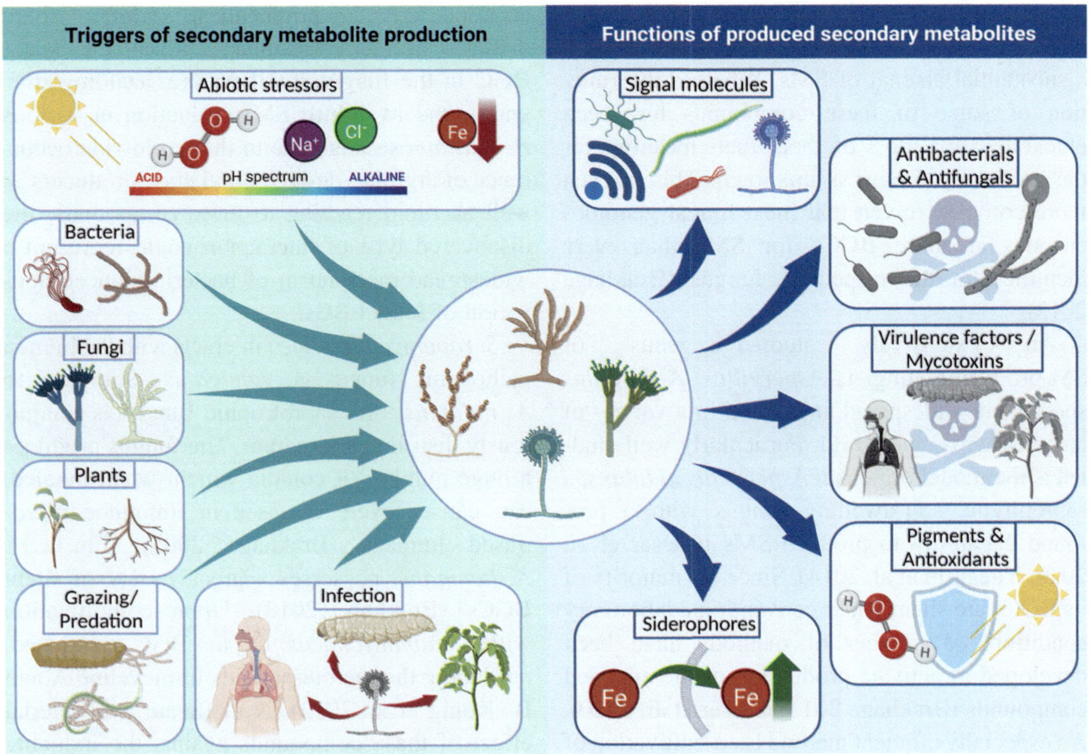

Fig. 11.1 Schematic depiction of factors and conditions that trigger fungal SM production as well as effects of the produced SMs. Figure created with BioRender.com

11.2 Induction of Fungal SM Production by Microbial Interaction

Fungi live in likely all habitats where they share their environment with a multitude of other (micro-)organisms. The composition of organisms in these ecosystems may vary, but bacteria will be conceivably encountered in all habitats. Similarly, fungi coexist, compete, or collaborate with each other. Therefore, they had to evolve strategies to either cooperate with their bacterial and fungal neighbors in beneficial interactions or reduce their competitive pressure, *e.g.*, by inhibiting growth of neighboring microorganisms or even killing them (Rangel et al. 2021; Jakubczyk and Dussart 2020; Venkatesh and Keller 2019).

In the following paragraphs, we present examples from various ecosystems, in which fungal SM production is induced by bacteria and other fungi, and summarize current knowledge on how the activation is regulated by the ecological context. We will discuss why the investigation of these processes is relevant for research, discovery of novel compounds, and industrial application.

11.2.1 Chemical Communication of Bacteria and Fungi in Terrestrial Ecosystems

A well-studied environment regarding microbial interactions is soil. It encompasses diverse habitats with an enormous number of ecological niches. In soil, fungi get in contact with a multitude of other organisms that compete for the same niche (Fig. 11.1). To prevail, fungi evolved a substantial arsenal of SMs. Whereas the function of some of these compounds has been elucidated, for most of them their meaning for the producing fungus is unknown. This is even more complex, given that most fungal genomes encode far more BGCs for SMs than were identified in the respective fungus (Brakhage 2013).

An extensively studied genus of SM-producing fungi is *Aspergillus*. *Aspergillus* species are widespread and found in a variety of habitats across the world. Particularly well studied is the model organism *Aspergillus nidulans*, a saprophytic, soil-dwelling fungus with a profound capability to produce SMs (Caesar et al. 2020; Yaegashi et al. 2014). Since the majority of its BGCs are silent under conventional laboratory conditions, a number of methods have been developed to activate production of the encoded compounds (Brakhage 2013; Caesar et al. 2020). An especially efficient method is co-cultivation of fungi with distinct microorganisms. It can apparently mimic the SM-producing ecological conditions of the natural habitat. The early establishment of this strategy led to the activation of the orsellinic acid (*ors*) BGC in *A. nidulans*. Co-cultivation of *A. nidulans* with both *Streptomyces rapamycinicus* and *Streptomyces iranensis* triggers the production of orsellinic acid and its derivatives lecanoric acid, F-9775A, and F-9775B (Schroeckh et al. 2009; Krespach et al. 2021). Whereas the effect of the produced compounds on other microorganisms remains elusive, remarkable findings regarding the molecular pathway of the induction have recently been reported. The bacteria induce an alteration of the fungal histone acetylation pattern, mediated by the Saga/Ada complex (Nützmann et al. 2011, 2013), and the activation of a distinct fungal transcription factor, BasR (Fischer et al. 2018). Both are required for the induction of the *ors* BGC. Recently, the bacterial inducer was identified as azalomycin F, a marginolactone produced by *S. rapamycinicus* and *S. iranensis* (Krespach et al. 2023). It was also demonstrated that a whole family of compounds, i.e., arginine-derived polyketides, which share distinct structural features, induce the silent BGC in the fungus. Furthermore, azalomycin F was found to induce SM production in various *Penicillium* strains. Due to the worldwide occurrence of arginine-derived polyketide producers as well as fungi reacting to these compounds, the discovered type of interaction could represent a widespread mechanism of bacteria-induced activation of fungal BGC.

S. rapamycinicus also interacts with the human pathogenic fungus *A. fumigatus*. Similarly to *A. nidulans*, this saprotrophic fungus is ubiquitously distributed in nature. The fungus produces a huge number of conidia which, when inhaled, can cause severe diseases in immunocompromised humans (Brakhage 2005) (Fig.11.1). *A. fumigatus* possesses a great variety of silent BGCs (Brakhage 2013). Upon co-cultivation with *S. rapamycinicus*, the *fcc* BGC is induced, producing the meroterpenoids fumicycline A and B (König et al. 2013). A moderate antibacterial effect of these compounds against the inducing bacterium was demonstrated, leading to the speculation that the compounds may be involved in the fungal defense against bacterial competitors.

Furthermore, co-cultivation of *S. rapamycinicus* with *A. fumigatus* induces the *fgn* BGC, resulting in the production of fumigermin (Stroe et al. 2020). Besides *S. rapamycinicus*, the closely related *S. iranensis* and the more distantly related *S. lividans* and *S. coelicolor* induce the production of fumigermin as well, albeit the two latter ones to a much lesser extent. Fumigermin was shown to reversibly inhibit spore germination of different *Streptomyces* strains and is therefore likely part of the chemical warfare of this fungus to inhibit bacteria in the same habitat. This hypothesis is strengthened by the fact that the two strong inducers of fumigermin production were more susceptible than the weak inducers *S. lividans* and *S. coelicolor* (Stroe et al. 2020).

Consistent with the activation of the *ors* BGC in *A. nidulans*, induction of fumigermin and fumicycline A/B production by *S. rapamycinicus* and *S. iranensis* is mediated by the arginine-derived polyketide azalomycin F

(Krespach et al. 2023). This indicates the importance of arginine-derived polyketides as widespread signaling molecules with the capability to induce fungal BGCs.

Interestingly, the *ors* BGC of *A. nidulans* can also be induced by *A. fumigatus* (Ninomiya et al. 2022). Production of the antibacterial diphenyl ethers violaceol I and II in *A. nidulans* is enhanced in co-cultivation with *A. fumigatus*. RNA-sequencing and gene disruption experiments confirmed that the production of these compounds depends on the upregulation of the *ors* BGC upon co-culture. This is in accordance with previous findings by Nielsen et al. (2011), which link violaceol production to one of the cluster genes, *orsA*. However, they also report constitutive production of violaceols by *A. nidulans*, without induction by *A. fumigatus*. Thus, it seems likely that culturing conditions (*e. g*., use of different media) affect violaceol production as well.

The above-described examples can be summarized as signaling via SMs. This is advantageous as it allows communication over short distances without the need for physical contact. Other compounds are so-called volatile organic compounds (VOCs). They are small molecules of low molecular weight (< 300 Da). Due to their volatile nature, they easily dissipate through air, water, and tissue (Schulz-Bohm et al. 2017; Netzker et al. 2020). Many VOCs have antibacterial or antifungal activity, while others are messenger molecules, able to alter the metabolic activity of the target organism. Plants are prolific producers of VOCs, since they play an important role in their communication with each other and their microbial environment (Ninkovic et al. 2021; Loreto et al. 2014; Minerdi et al. 2021). Here, we will focus on bacterial–fungal and fungal–fungal communication *via* VOCs.

Lutz et al. (2004) discovered that several *Pseudomonas fluorescens* strains influence the production of chitinases in *Trichoderma atroviride*. Both organisms are regarded as possible biocontrol agents, able to protect crop plants from infection by plant pathogenic microorganisms (Couillerot et al. 2009; Sood et al. 2020). *P. fluorescens* produces 2,4-diacetylphloroglucinol (DAPG) which has broad antifungal and antibacterial activity. Production of DAPG is increased upon exposure to VOCs produced by *T. atroviride*. In turn, exposure of *T. atroviride* to DAPG-producing *P. fluorescens* strains leads to an increased expression of the *nag1* gene and thus to higher amounts of the chitinase N-acetyl-β-D-glucosaminidase NAG1 (Lutz et al. 2004). Since NAG1 is of major importance for the biocontrol activity of *T. atroviride* (Brunner et al. 2003) and DAPG is one of the contributors to *P. fluorescens* biocontrol activity, this interplay indicates that mixtures of microorganisms can be more effective in biocontrol applications than axenic cultures (Lutz et al. 2004). The use of biocontrol agents in agriculture instead of classical treatments allows for more sustainable agriculture, as it reduces the need for pesticides and improves both health and yield of crop plants (Höfte and Altier 2010; De Silva et al. 2019).

VOCs also drive the interaction between the plant pathogenic fungus *Verticillium longisporum* and the bacterium *Paenibacillus polymyxa*. *V. longisporum* increases production of 1-butanol and 2-phenylethanol, two antimicrobial VOCs, when exposed to the various antifungal and plant growth-promoting VOCs emitted by *P. polymyxa* (Rybakova et al. 2017). As a consequence, *P. polymyxa* protects its host plant by inhibiting growth of *V. longisporum*, thereby reducing its detrimental effects on the host. *V. longisporum*-infected oil seed rape seedlings used as host plants showed increased growth when co-infected with *P. polymyxa* Sb3–1. This demonstrates how fungi and bacteria can limit each other's growth through production of VOCs, maintaining a balance between two organisms, which in this case also benefits the host plant.

A VOC-mediated two-way interaction was found between the plant pathogenic bacterium *Ralstonia solanacearum* and the likewise plant pathogenic fungus *Aspergillus flavus* (Spraker et al. 2014). The fungus reduces conidiation and increases aflatoxin production, while the bacterium lowers its extracellular polysaccharide production. However, as no specific VOCs responsible for the increased aflatoxin production could be identified, it was speculated that alterations in mycotoxin production could also be a secondary effect of the reduced growth and conidiation of the fungus in the presence of the bacterium.

An ecologically relevant co-culture of two wood-decaying fungi, *Eutypa lata* and *Botryosphaeria obtusa*, was studied by Azzollini et al. (2018). The authors simultaneously observed all SM-inducing processes for both volatile and non-volatile compounds. Using multivariate data analyses, they found that in co-culture the volatile compound 2-nonanone was overproduced over time. The compound 2-nonanone, an antifungal toward both fungi, was found to be co-produced with the known antifungal substance O-methylmellein. This co-production can be interpreted as generation of a synergistic activity for defense against competing fungi (Azzollini et al. 2018).

11.2.2 Fungal SM Production Induced by Endophytic Microorganisms

Endophytes inhabit the intercellular or intracellular spaces of plant tissue and either act as beneficial symbionts and neutral commensals or cause damage to the host as pathogens (Brader et al. 2017). Since plant tissues are inhabited by various microorganisms simultaneously, the dynamics and interactions between endophytes have a strong influence on growth and health of the host plant and therefore are of high interest to ecological and especially agricultural research (Khare et al. 2018). Furthermore, endophytic organisms represent a large pool of potential bioactive agents, making them a promising subject for drug discovery (Hagag et al. 2022; Hridoy et al. 2022; Singh et al. 2017).

A recent study found that the endophytic pathogen *Fusarium fujikuroi* produces several SMs in response to colonization of its chlamydospores by *R. solanacearum*. The bacterial compound ralsolamycin specifically triggers the induction of several BGCs in the fungus, resulting in production of the polyketide bikaverin and the non-ribosomal peptide beauvericin, among others. Together, the SMs possess antibacterial activity which may protect the chlamydospores from bacterial invasion by *R. solanacearum* since the compounds accumulate in chlamydospores (Spraker et al. 2018). Ralsolamycin also suppresses the biosynthesis of imizoquin, an antibacterial peptide, produced by the plant pathogen *A. flavus*. This compound plays a major role during the germination of the fungus and is suspected to give *A. flavus* a competitive advantage over the bacterium (Khalid et al. 2018). Therefore, suppression of imizoquin production in the fungus could be a strategy of *R. solanacearum* to defend its habitat against fungal invaders. These examples illustrate how endophytic organisms use SMs to regulate each other's growth in order to limit the spread of competing organisms and preserve their ecological niche.

The presence of bacteria can also alter expression of constitutively active fungal BGCs. The fungal endophyte *Fusarium tricinctum* was shown to increase its production of several SMs up to 78-fold upon co-cultivation with *Bacillus subtilis* (Ola et al. 2013). Among these compounds enniatin A1 and B1 showed antibacterial activity against the inducing *B. subtilis* as well as against *Staphylococcus aureus*, *Streptococcus pneumoniae*, and *Enterococcus faecalis*. Thus, these compounds might have a defensive function against bacterial competitors. In addition to enniatins, macrocarpon C, N-(carboxymethyl)-anthranilic acid, (−) citreoisocoumarin, and (−) citreoisocoumarinol were detected in the co-culture extracts. Their effects seem to be specific to *B. subtilis* as they were neither detectable in an axenic fungal culture nor in a co-culture with *Streptomyces lividans*. As none of these compounds were found to have antibacterial capabilities, their role remains elusive.

Co-cultivation experiments of the endophytic fungi *Alternaria tenuissima* and *Nigrospora sphaerica* revealed increased production of several polyketides by *A. tenuissima* (Chagas et al. 2013). The increase in production was particularly strong for the perylene quinone stemphyperylenol. The compound showed growth-inhibiting activity against *N. sphaerica*, indicating a competitive interaction between the two endophytic fungi. When applied to leaves of the host plant, no phytotoxic effect could be observed, which is remarkable since many other perylene quinones are reported to have phytotoxic activity. The authors argue a regulatory role of stemphyperylenol for maintaining the balance in the endophytic system of host plant and colonizers.

The endophytic fungus *Paraconiothyrium variabile* and the plant pathogen *Fusarium oxysporum* also exhibit a competitive interaction (Combès et al. 2012). *P. variabile*, isolated from the cowtail pine *Cephalotaxus harringtonia* showed antiphytopathogenic activity against *F. oxysporum*. Further investigation revealed that *P. variabile* produces twelve compounds, among them two oxylipins, 13-oxo-9,11-octadecadienoic acid (13-oxo-ODE) and 13-hydroperoxy-9,11-octadecadienoic acid (13-HPODE). In contrast, the amount of beauvericin, a mycotoxin produced by *F. oxysporum* in the inhibition zone between the two fungi, was strongly reduced. Since other studies indicated a signaling role for fungal oxylipins (Christensen and Kolomiets 2011), this change in metabolic activity might be mediated by oxylipins. Indeed, addition of the commercially available oxylipin 13-oxo-ODE reduced the amount of beauvericin detectable in the medium. A follow-up study by Bärenstrauch et al. (2020) showed that upon co-cultivation the production of PVLOX2, the enzyme responsible for production of oxylipin 13-HPODE, was strongly increased, but a repressive effect on the *beas* gene responsible for beauvericin production was not detected. This finding suggests that the previously reported effect of oxylipins on beauvericin production is not due to a transcriptional inhibition. Instead, exposure of *F. oxysporum* to *P. variabile* strongly induced *beas* expression indicating a specific response of *F. oxysporum* against the endophyte. Further experiments confirmed the antifungal activity of beauvericin against *P. variabile*, which supports the hypothesis that beauvericin production is part of a defensive response against the competing fungus. This finding contrasts those of Combès et al. (2012), who found that *P. variabile* is able to degrade or transform beauvericin, reducing the concentration of the mycotoxin in the surrounding medium. Since beauvericin is a known phytotoxin, a putative protective effect of the beauvericin-degrading *P. variabile* on the host plant was investigated. In plant infection experiments, *F. oxysporum* rapidly killed *Arabidopsis thaliana* when infected with the pathogen. When *A. thaliana* was colonized by *P. variabile* prior to infection with *F. oxysporum*, the endophytic fungus reduced the plant death rate up to 85%.

11.2.3 Aquatic Ecosystems: A Pool of Novel Compounds

It is fairly undisputed that the first lifeforms on our planet originated from water (Russell 2006; Damer and Deamer 2020; Osinski et al. 2020). Today, about 71% of our planet's surface is covered with water, over 96% being salt water of the oceans. Consequently, marine environments provide a multitude of habitats with highly variable conditions and billions of years of evolution lead to a huge biological diversity. As a result, aquatic ecosystems have proven a valuable source of SM-producing fungi (Hasan et al. 2015; Liu et al. 2020) and bacteria (Andryukov et al. 2019). In marine ecosystems, microbial life tends to accumulate on surfaces like those of aquatic plants and animals or sediment particles, since they provide constant access to nutrients and other resources (Dang and Lovell 2016). These surfaces are quickly covered by a diverse community of microorganisms, often forming biofilms. In these close associations, many marine microorganisms have evolved metabolic responses to the presence of other microbes or their produced compounds (Chen et al. 2020). This makes microorganisms from aquatic environments a fruitful topic in the endeavor to identify novel compounds and to study their activation conditions.

The antibiotic pestalone, a benzophenone, was found when the fungus *Pestalotia* sp. CNL-365, isolated from the surface of the brown alga *Rosenvingea* sp., was cultivated with the unclassified marine bacterium CNJ-328 (Cueto et al. 2001). The compound was only produced in co-cultivation, in the presence of bacterial competitors. Pestalone was not found after addition of bacterial culture extract or supernatant of the bacterial culture to the fungus. Pestalone exhibits strong antibacterial activity, *e.g.*, against methicillin-resistant *Staphylococcus aureus* and vancomycin-resistant *Enterococcus faecium*, indicating a role in antibacterial defense. However, the compound also exhibits moderate cytotoxicity against tumor cells. While initially considered as a promising precursor for antibiotic development, later studies could not further increase the antibacterial activity (Augner et al. 2013). Nevertheless, pestalone is a prominent example of how fungal–bacterial interactions can provide new candidates for antibiotic development.

The marine-derived fungus *Emericella* sp. was found to produce the cyclic depsipeptides emericellamide A and B in co-culture with the marine bacterium *Salinispora arenicola*. Emericellamides have modest antibacterial activities against methicillin-resistant *Staphylococcus aureus* (Oh et al. 2007). The production of emericellamides A and B increased 100-fold upon co-cultivation with the bacterium compared to the levels detected in fungal axenic culture, indicating that they might be part of an antibacterial defense.

Another example of a fungal–bacterial interaction involving two marine-derived organisms was discovered in the production of cytotoxic libertellenones A–D by the fungus *Libertella* sp. (Oh et al. 2005). The production of these diterpenoids was induced when co-cultivated with the α-proteobacterium *Thalassopia* sp. Libertellenone formation appears to rely on physical contact between the two microorganisms. Addition of bacterial supernatant or culture extract was insufficient to induce production of the SMs. The bacterial trigger as well as the function of the libertellenones remains elusive, as no inhibitory activity against the inducing bacterium was found.

Penicillium sp. DT-F29, isolated from marine sediment, was found to produce a number of 2,5-diketopiperazines (2,5-DKPs) when co-cultivated with *Bacillus* sp. B31 (Yu et al. 2017). In total, the authors found 12 2,5-DKPs in the fungal–bacterial co-culture that were absent in the fungal monoculture. Among the identified substances were 10 new compounds, but also the already known fumitremorgin A and 13-prenyl fumitremorgin B, all of which have bromodomain-containing protein 4 inhibitory activities. 2,5-DKPs have various biological activities, which makes them attractive research subjects especially for drug discovery and pharmaceutical applications (Borthwick 2012).

Berkeleylactones are macrolide SMs produced by a fungal co-cultivation of *Penicillium fuscum* and *Penicillium camemberti/clavigerum* isolated from the surface water of a lake (Stierle et al. 2017). An axenic cultivation of either fungus resulted in the production of known compounds like citrinin and patulin, but co-cultivation of the two fungal species activated an otherwise silent BGC leading to the production of berkeleylactones A–H. The produced SMs are active against four MRSA strains, *Bacillus anthracis*, *Streptococcus pyogenes*, *Candida albicans*, and *Candida glabrata*, indicating berkeleylactones may be produced as a means to protect a given habitat.

In another example, a co-cultivation of the two ascomycetes *Chaunopycnis* sp. with *Trichoderma hamatum* isolated from the inner tissue of the mollusk *Siphonaria* sp. led to the production of chaunopyran A (Shang et al. 2017). Interestingly, *Chaunopycnis* sp. was co-cultured with five other co-isolated fungal strains, but only co-cultivation with *T. hamatum* showed separate growth zones for both fungi. In a veritable competition scenario, *Chaunopycnis* sp. appears to produce the antifungal compound pyridoxatin to inhibit the growth of *T. hamatum*. As a response, *T. hamatum* biotransforms and modifies pyridoxatin to the inactive substance methyl-pyridoxatin. Finally, *Chaunopycnis* sp. activates the production of the antifungal chaunopyran A to gain a competitive advantage over a fungus inhabiting the same ecological niche.

Zhu et al. (2011) co-cultivated two marine-derived mangrove epiphytic *Aspergillus* sp. isolated from a rotten fruit of the *Avicennia marina* mangrove in the South China Sea. An extract of the mixed cultured mycelia revealed the presence of a new alkaloid designated as aspergicin, as well as a previously known SM, neoaspergillic acid. Significant antibacterial activity against the Gram-positive bacteria *Staphylococcus aureus*, *Staphylococcus epidermidis*, and *Bacillus subtilis* and the Gram-negative bacteria *Shigella dysenteriae*, *Proteus* sp., and *Escherichia coli* indicates bacterial competitors as target of both compounds.

An innovative approach to identify novel compounds *via* co-cultures was introduced by (Mandelare et al. 2018). The authors discovered that different developmental stages of the marine alga-derived *Aspergillus alliaceus* (teleomorph: *Petromyces alliaceus*), *i.e.*, the sexual and asexual form, produced different SMs. The vegetative stage produced the anthraquinone pigment nalgiovensin, the sexual morphotype mainly the mycotoxin ochratoxin. However, when both stages were combined, the metabolite profile drastically changed, becoming dominated by the cytotoxic bianthrone allianthrone A, its two diastereomers allianthrone B and C, and elevated levels of nalgiolaxin, the chlorinated congener of nalgiovensin.

Motivated by the need for structurally new bioactive compounds, Li et al. considered mangrove fungi as a source of SMs. In an initial study, the authors investigated a co-culture of *Phomopsis* sp. K38 and *Alternaria* sp. E33, which were both isolated from the coast of the South Chinese Sea (Li et al. 2006; Li et al. 2007; Li et al. 2008). The authors identified three new bioactive molecules: a diimide derivative and a nonadride derivative with weak cytotoxic activity, as well as a polysubstituted benzaldehyde with growth-inhibiting activity against several fungi (Li et al. 2010b, 2011; Wang et al. 2013). In a follow-up study, a new cyclopeptide, cyclo-(L-leucyl-trans-4-hydroxy-L-prolyl-D-leucyl-trans-4-hydroxy-L-proline) from the co-culture broth of the same fungi (Li et al. 2014) with antifungal activities against known fungal crop parasites such as *Gaeumannomyces graminis, Rhizoctonia cerealis, Helminthosporium sativum,* and *Fusarium graminearum,* was discovered.

An ecology-based approach of fungal–fungal interactions was applied by Oppong-Danquah et al. (2020). They classified eight fungi derived from marine sediment based on their inhibitory activity against a collection of phytopathogenic bacteria and fungi. Systematic co-cultivations of the eight fungal strains were conducted and their culture extracts tested against the collection of phytopathogens. A co-culture of *Plenodomus influorescens* and *Pyrenochaeta nobilis* showed an especially large inhibition zone between the two fungi. The extract obtained from this zone displayed strong inhibitory activity against the phytopathogenic bacterium *Xanthomonas campestris* and was further investigated. Five polyketide derivatives were isolated: the macrolides dendrolide E and dendrolide N, the azaphilones spiciferinone and 8a-hydroxy-spiciferinone, as well as the mycotoxin cephalochromin. Production of cephalochromin was specifically upregulated in the co-culture. In a screening against the eight phytopathogens, the compound showed the highest activity against *X. campestris* and the oomycete *Phytophthora infestans*. While its activity against *P. infestans* was already known, activity against *X. campestris* was reported for the first time, as well as its production by a fungus of the genus *Plenodomus*.

11.2.4 Microbial Interactions in the Human Host

Humans host a plethora of bacterial and fungal colonizers. In healthy humans, the homeostasis of the microbiome is regulated by both the human host and the interactions between the various microorganisms (Gilbert et al. 2018). Studies on the human microbiomes in the lung, gut, skin, or mouth revealed that fungal–bacterial interactions shaping these systems have a strong influence on the hosts' susceptibility to infections and their outcome (Krüger et al. 2019; Peleg et al. 2010). Whereas many studies link human microbiome and diseases (van Tilburg Bernardes et al. 2020;

Agus et al. 2021; Fobofou and Savidge 2022), only few interactions between human-colonizing microorganisms were traced back to specific SMs and their effects on other species of the microbiome yet. While the number of studies investigating SMs of the human microbiome is growing (Wilson et al. 2017; Milshteyn et al. 2018; Zipperer et al. 2016), not many microbial interactions have been examined to this level of detail. In this section, we report interactions of human-colonizing microorganisms. Activation of SM production in human pathogenic fungi during infection will be discussed in Sect. 11.4.1.

Candida albicans is an opportunistic fungal pathogen that can cause infections in immunocompromised patients. In healthy individuals it is a well-known member of the microbiome, *e.g.*, in the gut (Pérez 2021). Synergistic as well as antagonistic interactions with bacteria have been reported (Krüger et al. 2019; Peleg et al. 2010), but the molecular origin of these effects often remains obscure. Here, we focus on studies describing links to SMs. For example, dental plaque biofilms of *C. albicans* together with *Streptococcus mutans* showed higher pathogenicity than single-species biofilms, causing more aggressive caries (Falsetta et al. 2014). A later study attributed these findings to an increased production of the SM farnesol by the yeast upon co-cultivation with *S. mutans* (Kim et al. 2017). Farnesol, in the detected concentrations, had a growth-enhancing effect on *S. mutans*, which explains the reported increase in virulence. Pultar et al. (2021) discovered that the *S. mutans* product mutanobactin D, a non-ribosomal cyclic peptide, reduces yeast-to-hyphae transition and biofilm formation of *C. albicans*, which are essential mechanisms of the fungus to cause infection. Furthermore, it also inhibited the growth of the cariogenic bacteria *Actinomyces oris*, *Streptococcus oralis*, *Streptococcus anginosus*, *Streptococcus gordonii*, and *Veillonella dispar*, while it increased the growth of *Fusobacterium nucleatum*. The authors speculate that this may create a favorable environment for early colonization of dental surfaces by *S. mutans*, allowing it to outcompete other colonizers, thereby facilitating the transition from healthy to dysbiotic plaque. Thus, the interactions between *S. mutans* and *C. albicans* seem to switch from competitive in early colonization to beneficial in established, bipartite biofilms.

Considering that several fungal–bacterial interactions are reported, in which bacteria produce antifungal compounds, it is essential for fungi to recognize bacteria (Morales et al. 2010; Li et al. 2012; Islam et al. 2012). This is also the case within a human host. *Pseudomonas aeruginosa* is a pathogenic colonizer of the human lung and produces pyocyanin and phenazine, two SMs known to inhibit *A. fumigatus* as well as *C. albicans* (Kerr et al. 1999). Both fungi can cause invasive pulmonary infections in immunocompromised patients and have been found alongside *P. aeruginosa* in cystic fibrosis (CF) patients (Zhao et al. 2018; Haiko et al. 2019). This suggests that *P. aeruginosa* and the fungi compete for the same niche within human host. However, Scott et al. (2019) discovered that *P. aeruginosa* produces volatile sulfur compounds that promote growth of *A. fumigatus* and increases pathogenicity in co-infections of both organisms. Further studies confirmed that the nature of interactions between *P. aeruginosa* and *C. albicans* or *A. fumigatus* can vary strongly, from secretion of SMs inhibiting each other's growth to increased resistance against antimicrobials in mixed biofilms (Reece et al. 2021; Beswick et al. 2020). This indicates that interactions between bacteria and fungi in the human body are embedded in a complex regulatory network with shifting dynamics, which influences the composition of the microbiome. Elucidating these interaction networks will be of major importance for our understanding of many diseases and future advancement of therapies.

11.3 Induction of Fungal SM Production by Environmental Factors

Optimal adaptation to the environment significantly enhances fitness of fungi and is crucial for survival in nature. Fungi have therefore evolved the capability to sense their environment and to regulate transcriptional induction of BGCs accordingly. The diverse abiotic environmental factors that influence fungal secondary metabolism include light, oxidative and osmotic stress, temperature, and humidity. Others, such as pH and the availability of certain nutrients, can serve as indicators for potential microbial competition (Fig. 11.1). Altogether, regulation of metabolic pathways by the diverse abiotic factors is complex and regulatory cross-talk is often observed (Monroy et al. 2017).

11.3.1 Light and Oxygen-Dependent Induction of SM Production

Abiotic environmental factors such as light, temperature, and oxidative and osmotic stress trigger an adaptive cellular response (Brown et al. 2017). These factors have long been known as important stimuli for the induction of otherwise silent gene clusters in fungi and were used to increase the production of numerous SMs (Keller 2019). More recent studies elucidated some of the mechanisms behind this activation of BGCs. Light and temperature response in filamentous fungi is regulated by the trimeric velvet complex VeA-LaeA-VelB (Bok and Keller 2004; Lind et al. 2016). In the dark, VeA and VelB form a complex which interacts with LaeA in the nucleus. Light inhibits transport of VeA into the nucleus, which leads to reduced amounts of the trimeric complex in the nucleus and thus reduced activity of LaeA. VeA and LaeA have been shown to be global regulators of secondary metabolism and development in *Aspergillus* (Tisch and Schmoll 2010; Bayram et al. 2008). While in *A. nidulans* LaeA is required for the biosynthesis of sterigmatocystin and penicillin, the gliotoxin biosynthesis in *A. fumigatus* and the expression of the lovastatin BGC in *Aspergillus terreus* depend on LaeA as well (Caceres et al. 2020; Bok and Keller 2004; Bayram et al. 2008). When growing in soil under absence of light, filamentous fungi might encounter various competitors where the production of certain SMs is advantageous. Growth at the surface and thus exposure to light is a scenario that requires a different set of SMs and therefore might result in a changed SM arsenal.

Typical SMs produced under exposure to light are pigments. Just as in other organisms, fungal pigments often have photoprotective effects. Fungi produce a variety of pigment SMs, belonging to several chemical classes such as carotenoids, melanins, phenazines, quinones, monascin, and violacein (Braga et al. 2006; Lim et al. 2021; Langfelder et al. 2003). Carotenoids are a group of terpenoids and are even industrially manufactured due to their beneficial effects on humans and animals as antioxidants (Tisch and Schmoll 2010). Carotenoids can be found in species of all fungal phyla, and probably originated from archaea (Sandmann 2022). They were shown to be able to inactivate oxygen radicals and to quench singlet oxygen, generated by photosensitized reactions, thereby serving as a defensive measure against reactive oxygen species (ROS) (Sandmann 2022). The protective role of carotenoids against cell death caused by light and UV radiation was demonstrated with *Neurospora crassa* almost 50 years ago (Blanc et al. 1976). Although the regulation of carotenoid production and time needed to produce these compounds vary from species to species, carotenoids are commonly upregulated in blue light conditions since they are able to effectively absorb light of this wavelength (Tisch and Schmoll 2010). Melanins are typically found in spores and hyphae of fungi where they convey resistance to UV radiation (Langfelder et al. 2003) or assist the infection process (Heinekamp et al. 2015). There are several studies showing the inducing effect of light on melanization (Braga et al. 2006; Sun et al. 2021).

Other BGCs known to be induced under blue light conditions are the mycotoxins alternariol and altertoxin, produced by *Alternaria alternata* (Pruss et al. 2014). However, a photoprotective role has not yet been demonstrated for these SMs. In *Trichoderma reesei*, the blue light response is coupled to oxygen levels, and it is proposed that the oxidative stress response is coordinated with the response to blue light and metabolism to enhance growth and overall fitness under damaging levels of blue light (Lokhandwala et al. 2015).

Indeed, not only pigments but also other BGCs induced by light also react to oxidative stress. Especially highly oxygenated molecules such as the mycotoxin aflatoxin and its precursors are subject to redox regulation and thus their production is part of the oxidative stress response (Grintzalis et al. 2014; Kenne et al. 2018; Hong et al. 2013). Wild-type *Aspergillus parasiticus* is able to decrease the ROS concentration significantly faster than an aflatoxin deletion mutant (Kenne et al. 2018). Accordingly, oxidative stress inducing factors activate aflatoxin biosynthesis, while antioxidants block its production (Caceres et al. 2020; Hong et al. 2013). At least in some fungi, aflatoxin can be produced by the conversion of sterigmatocystin. Since this mycotoxin induces oxidative stress, an inverse regulation of these two metabolites is suggested (Zingales et al. 2020). Further examples of compounds produced under oxidative stress are fumonisins by *Fusarium fujikuroi* (Ferrigo et al. 2015) and the penitrems A–F by *Penicillium crustosum* (Kenne et al. 2018). However, studies regarding the involvement of these compounds in the oxidative stress response are still missing; hence, their physiological role remains elusive.

A recent study demonstrated the vital role of the fungal light desensitization protein VVD of *Podospora anserina* in response to white light (Shen et al. 2022). Besides severe growth defects and overproduction of spermatia, deletion of *vvd* resulted in increased pigmentation of the deletion mutant in white light. VVD negatively regulates melanin biosynthesis and production of yet uncharacterized further pink and orange pigments. Comparative transcriptomics also showed downregulation of the sterigmatocystin BGC in the Δ*vvd* mutant in white light, while no difference was detectable in the dark. This suggests a positive regulatory effect of VVD on the sterigmatocystin BGC, which was confirmed by HPLC analysis of extracts from wild-type and Δ*vvd P. anserina* cultures. Collectively, it was found that white light exposure represses production of sterigmatocystin, and VVD is able to attenuate this effect. Furthermore, growth of the Δ*vvd* mutant was strongly reduced when exposed to oxidative stress under constant illumination, confirming the previously indicated protective role of sterigmatocystin in response to oxidative stress (Shen et al. 2019).

11.3.2 Induction of Fungal BGCs in Response to Osmotic Stress

Besides illumination, water availability and temperature have been identified as two of the main abiotic factors modulating fungal growth (Li et al. 2020). Therefore, it can be assumed that these factors also influence the production of SMs. While there are several studies investigating how the association with fungi changes the effects of droughts on plants (Begum et al. 2021; Irankhah et al. 2020; Sheteiwy et al. 2021), the influence of drought on activation of fungal BGCs is understudied. There are only a few examples of fungal SMs produced under osmotic stress, resulting from reduced water availability (Mona et al. 2017; Overy et al. 2017). Osmolarity changes in the medium, caused for example by addition of sugars or NaCl, are transmitted by signal cascades. The most important signaling cascade for this environmental cue is the high osmolarity glycerol (HOG) pathway, which induces accumulation of osmolytes in the cell and provision of Na^+-ATPases (Duran et al. 2010). The involvement of the HOG signal transduction pathway in the regulation of BGCs has been described for various mycotoxins from several fungal species. Trichothecene biosynthesis in *F. graminearum* seems to be repressed in high osmolarity/ reduced water availability conditions while the alternariol biosynthesis in *Alternaria alternata* was shown to be induced (Graf et al. 2012; Medina et al. 2015). Deletion of the final MAP kinase in the MAPK HOG pathway in *A. alternata*, AaHOG, reduced fungal osmotic stress resistance and blocked alternariol production and colonization of tomato plants. Alternariol might be an important colonization factor, which is only produced under appropriate colonization conditions, for example in a tomato plant. However, the reduced osmotic stress resistance could also be due to other pleiotropic effects of the

Δaahog1 transformant and not directly linked to alternariol biosynthesis (Graf et al. 2012).

Penicillium nordicum, often found on dry cured ham, grows better at high salt concentrations (22% NaCl) compared to conditions where water is more available. Also, the production of ochratoxin A is significantly increased under these conditions (Schmidt-Heydt et al. 2012). The authors investigated the ecological reasons for the occurrence of several ochratoxin A-producing species in salt-rich habitats. They confirmed an adaptive role of *P. nordicum* ochratoxin A in NaCl-rich habitats. A mutant strain unable to produce ochratoxin A showed drastically reduced growth under high NaCl conditions. Another fungus, *Penicillium verrucosum*, shifts the production of citrinin, an SM structurally similar to ochratoxin A but not containing a chloride atom, to ochratoxin A at higher NaCl levels. These results suggest that biosynthesis and possibly excretion of the chloride-containing ochratoxin A support the maintenance of chloride homeostasis in the fungal cell (Schmidt-Heydt et al. 2012).

11.3.3 Influence of pH and Nutrient Starvation on Fungal Secondary Metabolism

Environmental pH, influenced by minerals and water entry, can also be a cue indicating the presence of neighboring microorganisms, since all microorganisms influence the surrounding pH by uptake and secretion of various molecules. The pH values significantly impact biological processes as they influence enzyme activities, proton gradients across membranes, as well as the synthesis and folding of proteins (Li et al. 2010a; Selvig and Alspaugh 2011; Hu et al. 2020). The initial pH in microbial cultures also influences the production of various compounds (Bizukojc et al. 2012). SMs produced upon a certain pH can have an antimicrobial effect (Ozcengiz and Demain 2013), suggesting that they were produced to inhibit the growth of a neighboring organism instead of simply changing the surrounding pH to a value more suitable for the producers. Similar to pH, the availability of nutrients is often influenced by microbial metabolic activity and can thus be indicative of a microbial neighbor (Bünemann et al. 2018). It is conceivable that fungi produce certain compounds in response to changes in nutrient availability. These compounds can be directly involved in resolving the nutritional shortage by making nutrients available, like siderophores, or they can be targeted at other organisms to reduce competition.

The most well-known example of an SM helping a fungus outcompete bacteria is penicillin. The discovery of penicillin by Alexander Fleming was one of the major milestones of modern medicine as it opened up the possibility to treat formerly untreatable diseases and infections. Penicillin, produced by various species of the genus *Aspergillus* and *Penicillium*, belongs to the β-lactam antibiotics, which hinder the peptidoglycan synthesis during cell division of bacteria, thereby leading to shedding of the bacterial cell wall (Brakhage 1998). Several factors influence the regulation of the penicillin BGC, some of which can also be an indicator for the presence of a microbial competitor. These factors, triggering major metabolic and developmental regulators, include availability of nutrients such as carbon or nitrogen and environmental pH. The latter might be connected to the observation that β-lactam antibiotics seem to exhibit increased toxicity on some bacterial species at alkaline pH (Brakhage 2013). Expression of the *acvA* gene, and thus penicillin production, in *Penicillium chrysogenum* is influenced by the major carbon catabolite repression transcription factor CreA and therefore dependent on sugar levels, especially glucose and sucrose (Cepeda-García et al. 2014). CreA in *A. flavus* has a major impact on aflatoxin production (Fasoyin et al. 2018). Similar as for penicillin, low availability of carbon resulted in a rise of production. A generally lowered availability of a carbon source appears to be an indicator of microbial competitors depleting available carbon sources. It can therefore be speculated that the induced production of antibiotics upon detection of low-carbon availability could be a measure to inhibit or kill the microbial competitors.

Besides carbon and oxygen, nitrogen is another important element that is needed for growth by all organisms. Low availability of nitrogen can hint toward a microbial competitor. Previous studies found hints that nitrogen availability might play a role in the regulation of the penicillin BGC in *P. chrysogenum* (Feng et al. 1994). Later, this observation was attributed to NRE, a transcription factor that mediates nitrogen metabolite repression by binding in the penicillin BGC (Haas and Marzluf 1995; Martin 2000). In *F. fujikuroi,* deletion of the *nre* ortholog *areA* resulted in the complete abolishment of gibberellin production (Tudzynski 2014). Furthermore, it was shown that AreA binds to the gibberellin biosynthesis gene promoters. Interestingly, gibberellins regulate plant growth and development (Davière and Achard 2013); hence, it can be speculated that the fungus manipulates the growth of the host plant through production of these phytohormones. In the opportunistic human pathogen *Penicillium marneffei*, AreA is important for virulence, since *area* deletion mutants failed to adapt to nitrogen-poor conditions during infection (Bugeja et al. 2012). Even though a direct link between NRE or its orthologs and other BGCs is not yet established in the case of *A. terreus*, it was shown that diverse SMs are differently regulated dependent on the amounts and the quality of the available nitrogen source. Production of diverse antibiotic and cytotoxic SMs as mevinolinic acid, geodin, terrein, asterric acid, and butyrolactone I was promoted during nitrogen limitation (Boruta and Bizukojc 2016).

As discussed above, the substrate pH can be a parameter for metabolic activity in the proximity of the fungus, as it is influenced by metabolites and degradation products secreted by the surrounding organisms. pH-dependent regulation is most prominently mediated by the transcription factor PacC that is described to be involved in the regulation of secondary metabolism in various *Aspergillus*, *Fusarium*, and *Penicillium* species (Arst and Peñalva 2003). Furthermore, PacC was shown to be involved in the virulence of pathogenic fungi, as deletion of *pacC* reduced virulence in several pathogenic fungi (Rascle et al. 2018; Barda et al. 2020; Zhang et al. 2013). However, this might also be due to the pleiotropic effects of reduced growth which was also reported for several fungi (Li et al. 2021a). Generally, PacC can directly regulate the expression of virulence factors. For example, it regulates the production of various mycotoxins, first reported for aflatoxin, where a PacC consensus motif was found in the promoter region of the aflatoxin transcription factor gene (Ehrlich et al. 1999).

In the cases presented above, the produced SMs are assumed to be directed against a microbial competitor or the host organism, either directly damaging them or inhibiting their growth. Siderophores, on the other hand, are known to chelate iron and other metals, making them better accessible. Iron is essential for fungal growth and development (Haas 2014). Under aerobic conditions, there is a constant competition among microorganisms for soluble iron. For iron uptake under low bioavailable conditions, many microorganisms evolved siderophores (Misslinger et al. 2021).

While the first fungal siderophore was identified from *Ustilago sphaerogena* by Neilands (1952), many other fungal species such as *Aspergillus, Penicillium, Candida, Alternaria*, and *Trichoderma* spp. were shown to produce siderophores as well (Pecoraro et al. 2021). Although various different mechanisms are involved in the regulation of siderophore biosynthesis (Misslinger et al. 2021), the most obvious trigger is the availability of iron. Expression of biosynthesis genes of the siderophores fusarinine C (FSC) and triacetylfusarinine C (TAFC) in *A. fumigatus* was induced upon co-cultivation with *A. nidulans* (Ninomiya et al. 2022). The authors speculate that the production of siderophores could be a reaction of *A. fumigatus* to microbial competition for available iron, as a similar increase in siderophore production was induced by phenazine production of *P. aeruginosa* (Moree et al. 2012). In *A. nidulans,* biosynthesis of siderophores is repressed by high concentrations of available iron, mediated by the GATA transcription factor

SreA (Oberegger et al. 2001). Additionally, siderophore production is influenced by pH, since SidA, catalyzing the initial step in TAFC biosynthesis, and the siderophore transporters MirA and MirB all contain PacC binding sites within their promotors (Eisendle et al. 2004). Acidic conditions repressed siderophore production, while neutral or alkaline conditions increased it. The authors offer several hypotheses: why concerted upregulation of siderophore biosynthesis genes during alkaline and low iron conditions is sensible. These include higher bioavailability of iron and lower efficiency of siderophores in acidic conditions as well as enhanced iron competition with bacteria during alkaline conditions. Furthermore, acquisition of iron is a major concern of fungi living in human or animal hosts. This is discussed in Sect. 11.4.1.

11.4 Induction of Fungal SMs in Infection

Co-evolution of fungi and mammals began, when the first mammals roamed the earth at least 160 million years ago, as evidenced from fossil records (Luo et al. 2011). An important result of this process is the ability of fungi to cause infections. A variety of factors can contribute to fungal virulence, but in many cases SMs produced by a particular fungus play a pivotal role in enabling or supporting an infection process. Typical examples for this are mycotoxins. Many fungal SMs have high value as therapeutics against bacterial infections, as cholesterol-lowering agents, or as immunosuppressants, just to name a few (Brakhage 2013). Conversely, fungal toxins have devastating effects on human health, as well as on crops, leading to high losses in agricultural food production (Almeida et al. 2019). Numerous mycotoxins are known that are virulence factors, causing diseases in animals and plants or accumulate in food (Scharf et al. 2014; Raffa and Keller 2019). Although many studies investigated the effects of mycotoxins in infection and disease (Richard 2007; Wild and Gong 2010; Fung and Clark 2004), the mechanisms by which the process of infection induces or enhances toxin production are less well studied. In the following, the current knowledge on fungal BGC induction during infection is summarized. We exemplify SMs that enable or support the infective process and whose biosynthesis is (co-)regulated with the course of infection in mammals, insects, and plants (Fig. 11.1).

11.4.1 Mycotoxins and Other Fungal SMs Activated Upon Infection of Humans

During infection of humans, fungal mycotoxins decrease the host's immune response, provide the fungus with immune evasion strategies, and induce tissue necrosis for nutrient acquisition and infection (Raffa and Keller 2019). We have already presented interactions of human-colonizing microorganisms resulting in fungal SM production in Sect. 11.2.4. Here, we focus on the few known examples, in which host factors or the environment generated by the host directly activates or enhances the production of mycotoxins and other SMs. Although there are many pathogenic fungi infecting humans, other fungi than *A. fumigatus* are understudied in this aspect. The presented examples focus on human hosts, but it is likely that similar mechanisms of SM induction exist during infection of other mammals.

Infection with *A. fumigatus* can lead to devastating diseases such as invasive aspergillosis. Lethality of invasive aspergillosis can be up to 95% due to its sophisticated immune evasion strategies (Brown et al. 2012; Schmidt et al. 2018). A microarray hybridization study revealed that at the onset of infection the BGCs for gliotoxin, fumagillin, fumitremorgin, and pseurotin biosynthesis are upregulated (McDonagh et al. 2008). Gliotoxin, an NRPS-derived mycotoxin, suppresses macrophage immune response, interferes with cytokine signaling, and directly damages host cells by the production of ROS (Arias et al. 2018; Scharf et al. 2016). The mycotoxin fumagillin inhibits the activity of phagocytes and amoeba, enabling the fungus to evade being killed by those cells

(Guruceaga et al. 2019), while fumitremorgin possesses activity against cancer cells and shows cell cycle inhibiting activity on mammalian cells (Cui et al. 1995). Pseurotin has neuritogenic and chitin synthase inhibitor function (Komagata et al. 1996; Wenke et al. 1993), moderate cytotoxicity against lung fibroblasts (Schmeda-Hirschmann et al. 2008), and the capacity to inhibit IgE production (Ishikawa et al. 2009). It is conceivable that these compounds impede the defensive capabilities of the host's immune system against the fungus and damage the surrounding tissue, thereby facilitating a successful infection of the host. Further studies were conducted to investigate the specific triggers for activation of the corresponding BGCs. Simulating the physiological conditions in the host during infection, Sueiro-Olivares et al. (2015) found that incubation of *A. fumigatus* at 37 °C instead of 24 °C induced the expression of *gliZ*, encoding a gliotoxin BGC transcription factor as well as *gliP*, coding for an NRPS that catalyzes the first step of gliotoxin biosynthesis. The hypoxic conditions in inflamed human tissue were also simulated (Grahl et al. 2012). Incubation at 0.2% O_2 instead of the 21% atmospheric O_2 concentration found in the air led to fumagillin and pseurotin production (Vödisch et al. 2011). Besides its role in disturbing the host's immune response, it is also conceivable that pseurotin A inhibits other microbes or nematodes, thereby enhancing the survival of *A. fumigatus* under natural hypoxic conditions, *e.g.*, in soil. Lastly, biofilm growth, which is often found in patients suffering from lung aspergilloma, triggers production of gliotoxin, fumitremorgin, and fumagillin further hinting at a potential importance of these SMs during infection (Bruns et al. 2010; Gibbons et al. 2012).

The immune system and hypoxic conditions in inflamed tissue are not the only challenges *A. fumigatus* faces during host infection. *A. fumigatus* can also penetrate epithelial tissue and disseminate *via* the blood stream (Hope et al. 2005). Here, *A. fumigatus* is not able to grow effectively due to hardly accessible iron and will try to escape into tissue. If this escape fails, the fungus reportedly shuts down most metabolic activity, while increasing siderophore production (Irmer et al. 2015) and expression of iron homeostasis genes. Along with *sidA*, the BGC for TAFC, an NRPS-derived siderophore, is overexpressed, suggesting an important role of TAFC for iron acquisition in blood (Irmer et al. 2015). The importance of siderophores for *A. fumigatus* lung infection was shown, as *sidA* was proven to be essential for *A. fumigatus* virulence (Schrettl et al. 2004; Hissen et al. 2005).

11.4.2 Fungal SM Activation in Response to Antifungal Treatment

When patients suffer from fungal infections, they are usually treated with antifungal therapeutics. However, the pharmaceutically applied antifungals are limited to few classes of antimycotics: azoles, polyenes, echinocandins, flucytosine, and terbinafine (Mattern et al. 2015). Interestingly, several studies showed that exposure of fungi to these drugs can trigger the production of fungal SMs. Caspofungin, an echinocandin-type antifungal, is used for treatment of aspergillosis. Exposure of *A. fumigatus* to caspofungin leads to increased production and release of gliotoxin and fumagillin (Eshwika et al. 2013; Conrad et al. 2018). It is speculated that the production of fumagillin could be an attack of the fungus on host cells or an attempt of communication, since echinocandin antifungals are products of other fungi (Conrad et al. 2018). The production of gliotoxin could be a countermeasure to mitigate the oxidative and osmotic stress induced by caspofungin, as gliotoxin also has antioxidative properties (Eshwika et al. 2013). Gliotoxin production is also increased upon treatment with the polyene macrolide amphotericin B. The increased release of gliotoxin after amphotericin B treatment might be due to membrane damage induced by amphotericin B (Reeves et al. 2004).

11.4.3 Activation of Fungal SM Biosynthesis During Infection of Insects

Insects have an even longer history of co-evolution with fungi than mammals, since they arose approximately 400 million years ago (Grimaldi 2010). Unsurprisingly, fungi have evolved specific interactions with insects that include infection and manipulation strategies. In addition to their ecological impact, these interactions are of interest as certain fungi and insects are major threats to crop plants and compounds involved could be potential candidates for future pesticides or biocontrol agents.

Beauveria bassiana is a well-studied pathogenic fungus infecting a wide range of insects (Uma Devi et al. 2008). During infection of *Triatoma infestans*, a blood-sucking bug and important vector of Chagas disease, expression of the core biosynthesis genes for tenellin and beauvericin was strongly increased, while genes for bassianolide were expressed to a lesser extent, suggesting specific roles of these compounds in the course of an infection (Lobo et al. 2015). Beauvericin and bassianolide are highly insecticidal virulence factors, while tenellin exhibits cytotoxic activity (Eley et al. 2007; Rohlfs and Churchill 2011). When larvae of the wax moth *Galleria mellonella* are infected by *B. bassiana*, oosporein, an orsellinic acid derivative, is produced. Oosporein is primarily formed after the insect died of the infection. Together, this suggests a suppressing effect of oosporein toward competing microorganisms to secure the insect carcass as food source for the fungus (Feng et al. 2015; Fan et al. 2017). The fungus *Metarhizium anisopliae* is a pathogen of arthropods. When cultured in a medium supplemented with tick cuticles, 15 SM BGCs are upregulated as revealed by RNA-Seq analysis. Among them are the biosynthetic machineries for a tropolone/citrinin-related compound, a pseurotin-related compound, the insecticidal and phytotoxic destruxin, and the siderophore ferricrocin (Sbaraini et al. 2016). Destruxins are known virulence factors for the infection of certain insects and have also found application in pharmacological research as a basis for the development of drugs against cancer and osteoporosis, as well as potential biocontrol agents (Liu and Tzeng 2012). The siderophore ferricrocin is of importance for virulence of *Metarhizium robertsii*, as it is required for germination and maintenance of iron homeostasis (Giuliano Garisto Donzelli et al. 2015). The functions of both the tropolone/citrinin-related compound and the pseurotin-related metabolite have not yet been identified; hence, their potential roles during infection are unclear.

11.4.4 Fungal SMs Induced Upon Infections of Plants

Fungi and plants have the longest history of co-evolution. Plants conquered land approximately 480 million years ago, possibly with the help of fungi that formed mycorrhiza (Lutzoni et al. 2018). Since then, fungi have also evolved the capability to infect plants and are among the biggest threats for crops and hence for the nutrition of humans. An estimated 10–23% of harvests are lost to fungal infections on the field and another 10–20% after harvesting (Fisher et al. 2012). Fungi employ a vast array of SM toxins that enable them to infect plants and cause diseases, as reviewed comprehensively in Evidente et al. (2019). Here, we focus on BGCs that are induced upon infection of a plant or throughout the infection process.

An important genus of fungi infecting major crop plants is *Fusarium*. *Fusarium* spp. colonize maize, peas, tobacco, and banana where they cause various diseases (Edel-Hermann and Lecomte 2019). *Fusarium fujikuroi* (teleomorph: *Gibberella fujikuroi*), for example, causes the ear rot disease in maize (*Zea mays*). When maize kernels are infected, the fungus produces fumonisins, a family of SM toxins. Fumonisins are predominantly produced during the dent stage of maize kernel development, characterized by a rise in internal pH and a high amylopectin content in the kernels (Picot et al. 2011). When

F. fujikuroi was cultivated in a medium in which the carbon source amylose was substituted by amylopectin the amounts of produced fumonisins increased (Bluhm and Woloshuk 2005). A positive transcription regulator of the fumonisin BGC in *F. fujikuroi* is ZFR1, also predominantly expressed in maize kernels (Flaherty and Woloshuk 2004), which leads to fumonisin accumulation in these structures. This is of concern since fumonisins are considered threats to human and animal health (Li et al. 2021b) and are potential human carcinogens (Riley et al. 1994). This needs to be observed because fumonisins are not entirely eliminated during food processing (Humpf and Voss 2004). Consequentially, efforts are undertaken to enhance the resistance of maize plants against *F. fujikuroi*, *e.g.*, by application of genetically engineered *Bacillus thuringiensis* on the maize plants (Munkvold et al. 1999). The transgenic bacteria produce CryIA(b) and other toxins to certain maize-feeding insects like the European corn borer *Ostrinia nubilalis*. Consequentially, this so-called Bt maize suffers less damage from feeding insects, which in turn makes it less vulnerable to infection by *F. fujikuroi*. Complementary to these approaches, improving our understanding of the regulatory mechanisms behind fumonisin production might help to design crop plants that accumulate less toxins, thereby reducing the health risk to humans and animals.

Another maize pathogen, *Cochliobolus carbonum,* causes the leaf spot disease. In early infection the fungus forms appressoria, swollen hyphal structures that adhere to plant leaves and facilitate penetration into the plant tissue. The formation of these infection structures coincides with the production of HC-toxin, a cyclic tetrapeptide and histone deacetylase inhibitor (Brosch et al. 1995). HC-toxin is of major importance for invading the host and therefore considered a virulence factor of *C. carbonum* (Yoder 1980). While it is still unclear how the inhibition of histone deacetylases induces disease in maize, several studies have confirmed that the activity of HC-toxin as HDAC inhibitor is not confined to this host. As reviewed in Walton (2006), it extends to other plants and even insects and mammals. An especially interesting aspect of this cyclic tetrapeptide is, that it only displays toxicity toward maize plants with specific genotypes. Presence of the dominant alleles *Hm1* or *Hm2* renders maize plants resistant, as they code for HC-toxin reductase, a detoxifying enzyme, speculated to be exclusively retained in the genome as a defense against HC-toxin and related compounds (Walton 2006). This exemplifies how the evolutionary pressure exerted by fungal pathogens and their toxins shaped the genetic equipment of host plants.

Botrytis cinerea causes gray mold disease in more than 200 agricultural plants and employs botcinic acid and botrydial as phytotoxic SMs. Both compounds induce chlorosis and plant tissue necrosis. Botcinic acid also exhibits antifungal activity. During tomato leaf infection the BGCs for both compounds are upregulated and double knock-out mutants exhibit markedly diminished virulence. A knock-out mutant deficient in only botcinic acid production retains a substantial virulence potential, suggesting an at least partially redundant role of both compounds during infection (Dalmais et al. 2011). Botcinic acid is also synthesized during infection of soybean plants with *Sclerotinia sclerotiorum*, causing Sclerotinia stem rot disease (Westrick et al. 2019). The suggested widespread role of this phytotoxin in plant infection and the broad range of crop plants susceptible to these pathogens emphasize the importance of understanding the underlying molecular mechanisms in this specific pathogen–host interaction.

Various *Penicillium* ser. *Corymbifera* strains are responsible for blue mold storage rot in plant bulbs (Overy et al. 2005). When these strains were cultivated on plant tissue-based agars, *e.g.*, with extracts of yellow onions (*Allium cepa*), the production of some corymbiferones and corymbiferan lactones was massively enhanced. In addition, these compounds were also identified in onions infected by *Penicillium* spp. (Overy et al. 2006). This finding indicates that conditioned media can activate SM production similar to the real infection of the plant. Through this method, the authors were able to amplify the production of several SMs compared to cultivation on commonly used media. Among them were the novel compound corymbiferone B and another yet unidentified but structurally similar compound. This approach could be employed in screening for novel compounds, as it enables detection of compounds that would have been overlooked when using standard media and will allow elucidation of a more complete metabolic profile of the used fungi (Overy et al. 2006).

Kobayashi et al. (1995) investigated plant calli, which are more easily infected by fungi than highly differentiated intact plants. A co-culture consisting of a fungus and a callus appears to be a promising strategy for initiation of new metabolite production. By infecting a *Phytolacca americana* callus with the fungus *Botrytis fabae* (Kobayashi et al. 1995), the authors identified the antifungal metabolite phytolaccoside B, which was likely produced as a defense mechanism against the fungal pathogen. In a follow-up study, Marfori et al. (2002) set up a co-culture of *Trichoderma harzianum* and *Catharanthus roseus* callus and isolated trichosetin, an antimicrobial compound with a remarkable activity against the Gram-positive bacteria *S. aureus* and *B. subtilis*. The compound was not produced in individual cultures of either *T. harzianum* or *C. roseus* callus. Due to the close structural analogy of trichosetin to the fungal toxin equisetin, the authors argue that the compound is likely produced by the fungus.

11.5 Induction of Fungal SMs During Non-infectious Interaction with Other Eukaryotes

Except of mycorrhiza fungi and infection-related interactions, reports of fungi interacting with multicellular eukaryotes are scarce. This is likely due to the complexity of such interactions which makes them more difficult to analyze than fungal–fungal or fungal–bacterial interactions. Nevertheless, the ecological context for and relationship between interacting partners can help to unravel the nature of a microbial interaction. Interactions may be classified as neutral (in which no species gains an advantage from the interaction), competitive (in which the partners antagonize each other to gain advantage over nutrients, territories, etc.), or synergistic (in which all species benefit from the interaction). With regard to SMs, in many cases their production was interpreted as useful for competition. However, it also needs to be taken into account that the amounts produced in the natural environment are often very low and a function as communication molecule should be also considered (Fig. 11.1).

11.5.1 Fungal–Nematode Interactions

Fungi form a heterogeneous group of species characterized by diverse lifestyles which are mainly driven by the nutritional mode of the fungus. Saprotrophic fungi obtain nutrients from dead organic matter; biotrophic species exploit living host cells, while necrotrophic fungi gain nutrients from killed host cells (Macauley and Thrower 1966). Some fungi may not be restricted to only one lifestyle and can switch from a saprotrophic existence to a predatory one. The latter is a fascinating area of microbial interaction: fungi able to catch and kill nematodes, called nematode-trapping fungi or NTFs. NTFs produce specific traps with which they capture nematodes. After entrapment, the nematodes are paralyzed and penetrated by the fungal hyphae, which proliferate and digest the worm. An intricate chemical warfare occurs between the pathogenic NTFs and the host nematode. NTFs eavesdrop on the *Caenorhabditis elegans* pheromones known as ascarosides, which are used by the fungus as a signal to switch from the saprotrophic lifestyle to a predatory one *via* induction of trap morphogenesis (Hsueh et al. 2013; Yu et al. 2021b). Moreover, NTFs secrete VOCs like butenoate derivatives and polyketide SMs like 6-methyl-salicylic acid to attract the worms into fungal colonies (Hsueh et al. 2017; Yu et al. 2021a). To enable penetration, NTFs produce virulence factors including protease inhibitors, chitin-binding proteins, small-secreted proteins, and, importantly, SMs (Stergiopoulos and de Wit 2009; Hsueh et al. 2017; Yu et al. 2021a, b) After penetration, NTF may use SMs to kill or paralyze nematodes, as shown for the most widely studied model NTF *Arthrobotrys oligospora* (Niu and Zhang 2011). It thus becomes evident that fungal SMs are crucial for the fungal–nematode interaction and the biosynthesis of a number of SMs is specifically activated by the presence of prey nematodes.

Recently, Yu et al. identified key SMs involved in the regulation of trap formation and attraction of the nematode *C. elegans* by the NTF *Duddingtonia flagrans*, a fungus phylogenetically closely related to *A. oligospora* (Yu et al. 2021a). The genome of *D. flagrans* (Youssar et al. 2019) encodes the PKS gene *artA*, which shows high similarity to a known PKS gene of *A. oligospora*. In *A. oligospora*, this PKS is responsible for the biosynthesis of arthrosporols, which were shown to inhibit the formation of nematode traps (Xu et al. 2015). In agreement, Yu et al. (2021a) found that by deleting *artA*, or the neighboring genes *artB* or *artC*, the ability of *D. flagrans* to form traps was increased. LC-MS analysis confirmed the absence of arthrosporols in *ΔartA*, *ΔartB*, and *ΔartC* mutants. However, the salicylic acid derivative 6-methyl-salicylic acid (6-MSA) accumulated in a *ΔartC* mutant strain. Furthermore, the heterologous expression of the PKS *artA* in the fungus *Aspergillus oryzae* led to the formation of 6-MSA, indicating that 6-MSA is the polyketide product of ArtA, and that ArtB and ArtC convert 6-MSA to arthrosporols. As shown for arthrosporols, 6-MSA was also found to inhibit trap formation. Interestingly, the expression of the three genes is uncoupled, in both time and space. At the hyphal tips, only *artA* was expressed, which led to accumulation of 6-MSA, while *artB* and *artC* were expressed in the rear of the hyphae, where arthrosporols accumulated. Intrigued by the fact that both 6-MSA and arthrosporols inhibit trap formation, the authors looked for another role for 6-MSA to be produced at the hyphal tip and found that it serves as a chemoattractant for *C. elegans*. Therefore, *D. flagrans* employs an interesting regulatory mechanism of trap formation based on the spatially resolved production of SMs: at the hyphal tips, 6-MSA is produced to attract *C. elegans*, while further back in the hyphae arthrosporols are produced to block trap formation until a sufficient number of nematodes are present. When enough nematodes are sensed, *D. flagrans* will switch off arthrosporol and 6-MSA production and proceed to form nematode traps. Overall, *D. flagrans* makes use of a multifunctional SM BGC, which through combinatorial biosynthesis produces multiple compounds (6-MSA, arthrosporols, and additionally pigments), each with a specific role. The fungal SM 6-MSA is produced when worm pheromones are sensed and serves a chemoattractant role for the nematode, while both 6-MSA and arthrosporols are involved in trap formation, which the fungus controls by modulation of *art* gene expression.

11.5.2 Leafcutter Ant–Fungus Cultivar: Interaction with Fungal Pathogen *Escovopsis weberi*

Another fascinating association is the mutualism between *Acromyrmex* leafcutter ants and their fungal cultivar (Currie 2001). As part of this symbiosis, the ants groom the basidiomycete *Leucoagaricus gongylophorus*, which then serves as their main food source. It was initially believed that the ants manage to maintain the fungus free of microbial pathogens. However, the ant gardens are also home to a highly virulent and pathogenic fungus belonging to the genus *Escovopsis*. To counteract the pathogen, the ants have evolved a beneficial association with actinomycetes (mostly *Streptomyces* and *Pseudonocardia* spp.), which produce antibiotics that suppress the growth of the ascomycete *Escovopsis weberi*. In turn, the pathogen *E. weberi* produces a variety of compounds in order to compete with the garden fungus and circumvent the mutualist-powered defense of the leafcutter ants. Since *E. weberi* can overwhelm *L. gongylophorus*, researchers assumed the effect was achieved through secretion of toxins and SMs. In agreement with this hypothesis, Dhodary et al. (2018) discovered that

E. weberi produces five shearinine-type indole triterpenoids including two novel derivatives, shearinine L and shearinine M, as well as the polyketides emodin and cycloarthropsone. The latter two SMs strongly inhibited the growth of the garden fungus *L. gongylophorous*, while shearinine L inhibited the ants, which after a two-day learning period quickly adapted to avoid the compound. These findings were confirmed by Heine et al. (2018) who found that in a mutually beneficial symbiosis of *Acromyrmex* leafcutter ants and the fungus *L. gongylophorus* and *Pseudonocardia* bacteria, the parasitic *Escovopsis* fungus upregulates the production of two specialized metabolites including shearinine D, which reduces ant behavioral defenses and is ultimately lethal when it accumulates in ant tissues.

11.5.3 Grazing Insect Larvae Induce the Chemical Defense of Fungi

Leafcutter ants are not the only insects feeding on fungi. Several other insects use fungi as a food source as well. It is not surprising that fungi evolved chemical defense mechanisms to fight fungivorous insects. Investigating these substances could lead to the discovery of new insecticidal compounds that could potentially be useful for agriculture.

Grazing of *Drosophila melanogaster* larvae on *A. nidulans* upregulates LaeA, a global regulator of SM biosynthesis (Caballero Ortiz et al. 2013). This leads to enhanced expression of four BGCs, encoding the mycotoxin sterigmatocystin, the insecticidal austinoids, the antibiotic emericellamides, and the prominent antibiotic penicillin (Caballero Ortiz et al. 2013; Valiante et al. 2017). Similar results were obtained from the soil arthropod *Folsomia candida* grazing on *A. nidulans* (Döll et al. 2013). In this case, the fungus overproduced emericellamides (Sanchez et al. 2012), as well as the meroterpenoids austinol and dehydroaustinol. The latter exhibit neurotoxic activity to insects (Kataoka et al. 2011). It was discovered that the shift in secondary metabolism is part of a defensive response increasing the resistance to fungivore grazers. The production of a combination of SMs might exert synergistic effects and make the chemical response more effective than producing a single compound (Döll et al. 2013). Additionally, a previous study on arthropod grazing behavior indicated that *F. candida* tends to distribute its feeding activity on different fungal colonies because it is speculated that feeding leads to resistance of the colonies against the fungivore (Stötefeld et al. 2012). This behavior could be evidence of the toxin concentration increasing over time in the fungal colony when the arthropods started feeding.

In response to grazing of *Drosophila melanogaster* larvae, *A. nidulans* produces the terpenoid juvenile hormone III (JH-III) (Nielsen et al. 2013), which is essential for larval development. This finding is particularly interesting, since JH-III was previously not reported to be a fungal SM and no genes homologous to the biosynthesis pathway of JH-III were found. The authors speculate that this could either be due to the long evolutionary distance between fungi and insects, which may have obscured a common origin of the biosynthesis genes, or that the fungus indeed evolved an alternative biosynthetic pathway for JH-III (Nielsen et al. 2013). In both cases, the fact that the fungus is able to produce an insect hormone might result from the long history of co-evolution between insects and fungi. JH-III is not excreted by the fungus and instead is retained inside the mycelium. When *D. melanogaster* larvae feed on the mycelium, they ingest the compound that disturbs the fragile balance of juvenile hormones in the animals, thereby having a devastating effect on the insects (Wilson 2004). These findings impressively demonstrate the intricate chemical defense mechanisms of the fungus against grazing insects.

11.6 Conclusion

Over the course of evolution, fungi became an incredibly diverse group, conquering niches in virtually all habitats on earth and surviving even the harshest conditions. A significant factor contributing to their adaptability is their capacity to produce secondary metabolites. These compounds allow them to shape the composition of their surrounding microorganisms, to fight their competitors, and to survive conditions that would otherwise be damaging or even lethal (Macheleidt et al. 2016). In general, secondary metabolites can be regarded as signal molecules that influence other (micro-)organisms either in a positive or negative way. The ecological meaning of most of these compounds has not been uncovered until today.

Understanding the ecological factors triggering fungal BGCs is of major importance because fungi encode an enormous number of silent BGCs (Keller 2019; Knowles et al. 2022; Brakhage 2013). Since these silent BGCs represent a pool of potentially new antibiotics and further useful compounds, it is of high scientific interest to investigate the mechanisms underlying their activation. In the last decade, many different approaches to mimic natural ecological conditions have been followed that lead to the activation of silent BGCs: changes in culturing conditions, genetic manipulation of the producing organism, or complex co-culture systems of two and more partners (Netzker et al. 2018; Brakhage 2013), to name a few. Besides new compounds potentially interesting for medical or industrial application, another aim to study the complex interactive networks is how microbial communities are shaped and what is the potential outcome for human health and a healthy environment. Consequentially, understanding the interplay of microorganisms and the roles SMs play in structuring microbiomes might enable us to manipulate these microbial systems to our benefit and treat diseases.

References

Agus A, Clément K, Sokol H (2021) Gut microbiota-derived metabolites as central regulators in metabolic disorders. Gut 70(6):1174–1182. https://doi.org/10.1136/gutjnl-2020-323,071

Almeida F, Rodrigues ML, Coelho C (2019) The still underestimated problem of fungal diseases worldwide. Front Microbiol 10:214. https://doi.org/10.3389/fmicb.2019.00214

Andryukov B, Mikhailov V, Besednova N (2019) The biotechnological potential of secondary metabolites from marine bacteria. J Mar Sci Eng 7(6):176. https://doi.org/10.3390/jmse7060176

Arias M, Santiago L, Vidal-Garcia M, Redrado S, Lanuza P, Comas L, Domingo MP, Rezusta A, Galvez EM (2018) Preparations for invasion: modulation of host lung immunity during pulmonary aspergillosis by gliotoxin and other fungal secondary metabolites. Front Immunol 9:2549. https://doi.org/10.3389/fimmu.2018.02549

Arst HN, Peñalva MA (2003) Regulation of gene expression by ambient pH in filamentous fungi and yeasts. Microbiol Mol Biol Rev 40(1):1–3. https://doi.org/10.1016/s1087-1845(03)00077-x

Augner D, Krut O, Slavov N, Gerbino DC, Sahl HG, Benting J, Nising CF, Hillebrand S, Krönke M, Schmalz HG (2013) On the antibiotic and antifungal activity of pestalone, pestalachloride A, and structurally related compounds. J Nat Prod 76(8):1519–1522. https://doi.org/10.1021/np400301d

Azzollini A, Boggia L, Boccard J, Sgorbini B, Lecoultre N, Allard PM, Rubiolo P, Rudaz S, Gindro K, Bicchi C, Wolfender JL (2018) Dynamics of metabolite induction in fungal co-cultures by metabolomics at both volatile and non-volatile levels. Front Microbiol 9:72. https://doi.org/10.3389/fmicb.2018.00072

Barda O, Maor U, Sadhasivam S, Bi Y, Zakin V, Prusky D, Sionov E (2020) The pH-responsive transcription factor PacC governs pathogenicity and ochratoxin A biosynthesis in Aspergillus carbonarius. Front Microbiol 11:210. https://doi.org/10.3389/fmicb.2020.00210

Bärenstrauch M, Mann S, Jacquemin C, Bibi S, Sylla O-K, Baudouin E, Buisson D, Prado S, Kunz C (2020) Molecular crosstalk between the endophyte Paraconiothyrium variabile and the phytopathogen Fusarium oxysporum – modulation of lipoxygenase activity and beauvericin production during the interaction. Fungal Genetics Biol 139:103383. https://doi.org/10.1016/j.fgb.2020.103383

Bayram O, Krappmann S, Ni M, Bok JW, Helmstaedt K, Valerius O, Braus-Stromeyer S, Kwon NJ, Keller NP, Yu JH, Braus GH (2008) VelB/VeA/LaeA complex coordinates light signal with fungal development and secondary metabolism. Science 320(5882):1504–1506. https://doi.org/10.1126/science.1155888

Begum N, Akhtar K, Ahanger MA, Iqbal M, Wang P, Mustafa NS, Zhang L (2021) Arbuscular mycorrhizal fungi improve growth, essential oil, secondary metabolism, and yield of tobacco (Nicotiana tabacum L.) under drought stress conditions. Environ Sci Pollut Res Int 28(33):45276–45295. https://doi.org/10.1007/s11356-021-13,755-3

Beswick E, Amich J, Gago S (2020) Factoring in the complexity of the cystic fibrosis lung to understand Aspergillus fumigatus and Pseudomonas aeruginosa interactions. Pathogens 9(8). https://doi.org/10.3390/pathogens9080639

Bizukojc M, Pawlak M, Boruta T, Gonciarz J (2012) Effect of pH on biosynthesis of lovastatin and other secondary metabolites by Aspergillus terreus ATCC 20542. J Biotechnol 162(2–3):253–261. https://doi.org/10.1016/j.jbiotec.2012.09.007

Blanc PL, Tuveson RW, Sargent ML (1976) Inactivation of carotenoid-producing and albino strains of Neurospora crassa by visible light, blacklight, and ultraviolet radiation. J Bacteriol 125(2):616–625. https://doi.org/10.1128/jb.125.2.616-625.1976

Bluhm BH, Woloshuk CP (2005) Amylopectin induces fumonisin B1 production by Fusarium verticillioides during colonization of maize kernels. Mol Plant Microbe Interact 18(12):1333–1339. https://doi.org/10.1094/MPMI-18-1333

Bok JW, Keller NP (2004) LaeA, a regulator of secondary metabolism in Aspergillus spp. Eukaryot Cell 3(2):527–535. https://doi.org/10.1128/EC.3.2.527-535.2004

Borthwick AD (2012) 2,5-Diketopiperazines: synthesis, reactions, medicinal chemistry, and bioactive natural products. Chem Rev 112(7):3641–3716. https://doi.org/10.1021/cr200398y

Boruta T, Bizukojc M (2016) Induction of secondary metabolism of Aspergillus terreus ATCC 20542 in the batch bioreactor cultures. Appl Microbiol Biotechnol 100(7):3009–3022. https://doi.org/10.1007/s00253-015-7157-1

Brader G, Compant S, Vescio K, Mitter B, Trognitz F, Ma LJ, Sessitsch A (2017) Ecology and genomic insights into plant-pathogenic and plant-nonpathogenic endophytes. Annu Rev Phytopathol 55:61–83. https://doi.org/10.1146/annurev-phyto-080516-035641

Braga GU, Rangel DE, Flint SD, Anderson AJ, Roberts DW (2006) Conidial pigmentation is important to tolerance against solar-simulated radiation in the entomopathogenic fungus Metarhizium anisopliae. Photochem Photobiol 82(2):418–422. https://doi.org/10.1562/2005-05-08-RA-52

Brakhage AA (1998) Molecular regulation of β-lactam biosynthesis in filamentous fungi. Microbiol Mol Biol Rev 62(3):547–585. https://doi.org/10.1128/MMBR.62.3.547-585.1998

Brakhage AA (2005) Systemic fungal infections caused by Aspergillus species: epidemiology, infection process and virulence determinants. Curr Drug Targets 6(8):875–886. https://doi.org/10.2174/138945005774912717

Brakhage AA (2013) Regulation of fungal secondary metabolism. Nat Rev Microbiol 11(1):21–32. https://doi.org/10.1038/nrmicro2916

Broom L (2015) Mycotoxins and the intestine. Anim Nutr 1(4):262–265. https://doi.org/10.1016/j.aninu.2015.11.001

Brosch G, Ransom R, Lechner T, Walton JD, Loidl P (1995) Inhibition of maize histone deacetylases by HC toxin, the host-selective toxin of Cochliobolus carbonum. Plant Cell 7(11):1941–1950. https://doi.org/10.1105/tpc.7.11.1941

Brown GD, Denning DW, Gow NA, Levitz SM, Netea MG, White TC (2012) Hidden killers: human fungal infections. Sci Transl Med 4(165):165rv113. https://doi.org/10.1126/scitranslmed.3004404

Brown AJP, Cowen LE, di Pietro A, Quinn J (2017) Stress adaptation. Microbiol Spectr 5(4). https://doi.org/10.1128/microbiolspec.FUNK-0048-2016

Brunner K, Peterbauer CK, Mach RL, Lorito M, Zeilinger S, Kubicek CP (2003) The Nag1 N-acetylglucosaminidase of Trichoderma atroviride is essential for chitinase induction by chitin and of major relevance to biocontrol. Curr Genet 43(4):289–295. https://doi.org/10.1007/s00294-003-0399-y

Bruns S, Seidler M, Albrecht D, Salvenmoser S, Remme N, Hertweck C, Brakhage AA, Kniemeyer O, Müller FM (2010) Functional genomic profiling of Aspergillus fumigatus biofilm reveals enhanced production of the mycotoxin gliotoxin. Proteomics 10(17):3097–3107. https://doi.org/10.1002/pmic.201000129

Bugeja HE, Hynes MJ, Andrianopoulos A (2012) AreA controls nitrogen source utilisation during both growth programs of the dimorphic fungus Penicillium marneffei. Fungal Biol 116(1):145–154. https://doi.org/10.1016/j.funbio.2011.10.009

Bünemann EK, Bongiorno G, Bai Z, Creamer RE, De Deyn G, de Goede R, Fleskens L, Geissen V, Kuyper TW, Mäder P, Pulleman M, Sukkel W, van Groenigen JW, Brussaard L (2018) Soil quality – a critical review. Soil Biol Biochem 120:105–125. https://doi.org/10.1016/j.soilbio.2018.01.030

Caballero Ortiz S, Trienens M, Rohlfs M (2013) Induced fungal resistance to insect grazing: reciprocal fitness consequences and fungal gene expression in the Drosophila-Aspergillus model system. PLoS One 8(8):e74951. https://doi.org/10.1371/journal.pone.0074951

Caceres I, Khoury AA, Khoury RE, Lorber S, Oswald IP, Khoury AE, Atoui A, Puel O, Bailly JD (2020) Aflatoxin biosynthesis and genetic regulation: a review. Toxins (Basel) 12(3). https://doi.org/10.3390/toxins12030150

Caesar LK, Kelleher NL, Keller NP (2020) In the fungus where it happens: History and future propelling Aspergillus nidulans as the archetype of natural products research. Fungal Genet Biol 144:103477. https://doi.org/10.1016/j.fgb.2020.103477

Cepeda-García C, Domínguez-Santos R, García-Rico RO, García-Estrada C, Cajiao A, Fierro F, Martín JF (2014) Direct involvement of the CreA transcription factor in penicillin biosynthesis and expression of the pcbAB gene in Penicillium chrysogenum. Appl Microbiol Biotechnol 98(16):7113–7124. https://doi.org/10.1007/s00253-014-5760-1

Chagas FO, Dias LG, Pupo MT (2013) A mixed culture of endophytic fungi increases production of antifungal polyketides. J Chem Ecol 39(10):1335–1342. https://doi.org/10.1007/s10886-013-0351-7

Chen J, Zhang P, Ye X, Wei B, Emam M, Zhang H, Wang H (2020) The structural diversity of marine microbial secondary metabolites based on co-culture strategy: 2009–2019. Mar Drugs 18(9). https://doi.org/10.3390/md18090449

Christensen SA, Kolomiets MV (2011) The lipid language of plant-fungal interactions. Fungal Genet Biol 48(1): 4–14. https://doi.org/10.1016/j.fgb.2010.05.005

Coban O, De Deyn GB, van der Ploeg M (2022) Soil microbiota as game-changers in restoration of degraded lands. Science 375(6584):abe0725. https://doi.org/10.1126/science.abe0725

Combès A, Ndoye I, Bance C, Bruzaud J, Djediat C, Dupont J, Nay B, Prado S (2012) Chemical communication between the endophytic fungus Paraconiothyrium variabile and the phytopathogen Fusarium oxysporum. PLoS One 7(10):e47313. https://doi.org/10.1371/journal.pone.0047313

Conrad T, Kniemeyer O, Henkel SG, Krüger T, Mattern DJ, Valiante V, Guthke R, Jacobsen ID, Brakhage AA, Vlaic S, Linde J (2018) Module-detection approaches for the integration of multilevel omics data highlight the comprehensive response of Aspergillus fumigatus to caspofungin. BMC Syst Biol 12(1):88. https://doi.org/10.1186/s12918-018-0620-8

Couillerot O, Prigent-Combaret C, Caballero-Mellado J, Moënne-Loccoz Y (2009) Pseudomonas fluorescens and closely-related fluorescent pseudomonads as biocontrol agents of soil-borne phytopathogens. Lett Appl Microbiol 48(5):505–512. https://doi.org/10.1111/j.1472-765X.2009.02566.x

Cueto M, Jensen PR, Kauffman C, Fenical W, Lobkovsky E, Clardy J (2001) Pestalone, a new antibiotic produced by a marine fungus in response to bacterial challenge. J Nat Prod 64(11):1444–1446. https://doi.org/10.1021/np0102713

Cui CB, Kakeya H, Okada G, Onose R, Ubukata M, Takahashi I, Isono K, Osada H (1995) Tryprostatins A and B, novel mammalian cell cycle inhibitors produced by Aspergillus fumigatus. J Antibiot (Tokyo) 48(11):1382–1384. https://doi.org/10.7164/antibiotics.48.1382

Currie CR (2001) A community of ants, fungi, and bacteria: a multilateral approach to studying symbiosis. Annu Rev Microbiol 55:357–380. https://doi.org/10.1146/annurev.micro.55.1.357

Dalmais B, Schumacher J, Moraga J, Tudzynski B, Collado IG, Viaud M (2011) The Botrytis cinerea phytotoxin botcinic acid requires two polyketide synthases for production and has a redundant role in virulence with botrydial. Mol Plant Pathol 12(6): 564–579. https://doi.org/10.1111/j.1364-3703.2010.00692.x

Damer B, Deamer D (2020) The hot spring hypothesis for an origin of life. Astrobiology 20(4):429–452. https://doi.org/10.1089/ast.2019.2045

Dang H, Lovell CR (2016) Microbial surface colonization and biofilm development in marine environments. Microbiol Mol Biol Rev 80(1):91–138. https://doi.org/10.1128/MMBR.00037-15

Davière JM, Achard P (2013) Gibberellin signaling in plants. Development 140 (6):1147–1151. doi:https://doi.org/10.1242/dev.087650.

De Silva NI, Brooks S, Lumyong S, Hyde KD (2019) Use of endophytes as biocontrol agents. Fungal Biol Rev 33(2):133–148. https://doi.org/10.1016/j.fbr.2018.10.001

Dhodary B, Schilg M, Wirth R, Spiteller D (2018) Secondary metabolites from Escovopsis weberi and their role in attacking the garden fungus of leaf-cutting ants. Chemistry 24(17):4445–4452. https://doi.org/10.1002/chem.201706071

Döll K, Chatterjee S, Scheu S, Karlovsky P, Rohlfs M (2013) Fungal metabolic plasticity and sexual development mediate induced resistance to arthropod fungivory. Proc Biol Sci 280(1771):20131219. https://doi.org/10.1098/rspb.2013.1219

Duran R, Cary JW, Calvo AM (2010) Role of the osmotic stress regulatory pathway in morphogenesis and secondary metabolism in filamentous fungi. Toxins (Basel) 2(4):367–381. https://doi.org/10.3390/toxins2040367

Edel-Hermann V, Lecomte C (2019) Current status of Fusarium oxysporum formae speciales and races. Phytopathology 109(4):512–530. https://doi.org/10.1094/PHYTO-08-18-0320-RVW

Ehrlich KC, Cary JW, Montalbano BG (1999) Characterization of the promoter for the gene encoding the aflatoxin biosynthetic pathway regulatory protein AFLR. Biochim Biophys Acta 1444(3):412–417. https://doi.org/10.1016/S0167-4781(99)00022-6

Eisendle M, Oberegger H, Buttinger R, Illmer P, Haas H (2004) Biosynthesis and uptake of siderophores is controlled by the PacC-mediated ambient-pH Regulatory system in Aspergillus nidulans. Eukaryot Cell 3(2): 561–563. https://doi.org/10.1128/EC.3.2.561-563.2004

Eley KL, Halo LM, Song Z, Powles H, Cox RJ, Bailey AM, Lazarus CM, Simpson TJ (2007) Biosynthesis of the 2-pyridone tenellin in the insect pathogenic fungus Beauveria bassiana. Chembiochem 8(3):289–297. https://doi.org/10.1002/cbic.200600398

Eshwika A, Kelly J, Fallon JP, Kavanagh K (2013) Exposure of Aspergillus fumigatus to caspofungin results in the release, and de novo biosynthesis, of gliotoxin. Med Mycol 51(2):121–127. https://doi.org/10.3109/13693786.2012.688180

Evidente A, Cimmino A, Masi M (2019) Phytotoxins produced by pathogenic fungi of agrarian plants. Phytochem Rev 18(3):843–870. https://doi.org/10.1007/s11101-019-09624-0

Falsetta ML, Klein MI, Colonne PM, Scott-Anne K, Gregoire S, Pai CH, Gonzalez-Begne M, Watson G, Krysan DJ, Bowen WH, Koo H (2014) Symbiotic relationship between Streptococcus mutans and Candida albicans synergizes virulence of plaque biofilms in vivo. Infect Immun 82(5):1968–1981. https://doi.org/10.1128/IAI.00087-14

Fan Y, Liu X, Keyhani NO, Tang G, Pei Y, Zhang W, Tong S (2017) Regulatory cascade and biological activity of Beauveria bassiana oosporein that limits bacterial growth after host death. Proc Natl Acad Sci U S A 114(9):E1578–E1586. https://doi.org/10.1073/pnas.1616543114

Fasoyin OE, Wang B, Qiu M, Han X, Chung KR, Wang S (2018) Carbon catabolite repression gene creA regulates morphology, aflatoxin biosynthesis and virulence in Aspergillus flavus. Fungal Genet Biol 115:41–51. https://doi.org/10.1016/j.fgb.2018.04.008

Feng B, Friedlin E, Marzluf GA (1994) A reporter gene analysis of penicillin biosynthesis gene expression in Penicillium chrysogenum and its regulation by nitrogen and glucose catabolite repression. Appl Environ Microbiol 60(12):4432–4439. https://doi.org/10.1128/aem.60.12.4432-4439.1994

Feng P, Shang Y, Cen K, Wang C (2015) Fungal biosynthesis of the bibenzoquinone oosporein to evade insect immunity. Proc Natl Acad Sci U S A 112(36): 11365–11370. https://doi.org/10.1073/pnas.1503200112

Ferrigo D, Raiola A, Bogialli S, Bortolini C, Tapparo A, Causin R (2015) In vitro production of Fumonisins by Fusarium verticillioides under oxidative stress induced by H_2O_2. J Agric Food Chem 63(19):4879–4885. https://doi.org/10.1021/acs.jafc.5b00113

Fischer J, Müller SY, Netzker T, Jäger N, Gacek-Matthews A, Scherlach K, Stroe MC, García-Altares M, Pezzini F, Schoeler H, Reichelt M, Gershenzon J, Krespach MKC, Shelest E, Schroeckh V, Valiante V, Heinzel T, Hertweck C, Strauss J, Brakhage AA (2018) Chromatin mapping identifies BasR, a key regulator of bacteria-triggered production of fungal secondary metabolites. Elife 7. https://doi.org/10.7554/eLife.40969

Fisher MC, Henk DA, Briggs CJ, Brownstein JS, Madoff LC, McCraw SL, Gurr SJ (2012) Emerging fungal threats to animal, plant and ecosystem health. Nature 484(7393):186–194. https://doi.org/10.1038/nature10947

Flaherty JE, Woloshuk CP (2004) Regulation of fumonisin biosynthesis in Fusarium verticillioides by a zinc binuclear cluster-type gene, ZFR1. Appl Environ Microbiol 70(5):2653–2659. https://doi.org/10.1128/AEM.70.5.2653-2659.2004

Fobofou SA, Savidge T (2022) Microbial metabolites: cause or consequence in gastrointestinal disease? Am J Physiol Gastrointest Liver Physiol 322(6):G535–G552. https://doi.org/10.1152/ajpgi.00008.2022

Fung F, Clark RF (2004) Health effects of mycotoxins: a toxicological overview. J Toxicol Clin Toxicol 42(2): 217–234. https://doi.org/10.1081/clt-120,030,947

Gibbons JG, Beauvais A, Beau R, McGary KL, Latge JP, Rokas A (2012) Global transcriptome changes underlying colony growth in the opportunistic human pathogen Aspergillus fumigatus. Eukaryot Cell 11(1):68–78. https://doi.org/10.1128/EC.05102-11

Gilbert JA, Blaser MJ, Caporaso JG, Jansson JK, Lynch SV, Knight R (2018) Current understanding of the human microbiome. Nat Med 24(4):392–400. https://doi.org/10.1038/nm.4517

Giuliano Garisto Donzelli B, Gibson DM, Krasnoff SB (2015) Intracellular siderophore but not extracellular siderophore is required for full virulence in Metarhizium robertsii. Fungal Genet Biol 82:56–68. https://doi.org/10.1016/j.fgb.2015.06.008

Graf E, Schmidt-Heydt M, Geisen R (2012) HOG MAP kinase regulation of alternariol biosynthesis in Alternaria alternata is important for substrate colonization. Int J Food Microbiol 157(3):353–359. https://doi.org/10.1016/j.ijfoodmicro.2012.06.004

Grahl N, Shepardson KM, Chung D, Cramer RA (2012) Hypoxia and fungal pathogenesis: to air or not to air? Eukaryot Cell 11(5):560–570. https://doi.org/10.1128/EC.00031-12

Grimaldi DA (2010) 400 million years on six legs: on the origin and early evolution of Hexapoda. Arthropod Struct Dev 39(2–3):191–203. https://doi.org/10.1016/j.asd.2009.10.008

Grintzalis K, Vernardis SI, Klapa MI, Georgiou CD (2014) Role of oxidative stress in Sclerotial differentiation and aflatoxin B1 biosynthesis in Aspergillus flavus. Appl Environ Microbiol 80(18):5561–5571. https://doi.org/10.1128/AEM.01282-14

Guruceaga X, Perez-Cuesta U, Abad-Diaz de Cerio A, Gonzalez O, Alonso RM, Hernando FL, Ramirez-Garcia A, Rementeria A (2019) Fumagillin, a mycotoxin of Aspergillus fumigatus: biosynthesis, biological activities, detection, and applications. Toxins (Basel) 12(1). https://doi.org/10.3390/toxins12010007

Haas H (2014) Fungal siderophore metabolism with a focus on Aspergillus fumigatus. Nat Prod Rep 31(10):1266–1276. https://doi.org/10.1039/c4np00071d

Haas H, Marzluf GA (1995) NRE, the major nitrogen regulatory protein of Penicillium chrysogenum, binds specifically to elements in the intergenic promoter regions of nitrate assimilation and penicillin biosynthetic gene clusters. Curr Genet 28(2):177–183. https://doi.org/10.1007/BF00315785

Hagag A, Abdelwahab MF, Abd El-Kader AM, Fouad MA (2022) The endophytic Aspergillus strains: a bountiful source of natural products. J Appl Microbiol 132(6):4150–4169. https://doi.org/10.1111/jam.15489

Haiko J, Saeedi B, Bagger G, Karpati F, Özenci V (2019) Coexistence of Candida species and bacteria in patients with cystic fibrosis. Eur J Clin Microbiol Infect Dis

38(6):1071–1077. https://doi.org/10.1007/s10096-019-03493-3

Hasan S, Ansari MI, Ahmad A, Mishra M (2015) Major bioactive metabolites from marine fungi: a review. Bioinformation 11(4):176–181. https://doi.org/10.6026/97320630011176

Heine D, Holmes NA, Worsley SF, Santos ACA, Innocent TM, Scherlach K, Patrick EH, Yu DW, Murrell JC, Vieria PC, Boomsma JJ, Hertweck C, Hutchings MI, Wilkinson B (2018) Chemical warfare between leafcutter ant symbionts and a co-evolved pathogen. Nat Commun 9(1):2208. https://doi.org/10.1038/s41467-018-04520-1

Heinekamp T, Schmidt H, Lapp K, Pähtz V, Shopova I, Köster-Eiserfunke N, Krüger T, Kniemeyer O, Brakhage AA (2015) Interference of Aspergillus fumigatus with the immune response. Semin Immunopathol 37(2):141–152. https://doi.org/10.1007/s00281-014-0465-1

Hissen AH, Wan AN, Warwas ML, Pinto LJ, Moore MM (2005) The Aspergillus fumigatus siderophore biosynthetic gene sidA, encoding L-ornithine N5-oxygenase, is required for virulence. Infect Immun 73(9): 5493–5503. https://doi.org/10.1128/IAI.73.9.5493-5503.2005

Höfte M, Altier N (2010) Fluorescent pseudomonads as biocontrol agents for sustainable agricultural systems. Res Microbiol 161(6):464–471. https://doi.org/10.1016/j.resmic.2010.04.007

Hong SY, Roze LV, Wee J, Linz JE (2013) Evidence that a transcription factor regulatory network coordinates oxidative stress response and secondary metabolism in aspergilli. Microbiologyopen 2(1):144–160. https://doi.org/10.1002/mbo3.63

Hope WW, Walsh TJ, Denning DW (2005) Laboratory diagnosis of invasive aspergillosis. Lancet Infect Dis 5(10):609–622. https://doi.org/10.1016/S1473-3099(05)70238-3

Hridoy M, Gorapi MZH, Noor S, Chowdhury NS, Rahman MM, Muscari I, Masia F, Adorisio S, Delfino DV, Mazid MA (2022) Putative anticancer compounds from plant-derived endophytic fungi: a review. Molecules 27(1). https://doi.org/10.3390/molecules27010296

Hsueh YP, Mahanti P, Schroeder FC, Sternberg PW (2013) Nematode-trapping fungi eavesdrop on nematode pheromones. Curr Biol 23(1):83–86. https://doi.org/10.1016/j.cub.2012.11.035

Hsueh YP, Gronquist MR, Schwarz EM, Nath RD, Lee CH, Gharib S, Schroeder FC, Sternberg PW (2017) Nematophagous fungus Arthrobotrys oligospora mimics olfactory cues of sex and food to lure its nematode prey. Elife 6. https://doi.org/10.7554/eLife.20023

Hu Y, Lian L, Xia J, Hu S, Xu W, Zhu J, Ren A, Shi L, Zhao MW (2020) Influence of PacC on the environmental stress adaptability and cell wall components of Ganoderma lucidum. Microbiol Res 230:126348. https://doi.org/10.1016/j.micres.2019.126348

Humpf HU, Voss KA (2004) Effects of thermal food processing on the chemical structure and toxicity of fumonisin mycotoxins. Mol Nutr Food Res 48(4): 255–269. https://doi.org/10.1002/mnfr.200400033

Irankhah S, Chitarra W, Nerva L, Antoniou C, Lumini E, Volpe V, Ganjeali A, Cheniany M, Mashreghi M, Fotopoulos V, Balestrini R (2020) Impact of an arbuscular mycorrhizal fungal inoculum and exogenous MeJA on fenugreek secondary metabolite production under water deficit. Environ Exp Bot 176: 104096. https://doi.org/10.1016/j.envexpbot.2020.104096

Irmer H, Tarazona S, Sasse C, Olbermann P, Loeffler J, Krappmann S, Conesa A, Braus GH (2015) RNAseq analysis of Aspergillus fumigatus in blood reveals a just wait and see resting stage behavior. BMC Genomics 16:640. https://doi.org/10.1186/s12864-015-1853-1

Ishikawa M, Ninomiya T, Akabane H, Kushida N, Tsujiuchi G, Ohyama M, Gomi S, Shito K, Murata T (2009) Pseurotin A and its analogues as inhibitors of immunoglobulin E [correction of immunoglobuline E] production. Bioorg Med Chem Lett 19(5):1457–1460. https://doi.org/10.1016/j.bmcl.2009.01.029

Islam MR, Jeong YT, Lee YS, Song CH (2012) Isolation and identification of antifungal compounds from Bacillus subtilis C9 inhibiting the growth of plant pathogenic fungi. Mycobiology 40(1):59–66. https://doi.org/10.5941/MYCO.2012.40.1.059

Jakubczyk D, Dussart F (2020) Selected fungal natural products with antimicrobial properties. Molecules 25(4). https://doi.org/10.3390/molecules25040911

Kardos N, Demain AL (2011) Penicillin: the medicine with the greatest impact on therapeutic outcomes. Appl Microbiol Biotechnol 92(4):677–687. https://doi.org/10.1007/s00253-011-3587-6

Kataoka S, Furutani S, Hirata K, Hayashi H, Matsuda K (2011) Three austin family compounds from Penicillium brasilianum exhibit selective blocking action on cockroach nicotinic acetylcholine receptors. Neurotoxicology 32(1):123–129. https://doi.org/10.1016/j.neuro.2010.10.003

Keller NP (2019) Fungal secondary metabolism: regulation, function and drug discovery. Nat Rev Microbiol 17(3):167–180. https://doi.org/10.1038/s41579-018-0121-1

Kenne GJ, Gummadidala PM, Omebeyinje MH, Mondal AM, Bett DK, McFadden S, Bromfield S, Banaszek N, Velez-Martinez M, Mitra C, Mikell I, Chatterjee S, Wee J, Chanda A (2018) Activation of aflatoxin biosynthesis alleviates total ROS in Aspergillus parasiticus. Toxins (Basel) 10(2). https://doi.org/10.3390/toxins10020057

Kerr JR, Taylor GW, Rutman A, Hoiby N, Cole PJ, Wilson R (1999) Pseudomonas aeruginosa pyocyanin and 1-hydroxyphenazine inhibit fungal growth. J Clin Pathol 52(5):385–387. https://doi.org/10.1136/jcp.52.5.385

Khalid S, Keller NP (2021) Chemical signals driving bacterial-fungal interactions. Environ Microbiol 23(3):1334–1347. https://doi.org/10.1111/1462-2920.15410

Khalid S, Baccile JA, Spraker JE, Tannous J, Imran M, Schroeder FC, Keller NP (2018) NRPS-derived isoquinolines and lipopetides mediate antagonism between plant pathogenic fungi and bacteria. ACS Chem Biol 13(1):171–179. https://doi.org/10.1021/acschembio.7b00731

Khare E, Mishra J, Arora NK (2018) Multifaceted interactions between endophytes and plant: developments and prospects. Front Microbiol 9:2732. https://doi.org/10.3389/fmicb.2018.02732

Kim D, Sengupta A, Niepa TH, Lee BH, Weljie A, Freitas-Blanco VS, Murata RM, Stebe KJ, Lee D, Koo H (2017) Candida albicans stimulates Streptococcus mutans microcolony development via cross-kingdom biofilm-derived metabolites. Sci Rep 7:41332. https://doi.org/10.1038/srep41332

Knowles SL, Raja HA, Roberts CD, Oberlies NH (2022) Fungal–fungal co-culture: a primer for generating chemical diversity. Nat Prod Rep. https://doi.org/10.1039/D1NP00070E

Kobayashi A, Hagihara K, Kajiyama S, Kanzaki H, Kawazu K (1995) Antifungal compounds induced in the dual culture with Phytolacca americana callus and Botrytis fabae. Z Naturforsch C J Biosci 50(5–6):398–402. https://doi.org/10.1515/znc-1995-5-610

Komagata D, Fujita S, Yamashita N, Saito S, Morino T (1996) Novel neuritogenic activities of pseurotin A and penicillic acid. J Antibiot (Tokyo) 49(9):958–959. https://doi.org/10.7164/antibiotics.49.958

König CC, Scherlach K, Schroeckh V, Horn F, Nietzsche S, Brakhage AA, Hertweck C (2013) Bacterium induces cryptic meroterpenoid pathway in the pathogenic fungus Aspergillus fumigatus. Chembiochem 14(8):938–942. https://doi.org/10.1002/cbic.201300070

Krespach MKC, Stroe MC, Flak M, Komor AJ, Nietzsche S, Sasso S, Hertweck C, Brakhage AA (2021) Bacterial marginolactones trigger formation of algal gloeocapsoids, protective aggregates on the verge of multicellularity. Proc Natl Acad Sci U S A 118(45). https://doi.org/10.1073/pnas.2100892118

Krespach MKC, Stroe MC, Netzker T, Rosin M, Zehner LM, Komor AJ, Beilmann JM, Krüger T, Scherlach K, Kniemeyer O, Schroeckh V, Hertweck C, Brakhage AA (2023) Streptomyces polyketides mediate bacteria-fungi interactions across soil environments. Nat Microbiol. Accepted. https://doi.org/10.1038/s41564-023-01382-2

Krüger W, Vielreicher S, Kapitan M, Jacobsen ID, Niemiec MJ (2019) Fungal-bacterial interactions in health and disease. Pathogens 8(2). https://doi.org/10.3390/pathogens8020070

Langfelder K, Streibel M, Jahn B, Haase G, Brakhage AA (2003) Biosynthesis of fungal melanins and their importance for human pathogenic fungi. Fungal Genet Biol 38(2):143–158. https://doi.org/10.1016/S1087-1845(02)00526-1

Li C, Yang R, Lin Y, Zhou S (2006) Isolation and crystal structure of (−)-byssochlamic acid from mangrove fungus (strain no. k38). Chem Nat Compd 42(3):290–293. https://doi.org/10.1007/s10600-006-0101-y

Li CY, Yang RY, Lin YC, She ZG, Zhou SN (2007) A new nonadride derivative from mangrove fungus (strain no. k38). J Asian Nat Prod Res 9(3–5):285–291. https://doi.org/10.1080/10286020600727418

Li C, Ding W, She Z, Lin Y (2008) A new biphenyl derivative from an unidentified marine fungus E33. Chem Nat Compd 44(2):163–165. https://doi.org/10.1007/s10600-008-9003-5

Li B, Lai T, Qin G, Tian S (2010a) Ambient pH stress inhibits spore germination of Penicillium expansum by impairing protein synthesis and folding: a proteomic-based study. J Proteome Res 9(1):298–307. https://doi.org/10.1021/pr900622j

Li CY, Ding WJ, Shao CL, She ZG, Lin YC (2010b) A new diimide derivative from the co-culture broth of two mangrove fungi (strain no. E33 and K38). J Asian Nat Prod Res 12(9):809–813. https://doi.org/10.1080/10286020.2010.497757

Li C, Zhang J, Shao C, Ding W, She Z, Lin Y (2011) A new xanthone derivative from the co-culture broth of two marine fungi (strain no. E33 and K38). Chem Nat Compd 47(3):382. https://doi.org/10.1007/s10600-011-9939-8

Li H, Liu L, Zhang S, Cui W, Lv J (2012) Identification of antifungal compounds produced by Lactobacillus casei AST18. Curr Microbiol 65(2):156–161. https://doi.org/10.1007/s00284-012-0135-2

Li C, Wang J, Luo C, Ding W, Cox DG (2014) A new cyclopeptide with antifungal activity from the co-culture broth of two marine mangrove fungi. Nat Prod Res 28(9):616–621. https://doi.org/10.1080/14786419.2014.887074

Li Y, Huang Z, Petropoulos E, Ma Y, Shen Y (2020) Humidity governs the wall-inhabiting fungal community composition in a 1600-year tomb of Emperor Yang. Sci Rep 10(1):8421. https://doi.org/10.1038/s41598-020-65,478-z

Li B, Chen Y, Tian S (2021a) Function of pH-dependent transcription factor PacC in regulating development, pathogenicity, and mycotoxin biosynthesis of phytopathogenic fungi. FEBS J. https://doi.org/10.1111/febs.15808

Li T, Su X, Qu H, Duan X, Jiang Y (2021b) Biosynthesis, regulation, and biological significance of fumonisins in fungi: current status and prospects. Crit Rev Microbiol 48:450–462. https://doi.org/10.1080/1040841X.2021.1979465

Lim S, Bijlani S, Blachowicz A, Chiang YM, Lee MS, Torok T, Venkateswaran K, Wang CCC (2021) Identification of the pigment and its role in UV resistance in Paecilomyces variotii, a Chernobyl isolate, using genetic manipulation strategies. Fungal Genet Biol

152:103567. https://doi.org/10.1016/j.fgb.2021.103567
Lind AL, Smith TD, Saterlee T, Calvo AM, Rokas A (2016) Regulation of secondary metabolism by the velvet complex is temperature-responsive in Aspergillus. G3 (Bethesda) 6(12):4023–4033. https://doi.org/10.1534/g3.116.033084
Liu BL, Tzeng YM (2012) Development and applications of destruxins: a review. Biotechnol Adv 30(6): 1242–1254. https://doi.org/10.1016/j.biotechadv.2011.10.006
Liu Z, Frank M, Yu X, Yu H, Tran-Cong NM, Gao Y, Proksch P (2020) Secondary metabolites from marine-derived fungi from China. Prog Chem Org Nat Prod 111:81–153. https://doi.org/10.1007/978-3-030-37,865-3_2
Lobo LS, Luz C, Fernandes EK, Juarez MP, Pedrini N (2015) Assessing gene expression during pathogenesis: use of qRT-PCR to follow toxin production in the entomopathogenic fungus Beauveria bassiana during infection and immune response of the insect host Triatoma infestans. J Invertebr Pathol 128:14–21. https://doi.org/10.1016/j.jip.2015.04.004
Lokhandwala J, Hopkins HC, Rodriguez-Iglesias A, Dattenböck C, Schmoll M, Zoltowski BD (2015) Structural biochemistry of a fungal LOV domain photoreceptor reveals an evolutionarily conserved pathway integrating light and oxidative stress. Structure 23(1): 116–125. https://doi.org/10.1016/j.str.2014.10.020
Loreto F, Dicke M, Schnitzler JP, Turlings TC (2014) Plant volatiles and the environment. Plant Cell Environ 37(8):1905–1908. https://doi.org/10.1111/pce.12369
Luo ZX, Yuan CX, Meng QJ, Ji Q (2011) A Jurassic eutherian mammal and divergence of marsupials and placentals. Nature 476(7361):442–445. https://doi.org/10.1038/nature10291
Lutz MP, Wenger S, Maurhofer M, Defago G, Duffy B (2004) Signaling between bacterial and fungal biocontrol agents in a strain mixture. FEMS Microbiol Ecol 48(3):447–455. https://doi.org/10.1016/j.femsec.2004.03.002
Lutzoni F, Nowak MD, Alfaro ME, Reeb V, Miadlikowska J, Krug M, Arnold AE, Lewis LA, Swofford DL, Hibbett D, Hilu K, James TY, Quandt D, Magallón S (2018) Contemporaneous radiations of fungi and plants linked to symbiosis. Nat Commun 9(1):5451. https://doi.org/10.1038/s41467-018-07849-9
Macauley BJ, Thrower LB (1966) Succession of fungi in leaf litter of Eucalyptus regnans. Trans Br Mycol Soc 49(3):509–520. https://doi.org/10.1016/S0007-1536(66)80097-9
Macheleidt J, Mattern DJ, Fischer J, Netzker T, Weber J, Schroeckh V, Valiante V, Brakhage AA (2016) Regulation and role of fungal secondary metabolites. Annu Rev Genet 50:371–392. https://doi.org/10.1146/annurev-genet-120,215-035203
Mandelare PE, Adpressa DA, Kaweesa EN, Zakharov LN, Loesgen S (2018) Coculture of two developmental stages of a marine-derived Aspergillus alliaceus results in the production of the cytotoxic Bianthrone Allianthrone A. J Nat Prod 81(4):1014–1022. https://doi.org/10.1021/acs.jnatprod.8b00024
Marfori EC, Kajiyama S, Fukusaki E, Kobayashi A (2002) Trichosetin, a novel tetramic acid antibiotic produced in dual culture of Trichoderma harzianum and Catharanthus roseus Callus. Z Naturforsch C J Biosci 57(5–6):465–470. https://doi.org/10.1515/znc-2002-5-611
Martin JF (2000) Molecular control of expression of penicillin biosynthesis genes in fungi: regulatory proteins interact with a bidirectional promoter region. J Bacteriol 182(9):2355–2362. https://doi.org/10.1128/JB.182.9.2355-2362.2000
Mattern DJ, Valiante V, Unkles SE, Brakhage AA (2015) Synthetic biology of fungal natural products. Front Microbiol 6:775. https://doi.org/10.3389/fmicb.2015.00775
McCormack PL, Perry CM (2005) Caspofungin: a review of its use in the treatment of fungal infections. Drugs 65(14):2049–2068. https://doi.org/10.2165/00003495-200,565,140-00009
McDonagh A, Fedorova ND, Crabtree J, Yu Y, Kim S, Chen D, Loss O, Cairns T, Goldman G, Armstrong-James D, Haynes K, Haas H, Schrettl M, May G, Nierman WC, Bignell E (2008) Sub-telomere directed gene expression during initiation of invasive aspergillosis. PLoS Pathog 4(9):e1000154. https://doi.org/10.1371/journal.ppat.1000154
Medina A, Schmidt-Heydt M, Rodríguez A, Parra R, Geisen R, Magan N (2015) Impacts of environmental stress on growth, secondary metabolite biosynthetic gene clusters and metabolite production of xerotolerant/xerophilic fungi. Curr Genet 61(3): 325–334. https://doi.org/10.1007/s00294-014-0455-9
Mendelson M, Sharland M, Mpundu M (2022) Antibiotic resistance: calling time on the 'silent pandemic'. JAC Antimicrob Resist 4(2):dlac016. https://doi.org/10.1093/jacamr/dlac016
Milshteyn A, Colosimo DA, Brady SF (2018) Accessing bioactive natural products from the human microbiome. Cell Host Microbe 23(6):725–736. https://doi.org/10.1016/j.chom.2018.05.013
Minerdi D, Maggini V, Fani R (2021) Volatile organic compounds: from figurants to leading actors in fungal symbiosis. FEMS Microbiol Ecol 97(5). https://doi.org/10.1093/femsec/fiab067
Misslinger M, Hortschansky P, Brakhage AA, Haas H (2021) Fungal iron homeostasis with a focus on Aspergillus fumigatus. Biochim Biophys Acta Mol Cell Res 1868(1):118885. https://doi.org/10.1016/j.bbamcr.2020.118885
Mona SA, Hashem A, Abd-Allah EF, Alqarawi AA, DWK S, Wirth S, Egamberdieva D (2017) Increased resistance of drought by Trichoderma harzianum fungal treatment correlates with increased secondary metabolites and proline content. J Integr Agric 16(8):

1751–1757. https://doi.org/10.1016/S2095-3119(17)61695-2

Monroy AA, Stappler E, Schuster A, Sulyok M, Schmoll M (2017) A CRE1-regulated cluster is responsible for light dependent production of dihydrotrichotetronin in Trichoderma reesei. PLoS One 12(8):e0182530. https://doi.org/10.1371/journal.pone.0182530

Morales DK, Jacobs NJ, Rajamani S, Krishnamurthy M, Cubillos-Ruiz JR, Hogan DA (2010) Antifungal mechanisms by which a novel Pseudomonas aeruginosa phenazine toxin kills Candida albicans in biofilms. Mol Microbiol 78(6):1379–1392. https://doi.org/10.1111/j.1365-2958.2010.07414.x

Moree WJ, Phelan VV, Wu CH, Bandeira N, Cornett DS, Duggan BM, Dorrestein PC (2012) Interkingdom metabolic transformations captured by microbial imaging mass spectrometry. Proc Natl Acad Sci U S A 109(34): 13811–13816. https://doi.org/10.1073/pnas.1206855109

Munkvold GP, Hellmich RL, Rice LG (1999) Comparison of Fumonisin concentrations in kernels of transgenic Bt maize hybrids and nontransgenic hybrids. Plant Dis 83(2):130–138. https://doi.org/10.1094/PDIS.1999.83.2.130

Narsing Rao MP, Xiao M, Li WJ (2017) Fungal and bacterial pigments: secondary metabolites with wide applications. Front Microbiol 8:1113. https://doi.org/10.3389/fmicb.2017.01113

Neilands JB (1952) A crystalline organo-iron pigment from a rust fungus (Ustilago sphaerogena)1. J Am Chem Soc 74(19):4846–4847. https://doi.org/10.1021/ja01139a033

Netzker T, Fischer J, Weber J, Mattern DJ, König CC, Valiante V, Schroeckh V, Brakhage AA (2015) Microbial communication leading to the activation of silent fungal secondary metabolite gene clusters. Front Microbiol 6:299. https://doi.org/10.3389/fmicb.2015.00299

Netzker T, Flak M, Krespach MK, Stroe MC, Weber J, Schroeckh V, Brakhage AA (2018) Microbial interactions trigger the production of antibiotics. Curr Opin Microbiol 45:117–123. https://doi.org/10.1016/j.mib.2018.04.002

Netzker T, Shepherdson EMF, Zambri MP, Elliot MA (2020) Bacterial volatile compounds: functions in communication, cooperation, and competition. Annu Rev Microbiol 74:409–430. https://doi.org/10.1146/annurev-micro-011320-015542

Nielsen ML, Nielsen JB, Rank C, Klejnstrup ML, Holm DK, Brogaard KH, Hansen BG, Frisvad JC, Larsen TO, Mortensen UH (2011) A genome-wide polyketide synthase deletion library uncovers novel genetic links to polyketides and meroterpenoids in Aspergillus nidulans. FEMS Microbiol Lett 321(2):157–166. https://doi.org/10.1111/j.1574-6968.2011.02327.x

Nielsen MT, Klejnstrup ML, Rohlfs M, Anyaogu DC, Nielsen JB, Gotfredsen CH, Andersen MR, Hansen BG, Mortensen UH, Larsen TO (2013) Aspergillus nidulans synthesize insect juvenile hormones upon expression of a heterologous regulatory protein and in response to grazing by Drosophila melanogaster larvae. PLoS One 8(8):e73369. https://doi.org/10.1371/journal.pone.0073369

Ninkovic V, Markovic D, Rensing M (2021) Plant volatiles as cues and signals in plant communication. Plant Cell Environ 44(4):1030–1043. https://doi.org/10.1111/pce.13910

Ninomiya A, Urayama SI, Hagiwara D (2022) Antibacterial diphenyl ether production induced by co-culture of Aspergillus nidulans and Aspergillus fumigatus. Appl Microbiol Biotechnol 106(11): 4169–4185. https://doi.org/10.1007/s00253-022-11,964-5

Niu X-M, Zhang K-Q (2011) Arthrobotrys oligospora: a model organism for understanding the interaction between fungi and nematodes. Mycology 2(2):59–78. https://doi.org/10.1080/21501203.2011.562559

Nützmann HW, Reyes-Dominguez Y, Scherlach K, Schroeckh V, Horn F, Gacek A, Schumann J, Hertweck C, Strauss J, Brakhage AA (2011) Bacteria-induced natural product formation in the fungus Aspergillus nidulans requires Saga/Ada-mediated histone acetylation. Proc Natl Acad Sci U S A 108(34): 14282–14287. https://doi.org/10.1073/pnas.1103523108

Nützmann HW, Fischer J, Scherlach K, Hertweck C, Brakhage AA (2013) Distinct amino acids of histone H3 control secondary metabolism in Aspergillus nidulans. Appl Environ Microbiol 79(19):6102–6109. https://doi.org/10.1128/AEM.01578-13

Oberegger H, Schoeser M, Zadra I, Abt B, Haas H (2001) SREA is involved in regulation of siderophore biosynthesis, utilization and uptake in Aspergillus nidulans. Mol Microbiol 41(5):1077–1089. https://doi.org/10.1046/j.1365-2958.2001.02586.x

Oh DC, Jensen PR, Kauffman CA, Fenical W (2005) Libertellenones A-D: induction of cytotoxic diterpenoid biosynthesis by marine microbial competition. Bioorg Med Chem 13(17):5267–5273. https://doi.org/10.1016/j.bmc.2005.05.068

Oh DC, Kauffman CA, Jensen PR, Fenical W (2007) Induced production of emericellamides A and B from the marine-derived fungus Emericella sp. in competing co-culture. J Nat Prod 70(4):515–520. https://doi.org/10.1021/np060381f

Ola AR, Thomy D, Lai D, Brotz-Oesterhelt H, Proksch P (2013) Inducing secondary metabolite production by the endophytic fungus Fusarium tricinctum through coculture with Bacillus subtilis. J Nat Prod 76(11): 2094–2099. https://doi.org/10.1021/np400589h

Oppong-Danquah E, Budnicka P, Blümel M, Tasdemir D (2020) Design of fungal co-cultivation based on comparative metabolomics and bioactivity for discovery of marine fungal agrochemicals. Mar Drugs 18(2). https://doi.org/10.3390/md18020073

Osinski GR, Cockell CS, Pontefract A, Sapers HM (2020) The role of meteorite impacts in the origin of life.

Astrobiology 20(9):1121–1149. https://doi.org/10.1089/ast.2019.2203

Overy DP, Frisvad JC, Steinmeier U, Thrane U (2005) Clarification of the agents causing blue mold storage rot upon various flower and vegetable bulbs: implications for mycotoxin contamination. Postharvest Biol Technol 35(2):217–221. https://doi.org/10.1016/j.postharvbio.2004.08.001

Overy DP, Smedsgaard J, Frisvad JC, Phipps RK, Thrane U (2006) Host-derived media used as a predictor for low abundant, in planta metabolite production from necrotrophic fungi. J Appl Microbiol 101(6):1292–1300. https://doi.org/10.1111/j.1365-2672.2006.03037.x

Overy D, Correa H, Roullier C, Chi WC, Pang KL, Rateb M, Ebel R, Shang Z, Capon R, Bills G, Kerr R (2017) Does osmotic stress affect natural product expression in fungi? Mar Drugs 15(8). https://doi.org/10.3390/md15080254

Ozcengiz G, Demain AL (2013) Recent advances in the biosynthesis of penicillins, cephalosporins and clavams and its regulation. Biotechnol Adv 31(2):287–311. https://doi.org/10.1016/j.biotechadv.2012.12.001

Park SH, Kim D, Kim J, Moon Y (2015) Effects of Mycotoxins on mucosal microbial infection and related pathogenesis. Toxins (Basel) 7(11):4484–4502. https://doi.org/10.3390/toxins7114484

Pecoraro L, Wang X, Shah D, Song X, Kumar V, Shakoor A, Tripathi K, Ramteke PW, Rani R (2021) Biosynthesis pathways, transport mechanisms and biotechnological applications of fungal siderophores. J Fungi (Basel) 8(1). https://doi.org/10.3390/jof8010021

Peleg AY, Hogan DA, Mylonakis E (2010) Medically important bacterial-fungal interactions. Nat Rev Microbiol 8(5):340–349. https://doi.org/10.1038/nrmicro2313

Pérez JC (2021) The interplay between gut bacteria and the yeast Candida albicans. Gut Microbes 13(1):1979877. https://doi.org/10.1080/19490976.2021.1979877

Picot A, Barreau C, Pinson-Gadais L, Piraux F, Caron D, Lannou C, Richard-Forget F (2011) The dent stage of maize kernels is the most conducive for fumonisin biosynthesis under field conditions. Appl Environ Microbiol 77(23):8382–8390. https://doi.org/10.1128/AEM.05216-11

Pruss S, Fetzner R, Seither K, Herr A, Pfeiffer E, Metzler M, Lawrence CB, Fischer R (2014) Role of the Alternaria alternata blue-light receptor LreA (white-collar 1) in spore formation and secondary metabolism. Appl Environ Microbiol 80(8):2582–2591. https://doi.org/10.1128/AEM.00327-14

Pultar F, Hansen ME, Wolfrum S, Böselt L, Fróis-Martins R, Bloch S, Kravina AG, Pehlivanoglu D, Schäffer C, LeibundGut-Landmann S, Riniker S, Carreira EM (2021) Mutanobactin D from the human microbiome: total synthesis, configurational assignment, and biological evaluation. J Am Chem Soc 143(27):10389–10402. https://doi.org/10.1021/jacs.1c04825

Raffa N, Keller NP (2019) A call to arms: Mustering secondary metabolites for success and survival of an opportunistic pathogen. PLoS Pathog 15(4):e1007606. https://doi.org/10.1371/journal.ppat.1007606

Rangel LI, Hamilton O, de Jonge R, Bolton MD (2021) Fungal social influencers: secondary metabolites as a platform for shaping the plant-associated community. Plant J 108(3):632–645. https://doi.org/10.1111/tpj.15490

Rascle C, Dieryckx C, Dupuy JW, Muszkieta L, Souibgui E, Droux M, Bruel C, Girard V, Poussereau N (2018) The pH regulator PacC: a host-dependent virulence factor in Botrytis cinerea. Environ Microbiol Rep 10(5):555–568. https://doi.org/10.1111/1758-2229.12663

Reece E, Bettio PHA, Renwick J (2021) Polymicrobial interactions in the cystic fibrosis airway microbiome impact the antimicrobial susceptibility of Pseudomonas aeruginosa. Antibiotics (Basel) 10(7). https://doi.org/10.3390/antibiotics10070827

Reeves EP, Murphy T, Daly P, Kavanagh K (2004) Amphotericin B enhances the synthesis and release of the immunosuppressive agent gliotoxin from the pulmonary pathogen Aspergillus fumigatus. J Med Microbiol 53(Pt 8):719–725. https://doi.org/10.1099/jmm.0.45626-0

Richard JL (2007) Some major mycotoxins and their mycotoxicoses: an overview. Int J Food Microbiol 119(1–2):3–10. https://doi.org/10.1016/j.ijfoodmicro.2007.07.019

Riley RT, Voss KA, Yool HS, Gelderblom WCA, Merrill AH Jr (1994) Mechanism of fumonisin toxicity and carcinogenesis. J Food Prot 57(6):528–535. https://doi.org/10.4315/0362-028X-57.6.528

Rohlfs M, Churchill AC (2011) Fungal secondary metabolites as modulators of interactions with insects and other arthropods. Fungal Genet Biol 48(1):23–34. https://doi.org/10.1016/j.fgb.2010.08.008

Russell M (2006) First life: billions of years ago, deep under the ocean, the pores and pockets in minerals that surrounded warm, alkaline springs catalyzed the beginning of life. Am Sci 94(1):32–39

Rybakova D, Rack-Wetzlinger U, Cernava T, Schaefer A, Schmuck M, Berg G (2017) Aerial warfare: a volatile dialogue between the plant pathogen Verticillium longisporum and Its Antagonist Paenibacillus polymyxa. Front Plant Sci 8:1294. https://doi.org/10.3389/fpls.2017.01294

Sanchez JF, Somoza AD, Keller NP, Wang CC (2012) Advances in Aspergillus secondary metabolite research in the post-genomic era. Nat Prod Rep 29(3):351–371. https://doi.org/10.1039/c2np00084a

Sandmann G (2022) Carotenoids and their biosynthesis in fungi. Molecules 27(4). https://doi.org/10.3390/molecules27041431

Sbaraini N, Guedes RL, Andreis FC, Junges A, de Morais GL, Vainstein MH, de Vasconcelos AT, Schrank A

(2016) Secondary metabolite gene clusters in the entomopathogen fungus Metarhizium anisopliae: genome identification and patterns of expression in a cuticle infection model. BMC Genomics 17(Suppl 8): 736. https://doi.org/10.1186/s12864-016-3067-6

Scharf DH, Heinekamp T, Brakhage AA (2014) Human and plant fungal pathogens: the role of secondary metabolites. PLoS Pathog 10(1):e1003859. https://doi.org/10.1371/journal.ppat.1003859

Scharf DH, Brakhage AA, Mukherjee PK (2016) Gliotoxin--bane or boon? Environ Microbiol 18(4): 1096–1109. https://doi.org/10.1111/1462-2920.13080

Schmeda-Hirschmann G, Hormazabal E, Rodriguez JA, Theoduloz C (2008) Cycloaspeptide A and pseurotin A from the endophytic fungus Penicillium janczewskii. Z Naturforsch C J Biosci 63(5–6):383–388. https://doi.org/10.1515/znc-2008-5-612

Schmidt H, Vlaic S, Krüger T, Schmidt F, Balkenhol J, Dandekar T, Guthke R, Kniemeyer O, Heinekamp T, Brakhage AA (2018) Proteomics of Aspergillus fumigatus conidia-containing phagolysosomes identifies processes governing immune evasion. Mol Cell Proteomics 17(6):1084–1096. https://doi.org/10.1074/mcp.RA117.000069

Schmidt-Heydt M, Graf E, Stoll D, Geisen R (2012) The biosynthesis of ochratoxin A by Penicillium as one mechanism for adaptation to NaCl rich foods. Food Microbiol 29(2):233–241. https://doi.org/10.1016/j.fm.2011.08.003

Schrettl M, Bignell E, Kragl C, Joechl C, Rogers T, Arst HN Jr, Haynes K, Haas H (2004) Siderophore biosynthesis but not reductive iron assimilation is essential for Aspergillus fumigatus virulence. J Exp Med 200(9): 1213–1219. https://doi.org/10.1084/jem.20041242

Schroeckh V, Scherlach K, Nützmann HW, Shelest E, Schmidt-Heck W, Schuemann J, Martin K, Hertweck C, Brakhage AA (2009) Intimate bacterial-fungal interaction triggers biosynthesis of archetypal polyketides in Aspergillus nidulans. Proc Natl Acad Sci U S A 106(34):14558–14563. https://doi.org/10.1073/pnas.0901870106

Schulz-Bohm K, Martin-Sanchez L, Garbeva P (2017) Microbial volatiles: small molecules with an important role in intra- and inter-kingdom interactions. Front Microbiol 8:2484. https://doi.org/10.3389/fmicb.2017.02484

Scott J, Sueiro-Olivares M, Ahmed W, Heddergott C, Zhao C, Thomas R, Bromley M, Latgé JP, Krappmann S, Fowler S, Bignell E, Amich J (2019) Pseudomonas aeruginosa-derived volatile sulfur compounds promote distal Aspergillus fumigatus growth and a synergistic pathogen-pathogen interaction that increases pathogenicity in co-infection. Front Microbiol 10:2311. https://doi.org/10.3389/fmicb.2019.02311

Selvig K, Alspaugh JA (2011) pH response pathways in fungi: adapting to host-derived and environmental signals. Mycobiology 39(4):249–256. https://doi.org/10.5941/MYCO.2011.39.4.249

Shang Z, Salim AA, Capon RJ (2017) Chaunopyran A: co-cultivation of marine mollusk-derived fungi activates a rare class of 2-alkenyl-tetrahydropyran. J Nat Prod 80(4):1167–1172. https://doi.org/10.1021/acs.jnatprod.7b00144

Shen L, Porée FH, Gaslonde T, Lalucque H, Chapeland-Leclerc F, Ruprich-Robert G (2019) Functional characterization of the sterigmatocystin secondary metabolite gene cluster in the filamentous fungus Podospora anserina: involvement in oxidative stress response, sexual development, pigmentation and interspecific competitions. Environ Microbiol 21(8):3011–3026. https://doi.org/10.1111/1462-2920.14698

Shen L, Chapeland-Leclerc F, Ruprich-Robert G, Chen Q, Chen S, Adnan M, Wang J, Liu G, Xie N (2022) Involvement of VIVID in white light-responsive pigmentation, sexual development and sterigmatocystin biosynthesis in the filamentous fungus Podospora anserina. Environ Microbiol 24(7):2907–2923. https://doi.org/10.1111/1462-2920.15978

Sheteiwy MS, Ali DFI, Xiong YC, Brestic M, Skalicky M, Hamoud YA, Ulhassan Z, Shaghaleh H, AbdElgawad H, Farooq M, Sharma A, El-Sawah AM (2021) Physiological and biochemical responses of soybean plants inoculated with Arbuscular mycorrhizal fungi and Bradyrhizobium under drought stress. BMC Plant Biol 21(1):195. https://doi.org/10.1186/s12870-021-02949-z

Singh M, Kumar A, Singh R, Pandey KD (2017) Endophytic bacteria: a new source of bioactive compounds. 3 Biotech 7(5):315. https://doi.org/10.1007/s13205-017-0942-z

Sood M, Kapoor D, Kumar V, Sheteiwy MS, Ramakrishnan M, Landi M, Araniti F, Sharma A (2020) Trichoderma: the "secrets" of a multitalented biocontrol agent. Plants (Basel) 9(6). https://doi.org/10.3390/plants9060762

Spraker JE, Jewell K, Roze LV, Scherf J, Ndagano D, Beaudry R, Linz JE, Allen C, Keller NP (2014) A volatile relationship: profiling an inter-kingdom dialogue between two plant pathogens, Ralstonia Solanacearum and Aspergillus Flavus. J Chem Ecol 40(5):502–513. https://doi.org/10.1007/s10886-014-0432-2

Spraker JE, Wiemann P, Baccile JA, Venkatesh N, Schumacher J, Schroeder FC, Sanchez LM, Keller NP (2018) Conserved responses in a war of small molecules between a plant-pathogenic bacterium and fungi. mBio 9(3). https://doi.org/10.1128/mBio.00820-18

Stergiopoulos I, de Wit PJ (2009) Fungal effector proteins. Annu Rev Phytopathol 47:233–263. https://doi.org/10.1146/annurev.phyto.112408.132637

Stierle AA, Stierle DB, Decato D, Priestley ND, Alverson JB, Hoody J, McGrath K, Klepacki D (2017) The berkeleylactones, antibiotic macrolides from fungal coculture. J Nat Prod 80(4):1150–1160. https://doi.org/10.1021/acs.jnatprod.7b00133

Stötefeld L, Scheu S, Rohlfs M (2012) Fungal chemical defence alters density-dependent foraging behaviour and success in a fungivorous soil arthropod. Ecol Entomol 37:323–329. https://doi.org/10.1111/j.1365-2311.2012.01373.x

Stroe MC, Netzker T, Scherlach K, Kruger T, Hertweck C, Valiante V, Brakhage AA (2020) Targeted induction of a silent fungal gene cluster encoding the bacteria-specific germination inhibitor fumigermin. Elife 9. https://doi.org/10.7554/eLife.52541

Sueiro-Olivares M, Fernandez-Molina JV, Abad-Diaz-de-Cerio A, Gorospe E, Pascual E, Guruceaga X, Ramirez-Garcia A, Garaizar J, Hernando FL, Margareto J, Rementeria A (2015) Aspergillus fumigatus transcriptome response to a higher temperature during the earliest steps of germination monitored using a new customized expression microarray. Microbiology (Reading) 161(Pt 3):490–502. https://doi.org/10.1099/mic.0.000021

Sun W, Yu Y, Chen J, Yu B, Chen T, Ying H, Zhou S, Ouyang P, Liu D, Chen Y (2021) Light signaling regulates Aspergillus niger biofilm formation by affecting melanin and extracellular polysaccharide biosynthesis. mBio 12(1). https://doi.org/10.1128/mBio.03434-20

Tisch D, Schmoll M (2010) Light regulation of metabolic pathways in fungi. Appl Microbiol Biotechnol 85(5): 1259–1277. https://doi.org/10.1007/s00253-009-2320-1

Tudzynski B (2014) Nitrogen regulation of fungal secondary metabolism in fungi. Front Microbiol 5:656. https://doi.org/10.3389/fmicb.2014.00656

Uma Devi K, Padmavathi J, Uma Maheswara Rao C, Khan AAP, Mohan MC (2008) A study of host specificity in the entomopathogenic fungus Beauveria bassiana (Hypocreales, Clavicipitaceae). Biocontrol Sci Technol 18(10):975–989. https://doi.org/10.1080/09583150802450451

Valiante V, Mattern DJ, Schuffler A, Horn F, Walther G, Scherlach K, Petzke L, Dickhaut J, Guthke R, Hertweck C, Nett M, Thines E, Brakhage AA (2017) Discovery of an extended austinoid biosynthetic pathway in Aspergillus calidoustus. ACS Chem Biol 12(5): 1227–1234. https://doi.org/10.1021/acschembio.7b00003

van Tilburg Bernardes E, Pettersen VK, Gutierrez MW, Laforest-Lapointe I, Jendzjowsky NG, Cavin J-B, Vicentini FA, Keenan CM, Ramay HR, Samara J, MacNaughton WK, Wilson RJA, Kelly MM, McCoy KD, Sharkey KA, Arrieta M-C (2020) Intestinal fungi are causally implicated in microbiome assembly and immune development in mice. Nat Commun 11(1): 2577. https://doi.org/10.1038/s41467-020-16,431-1

Venkatesh N, Keller NP (2019) Mycotoxins in conversation with bacteria and fungi. Front Microbiol 10:403. https://doi.org/10.3389/fmicb.2019.00403

Vödisch M, Scherlach K, Winkler R, Hertweck C, Braun HP, Roth M, Haas H, Werner ER, Brakhage AA, Kniemeyer O (2011) Analysis of the Aspergillus fumigatus proteome reveals metabolic changes and the activation of the pseurotin A biosynthesis gene cluster in response to hypoxia. J Proteome Res 10(5): 2508–2524. https://doi.org/10.1021/pr1012812

Walton JD (2006) HC-toxin. Phytochemistry 67(14): 1406–1413. https://doi.org/10.1016/j.phytochem.2006.05.033

Wang J, Ding W, Li C, Huang S, She Z, Lin Y (2013) A new polysubstituted benzaldehyde from the co-culture broth of two marine fungi (strains nos. E33 and K38). Chem Nat Compd 49(5):799–802. https://doi.org/10.1007/s10600-013-0751-5

Wenke J, Anke H, Sterner O (1993) Pseurotin A and 8-O-Demethylpseurotin A from Aspergillus fumigatus and their inhibitory activities on chitin synthase. Biosci Biotechnol Biochem 57(6):961–964. https://doi.org/10.1271/bbb.57.961

Westrick NM, Ranjan A, Jain S, Grau CR, Smith DL, Kabbage M (2019) Gene regulation of Sclerotinia sclerotiorum during infection of Glycine max: on the road to pathogenesis. BMC Genomics 20(1):157. https://doi.org/10.1186/s12864-019-5517-4

Wild CP, Gong YY (2010) Mycotoxins and human disease: a largely ignored global health issue. Carcinogenesis 31(1):71–82. https://doi.org/10.1093/carcin/bgp264

Wilson TG (2004) The molecular site of action of juvenile hormone and juvenile hormone insecticides during metamorphosis: how these compounds kill insects. J Insect Physiol 50(2–3):111–121. https://doi.org/10.1016/j.jinsphys.2003.12.004

Wilson MR, Zha L, Balskus EP (2017) Natural product discovery from the human microbiome. J Biol Chem 292(21):8546–8552. https://doi.org/10.1074/jbc.R116.762906

Xu ZF, Wang BL, Sun HK, Yan N, Zeng ZJ, Zhang KQ, Niu XM (2015) High trap formation and low metabolite production by disruption of the polyketide synthase gene involved in the biosynthesis of arthrosporols from nematode-trapping fungus Arthrobotrys oligospora. J Agric Food Chem 63(41):9076–9082. https://doi.org/10.1021/acs.jafc.5b04244

Yaegashi J, Oakley BR, Wang CC (2014) Recent advances in genome mining of secondary metabolite biosynthetic gene clusters and the development of heterologous expression systems in Aspergillus nidulans. J Ind Microbiol Biotechnol 41(2):433–442. https://doi.org/10.1007/s10295-013-1386-z

Yoder OC (1980) Toxins in Pathogenesis. Annu Rev Phytopathol 18(1):103–129. https://doi.org/10.1146/annurev.py.18.090180.000535

Youssar L, Wernet V, Hensel N, Yu X, Hildebrand HG, Schreckenberger B, Kriegler M, Hetzer B, Frankino P, Dillin A, Fischer R (2019) Intercellular communication is required for trap formation in the nematode-trapping fungus Duddingtonia flagrans. PLoS Genet 15(3): e1008029. https://doi.org/10.1371/journal.pgen.1008029

Yu L, Ding W, Wang Q, Ma Z, Xu X, Zhao X, Chen Z (2017) Induction of cryptic bioactive 2,5-diketopiperazines in fungus Penicillium sp. DT-F29 by microbial co-culture. Tetrahedron 73(7):907–914. https://doi.org/10.1016/j.tet.2016.12.077

Yu X, Hu X, Pop M, Wernet N, Kirschhöfer F, Brenner-Weiss G, Keller J, Bunzel M, Fischer R (2021a) Fatal attraction of Caenorhabditis elegans to predatory fungi through 6-methyl-salicylic acid. Nat Commun 12(1): 5462. https://doi.org/10.1038/s41467-021-25,535-1

Yu Y, Zhang YK, Manohar M, Artyukhin AB, Kumari A, Tenjo-Castano FJ, Nguyen H, Routray P, Choe A, Klessig DF, Schroeder FC (2021b) Nematode signaling molecules are extensively metabolized by animals, plants, and microorganisms. ACS Chem Biol 16(6): 1050–1058. https://doi.org/10.1021/acschembio.1c00217

Zhang T, Sun X, Xu Q, Candelas LG, Li H (2013) The pH signaling transcription factor PacC is required for full virulence in Penicillium digitatum. Appl Microbiol Biotechnol 97(20):9087–9098. https://doi.org/10.1007/s00253-013-5129-x

Zhao J, Cheng W, He X, Liu Y (2018) The co-colonization prevalence of Pseudomonas aeruginosa and Aspergillus fumigatus in cystic fibrosis: a systematic review and meta-analysis. Microb Pathog 125:122–128. https://doi.org/10.1016/j.micpath.2018.09.010

Zhu F, Chen G, Chen X, Huang M, Wan X (2011) Aspergicin, a new antibacterial alkaloid produced by mixed fermentation of two marine-derived mangrove epiphytic fungi. Chem Nat Compd 47(5):767–769. https://doi.org/10.1007/s10600-011-0053-8

Zingales V, Fernández-Franzón M, Ruiz MJ (2020) Sterigmatocystin: occurrence, toxicity and molecular mechanisms of action - a review. Food Chem Toxicol 146:111802. https://doi.org/10.1016/j.fct.2020.111802

Zipperer A, Konnerth MC, Laux C, Berscheid A, Janek D, Weidenmaier C, Burian M, Schilling NA, Slavetinsky C, Marschal M, Willmann M, Kalbacher H, Schittek B, Brötz-Oesterhelt H, Grond S, Peschel A, Krismer B (2016) Human commensals producing a novel antibiotic impair pathogen colonization. Nature 535(7613):511–516. https://doi.org/10.1038/nature18634

Evolution and Diversity of Bioluminescent Fungi

12

Huei-Mien Ke, Min R. Lu, Chiung-Chih Chang, Chen Hsiao, Jie-Hao Ou, Yuichi Taneyama, and Isheng Jason Tsai

Abstract

Bioluminescence is one of the most fascinating features of basidiomycetes. In this chapter, we have reviewed a total of 105 bioluminescent fungi that have been documented to date. We show that these species are restricted to three major lineages, including mycenoid, Armillaria, and Omphalotus, and display different bioluminescent patterns and geographical distributions. A global phylogeny shows that these species also exhibit patchy phylogenetic placement with their non-bioluminescent relatives. We review previous work on their genomes, which contain the fungal luciferin biosynthetic cluster. The dynamics of this cluster provide an evolutionary scenario since the last common ancestor of the family Mycenaceae and the marasmioid clade of Agaricales. Finally, we discuss potential functions and possible research directions of bioluminescence in these fungi.

Keywords

Basidiomycetes, Fungal evolution, Bioluminescence, Luciferase, Mycena

H.-M. Ke (✉)
Department of Microbiology, Soochow University, Taipei, Taiwan
e-mail: hmke@scu.edu.tw

M. R. Lu · C. Hsiao · I. J. Tsai (✉)
Biodiversity Research Center, Academia Sinica, Taipei, Taiwan
e-mail: ijtsai@gate.sinica.edu.tw

C.-C. Chang
Biodiversity Research Center, Academia Sinica, Taipei, Taiwan

Biodiversity Program, Taiwan International Graduate Program, Academia Sinica and National Taiwan Normal University, Taipei, Taiwan

Department of Life Science, National Taiwan Normal University, Taipei, Taiwan

J.-H. Ou
Department of Plant Pathology, National Chung Hsing University, Taichung, Taiwan

Y. Taneyama
Nagano, Japan

12.1 Introduction

Bioluminescence is the emission of light by a living organism, generated by a chemical reaction that involves proteins called luciferases that catalyze the oxidation of substrates called luciferins. It is a trait that has evolved at least 94 times independently across the tree of life (Lau and Oakley 2021), with over 700 genera of bioluminescent organisms recorded from freshwater, terrestrial, and marine environments (Haddock et al. 2010; Widder 2010; Lau and Oakley 2021; Delroisse et al. 2021). Bioluminescence can be further categorized into two types, autogenic and bacteriogenic, in which light is produced either by the organism itself or bioluminescent bacterial symbionts, respectively (Lau and Oakley 2021).

© The Author(s), under exclusive license to Springer Nature Switzerland AG 2023

S. Pöggeler, T. James (eds.), *Evolution of Fungi and Fungal-Like Organisms*, The Mycota 14,
https://doi.org/10.1007/978-3-031-29199-9_12

Bioluminescence in fungi is autogenic, which means fungi emit light using their own luciferin and luciferase (Oliveira et al. 2012; Oliveira and Stevani 2009; Airth and Mc 1959). Fungal luciferin is different from the other 13 known luciferins (Lau and Oakley 2021; Purtov et al. 2015) and all fungal luciferases are homologous and distinct from other known luciferases used by other organisms based on sequence similarity (Delroisse et al. 2021; Kotlobay et al. 2018).

All bioluminescent fungi belong to the Agaricales of Basidiomycetes, which is a major order comprising around 13,000 gilled mushrooms (Kirk et al. 2008). These fungi belong to ten different unrelated genera although some of which have taxonomic synonyms. The apparent patchy phylogenetic distribution of bioluminescence in fungi and the conserved genetic mechanisms underlying phenotype can be explained by either convergent evolution or an intricate loss/gain history. Recent phylogenomic progress has resolved the phylogenetic relationships among species and helps us answer those questions.

This review focuses on the evolution of bioluminescence in mushrooms in the Agaricales order. The recent identification of the luciferin biosynthetic pathway genes and new phylogenomic analyses of bioluminescent fungi and their sister taxa have shown that fungal bioluminescence occurred in the last common ancestor of the mycenoid and the marasmioid clade. We will review the features of bioluminescent fungi, the known genes involved in the bioluminescence, and their evolutionary history.

12.2 Diversity of Bioluminescent Fungi

Bioluminescence has long captured the imagination of scientists. The taxonomy and evolution of bioluminescent fungi was first compiled and discussed 70 years ago (Wassink 1978; Harvey 1952). The number of taxonomically reported bioluminescent fungi continues to increase with improvement over higher sensitivity in light detection initially from the naked eye to digital camera, luminometer, or photomultiplier. Two reviews have updated the accounts to 64 luminescent species in 2008 (Desjardin et al. 2008) and 81 species in 2015 (Chew et al. 2015). We have removed the phylogenetically uncertain species and merged synonyms where possible, as well as compiled and assigned a list of 28 additional bioluminescent species (Mihail 2015; Desjardin et al. 2016; Cortes-Perez et al. 2019; Chang et al. 2020; Karunarathna et al. 2020; Dauner et al. 2021; Terashima et al. 2016). The current number of recorded bioluminescent fungi is 105. In recent years, molecular data were generated using barcode DNA sequences including rRNA ITS (Internal transcribed spacer), large subunit rRNA (LSU), and RNA polymerase II (*rpb2*) to revise the phylogenetic positions of these species, which can be classified into three lineages (Fig. 12.1): 80 in the mycenoid (78 Mycenaceae and two from the Lucentipes lineage) (Table 12.1), 11 in the *Armillaria* lineage and 13 in the *Omphalotus* lineage, and one species (*Pleurotus nitidus* Har. Takah. & Taneyama) with uncertain position (Table 12.2). With the availability of molecular markers, the term mycenoid was originally intended to describe various morphological features that are now coherently grouped in species belonging to the Mycenaceae family. A fourth lineage (Lucentipes) containing the bioluminescent *Gerronema viridilucens* Desjardin, Capelari & Stevani (Desjardin et al. 2005) and *Mycena lucentipes* Desjardin, Capelari & Stevani (Desjardin et al. 2007) which were classified as mycenoid species is excluded in this review because of the lack of available marker sequences. The three lineages shared a common origin in the last common ancestor of Mycenaceae and the marasmioid clade. For more details on the relationships, see Ke et al. 2020. Here, we summarize the properties of these three major bioluminescent fungal lineages.

12.2.1 Mycenaceae Family

The Mycenaceae family contains most species originally part of the mycenoid lineage (Desjardin et al. 2008), which are a type of mushrooms

Fig. 12.1 Overview of bioluminescent fungi lineages and their sister groups. Basidiomes of *Mycena kentingensis* in light (a-1) and dark (a-2) exposures and its mycelium in light (a-3) and dark (a-4) cultured in compost. Basidiomes of *Mycena luguensis* (b-1) and its mycelium in light (b-2) and dark (b-3) grown on wood. Basidiomes of *Armillaria mellea* (c-1) and its cultured mycelium in light (c-2) and dark (c-3). Basidiomes of *Omphalotus japonicus* on natural host tree (d-1) and the fruiting body in light (d-2) and dark (d-3)

whose gills adhere to cartilaginous and central stipes and lack a ring or volva (Ulloa and Hanlin 2017). The mycenoid lineage includes the majority of recorded bioluminescent fungi (80) belonging to the Mycenaceae family with the exception of two species (*Gerronema viridilucens* and *Mycena lucentipes*; Desjardin et al. 2008). In this review, we focus on the diversity of Mycenaceae. Mycenaceae contains over 600 species including *Atheniella, Dictyopanus*, *Favolaschia*, *Mycena*, *Mycenoporella*, *Panellus*, *Resinomycena*, *Roridomyces,* and *Xeromphalina* (Redhead et al. 2012; Kirk et al. 2008). Members of this family are known primarily to colonize leaves or wood as litter- or wood-decaying saprophytes (Moncalvo et al. 2002). Still, other members of this family were reported to use different nutritional strategies such as orchid mycorrhizae (Ogura-Tsujita et al. 2009; Martos et al. 2009; Zhang et al. 2012), the coffee pathogen *Mycena citricolor* (Rao and Tewari 1987; Quesada-Chanto and Jimenez-Ulate 1996), a plant root invader (Harder et al. 2021), and tissues in moss and root system in healthy plant (Davey et al. 2013; Liao et al. 2014; Botnen et al. 2014; Lorberau et al. 2017; Kernaghan and Patriquin 2011). ITS is the most common marker used in this lineage. We downloaded available ITS sequences (>500 bp) from the National Center for Biotechnology Information (NCBI) according to taxId including Mycenaceae, *Favolaschia, Filoboletus, Dictyopanus*, and *Roridomyces*. We constructed a phylogentic tree of 287 non-redundant sequences with species information using IQ-TREE (Nguyen et al. 2015). An alignment length of 1220 bp (last updated 2022.09.11; https://github.com/HueiMien/bioluminescent-fungi.git) was produced with MAFFT (Katoh and Standley 2013) and trimming with trimal (Capella-Gutierrez et al. 2009). (Fig. 12.2).

Table 12.1 Bioluminescent fungi recorded in the Mycenaceae

Section[a]	Scientific name[b]	Mycelium[c]	Fruiting body[c]	Stipe[c]	Cap[c]	Spore[c]	Reference[d]
Aspratiles	*Mycena lacrimans* Singer	–	+	+	–	–	(Desjardin and Braga-Neto 2007)
Aspratiles	*Mycena aspratilis* Maas Geest. & de Meijer	?	+	–	+	?	(Desjardin et al. 2010)
Basipedes	*Mycena illuminans* Henn.	+	+	–	+	–	(Corner 1954; Chew et al. 2015; Chew et al. 2013)
Basipedes	*Mycena stylobates* (Pers.: Fr.) P.Kumm.	+	–	–	–	–	(Bothe 1931)
Basipedes	*Mycena kentingensis* Y.S. Shih, C.Y. Chen, W.W. Lin & H.W. Kao	+	+	–	+	–	(Shih et al. 2014)
Basipedes	*Mycena nocticaelum* A.L.C. Chew & Desjardin	+	+	–	+	?	(Chew et al. 2015)
Calodontes	*Mycena pura* (Pers.: Fr.) P. Kumm.	+	–	–	–	–	(Treu and Agerer 1990)
Calodontes	*Mycena rosea* (Bull.) Gramberg	+	–	–	–	–	(Treu and Agerer 1990)
Calodontes	*Mycena cahaya* A.L.C. Chew & Desjardin	+	+	+	+	?	(Chew et al. 2014)
Calodontes	*Mycena seminau* A.L.C. Chew & Desjardin	+	+	–	+	?	(Chew et al. 2014)
Calodontes	*Mycena sinar* A.L.C.Chew & Desjardin	+	+	+	+	–	(Chew et al. 2014)
Calodontes	*Mycena sinar* var. tangkaisinar A.L.C.Chew & Desjardin	?	+	+	+	?	(Chew et al. 2014)
Citricolores	*Mycena citricolor* (Berk. & M.A.Curtis) Sacc.	+	–	–	–	–	(Berliner 1961a)
Crocatae	*Mycena luxfoliicola* Cortés-Pérez, Desjardin & Ram.-Cruz	+	+	+	+	?	(Cortes-Perez et al. 2019)
Exornatae	*Mycena chlorophos* (Berk. & M.A.Curtis) Sacc.	+	+	+	+	–	(Corner 1954)
Exornatae	*Mycena discobasis* Metrod	?	+	+	+	–	(Desjardin et al. 2007)
Exornatae	*Mycena margarita* (Murrill) Murrill	?	+	+	+	?	(Desjardin et al. 2010)
Exornatae	*Mycena deeptha* Aravind. & Manim.	+	–	–	–	–	(Aravindakshan et al. 2012)
Fragilipedes	*Mycena polygramma* (Bull.: Fr.) S.F.Gray	+	–	–	–	–	(Berliner 1961a; Treu and Agerer 1990)
Fragilipedes	*Mycena zephirus* (Fr.: Fr.) P.Kumm.	+	–	–	–	–	(Bothe 1931; Treu and Agerer 1990)
Fragilipedes	*Mycena silvaelucens* B.A. Perry & Desjardin	?	+	+	+	?	(Desjardin et al. 2010)
Fragilipedes	*Mycena flammifera* Har. Takah.& Taneyama	+	+	+	+	?	(Terashima et al. 2016)

(continued)

Table 12.1 (continued)

Section[a]	Scientific name[b]	Mycelium[c]	Fruiting body[c]	Stipe[c]	Cap[c]	Spore[c]	Reference[d]
Fragilipedes	*Mycena stellaris* Har. Takah., Taneyama & A. Hadano	+	+	+	+	?	(Terashima et al. 2016)
Fragilipedes	*Mycena jingyinga* C.C. Chang, C.Y. Chen, W.W. Lin & H.W. Kao	+	–	–	–	–	(Chang et al. 2020)
Fragilipedes	*Mycena luguensis* C.C. Chang, C.Y. Chen, W.W. Lin & H.W. Kao	+	–	–	–	–	(Chang et al. 2020)
Fragilipedes	*Mycena venus* C.C. Chang, C.Y. Chen, W.W. Lin & H.W. Kao	+	–	–	–	–	(Chang et al. 2020)
Galactopoda	*Mycena haematopus* (Pers.: Fr.) P.Kumm.	+	+	?	?	?	(Bermudes et al. 1992; Treu and Agerer 1990)
Hygrocyboideae	*Mycena epipterygia* (Scop.: Fr.) S.F.Gray	+	–	–	–	–	(Bothe 1931)
Incertae Sedis	*Mycena daisyogunensis* Kobayasi	?	+	?	?	?	(Kobayasi 1951)
Incertae Sedis	*Mycena pseudostylobates* Kobayasi	+	?	?	?	?	(Kobayasi 1951)
Lactipedes	*Mycena galopus* (Pers.: Fr.) P.Kumm.	+	–	–	–	–	(Berliner 1961a; Bothe 1931; Treu and Agerer 1990)
Manipularis-group	*Filoboletus pallescens* (Boedijn) Maas Geest.	?	+	?	?	?	(Maas Geesteranus 1992)
Manipularis-group	*Panellus yunnanensis* P.G. Liu = *Filoboletus yunnanensis*	?	+	?	?	?	(Liu 1995)
Manipularis-group	*Filoboletus manipularis* (Berk.) Teng = *Favolaschia manipularis*	+	+	+	?	?	(Corner 1954)
Manipularis-group	*Filoboletus manipularis* var. microporus Kawam. ex Corner nom. inval. = *Polyporus microporus* Kawam. nom. inval.	?	+	+	?	?	(Corner 1954)
Manipularis-group	*Filoboletus Hanedae* Kobayasi = *Poromycena* hanedai	?	+	+	?	?	(Kobayasi 1951)
Mycena	*Mycena inclinata* (Fr.) Quél.	+	–	–	–	–	(Wassink 1948)
Mycena	*Mycena maculata* P.Karst.	+	–	–	–	–	(Treu and Agerer 1990)
Mycena	*Mycena tintinnabulum* (Fr.) Quél.	+	–	–	–	–	(Bothe 1930)
Mycena	*Mycena gombakensis* A.L.C. Chew & Desjardin	+	+	+	+	?	(Chew et al. 2015)

(continued)

Table 12.1 (continued)

Section[a]	Scientific name[b]	Mycelium[c]	Fruiting body[c]	Stipe[c]	Cap[c]	Spore[c]	Reference[d]
Nodosae	*deformis* Maas Geest. & de Meijer	+	–	–	–	–	(Desjardin et al. 2016)
Panellus	*Panellus luxfilamentus* A.L.C. Chew & Desjardin	+	–	–	–	–	(Chew et al. 2015)
Panellus	*Dictyopanus foliicola* Kobayasi	+	+	?	?	?	(Kobayasi 1951)
Panellus	*Panellus pusillus* var. sublamellatus Corner	?	+	?	?	?	(Corner 1954)
Panellus	*Panellus luminescens* (Corner) Corner	+	+	+	+	+	(Chew et al. 2015; Corner 1950)
Panellus	*Panellus pusillus* (Pers. ex Lév.) Burdsall & O.K.Mill	+	+	?	+	?	(Corner 1954; Desjardin et al. 2008)
Panellus	*Panellus stipticus* (Bull.: Fr.) P.Karst.	+	+	–	+	?	(Berliner 1961a; Wassink 1948)
Resinomycena	*Resinomycena fulgens* Har. Takah.,Taneyama & Oba	?	+	+	+	?	(Terashima et al. 2016)
Resinomycena	*Resinomycena petarensis* Desjardin, B.A. Perry & Stevani	+	–	–	–	–	(Desjardin et al. 2016)
Roridae	*Roridomyces irritans* E. Horak = *Mycena irritans*)	–	+	–	+	–	(Horak 1978)
Roridae	*Roridomyces lamprosporus* (Corner) Rexer = *Mycena lamprospora* (Corner) E. Horak	–	+	–	–	+	(Horak 1978; Corner 1950)
Roridae	*Roridomyces pruinosoviscida* Corner = *Roridomyces pruinosoviscidus* (Corner) A.L.C. Chew & Desjardin	+	+	+	+	?	(Chew et al. 2015; Corner 1954)
Roridae	*Roridomyces pruinosoviscidus* var. rabaulensis (Corner) Blanco-Dios	?	+	?	?	?)	(Corner 1954)
Roridae	*Roridomyces roridus* (Fr.) Rexer = *rorida* (Fr.) Quel.	+	–	–	–	–	(Josserand 1953)
Roridae	*Roridomyces sublucens* Corner	–	+	+	+	–	(Corner 1954)
Roridae	*Roridomyces phyllostachydis* Karun., Mortimer and Axford	+	+	+	–	?	(Karunarathna et al. 2020)
Roridae	*Roridomyces viridiluminus* L.A.P. Dauner, Karunarathna & P.E. Mortimer	+	+	+	+	?	(Dauner et al. 2021)
Rubromarginatae	*Mycena lux-coeli* Corner	?	+	+	+	–	(Corner 1954)
Rubromarginatae	*Mycena noctilucens* Kawam. ex Corner = *Mycena noctilucens* var. magnispora Corner	+	+	+	+	?	(Chew et al. 2015; Corner 1954)

(continued)

Table 12.1 (continued)

Section[a]	Scientific name[b]	Mycelium[c]	Fruiting body[c]	Stipe[c]	Cap[c]	Spore[c]	Reference[d]
Rubromarginatae	*Mycena olivaceomarginata* (Massee apud Cooke)	+	–	–	–	–	(Wassink 1979)
Rubromarginatae	*Mycena singeri* Lodge	?	+	+	+	–	(Desjardin et al. 2007)
Rubromarginatae	*Mycena coralliformis* A.L.C. Chew & Desjardin	+	–	–	–	–	(Chew et al. 2015)
Rubromarginatae	*Mycena fulgoris* Cortés-Pérez, Desjardin	–	+	+	–	?	(Desjardin et al. 2016)
Rubromarginatae	*Mycena lumina* Cortés-Pérez, Desjardin	+	+	+	+	?	(Cortes-Perez et al. 2019)
Sacchariferae	*Mycena asterina* Desjardin, Capelari & Stevani	?	+	–	+	–	(Desjardin et al. 2007; Desjardin et al. 2008)
Sacchariferae	*Mycena lazulina* Har. Takah.,Taneyama, Terashima & Oba	+	+	+	+	?	(Terashima et al. 2016)
Sacchariferae	*Mycena perlae* Cortés-Pérez, Desjardin & Rockefeller	–	+	–	+	?	(Cortes-Perez et al. 2019)
Sanguinolentae	*Mycena sanguinolenta* (Alb. & Schwein.: Fr.) P. Kumm.	+	–	–	–	–	(Desjardin et al. 2007)
Sanguinolentae	*Mycena nebula* Cortés-Pérez, Desjardin & Rockefeller	?	+	+		?	(Cortes-Perez et al. 2019)
Supinae	*Mycena fera* Maas Geest. & de Meijer	?	+	+	+	–	(Desjardin et al. 2007)
Supinae	*Mycena luxarboricola* B. A.Perry & Desjardin	?	+	+	+	?	(Desjardin et al. 2010)
Supinae	*Mycena globulispora* Maas Geest. & de Meijer	?	+	+	–	?	(Desjardin et al. 2016)
Similar to Euspeireae/ Aspratiles	*Mycena guzmanii* Cortés-Pérez, Desjardin	+	+	+	+	?	(Cortes-Perez et al. 2019)
Similar to Hygrocyboideae/ Insignes/ Euspeireae	*Mycena luxaeterna* B.A. Perry & Desjardin	+	+	+	–	?	(Desjardin et al. 2010)
Similar to Insignes/ Euspeireae	*Mycena luxperpetua* B.A. Perry & Desjardin	+	+	+	+	?	(Desjardin et al. 2010)
Uncertain	*Mycena* aff. *abieticola* Singer	?	+	+	+	?	(Desjardin et al. 2010)
Uncertain	*Mycena luxfoliata* Har. Takah., Taneyama & Terashima	+	–	–	–	–	(Terashima et al. 2016)
Uncertain	*Mycena oculisnymphae* Desjardin, B.A. Perry & Stevanir	?	+	+	+	?	(Desjardin et al. 2016)

[a] The "Uncertain" section means there is no assignment according to the reference
[b] Use current name of the species in Index Fungorum, and their synonyms are listed as well.
[c] +, –, and? denote the structure with, without, and uncertain illumination, respectively
[d] Citations which report the bioluminescence feature

Table 12.2 Bioluminescent fungi in *Armillaria* and *Omphalotus* lineage

Lineage	Taxon	Mycelium	Fruiting body	Reference
Armillaria lineage	*Armillaria fuscipes* Petch	+	–	(Wassink 1978; Wassink 1948; Berliner 1961a)
	Armillaria gallica Marxm. & Romagn	+	–	(Mihail and Bruhn 2007)
	Armillaria mellea (Valh.) P.Kumm.	+	–	(Mihail and Bruhn 2007)
	Armillaria ostoyae (Romagn.) Henrik	+	–	(Rishbeth 1986)
	Armillaria tabescens (Scop.) Emel	+	–	(Mihail and Bruhn 2007)
	Armillaria calvescens Bérubé & Dessur.	+	–	(Mihail 2015)
	Armillaria cepistipes Velen.	+	–	(Mihail 2015)
	Armillaria gemina Bérubé & Dessur.	+	–	(Mihail 2015)
	Armillaria nabsnona T.J. Volk & Burds.	+	–	(Mihail 2015)
	Armillaria sinapina Bérubé & Dessur.	+	–	(Mihail 2015)
	Desarmillaria ectypa (Fr.) R.A. Koch & Aime = *Armillaria ectypa* (Fr.) Lamoure	+	+	(Ainsworth 2004)
Omphalotus lineage	*Omphalotus guepiniiformis* (Berk.) Neda = *Lampteromyces luminescens* M. Zang = *Omphalotus japonicus* (Kawam.) Kirchm. & O.K. Mill.	+	?	(Zang 1979; Kawamura 1910; Bermudes et al. 1992; Singer 1947)
	Neonothopanus nambi (Speg.) Petersen & Krisai-Greilhuber	+	?	(Corner 1981; Chew et al. 2015)
	Nothopanus noctilucens (Le´v.) Singer = *Pleurotus noctilucens* Le´v.	+	?	(Léveillé 1844; Haneda 1955)
	Omphalotus illudens (Schwein.) Bresinsky & Besl. = *Clitocybe illudens* Schwein. = *Panus illudens* (Schwein.) Fr. = *Pleurotus facifer* Berk. & M. A. Curtis	+	+	(Berliner 1961b, a; Wassink 1948)
	Omphalotus mangensis (J. Li & X. Hu) Kirchm. & O. K. Mill. = *Lampteromyces mangensis* J. Li & X. Hu	+	?	(Li and Hu 1993)
	Omphalotus nidiformis (Berk.) O. K. Mill. = *Pleurotus nidiformis* (Berk.) Sacc. = *Pleurotus candescens* (F. Muell. & Berk.) Sacc. = *Pleurotus illuminans* (Berk.) Sacc. = *Pleurotus lampas* (Berk.) Sacc. = *Pleurotus phosphorus* (Berk.) Sacc.	+	?	(Berkeley 1844; Miller Jr. 1994)
	Omphalotus olearius (DC.) Singer = *Pleurotus olearius* (DC.) Gillet	+	+	(Wassink 1948)
	Omphalotus olivascens H.E.Bigelow, O.K.Mill. & Thiers	+	–	(Bigelow et al. 1976)
	Pleurotus decipiens Corner	+	?	(Corner 1981)
	Nothopanus eugrammus (Mont.) Singer 1944 = Pleurotus eugrammus var. radicicolus Corner	+	?	(Corner 1981)
	Neonothopanus gardneri (Berk.) Capelari, Desjardin, B.A. Perry, T. Asai & Stevani	+	?	(Capelari et al. 2011; Saccardo 1887)
	Marasmiellus lucidus Har. Takah., Taneyama & S. Kurogi	+	?	(Terashima et al. 2016)
	Marasmiellus venosus Har. Takah., Taneyama & A. Hadano	+	+	(Terashima et al. 2016)
Uncertain	*Pleurotus nitidus* Har. Takah. & Taneyama	+	?	(Terashima et al. 2016)

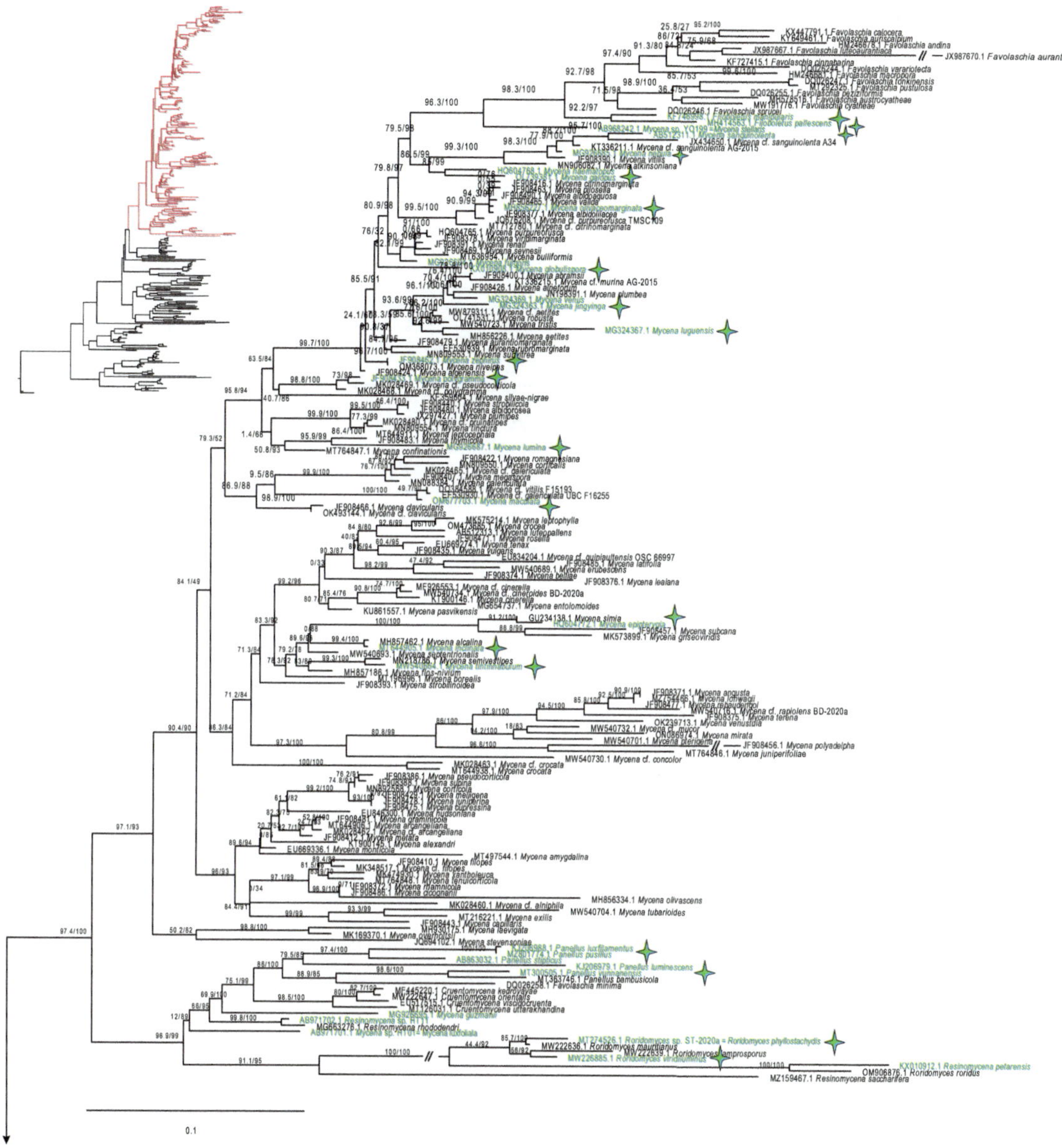

Fig. 12.2 ITS tree of 287 Mycenaceae sequences with two *Armillaria* species as outgroup which was constructed by IQ-TREE (Nguyen et al. 2015) with an alignment length of 1220 bp without redundant species downloaded from NCBI. Numbers on each branch denote support values (SH-aLRT support (%) / ultrafast bootstrap support (%)). Dark green nodes annotated with a star denote the bioluminescent species. Note that the sequences are not necessary the same as the holotype, and not all bioluminescent Mycenaceae species were listed here since not all of them have been recorded in NCBI. The inset indicates the lineage with species belong to sect. Calodontes

We found that the bioluminescent trait was scattered along the Mycenaceae phylogeny and did not group to a specific genus. This observation is consistent with previous patchy distributions observed in smaller datasets using different molecular markers and focusing on

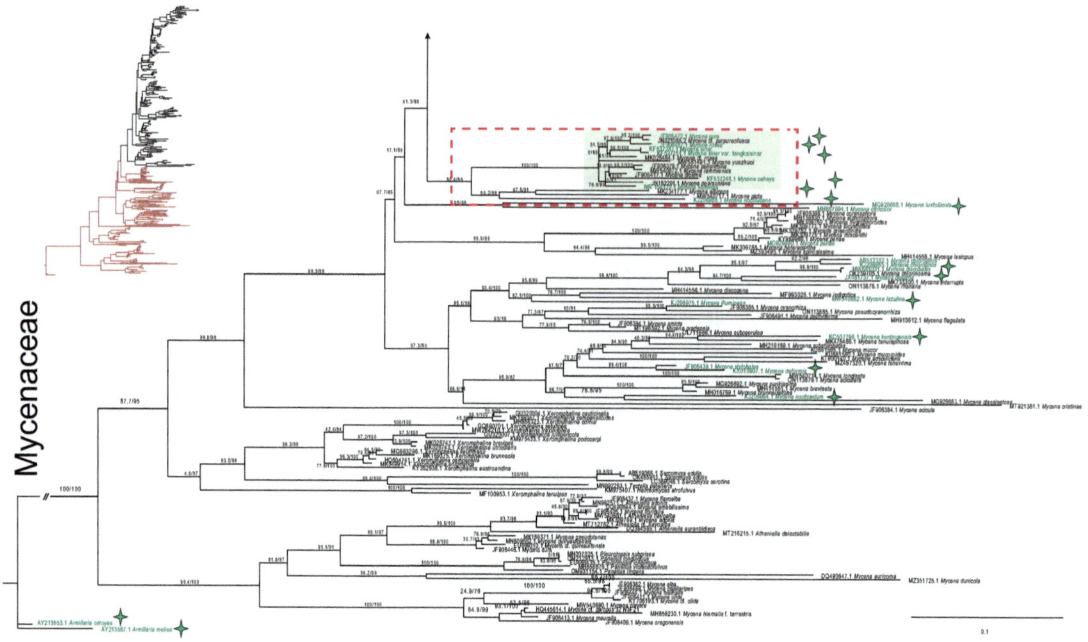

Fig. 12.2 (continued)

different closely related species (Chew et al. 2015; Karunarathna et al. 2020; Dauner et al. 2021; Shih et al. 2014; Chang et al. 2020).

Among different species, luminescence also showed diverse patterns in different tissues, including bioluminescent mycelia, caps and stipes of basidiomes (Table 12.1).

Our revision was consistent with the observations of Desjardin et al. 2008. Overall, the species with bioluminescent basidiomes mostly have bioluminescent mycelia (26/32) but some have non-bioluminescent mycelia (6/32). The variations can usually be found between closely related species, indicating that the expression of bioluminescent is not restricted to specific tissues. An example was the *Mycena* sect. Calodontes shown in Fig. 12.2. These 13 species with >85.8 ITS nucleotide identity form a monophyletic clade with seven non-bioluminescent species and six bioluminescent species, including two species with bioluminescent mycelia but non-bioluminescent fruiting bodies, as well as four species with bioluminescent fruiting bodies with different intensity in the stipes and caps. Different species also displayed different degrees of endemism (Fig. 12.3). *M. pura* is the most widely distributed across different continents. By contrast, *M. chlorophos* is restricted in Asia, and *M. kentingensis* has only been found in Taiwan so far.

12.2.2 *Armillaria* lineage

This lineage includes all members of genus *Armillaria* comprising more than 40 species (Baumgartner et al. 2011; Anderson and Stasovski 1992; Piercey-Normore et al. 1998; Coetzee et al. 2001; Koch et al. 2017). The *Armillaria* lineage and its sister group, the physalacrioid lineage (all of which are non-bioluminescent), form the family Physalacriaceae (Moncalvo et al. 2002), which includes 11 genera and 169 species (Kirk et al. 2008). Most of them can decompose organic material as saprotrophs and infect plants as necrotrophic pathogens when they are in a suitable environment (Sipos et al. 2018; Shaw et al. 1991; Smith et al. 1992; Kedves et al. 2021). Some of them are associated with orchid mycorrhizas (Martos et al. 2009; Cha and Igarashi 1995; Sekizaki et al. 2008; Guo et al. 2016).

Eleven species in the *Armillaria* lineage have been confirmed to have a luminescent property in

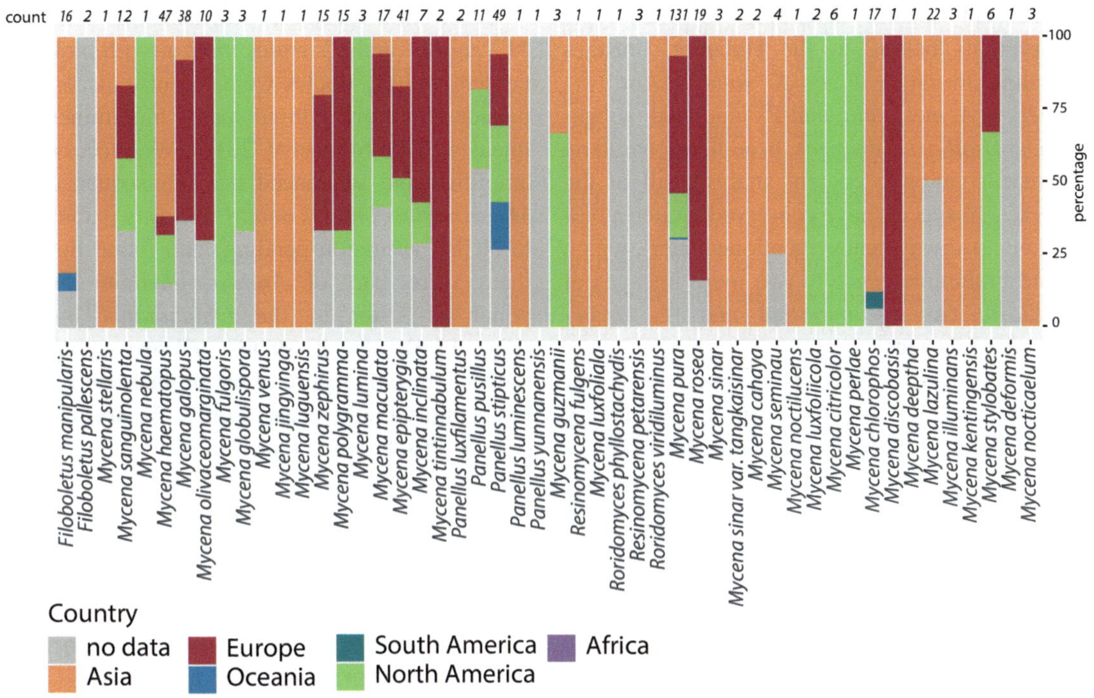

Fig. 12.3 The geographic distribution of bioluminescent fungi in Mycenaceae lineage recorded by extracting ITS (Internal transcribed spacer) from NCBI and converting the collected country to six continents

their mycelium (Table 12.2), and it has been believed that the mycelium of most *Armillaria* species is luminescent (Desjardin et al. 2008). Among them, only *Desarmillaria ectypa* (Synonym *A. ectypa*) emit light in both mycelium and fruitbodies (mainly from the gills) so far in report. Based on the classification of a 2018 review using a translation elongation factor subunit 1-alpha (tef-1α) (Coetzee et al. 2018), nine sub-lineages were proposed (Fig. 12.4). Most of the known

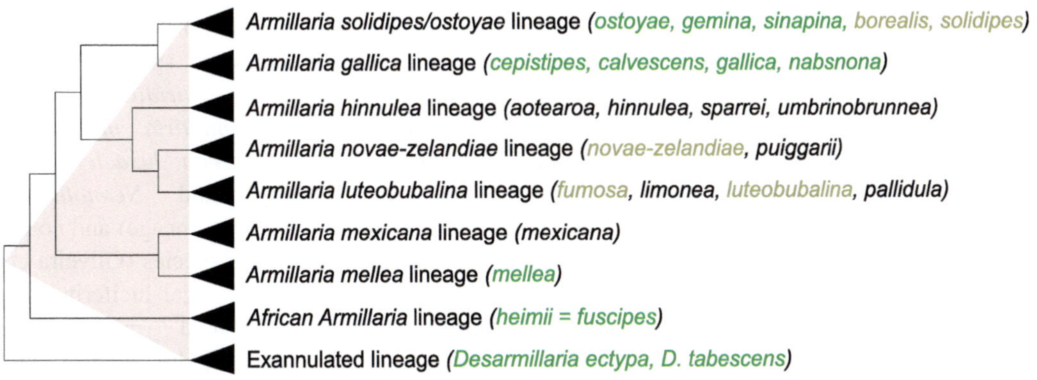

Fig. 12.4 The abbreviation of a phylogeny from (Coetzee et al. 2018), created using tef-1α sequences. Species with dark green denote the bioluminescent fungi according to observation of light emission, those with light green denotes the bioluminescent fungi with prediction of luciferase cluster in genome. The figure was adapted with permission and redrawn

luminescent species according to phenotypic observations belong to the Holarctic lineages including the *A. gallica*, *A. solidipes/ostoyae*, *A. mellea*, *A. mexicana*, and Exannulated (subgenus *Desarmillaria*) lineages. In addition, based on the recent genomic content analysis, all of the analyzed 14 species have conserved luciferin biosynthetic pathway genes (mentioned later), which extend bioluminescence to two more lineages of *Armillaria* (*A. luteobubalina* and *A. novae-zelandiae* Lineages).

Even when based just on those recorded bioluminescent fungi, it does not seem that the ability to emit light is related to geographic distribution. For example, the known luminescent species include *A. gemina* which is confined to North America, *A. fuscipes* which is confined to Africa, and *A. ostoyae* which has transcontinental distribution across Europe, North America, and Asia. Light emission was recorded in *Desarmillaria ectypa* and *D. tabescens*, which belongs to the monophyletic group of the Exannulated lineage basal to all *Armillaria* species, but no light emission has been recorded in the closely related genus *Guyanagaster*. Interestingly, the genome of *G. necrorhizus* maintains the luciferase cluster but with a truncated luciferase gene (Sect. 12.3).

12.2.3 *Omphalotus* Lineage

The *Omphalotus* lineage includes the genera *Omphalotus* and *Neonothopanus* (Kirchmair et al. 2004). According to a 2019 review (Oliveira et al. 2019) and Mycobank and Index Fungorum, in addition to *Omphalotus* and *Neonothopanus*, Omphalotaceae includes the other nine genera (*Anthracophyllum*, *Connopus*, *Gymnopanella*, *Gymnopus*, *Lentinula*, *Marasmiellus*, *Marasmius*, *Mycetinis*, *Pusillomyces*, and *Rhodocollybia*). It has been suspected that all species from the *Omphalotus* lineage form luminescent fruiting bodies but that their mycelia can be either luminescent or non-luminescent. They are known for large mushrooms and *Neonothopanus gardneri* and *N. nambi* were well-characterized by bioluminescent mechanisms (Oliveira et al. 2015) and the relevant enzyme activity (Kotlobay et al. 2018).

Three new species, *Marasmiellus lucidus*, *M. venosus*, and *Pleurotus nitidus*, with bioluminescent fruiting bodies were recently discovered outside the current three major bioluminescent lineages (Terashima et al. 2016). We successfully amplified and sequenced the ITS sequences of *Marasmiellus lucidus* and *M. venosus* (submitted to NCBI; accession pending) and found both of them were part of *Anthracophyllum* lineage which is sister to the *Omphalotus* lineage (Fig. 12.5). As it stands, *Pleurotus nitidus* remains the only record of bioluminescent fungus in the entire Pleurotaceae family. We speculate it may be misclassified that warrants further investigation as this family often was confused with species belonging in the *Omphalotus* lineage.

12.3 Genes Responsible for Bioluminescence; Luciferase Cluster: Luz and Luciferin Biosynthesis Genes

Green light (~530 nm) emission in fungi was confirmed to be a result of an enzyme-mediated reaction in which a hot extract containing heat-stable substrates (e.g., luciferin) mixed with a cold extract containing enzymes (e.g., luciferase) (Oliveira and Stevani 2009).

Later, one single bioluminescent system shared by all known bioluminescent fungal lineages was validated by detecting bioluminescence from combinations of substrate/enzyme extract from bioluminescent species from four lineages (*Gerronema viridilucens* from Lucentipes lineage, *Armillaria mellea* from *Armillaria* lineage, *Mycena luxaeterna* from Mycenaceae lineage, and *Neonothopanus gardneri* from *Omphalotus* lineage) and comparing them to nonluminous species (Oliveira et al. 2012). The structure of fungal luciferin and its precursor were identified in 2015 as 3-hydroxyhispidin and hispidin in the bioluminescent *Neonothopanus nambi*, *Mycena citricolor*, *Panellus stipticus*, and *Armillaria borealis* (Purtov et al. 2015). In 2017, the light-emitting process was demonstrated by reporting the structure of the emitter, oxyluciferin, and oxyluciferin is enzymatically hydrolyzed yielding

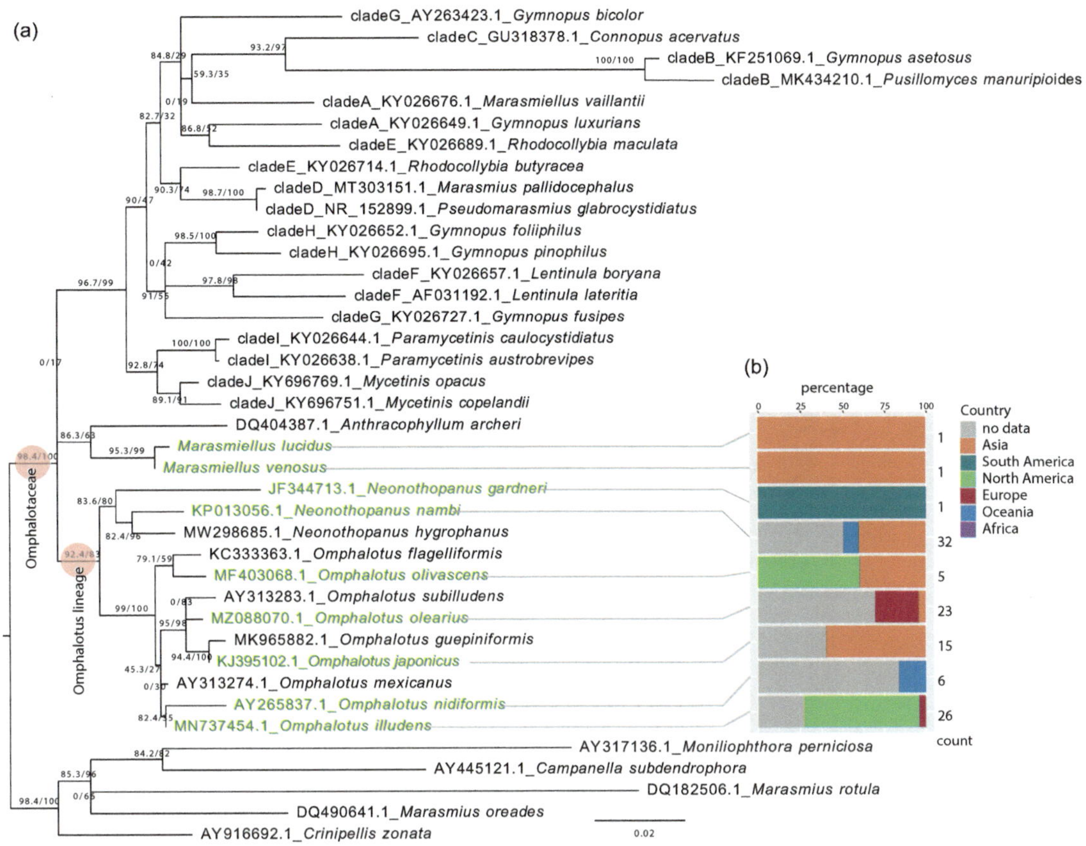

Fig. 12.5 (**a**), ITS tree of 34 Omphalotaceae sequences with five species of Marasmiaceae as an outgroup. The tree was constructed by IQ-TREE (Nguyen et al. 2015) with an alignment length of 1373 bp and without redundant species downloaded from NCBI. Numbers on each branch denote support values (SH-aLRT support (%) / ultrafast bootstrap support (%)). Nodes with dark green color denote bioluminescent species. Note that the sequences not necessary the same as the holotype, and not all bioluminescent species were listed here since not all of them have been recorded in NCBI. The clade name in the front of species name according to previous review (Oliveira et al. 2019). (**b**), The geographic distribution of bioluminescent fungi recorded by extracting ITS (Internal transcribed spacer) from NCBI and converting the collected country to six continents

caffeic acid (Kaskova et al. 2017). The biosynthetic cycle was finalized and the four enzymes involved were identified in *N. nambi* (Fig. 12.6a): Hispidin synthase (*hisps*) converts caffeic acid to hispidin; hispidin-3-hydroxylase (*h3h*) hydroxylates hispidin to give 3-hydroxyhispidin, the fungal luciferin; and the oxidation of luciferin by luciferase (*luz*) yields an unstable high-energy intermediate, which emits light while turning into oxyluciferin, a substance that can be recycled to caffeic acid by caffeylpyruvate hydrolase (*cph*) (Kotlobay et al. 2018). The *luz*, *h3h*, and *hispS* were introduced in yeast *Pichia pastoris* for autoluminescence.

12.4 The Origin of Fungal Bioluminescence

Like other bioluminescent species, the discovery of bioluminescent fungi, with their minor presence and patchy phylogenetic placement in gilled mushrooms, initially suggested they evolved independently multiple times in a convergent manner. However, recent advances in the genome

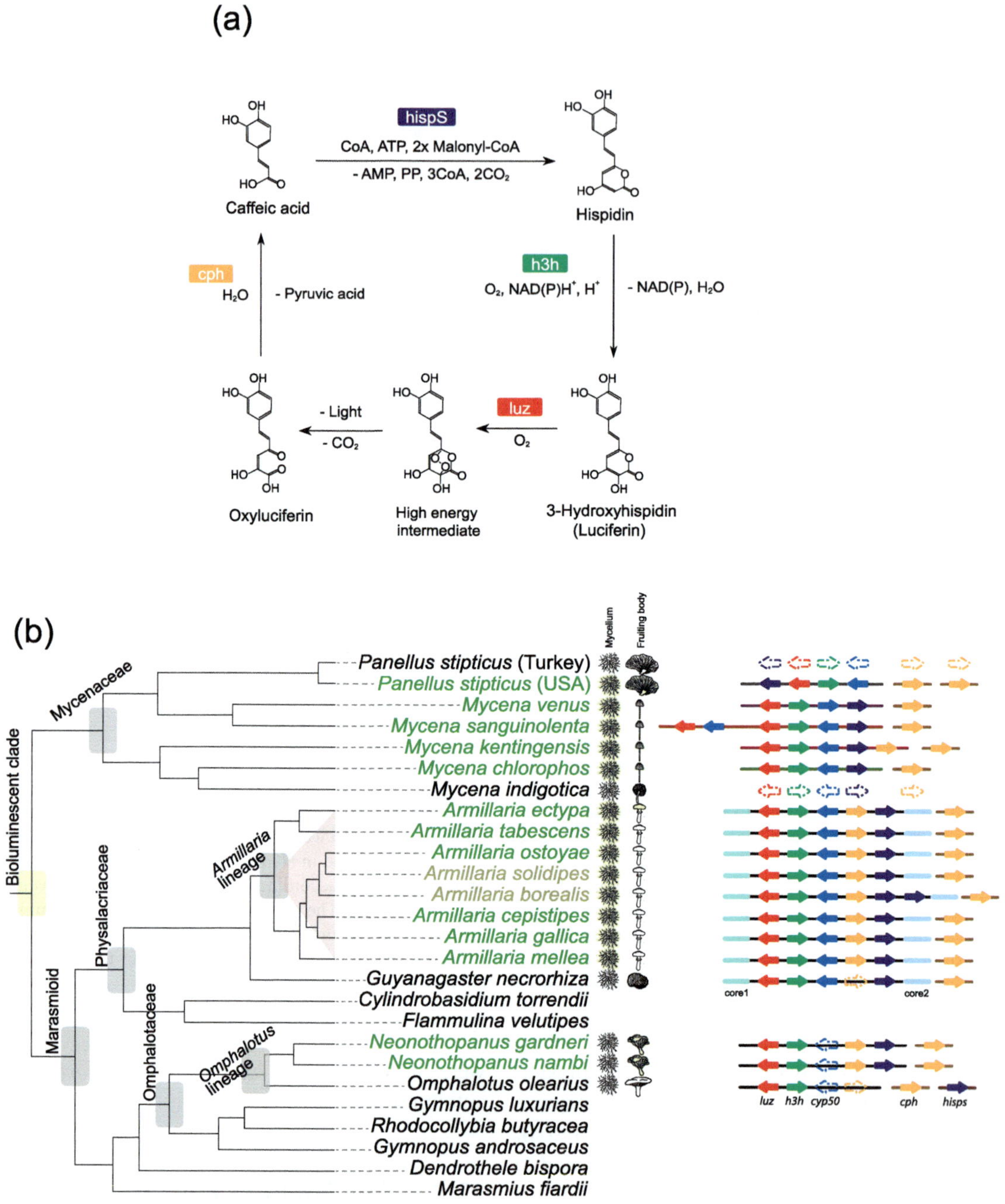

Fig. 12.6 (**a**), Fungal luciferin biosynthetic pathway proposed by (Kotlobay et al. 2018). (**b**), Tree topology from species with genomes and their gene organization of the luciferase cluster inferred from (Ke et al. 2020) and (Kotlobay et al. 2018). Block arrows indicate genes and their orientation. The dashed block arrows denote the loss of a gene. All the species with dark green are bioluminescent species with records in the literature. The species with light green are the species believed to illuminate

sequencing of Agaricales have pointed to an alternative scenario. One study found that the genes responsible for luciferin biosynthetic cycles were physically adjacent in genomes (luciferase cluster; Fig. 12.6b) by comparing genomes from ten bioluminescent fungi and their sister species (Kotlobay et al. 2018). By generating reference genomes of bioluminescent fungal genomes from five Mycenaceae (Ke et al. 2020), eight *Armillaria* lineages, and two *Omphalotus* lineages, we were able to investigate the evolutionary dynamics around the luciferase cluster. Members of the luciferase cluster were clearly homologous across distant bioluminescent fungi, suggesting that bioluminescence was not independently evolved. Both studies concluded that bioluminescence originated from the last common ancestor of the Mycenaceae family and marasmioid clade of Agaricales, although the species trees in these two studies were not congruent. The latter showed that the Schizophyllaceae was the closest lineage with the bioluminescent clade and had a genome encoding a homologous luciferase. Through phylogenetic reconciliation, it was found that the species tree was congruent to the gene trees in the luciferase cluster, suggesting the patchy distribution was a result of a loss of the luciferase cluster in the majority of Agaricales rather than horizontal gene transfer (Ke et al. 2020). However, the origin of these luciferin biosynthetic pathway genes within Schizophyllaceae is still unknown. Our experiment of adding hispidin (luciferin precursor) to *Schizophyllum commune* (unpublished data) showed that it could not be converted to light.

A fungal gene cluster can be differentially maintained due to differences in genome plasticity (Marcet-Houben and Gabaldon 2019). The luciferase cluster was located in different low synteny regions of *Mycena* genomes, which may suggest that they were dispensable and could be lost in most species in the Mycenaceae lineage. In contrast, the luciferase cluster was located in the conserved syntenic region in all eight of the bioluminescent fungal genomes from the *Armillaria* lineage. Ke et al. proposed an evolutionary scenario that the luciferase cluster was originally located at the dispensable region of the last common ancestor and the cluster was translocated to different genomic locations through rearrangement (Ke et al. 2020). In the *Armillaria* lineage, the luciferase cluster was translocated to the core region and became more conserved.

12.5 Biological Function of Bioluminescent Fungi

The roles of bioluminescence in various organisms have been extensively studied (for review please see (Lau and Oakley 2021)), and often involve mating, luring prey or defense. At the molecular level, negative selection was detected on the luciferase cluster, and it could explain why the luciferase cluster had been maintained across hundreds of millions of years (Ke et al. 2020) suggesting that bioluminescence may play a role in environmental adaptation but in a species-specific way. Several pieces of evidence support different possible hypotheses for the function of bioluminescence. First, several studies proposed that fungal bioluminescence may facilitate spore dispersal by attracting potential vectors. More recently, in the Brazilian forest, a biological circadian control of luminescence was found by measuring the profiles of the luciferase-, reductase-, and luciferin-rich extract from mycelia of *N. gardneri*, as well as insects captured using an artificial mushroom with light similar to the bioluminescence of dark mushrooms (Oliveira et al. 2015). It was proposed that the bioluminescent *N. gardneri* of the *Omphalotus* lineage increased spore dispersal by attracting arthropods in the evening. However, the insects' attraction has not been proven in other bioluminescent species in the same lineage, such as *Omphalotus nidiformis* (Weinstein et al. 2016).

Second, fungal bioluminescence is a by-product of an unknown biological process. This hypothesis has been proposed two decades ago according to the existence of luminous and nonluminous populations of the same species or

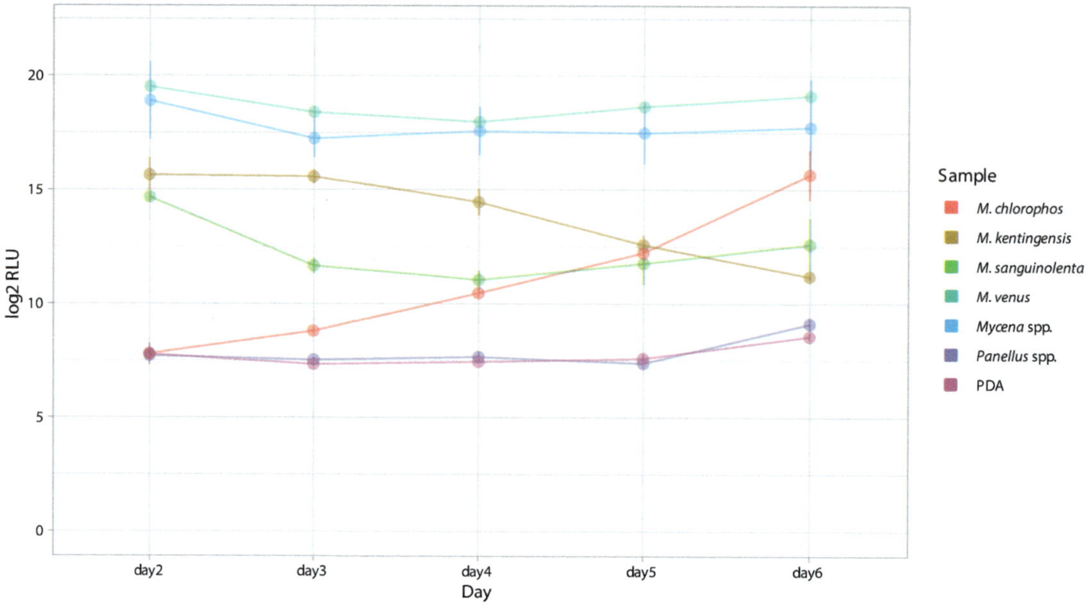

Fig. 12.7 Bioluminescence patterns were measured for 6 days after inoculation of mycelium on potato dextrose agar (PDA). The control was PDA only

the continuous light production in bioluminescent fungi (Herring 1994; Deheyn and Latz 2007). Recent studies revealing the expression of bioluminescence also indicate a biological function. It has been found that the synthesis of luciferin precursors and H3H was inhibited during the *Armillaria* development from mycelium to mushroom (Purtov et al. 2017; Mihail et al. 2018; Puzyr et al. 2017). This suppression resulted in decreased luminescence indicating a possible form of regulation and biological role. There is a divergent intensity in cap, stipe, or mycelium of species of Mycenaceae lineage. Several genes that could be responsible for this regulation were enriched when investigating the differential expression of genes among different stages of *Mycena kentingensis* (Ke et al. 2020). In addition, a diversity of intensities was also found in cultured mycelium among species in the Mycenaceae lineage (Fig. 12.7). Since regulation implies an adaptive function, these observations and analyses provide indirect evidence for a biological role. However, further validation will be necessary to clear up the biological role for each species.

12.6 Conclusions

Bioluminescence in fungi is an intriguing feature restricted to gilled mushrooms that utilizes a single type of biochemical reaction. We have been able to understand the evolution of bioluminescent mushrooms better recently thanks to the accumulated record of which fungi are bioluminescent, molecular studies on luciferin biosynthetic pathways, and genome research. When the first fungal luciferase gene cluster of *Neonothopanus nambi* was identified in 2018, this successfully heterogeneously expressed biosynthetic pathway advanced the further study and application of fungal bioluminescence ((Kotlobay et al. 2018); for a detailed review on potential applications see (Ke and Tsai 2021)). The homologous luciferase found from the genomic sequencing of several bioluminescent fungi and their sister groups allowed us to trace the evolutionary history to a last common ancestor of the Mycenaceae family and the marasmioid clade within Agaricales. Despite a great deal of progress, many important questions still remain. Important goals for future work include developing a more complete understanding of bioluminescent features, ecological function and its gene regulation.

Acknowledgment We thank Yu-Shen Shih and Hua-Te Fang for providing photos of bioluminescent fungi, and Eiji Hadano and Atsuko Hadano for specimen collection to validate the phylogenetic position of *Marasmiellus lucidus* and *M. venosus*. I.J.T was supported by Career Development Award AS-CDA-107-L01, Academia Sinica, Taiwan, and the Ministry of Science and Technology, Taiwan under Grant 110-2628-B-001-027.

References

Ainsworth M (2004) Searching for luminous mushrooms of the marsh fungus Armillaria ectypa. Field Mycol 5(4):142–144

Airth RL, Mc EW (1959) Light emission from extracts of luminous fungi. J Bacteriol 77(2):249–250. https://doi.org/10.1128/jb.77.2.249-250.1959

Anderson JB, Stasovski E (1992) Molecular phylogeny of northern-hemisphere species of Armillaria. Mycologia 84(4):505–516. https://doi.org/10.2307/3760315

Aravindakshan DM, Kumar TKA, Manimohan P (2012) A new bioluminescent species of Mycena sect. Exornatae from Kerala State, India. Mycosphere 3(5):556–561. https://doi.org/10.5943/mycosphere/3/5/4

Baumgartner K, Coetzee MP, Hoffmeister D (2011) Secrets of the subterranean pathosystem of Armillaria. Mol Plant Pathol 12(6):515–534. https://doi.org/10.1111/j.1364-3703.2010.00693.x

Berkeley MJ (1844) Decades of fungi London J Bot 3:185–194

Berliner MD (1961a) Diurnal periodicity of luminescence in three basidiomycetes. Science 134(3481):740. https://doi.org/10.1126/science.134.3481.740

Berliner MD (1961b) Studies in fungal luminescence. Mycologia 53(1):84–90. https://doi.org/10.2307/3756133

Bermudes D, Petersen RH, Nealson KH (1992) Low-level bioluminescence detected in Mycena-Haematopus Basidiocarps. Mycologia 84(5):799–802. https://doi.org/10.2307/3760392

Bigelow HE, Miller JOK, Thiers HD (1976) A new species of Omphalotus. Mycotaxon 3(3):363–372

Bothe F (1930) Ein neuer einheimischer Leuchtpilz, Mycena tintinnabulum. Ber Deut Bot Ges 48:394–399

Bothe F (1931) Über das Leuchten verwesender Blätter und seine Erreger. Planta 14:752–765. https://doi.org/10.1007/bf01917160

Botnen S, Vik U, Carlsen T, Eidesen PB, Davey ML, Kauserud H (2014) Low host specificity of root-associated fungi at an Arctic site. Mol Ecol 23(4): 975–985. https://doi.org/10.1111/mec.12646

Capelari M, Desjardin DE, Perry BA, Asai T, Stevani CV (2011) Neonothopanus gardneri: a new combination for a bioluminescent agaric from Brazil. Mycologia 103(6):1433–1440. https://doi.org/10.3852/11-097

Capella-Gutierrez S, Silla-Martinez JM, Gabaldon T (2009) trimAl: a tool for automated alignment trimming in large-scale phylogenetic analyses. Bioinformatics 25(15):1972–1973. https://doi.org/10.1093/bioinformatics/btp348

Cha JY, Igarashi T (1995) Armillaria species associated with Gastrodia elata in Japan. Eur J For Pathol 25(6–7): 319–326

Chang CC, Chen CY, Lin WW, Kao HW (2020) Mycena jingyinga, Mycena luguensis, and Mycena venus: three new species of bioluminescent fungi from Taiwan. Taiwania 65(3):396–406. https://doi.org/10.6165/tai.2020.65.396

Chew ALC, Desjardin DE, Tan YS, Musa MY, Sabaratnam V (2015) Bioluminescent fungi from Peninsular Malaysia-a taxonomic and phylogenetic overview. Fungal Divers 70(1):149–187. https://doi.org/10.1007/s13225-014-0302-9

Chew ALC, Tan YS, Desjardin DE, Musa MY, Sabaratnam V (2013) Taxonomic and phylogenetic re-evaluation of Mycena illuminans. Mycologia 105(5):1325–1335. https://doi.org/10.3852/13-009

Chew AL, Tan YS, Desjardin DE, Musa MY, Sabaratnam V (2014) Four new bioluminescent taxa of Mycena sect. Calodontes from Peninsular Malaysia. Mycologia 106(5):976–988. https://doi.org/10.3852/13-274

Coetzee MPA, Wingfield BD, Bloomer P, Ridley GS, Kile GA, Wingfield MJ (2001) Phylogenetic relationships of Australian and New Zealand Armillaria species. Mycologia 93(5):887–896. https://doi.org/10.2307/3761754

Coetzee MPA, Wingfield BD, Wingfield MJ (2018) Armillaria root-rot pathogens: species boundaries and global distribution. Pathogens 7(4):83. https://doi.org/10.3390/pathogens7040083

Corner EJH (1950) Descriptions of two luminous tropical agarics (Dictyopanus and Mycena). Mycologia 42: 423–431

Corner EJH (1954) Further descriptions of luminous agarics. Trans Br Mycol Soc 37:256–271

Corner EJH (1981) The agaric genera Lentinus, Panus, and Pleurotus: with particular reference to Malaysian species. Beihefte zur Nova Hedwigia; Heft 69. J. Cramer, Vaduz

Cortes-Perez A, Desjardin DE, Perry BA, Ramirez-Cruz V, Ramirez-Guillen F, Villalobos-Arambula AR, Rockefeller A (2019) New species and records of bioluminescent Mycena from Mexico. Mycologia 111(2):319–338. https://doi.org/10.1080/00275514.2018.1554172

Dauner LAP, Karunarathna SC, Tibpromma S, Xu JC, Mortimer PE (2021) Bioluminescent fungus Roridomyces viridiluminus sp. nov. and the first Chinese record of the genus Roridomyces, from

Southwestern China. Phytotaxa 487(3):233–250. https://doi.org/10.11646/phytotaxa.487.3.4
Davey ML, Heimdal R, Ohlson M, Kauserud H (2013) Host- and tissue-specificity of moss-associated Galerina and Mycena determined from amplicon pyrosequencing data. Fungal Ecol 6(3):179–186. https://doi.org/10.1016/j.funeco.2013.02.003
Deheyn DD, Latz MI (2007) Bioluminescence characteristics of a tropical terrestrial fungus (Basidiomycetes). Luminescence 22(5):462–467. https://doi.org/10.1002/bio.985
Delroisse J, Duchatelet L, Flammang P, Mallefet J (2021) Leaving the dark side? Insights into the evolution of luciferases. Front Mar Sci 8:ARTN 673620. https://doi.org/10.3389/fmars.2021.673620
Desjardin DE, Braga-Neto R (2007) Mycena lacrimans, a rare species from Amazonia, is bioluminescent. Edinb J Bot 64(3):275–281
Desjardin DE, Capelari M, Stevani CV (2005) A new bioluminescent agaric from Sao Paulo, Brazil. Fungal Divers 18:9–14
Desjardin DE, Capelari M, Stevani C (2007) Bioluminescent Mycena species from Sao Paulo, Brazil. Mycologia 99(2):317–331. https://doi.org/10.3852/mycologia.99.2.317
Desjardin DE, Oliveira AG, Stevani CV (2008) Fungi bioluminescence revisited. Photochem Photobiol Sci 7(2):170–182. https://doi.org/10.1039/b713328f
Desjardin DE, Perry BA, Lodge DJ, Stevani CV, Nagasawa E (2010) Luminescent Mycena: new and noteworthy species. Mycologia 102(2):459–477
Desjardin DE, Perry BA, Stevani CV (2016) New luminescent mycenoid fungi (Basidiomycota, Agaricales) from Sao Paulo State, Brazil. Mycologia 108(6): 1165–1174. https://doi.org/10.3852/16-077
Guo T, Wang HC, Xue WQ, Zhao J, Yang ZL (2016) Phylogenetic analyses of Armillaria reveal at least 15 phylogenetic lineages in China, seven of which are associated with cultivated Gastrodia elata. PLoS One 11(5):e0154794. https://doi.org/10.1371/journal.pone.0154794
Haddock SH, Moline MA, Case JF (2010) Bioluminescence in the sea. Annu Rev Mar Sci 2:443–493. https://doi.org/10.1146/annurev-marine-120308-081028
Haneda Y (1955) Luminous organisms of Japan and the Far East. The Luminescence of Biological Systems. American Association for the Advancement of Science, Washington DC
Harder CB, Hesling E, Botnen SS, Dima B, von Bonsdorff-Salminen T, Niskanen T, Jarvis SG, Lorberau KE, Ouimette A, Hester A, Hobbie EA, Taylor AFS, Kauserud H (2021) Mycena species can be opportunist-generalist plant root invaders. bioRxiv. https://doi.org/10.1101/2021.03.23.436563
Harvey EN (1952) Bioluminescence. Academic Press, New York
Herring PJ (1994) Luminous fungi. s.n., S.l
Horak E (1978) Mycena rorida (Fr.) Quél. And related species from the southern hemisphere. Ber Schweiz Bot Ges 88(1/2):20–29
Josserand M (1953) Sur la luminescence de "Mycena rorida" en Europe occidentale. Bull Mens Soc Linn Lyon 22(4):99–102
Karunarathna SC, Mortimer PE, Tibpromma S, Dutta AK, Paloi S, Hu YW, Baurah G, Axford S, Marciniak C, Luangharn T, Madawala S, Lin C, Chen JZ, Acharya K, Kobmoo N, Samarakoon MC, Karunarathna A, Gao SY, Xu JC, Lumyong S (2020) Roridomyces phyllostachydis (Agaricales, Mycenaceae), a new bioluminescent fungus from Northeast India. Phytotaxa 459(2):155–167. https://doi.org/10.11646/phytotaxa.459.2.6
Kaskova ZM, Dorr FA, Petushkov VN, Purtov KV, Tsarkova AS, Rodionova NS, Mineev KS, Guglya EB, Kotlobay A, Baleeva NS, Baranov MS, Arseniev AS, Gitelson JI, Lukyanov S, Suzuki Y, Kanie S, Pinto E, Di Mascio P, Waldenmaier HE, Pereira TA, Carvalho RP, Oliveira AG, Oba Y, Bastos EL, Stevani CV, Yampolsky IV (2017) Mechanism and color modulation of fungal bioluminescence. Sci Adv 3(4): e1602847. https://doi.org/10.1126/sciadv.1602847
Katoh K, Standley DM (2013) MAFFT multiple sequence alignment software version 7: improvements in performance and usability. Mol Biol Evol 30(4):772–780. https://doi.org/10.1093/molbev/mst010
Kawamura S (1910) Studies on the luminous fungus, Pleurotus japonicus,sp. nov. 24:275–281
Ke HM, Lee HH, Lin CI, Liu YC, Lu MR, Hsieh JA, Chang CC, Wu PH, Lu MJ, Li JY, Shang G, Lu RJ, Nagy LG, Chen PY, Kao HW, Tsai IJ (2020) Mycena genomes resolve the evolution of fungal bioluminescence. Proc Natl Acad Sci U S A 117(49): 31267–31277. https://doi.org/10.1073/pnas.2010761117
Ke HM, Tsai IJ (2021) Understanding and using fungal bioluminescence – Recent progress and future perspectives. Curr Opin Green Sustain Chem 33(100570):1–6. https://doi.org/10.1016/j.cogsc.2021.100570
Kedves O, Shahab D, Champramary S, Chen L, Indic B, Boka B, Nagy VD, Vagvolgyi C, Kredics L, Sipos G (2021) Epidemiology, biotic interactions and biological control of Armillarioids in the northern hemisphere. Pathogens 10(1):76. https://doi.org/10.3390/pathogens10010076
Kernaghan G, Patriquin G (2011) Host associations between fungal root endophytes and boreal trees. Microb Ecol 62(2):460–473. https://doi.org/10.1007/s00248-011-9851-6
Kirchmair M, Morandell S, Stolz D, Poder R, Sturmbauer C (2004) Phylogeny of the genus Omphalotus based on nuclear ribosomal DNA-sequences. Mycologia 96(6): 1253–1260. https://doi.org/10.2307/3762142
Kirk PM, Ainsworth GC, Bisby GR, Bioscience C (2008) Dictionary of the fungi. 10th / prepared by CABI Bioscience, 10th edn. CAB International, Wallingford

Kobayasi Y (1951) Contributions to the luminous fungi from Japan. J Hattori Bot Lab 5:1–6

Koch RA, Wilson AW, Sene O, Henkel TW, Aime MC (2017) Resolved phylogeny and biogeography of the root pathogen Armillaria and its gasteroid relative, Guyanagaster. BMC Evol Biol 17(1):33. https://doi.org/10.1186/s12862-017-0877-3

Kotlobay AA, Sarkisyan KS, Mokrushina YA, Marcet-Houben M, Serebrovskaya EO, Markina NM, Gonzalez Somermeyer L, Gorokhovatsky AY, Vvedensky A, Purtov KV, Petushkov VN, Rodionova NS, Chepurnyh TV, Fakhranurova LI, Guglya EB, Ziganshin R, Tsarkova AS, Kaskova ZM, Shender V, Abakumov M, Abakumova TO, Povolotskaya IS, Eroshkin FM, Zaraisky AG, Mishin AS, Dolgov SV, Mitiouchkina TY, Kopantzev EP, Waldenmaier HE, Oliveira AG, Oba Y, Barsova E, Bogdanova EA, Gabaldon T, Stevani CV, Lukyanov S, Smirnov IV, Gitelson JI, Kondrashov FA, Yampolsky IV (2018) Genetically encodable bioluminescent system from fungi. Proc Natl Acad Sci U S A 115(50): 12728–12732. https://doi.org/10.1073/pnas.1803615115

Lau ES, Oakley TH (2021) Multi-level convergence of complex traits and the evolution of bioluminescence. Biol Rev Camb Philos Soc 96(2):673–691. https://doi.org/10.1111/brv.12672

Léveillé JH (1844) Champignons exotiques. Annales des Sciences Naturelles Botanique 3:167–221

Li J, Hu X (1993) A new species of Lampteromyces from Hunan. Act Sci Nat Univ Norm Hunan 16(2):188–189

Liao HL, Chen Y, Bruns TD, Peay KG, Taylor JW, Branco S, Talbot JM, Vilgalys R (2014) Metatranscriptomic analysis of ectomycorrhizal roots reveals genes associated with Piloderma-Pinus symbiosis: improved methodologies for assessing gene expression in situ. Environ Microbiol 16(12): 3730–3742. https://doi.org/10.1111/1462-2920.12619

Liu PG (1995) Luminous fungi. Biodivers Sci 3(2): 109–112

Lorberau KE, Botnen SS, Mundra S, Aas AB, Rozema J, Eidesen PB, Kauserud H (2017) Does warming by open-top chambers induce change in the root-associated fungal community of the arctic dwarf shrub Cassiope tetragona (Ericaceae)? Mycorrhiza 27(5):513–524. https://doi.org/10.1007/s00572-017-0767-y

Maas Geesteranus RA (1992) Filoboletus manipularis and some related species. Proceedings of the Koninklijke Nederlandse Akademie van Wetenschappen 95(2): 267–274

Marcet-Houben M, Gabaldon T (2019) Evolutionary and functional patterns of shared gene neighbourhood in fungi. Nat Microbiol 4(12):2383–2392. https://doi.org/10.1038/s41564-019-0552-0

Martos F, Dulormne M, Pailler T, Bonfante P, Faccio A, Fournel J, Dubois MP, Selosse MA (2009) Independent recruitment of saprotrophic fungi as mycorrhizal partners by tropical achlorophyllous orchids. New Phytol 184(3):668–681. https://doi.org/10.1111/j.1469-8137.2009.02987.x

Mihail JD (2015) Bioluminescence patterns among North American Armillaria species. Fungal Biol 119(6): 528–537. https://doi.org/10.1016/j.funbio.2015.02.004

Mihail JD, Bilyeu L, Lalk SR (2018) Bioluminescence expression during the transition from mycelium to mushroom in three north American Armillaria and Desarmillaria species. Fungal Biol 122(11): 1064–1068. https://doi.org/10.1016/j.funbio.2018.08.007

Mihail JD, Bruhn JN (2007) Dynamics of bioluminescence by Armillaria gallica, A mellea and A tabescens. Mycologia 99(3):341–350

Miller OK Jr (1994) Observations on the genus Omphalotus in Australia. Mycol Helv 6:91–100

Moncalvo JM, Vilgalys R, Redhead SA, Johnson JE, James TY, Catherine Aime M, Hofstetter V, Verduin SJ, Larsson E, Baroni TJ, Greg Thorn R, Jacobsson S, Clemencon H, Miller OK Jr (2002) One hundred and seventeen clades of euagarics. Mol Phylogenet Evol 23(3):357–400. https://doi.org/10.1016/S1055-7903(02)00027-1

Nguyen LT, Schmidt HA, von Haeseler A, Minh BQ (2015) IQ-TREE: a fast and effective stochastic algorithm for estimating maximum-likelihood phylogenies. Mol Biol Evol 32(1):268–274. https://doi.org/10.1093/molbev/msu300

Ogura-Tsujita Y, Gebauer G, Hashimoto T, Umata H, Yukawa T (2009) Evidence for novel and specialized mycorrhizal parasitism: the orchid Gastrodia confusa gains carbon from saprotrophic Mycena. Proc Biol Sci 276(1657):761–767. https://doi.org/10.1098/rspb.2008.1225

Oliveira AG, Desjardin DE, Perry BA, Stevani CV (2012) Evidence that a single bioluminescent system is shared by all known bioluminescent fungal lineages. Photochem Photobiol Sci 11(5):848–852. https://doi.org/10.1039/c2pp25032b

Oliveira AG, Stevani CV (2009) The enzymatic nature of fungal bioluminescence. Photochem Photobiol Sci 8(10):1416–1421. https://doi.org/10.1039/b908982a

Oliveira AG, Stevani CV, Waldenmaier HE, Viviani V, Emerson JM, Loros JJ, Dunlap JC (2015) Circadian control sheds light on fungal bioluminescence. Curr Biol 25(7):964–968. https://doi.org/10.1016/j.cub.2015.02.021

Oliveira JJS, Vargas-Isla R, Cabral TS, Rodrigues DP, Ishikawa NK (2019) Progress on the phylogeny of the Omphalotaceae: Gymnopus s. str., Marasmiellus s. str., Paragymnopus gen. nov. and Pusillomyces gen. nov. Mycol Prog 18(5):713–739. https://doi.org/10.1007/s11557-019-01483-5

Piercey-Normore MD, Egger KN, Berube JA (1998) Molecular phylogeny and evolutionary divergence of North American biological species of Armillaria. Mol Phylogenet Evol 10(1):49–66. https://doi.org/10.1006/mpev.1997.0485

Purtov KV, Petushkov VN, Baranov MS, Mineev KS, Rodionova NS, Kaskova ZM, Tsarkova AS, Petunin AI, Bondar VS, Rodicheva EK, Medvedeva SE, Oba Y, Oba Y, Arseniev AS, Lukyanov S, Gitelson JI, Yampolsky IV (2015) The chemical basis of fungal bioluminescence. Angew Chem Int Ed Engl 54(28): 8124–8128. https://doi.org/10.1002/anie.201501779

Purtov KV, Petushkov VN, Rodionova NS, Gitelson JI (2017) Why does the bioluminescent fungus Armillaria mellea have luminous mycelium but nonluminous fruiting body? Dokl Biochem Biophys 474(1): 217–219. https://doi.org/10.1134/S1607672917030176

Puzyr AP, Medvedeva SE, Bondar VS (2017) Biochemical changes causes lack of bioluminescence in fruiting bodies of Armillaria. Mycosphere 8(1):9–17. https://doi.org/10.5943/mycosphere/8/1/2

Quesada-Chanto A, Jimenez-Ulate F (1996) Short communication: in vitro evaluation of a Bacillus sp. for the biological control of the coffee phytopathogen Mycena citricolor. World J Microbiol Biotechnol 12(1):97–98. https://doi.org/10.1007/BF00327810

Rao DV, Tewari JP (1987) Production of oxalic acid by Mycena citricolor, causal agent of the American leaf spot of coffee. Phytopathology 77(6):780–785. https://doi.org/10.1094/Phyto-77-780. Index Fungorum 14: 1–1 (2012)

Redhead SA, Moncalvo JM, Vilgalys R, Desjardin DE, Perry BA (2012) Index Fungorum 14:1–1

Rishbeth J (1986) Some characteristics of English Armillaria species in culture. T Brit Mycol Soc 86: 213–218. https://doi.org/10.1016/S0007-1536(86)80147-4

Saccardo PA (1887) Sylloge Hymenomycetum, Vol. I. Agaricineae. Sylloge Fungorum 5:1–1146

Sekizaki H, Kuninaga S, Yamamoto M, Asazu SN, Sawa S, Kojoma M, Yokosawa R, Yoshida N (2008) Identification of Armillaria nabsnona in gastrodia tubers. Biol Pharm Bull 31(7):1410–1414. https://doi.org/10.1248/bpb.31.1410

Shaw CG, Kile GA, United States. Forest Service (1991) Armillaria root disease. Agriculture handbook, vol 691. Forest Service, U.S. Dept. of Agriculture, Washington, DC

Shih YS, Chen CY, Lin WW, Kao HW (2014) Mycena kentingensis, a new species of luminous mushroom in Taiwan, with reference to its culture method. Mycol Prog 13(2):429–435. https://doi.org/10.1007/s11557-013-0939-x

Singer R (1947) New genera of Fungi.3. Mycologia 39(1): 77–89. https://doi.org/10.2307/3755289

Sipos G, Anderson JB, Nagy LG (2018) Armillaria. Curr Biol 28(7):R297–R298. https://doi.org/10.1016/j.cub.2018.01.026

Smith ML, Bruhn JN, Anderson JB (1992) The fungus Armillaria-Bulbosa is among the largest and oldest living organisms. Nature 356(6368):428–431. https://doi.org/10.1038/356428a0

Terashima Y, Takahashi H, Taneyama Y (2016) The fungal flora in southwestern Japan: agarics and boletes. Tokai University Press, Japan

Treu R, Agerer R (1990) Culture characteristics of some Mycena species. Mycotaxon 38:279–309

Ulloa M, Hanlin RT (2017) Illustrated dictionary of mycology, 2nd edn. APS Press, St. Paul, MN

Wassink EC (1948) Observations on the luminescence in fungi, 1, including a critical review of the species mentioned as luminescent in literature. Recueil des travaux botaniques néerlandais 41(1):150–212

Wassink EC (1978) Luminescence in fungi. In: Herring PJ (ed) Bioluminescence in action. Academic Press, London, pp 171–197

Wassink EC (1979) On fungus luminescence. Meded Landbouwhogeschool Wageningen 79 (5)

Weinstein P, Delean S, Wood T, Austin AD (2016) Bioluminescence in the ghost fungus Omphalotus nidiformis does not attract potential spore dispersing insects. IMA Fungus 7(2):229–234. https://doi.org/10.5598/imafungus.2016.07.02.01

Widder EA (2010) Bioluminescence in the ocean: origins of biological, chemical, and ecological diversity. Science 328(5979):704–708. https://doi.org/10.1126/science.1174269

Zang M (1979) Some new species of higher fungi from Xizang (Tibet) of China. Acta Bot Yunnanica 1(2): 101–105

Zhang L, Chen J, Lv Y, Gao C, Guo S (2012) *Mycena* sp., a mycorrhizal fungus of the orchid Dendrobium officinale. Mycol Prog 11:395–401. https://doi.org/10.1007/s11557-011-0754-1

Paedomorphosis and Evolution of Sequestrate Basidiomycetes

13

Francisco Kuhar, Eduardo Nouhra, Donald H. Pfister, and Matthew E. Smith

Abstract

Current theories on the evolution of sequestrate (enclosed) basidiomes explain the origin of these forms in a gradualist adaptational process led by selective forces, such as drought and animal consumption. Paedomorphosis (the retention of juvenile traits) has been invoked as the phenomenon underlying sequestration, but many consequences of this process have not yet been explored. Our present interpretation of sequestrate morphologies, in light of Stephen Jay Gould's characterization of neoteny (retention of juvenile features in an adult stage with mature reproductive structures) and progenesis (the onset of sexual maturity in a morphologically immature stage that does not reach the mature morphology observed in the ancestral form), both involved in paedomorphosis, implies that the origin of sequestrate basidiomes might constitute two distinct evolutionary processes. These two processes could be recognized among fungi by contrasting their morphological plasticity, phylogenetic diversification, and ecological patterns. The hypotheses discussed here provide new insights for interpreting and studying the evolution of sequestrate fungi.

Keywords

Gasteroid · Secotioid · Progenesis · Neoteny · Heterochrony

13.1 Introduction

Sequestrate Agaricomycotina (Basidiomycota) is an artificial (polyphyletic) group of taxa that develops more or less enclosed spore-producing structures (basidiomes) that lack forcibly discharged basidiospores. Sequestrate is a term used to describe fungi with a more-or-less enclosed fruiting bodies, including a wide range of fungal taxa. Sequestrate basidiomes exhibit various morphologies ranging from veiled mushrooms in which the veil remains in place, to bird's nest fungi, earthballs, earthstars, puffballs, truffle-like fungi, and stinkhorns. Here we use the term gasteroid to refer to enclosed basidiomes that do not open or open only late in the maturation process. We use the term secotioid to refer to a subset of these sequestrate taxa that resemble agaricoid relatives in having a stalk (which may be vestigial) and cap, but in which the cap never completely opens. The term

F. Kuhar (✉) · E. Nouhra
Instituto Multidisciplinario de Biología Vegetal (CONICET), FCEFyN, Universidad Nacional de Córdoba, Córdoba, Argentina

D. H. Pfister
Department of Organismic and Evolutionary Biology, Harvard University, Cambridge, MA, USA

M. E. Smith
Department of Plant Pathology, University of Florida, Gainesville, FL, USA

© The Author(s), under exclusive license to Springer Nature Switzerland AG 2023

S. Pöggeler, T. James (eds.), *Evolution of Fungi and Fungal-Like Organisms*, The Mycota 14,
https://doi.org/10.1007/978-3-031-29199-9_13

sequestrate is used here following Peintner et al. (2001), Justo et al. (2010), and Caiafa et al. (2021a) to refer to secotioid and gasteroid basidiomes as well as taxa such as bird's nest fungi and stinkhorns.

Historically the evolution of sequestrate forms within Basidiomycota has been a subject of great interest among mycologists. Early evolutionary hypotheses interpreted the sequestrate morphology as primitive in respect to the typical mushroom morphology. This idea was held until the mid-twentieth century (Fig. 13.1a). For example, Singer (1958) regarded the evolution of sequestrate fungi from agaricoid ancestors as "degradationist." This idea recalled the Haeckelian paradigm of recapitulation, which maintained that "ontogeny recapitulates phylogeny" (Haeckel 1866). Since many agaricoid basidiomes show enclosed stages during their development, it seemed logical to assume that this developmental sequence could be applied to discussions of phylogeny. This is evident from Singer's (1958) words, "According to Haeckel's rule, one might then assume a progressive differentiation." Singer's implication was that agaricoid basidiomes develop from enclosed forms, so agaricoid basidiomes must have evolved from gasteroid ancestors (and that secotioid fungi must be an intermediate stage between the two forms).

The idea of heterochrony (proposed by Haeckel (1866) and reinterpreted by De Beer (1951)) is that changes in the timing of embryonic development are the basis for many evolutionary innovations. Heterochrony challenged the concept of recapitulation and also played a decisive role in the biological sciences during the twentieth century (Shea 1989). Although basidiome fruiting does not represent the development of the entire organism (and is therefore different from embryonic development), this concept has nonetheless been applied to sequestrate basidiome evolution by many authors, as summarized by Hibbett et al. (1994). Gerber and Hopkins (2011) championed the idea that heterochronic changes may often affect only a part of the organism (e.g., only the development of basidiomes), a phenomenon that they call "local or dissociated heterochrony."

Sequestrate basidiomycetes have been interpreted as paedomorphic, i.e. as evolutionary lines in which the development remains arrested at an enclosed "juvenile" stage (Thiers 1984; Bruns et al. 1989). Gube (2009) suggests that paedomorphosis is "the disturbance of the succession of developmental regulation" and assumed it to be "the triggering process of gasteromycetation." As a logical consequence, we now accept that the sequestrate state is often (but not always) evolutionarily derived from the agaricoid morphology (Fig. 13.1b), an idea that was first outlined by Anton de Bary (1884). A recent review by Virágh et al. (2021) explains that "from normal pileate-stipitate species, the road seems to lead through agaricoid forms with permanently closed caps and contorted gills to closed (angiocarpic) fruiting bodies in which the stipe starts to degenerate and eventually disappear (or remain as a columella) and hymenial tissue shows transitions to a gleba-like structure." This statement highlights how this "sequential" hypothesis has permeated the literature and is still the most accepted and influential concept to explain the development of sequestrate fungi.

In 1977, the paleontologist Stephen Jay Gould published "Ontogeny and Phylogeny," a book in which he extensively revised the major ideas developed around heterochrony. In this work, Gould (1977) not only discussed the processes involved but also summarized the implications of these processes at levels ranging from morphological considerations, to phylogenetic, taxonomic, and even ecological considerations. Although the paedogenetic origin of sequestrate taxa has been proposed by many mycologists, the implications of this hypothesis have not been fully explored. This discontinuity in the study of heterochronic processes has also been the case for many animal groups; Gould (1977) attributed this neglect to the fact that much of the research on this subject has been published in German (e.g., Müller 1864; Wirz 1950; Mangold-Wirz 1966).

Gould dedicated the ninth chapter of his book to the distinction between two different processes

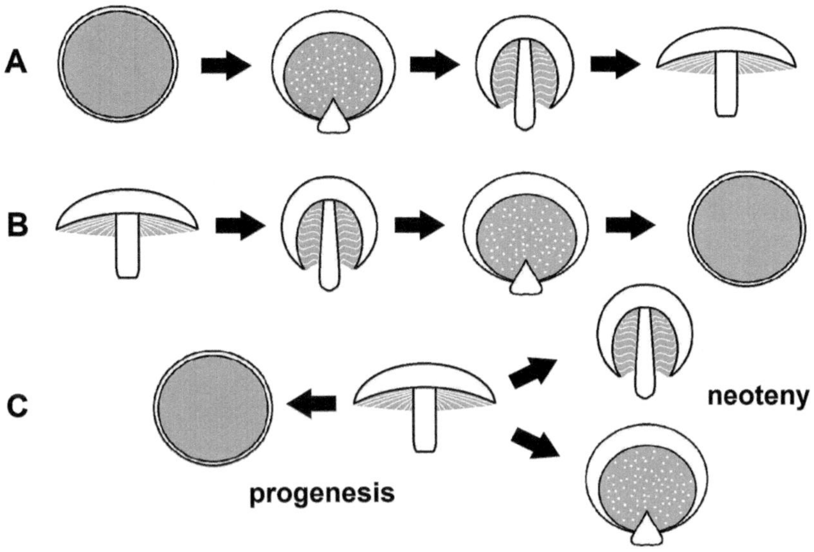

Fig. 13.1 Hypotheses on the morphological evolution of sequestrate basidiomes (**a**) obsolete concept based on the Hackelian "recapitulation paradigm." (**b**) modern gradualist-adaptationist conception. (**c**) present proposal considering Gould's concept of paedomorphosis that may be the result of two distinct processes, either neoteny or progenesis

that both fit the definition of paedomorphosis: progenesis and neoteny. He defines neoteny as the retention of juvenile features in an adult stage with mature reproductive structures. In contrast, progenesis is the onset of sexual maturity in a morphologically immature stage (relative to the ancestral developmental sequence) that does not reach the mature morphology observed in the ancestral form. Based on these delimitations, he suggested a series of features and patterns in development that would be associated with either progenesis or neoteny.

If we apply these concepts to sequestrate fungi (Fig. 13.2), we can interpret basidiomes as a product of either progenesis or neoteny. Neotenic basidiomes are those where the fertile structure results from the contortion and anastomosis of an adult-like hymenophore (spore-producing layers—the tissues supporting the hymenium, e.g. gills, pores, teeth, etc.). In the case of neotenic basidiomes, it should be possible to identify a true hymenium. In contrast, progenetic basidiomes are those in which the ancestral hymenophoral configuration is not even insinuated during the ontogeny and therefore no true hymenium should be present. Although in most cases this distinction is straightforward, the presence of true hymenia is difficult to assess in some taxa. In some cases, fertile hyphae can secondarily invade the surface of preexistent glebal cavities (gleba is the term used in sequestrate fungi for the mass of basidiospores produced in the enclosed basidiome). For example, this seems to be the case in species of *Rhizopogon* (Reijnders 2000, see discussion). Other macromorphological features highlight this distinction: in the basidiomes of neotenic fungi such as those occurring in genera such as *Amanita, Cortinarius, Descolea, Entoloma,* and *Russula*, the "adult" ancestral structures often can be recognized. In cases such as these, the differences between the gasteroid morphology and that of the ancestral stage are the consequence of the retention of the enclosed hymenophore. At the opposite end of the spectrum, progenetic forms are those where the ancestral, mature morphology is not clear and the reproductive stage is reached in an "egg-like" basidiome and related "transitional forms" are lacking. Examples of this progenetic type are the saprobic Geastrales, *Lycoperdon* (and allied Agaricales), and Phallales, as well as the ectomycorrhizal Hysterangiales and Sclerodermatineae. The presence of a stipe has proven to be misleading and is apparently unrelated to hymenophore development, since some progenetic taxa (e.g., *Tulostoma* and relatives) retain a robust stipe (Reijnders 2000). Accordingly, the ontogeny of the fertile

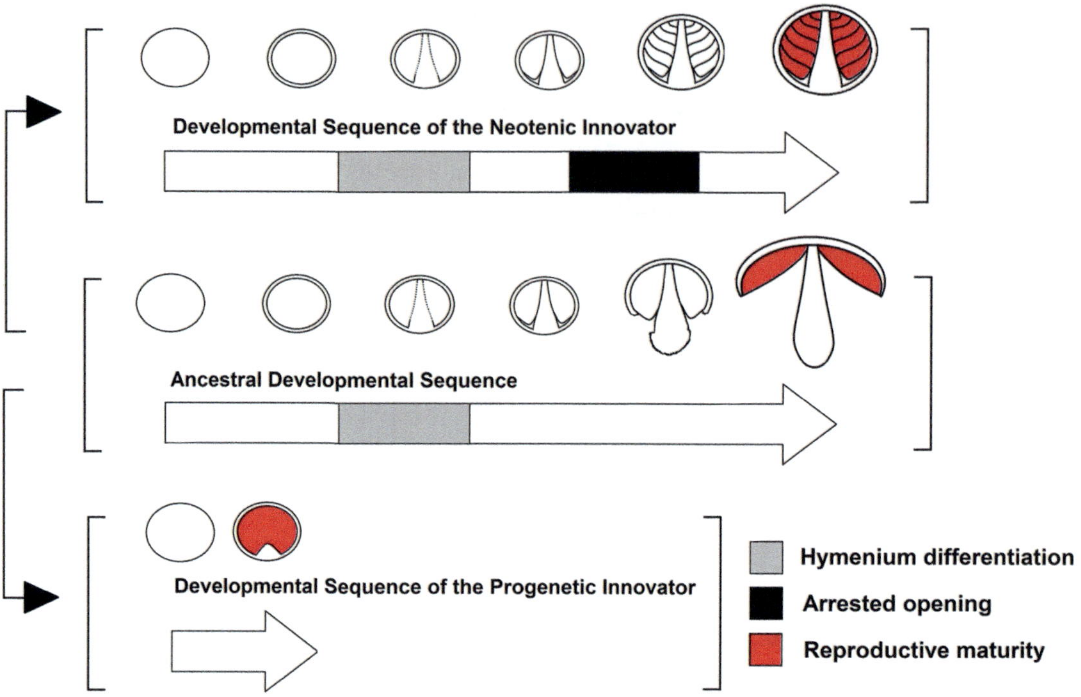

Fig. 13.2 Neotenic and progenetic development compared with the ancestral sequence. The example provided here shows the idealized sequence for a typical agaricoid mushroom and related forms, but the concept could also be applied to other sequestrate morphological forms

tissue is the most reliable criterion to differentiate neotenic versus progenetic forms.

Progenetic and neotenic fungi have an interesting taxonomic history. Glebal development without traces of a true hymenium was the critical character used by Schröter (1876) to define the order Plectobasidiales. Rauschert (1964) indicated that other mycologists had already recognized these significant differences in the ontogeny of the spore-producing layers. The fungi we treat here as progenetic and neotenic were delimited by Malençon (1955) as Exogastrineae (with a true hymenium) and Endogastrineae (lacking a true hymenium). Pilát (1958) used the similar terms Exogasteromycetidae and Endogasteromycetidae, whereas Heim (1971) used *Gastérales agaricoides* and *Gastérales vraies* (roughly equivalent to "agaricoid gasteromycetes" and "true gasteromycetes"). These three authors employed similar groupings and included mostly the same taxa (Reijnders 2000). Based on current molecular data we now know that these groups are not monophyletic taxonomic units but instead represent several independently derived lineages. These legacy classifications are informative since they reflect different developmental and potentially evolutionary processes. The first group, Exogastrineae, more closely resembles the agaricoid fungi than the other group, Endogastrineae, due to the presence of recognizable hymenia (Reijnders 2000). We hypothesize that this similarity to agaricoid fungi is due to the neotenic origin of the gasteroid morphology and can be attributed to the retention of the enclosed basidiome at maturity (neotenic sequestration, Fig. 13.2). The second group of species (Endogastrineae) lacks true hymenia because the progenetic enclosed structure reaches sexual maturity while arrested in an ontogenetic stage prior to hymenophoral differentiation (progenetic sequestration, Fig. 13.2).

Some authors have suggested a hymenium is present in groups of gasteroid fungi that have a more-or-less unorganized gleba, such as *Lycoperdon* (Rehsteiner 1892), members of the Geastrales (Sunhede 1993), and the Phallales (Dring 1973). The organization of the spore-producing tissues in these groups has been discussed extensively in the past (e.g., Lohwag 1926; Fischer 1933; Malençon 1955; Dring 1973). This line of work was based on the idea that spaces within the developing gleba were remnants of an ancestral hymenial surface. However, these concepts have mostly been ignored in recent literature on sequestrate fungi and were regarded as unproductive and potentially misleading by Reijnders (2000) who stated that "this distinction of glebal development lacks much value," and that "the facts concerning the question of hymenia in gasteromycetes is disputable." For the present discussion, we follow Reijnders (2000) regarding the definition and phylogenetic distribution of true hymenia with the caveat that some taxa likely require additional morphological study to better characterize their development.

13.2 Phylogenetic Trends

The fact that sequestrate forms are prolific and phylogenetically dispersed among Agaricomycotina has been ascribed to adaptive forces such as drought or mycophagy (Thiers 1984). Accelerated differentiation of features associated with sequestration has been demonstrated by Bruns et al. (1989) for the genus *Rhizopogon*. These authors mention paedomorphosis as a possible explanation, but they also indicate that "sparsity of intermediates between *Rhizopogon* and *Suillus* provides indirect evidence of intense selective pressure" making clear that the resulting sequestrate morphology is driven by this force. If the paedomorphic hypothesis proposed here is corroborated, no intensified selective pressure is needed to explain the sudden morphological reconfiguration. Although selective pressures are the basis for gradual evolution, heterochrony (e.g., paedomorphosis) leads to saltational evolution (a process of sudden, abrupt changes that give rise to new taxa). The concept of saltation can be traced back to E. G. Saint-Hilaire (1837). Goldschmidt (1933) later delimited saltation by explaining that macromutations (mutations involving dramatic rearrangements) would result in "hopeful monsters" which, if successful, might be perpetuated in the population. Gould (1977) remarked that progenesis would provide the basis for the rise of "hopeful monsters," citing Goldschmidt: "The mutant gene produces its effect... by changing the rates of partial processes of development." Sequestration might constitute a great example of saltation (Baura et al. 1992); rather than evolving towards enclosed forms through long selective processes, basidiome evolution could come about by exploiting the library of immature stages of the organism in one simple evolutionary step. Evolving towards an "immature" form would explain this acceleration (saltation) of the evolution of truffle-like forms suggested by Bruns et al. (1989). This is not unlike the rampant convergence in form exhibited by sequestrate fungi, and is reflected in the work of Wake (1991) in paedomorphic salamanders conceptualized as "paedomorphic homoplasy." Further changes following the sequestration event, such as the loss of active spore discharge, can be ascribed to the suggestion by Bruns et al. (1989) that "loss of function" mutations occur more frequently than acquisition of complex features. An alternative model to explain the emergence of "highly derived hypogeous lineages" was proposed by Albee-Scott (2007) which he called, "secotioid inertia." This model proposes extinction of epigeous (above ground) sister taxa after the emergence of gasteroid taxa due to selection pressures (such as drought or animal dispersal). In this scenario, selection would play a key role and the loss of epigeous taxa via extinction could make evolutionary relationships difficult to discern. Gube (2009) suggested that paedomorphosis offers a more plausible hypothesis for sequestration than does secotioid inertia, since it does not imply orthogenesis (an intrinsic or innate tendency to evolve toward some goal). However, the two concepts are not mutually exclusive.

This apparently directional evolution (from exposed to enclosed fertile tissues) is indeed a likely result from heterochronic evolution. McNamara (1982) suggests that heterochronicity is one of the main causes of directional change, since related lineages evolving immature morphology would independently reach similar body plans. In simple words, if fruiting bodies of a certain genus evolve toward forms that resemble immature stages, the products are likely to look similar since the immature stages also tend to look similar for taxa within a genus. This also means that the final morphology is not chosen among many potential possibilities by natural selection (Gould 1977). Another consequence suggested by McNamara (1982) is that the coexistence of the neotenic innovator with the ancestral morphology would secure the irreversibility of this change via competition, constituting an alternative explanation to the "non-reversibility" of sequestrate forms due to the loss of the ballistic discharge of basidiospores.

Another aspect of paedomorphosis discussed by Gould (1977) is the probability that heterochronic changes are retained by natural selection. Whereas neoteny implies more subtle modifications with a higher likelihood for survival, the dramatic changes provoked by progenesis are expected to have lower fitness on average and therefore lower chances of success in nature. This is evident in the numerous origins of neotenic sequestrate forms that have given rise to polyphyletic entities such as *Thaxterogaster* (sequestrate *Cortinarius*), *Gastroboletus* (sequestrate *Boletus* relatives), and *Zelleromyces* (sequestrate *Lactarius*), among others.

13.3 Diversification Patterns

One of the most interesting predictions that Gould outlines in Ontogeny and Phylogeny (1977) is that progenesis and neoteny play different roles in macroevolution. Gould predicts that neoteny would be an escape from a highly competitive niche for a single species, whereas progenesis would provide the foundation for new lineages to arise and be tested by selection. However, he predicted that progenesis would be only rarely successful because of the new and dramatic morphological changes that could be deleterious (e.g., the "hopeful monsters").

The diversification patterns of sequestrate fungi follow the predictions of Gould (1977): the progenetic puffballs constitute clades in which numerous taxa arise from a single ancestor and the neotenic examples are the result of multiple sequestration events within otherwise agaricoid genera (Sanchez García et al. 2020). This was proposed by Baura et al. (1992) when they stated that "secotioid forms are easy to produce." According to Gould's reasoning, we find that groups originating by neoteny will necessarily result in one or a few sequestrate taxa nested among agaricoid relatives, whereas progenetic innovations will produce monophyletic radiations with many species (Fig. 13.3). Evidence of multiple neotenic sequestration events has been well established in many of the major neotenic groups such as *Amanita* (Truong et al. 2017a), *Chlorophyllum* (Loizides et al. 2020), *Cortinarius* (Peintner et al. 2001), *Descolea* (Kuhar et al. 2017), *Entoloma* (Kinoshita et al. 2012), *Inocybe* (Caiafa et al. 2021a), and *Laccaria* (Sheedy et al. 2013). In contrast, a single origin with monophyly is documented for many progenetic groups, such as the formerly circumscribed Lycoperdales inc. *Tulostoma* (Larsson and Jeppson 2008), in the Sclerodermatineae (Sato and Toju 2019) and in the Hysterangiales, among others (Hosaka et al.

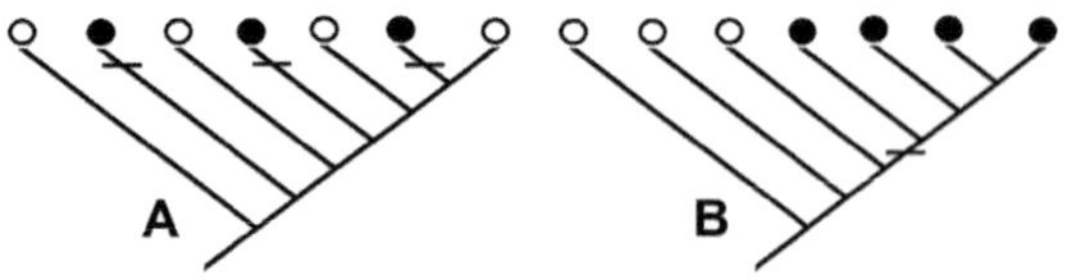

Fig. 13.3 Diversification patterns interpreted from Gould (1977) depicting (**a**) frequent neotenic innovations resulting in the polyphyletic arrangement observed in secotioid taxa (i.e., neotenic evolution of sequestrate taxa), or (**b**) rare progenetic innovation resulting in a prolific monophyletic clade, as observed in many gasteroid fungi (i.e., progenetic evolution of sequestrate taxa). White circles represent the ancestral agaricoid morphology and black circles the derived sequestrate (gasteroid or secotioid) stages. Horizontal lines represent sequestration events

2006). Gube and Dörfelt (2011) pointed out that within the Agaricaceae, tulostomatoid and lycoperdoid fungi (progenetic gasteroid taxa) constitute monophyletic lineages whereas the secotioid and truffle-like taxa (neotenic gasteroid taxa) within this family are polyphyletic and are closely related to agaricoid sister taxa (Lebel and Syme 2012).

The question of whether sequestrate fungi constitute "evolutionary dead-ends" was addressed in two presentations at the annual meeting of the Mycological Society of America, first by Wilson et al. (2008, later published in Wilson et al. 2011) and then by Harrower et al. (2016, published in Harrower 2017). Interestingly, the two studies arrived at opposite conclusions. Harrower (2017) concluded that no further diversification occurs after a sequestration event, while Wilson et al. (2011) detected an increased diversification rate in sequestrate lineages.

This apparent contradiction is interesting in the light of the two different paedogenetic evolutionary processes that are described here. Harrower (2017) focused on the neotenic genus *Cortinarius* whereas Wilson et al. (2011) studied progenetic groups, including Sclerodermatineae and Lycoperdales. Moreover, Wilson et al. (2011) concluded that no increase in the extinction rates could be detected, which also supports Gube's (2009) idea that paedomorphosis is more likely than secotioid inertia to explain the evolutionary patterns seen in gasteroid fungi. A progenetic origin of the "highly derived forms" found in Sclerodermatineae and Lycoperdaceae would help to explain the absence of secotioid "missing links" in those clades. This is because no stepwise gradual evolution is needed in the progenetic model, only the heterochronic maturation of basidia without hymenial differentiation (Fig. 13.2).

Another argument that supports the neotenic nature of the secotioid taxa is the co-occurrence of neotenic and agaricoid individuals within a species. This aligns with the paradigmatic case of the paedomorphic salamanders, whose "transformation" might occur or not, depending on environmental factors (Tilley 1973). Gould emphasizes that stable environments with crowded niches would favor neoteny as a way to escape the "adult" niche by slight modifications of the body plan (Bauplan). In salamanders, this implies the ability to avoid competition with adult salamanders by remaining in the aquatic environment and reproducing there. Among neotenic mushrooms, this occurrence of both "typical" (agaricoid) and neotenic (secotioid) individuals has been observed in many genera, including *Descolea* (Neville et al. 2004), *Gastrosuillus* (Kretzer and Bruns 1997), *Inocybe* (Braaten et al. 2014), and *Russula* (Vidal et al. 2019). As expected based on the conceptual model presented here, this intraspecific alternation of forms has not been observed in any of the progenetic fungi.

13.4 Morphological Plasticity

Gould (1977) concluded that progenesis might have played a major role in the diversification of body plans in the animal kingdom. The example used by Gould to illustrate this feature are the progenetic females of the scale insects, whose sessile bodies enlarge and remain attached to leaves and stems without developing wings or even limbs and they instead produce a large number of eggs. Neoteny, on the other hand, would mean only a step back from a specialized adult morphology that is unsuccessful or non-competitive under certain conditions, and would therefore prevent the neotenic innovator from entering the occupied reproductive niche of the "normal adults." Gould's favorite example of neotenic animals are aquatic salamanders such as axolotls, who retain gills throughout their lives and are able to reproduce in the "larval" aquatic niche. Gould argues that the success of progenetic changes relies on their explosive reproductive ability and not in the new morphology, thus liberating the morphology for structural experiments. He also cites De Beer's (1951) opinion that genes related to the ancestral adult form would be absent in the progenetic innovator and ready to undertake transformational processes (such as gene duplication). Surprisingly, a comparative genomics study by Nagy et al. (2016), which included two progenetic species (*Scleroderma citrinum* and *Sphaerobolus stellatus*), showed that they have significantly larger genomes than the closest sampled taxa. However, studies based on larger taxon sampling and including neotenic forms are needed to adequately test De Beer's (1951) prediction.

The evolutionary potential of this process was also highlighted by Takhtajan (1969), who suggested that heterochrony played a principal role in angiosperm diversification, e.g. in structures such as the folded carpels of the flowering plants, which would result from a neoteny event recovering the folded disposition of immature leaves (Takhtajan 1976). In zoology, this idea has also been supported by Britz & Conway (2016), who proposed that progenesis allowed some fishes in the order Cypriniformes to escape the characteristic body plan of this group. Gube and Dörflet (2011) indicated that the "profound novelties" produced during gasteromycetation correspond to the scheme of saltational evolution as explained by Theißen (2009) and that heterochrony may be involved in these processes via mutations in key regulatory factors of developmental processes. Gould's idea that progenesis triggers structural innovation is supported in sequestrate fungi by the work of Wilson et al. (2011) who focused on progenetic taxa. Wilson et al. (2011) emphasize that "gasteroid fungi represent a small fraction of the total number of species of Agaricomycotina, but they encompass a tremendous range of morphological diversity." On the other hand, the neotenic taxa (e.g., secotioid fungi and their relatives that are recently derived from agaricoid ancestors, Table 13.1) display a limited level of structural plasticity, as evidenced by the fact that they often maintain a recognizable similarity to their agaricoid relatives. We hypothesize that the different degrees of lamellar anastomosis might be a consequence of hymenophore development within an enclosed basidiome following complex folding dynamics (Kuhar et al. 2022) or in the case of hypogeous species it could be due to the pressure exerted by the surrounding soil. In contrast to these limited morphologies present in neotenic sequestrate taxa, we see a great diversity of secondary developments in progenetic forms that range from earth stars to cannonballs to bird's nest fungi (Fig. 13.4). These species can even overcome sequestration and develop novel

stipitate basidiomes (such as in the Phallales or *Calostoma*) or even develop unique dispersal mechanisms (such as in the independent ballistospory found in *Sphaerobolus*) (Hosaka et al. 2006). The overwhelming morphological plasticity seen in the progenetic gasteroid forms illustrates better than any other group Gould's characterization of progenesis as "the working laboratory of evolution." Even if in some cases the morphological diversity is low (as in the genus *Rhizopogon*), we can generalize that the gasteroid taxa with the highest morphological diversification are generally progenetic.

13.5 Selection Strategies

The r/K selection theory attempts to explain trait selection that is correlated with reproductive strategies. K-selected traits would be related to taxa that have fewer offspring but allocate more resources to parental care (in a broad sense), while r-strategies would lead to higher production of offspring (either individuals or propagules) with less investment in their survival. This theory has been challenged in recent decades by more complex models (Reznick et al. 2002). However, modern approaches using the r/K selection theory (while taking into account population density or internal stochasticity) still have high explanatory value (Oizumi et al. 2016). Gould (1977) suggested that r- or K-selection strategies were linked to the different processes of paedomorphosis. Gould's (1977) predictions are not based on observations but instead a series of cause–effect deductions. He suggested that the retention of an immature morphology (as in neotenic salamanders) would provide an escape from a crowded adult niche. Thus, if the larval niche has less competition than the adult niche, then selection would favor the retention of juvenile features. On the contrary, the reproduction in larval stages involving a simplification of the lifecycle and the higher allocation of biomass to

Table 13.1 Origin, examples, and predictions of the progenetic and neotenic processes as interpreted from Gould (1977) and adapted to the evolution of sequestrate basidiomes

	Progenetic Sequestration	Neotenic Sequestration
Origin	Onset of reproductive maturity prior to hymenium differentiation	Retention of the enclosed basidiome throughout the development
Phylogenetic consequences	Increased speciation, monophyletic (i.e., one event leads to a sequestrate ancestor that subsequently diversifies to produce descendant sequestrate taxa)	Limited speciation, polyphyletic (i.e., sequestrate taxa evolve multiple times independently within an evolutionary lineage)
Morphological consequences	Enhanced morphoanatomical diversification	Limited morphoanatomical diversification
Selection strategy	r - strategy	K - strategy
Frequency of sequestration events	Rare (15)[a]	Frequent (102)[a]
Example taxa	Geastrales - Hysterangiales - "Lycoperdales" - Phallales - Sclerodermatineae	[b]Cystangium (*Russula*) - *Descomyces* (*Descolea*) - *Endoptychum* (*Chlorophyllum*) *Thaxterogaster* (*Cortinarius*) *Torrendia* (*Amanita*)-

[a]estimated number of sequestration events in the taxon sampling of Sanchez García et al. (2020). These numbers constitute conservative calculations that are likely significant underestimations of the total number of sequestration events in Agaricomycotina. Genera of fungi are categorized as either progenetic or neotenic based on the documented morphology and development in known species according to published literature (see reference list)

[b] Here we use generic names as traditionally used for sequestrate taxa with modern placement indicated by names in parentheses. This illustrates the polyphyletic nature of the classification resulting from the multiple origins of sequestrate taxa

propagation implies a deep change in the body plan (Bauplan). In the case of scale insects, progenetic females become enlarged "casings" for their eggs, and in the case of fungi, the progenetic result is fruiting bodies consisting of sacs containing a large number of spores. Progenesis is only successful, as Gould sees it, in a niche that is available for fast colonization, since it allows an explosive/aggressive propagule delivery. Based on this logic, Gould suggested that progenesis would be linked to r-selection strategies, whereas neoteny would be a result of K-selective regimes. The state of our knowledge of basidiome ecology is still too limited to speculate if these predictions fit with the sequestration scheme proposed here. However, given the importance of this ecological framework to Gould's idea, it is important to mention these hypotheses so that future researchers can evaluate new data in light of Gould's proposal.

It is well known that a simplification of the lifecycle and the massive production of spores are associated with the colonization of highly available (r-selective) niches, as happens with the high prevalence of anamorphic stages in Ascomycota that colonize fruits or food products (Hornok 2007). We could provisionally reason that puffballs, for example, would also follow this pattern of life cycle shortening and modification of the structural plan to produce high numbers of spores (Table 13.1). Along the same line, we could argue that the retention of the closed morphology could be a successful way to avoid the overcrowded (and thus competitive; K-selected) agaricoid reproductive niche, thus introducing novel colonization strategies such as dispersal by mycophagous animals. A preliminary observation pointing in this direction is that basidia with increased number of sterigmata (and thus basidiospores) are more likely to occur in progenetic genera like *Alpova* (Nouhra et al. 2005), *Astraeus* (Coker and Couch 1974), *Geastrum* (Sunhede 1993), *Laternea* (Sáenz et al. 1983), *Phallus* (Gull 1981), and *Scleroderma* (Reijnders 1999). However, reduction in the number of sterigmata and spores is more

Fig. 13.4 A-C, Neotenic basidiomes showing a low degree of morphological modification from the agaricoid architecture (A and B *Cortinarius* spp., C, *Inocybe* sp.; D-L, progenetic basidiomes illustrating the high level of morphological differentiation, D, *Cyathus* sp.: E and F, *Phallus* spp.; G, *Scleroderma* sp.; H, *Pisolithus* sp.; I, *Astraeus* sp.; J, *Calvatia* sp.: K, *Geastrum* sp.; and L, *Tulostoma* sp.

frequent in neotenic genera, including *Chlorophyllum* (Loizides et al. 2020), *Cortinarius* (Bougher and Castellano 1993 as *Cortinomyces*), *Descolea* (Kuhar et al. 2017), *Entoloma* (Kinoshita et al. 2012), *Hymenogaster* (Stewart and Trappe 1975), and *Inocybe* (Caiafa et al. 2021a).

If Gould's hypothesis is correct and can be applied to the fungal model, stable environments that favor K-selective conditions would often trigger the emergenge of neotenic taxa whereas environments with r- selective conditions would favor progenetic innovators. However, these ideas are only speculative and further research on the selective strategies and any possible link to basidiome morphology are needed to propose a plausible hypothesis.

13.6 Outline of Neoteny and Progenesis within the Agaricomycotina

Within the Agaricomycotina, sequestration is a well-known evolutionary phenomenon that took place in many modern taxonomic orders. Although Reijnders' (2000) revision serves as the basis of the scheme we propose here, further work is needed to fully characterize the developmental changes leading to sequestrate morphologies before testing the applicability of Gould's proposal to this framework. Many taxa still need to be studied in more detail. Understanding their phylogenetic position, speciation patterns, and the ontogeny of their fertile tissues are of paramount importance in order to understand where they fit in the proposed evolutionary scenarios discussed here.

Gasteroid taxa in the order Russulales have been recognized for many decades (e.g., Singer and Smith 1960; Lebel and Castellano 2002). Although different levels of modification of the agaricoid structure have been documented, most described taxa have true hymenia (e.g., Pegler and Young 1979). Molecular data (e.g., Vidal et al. 2019) confirm a pattern of multiple origins of sequestrate taxa with limited radiation. This pattern is also supported by the paraphyletic nature of *Russula*, which now contains the obsolete genera *Bucholtzia*, *Cystangium*, *Elasmomyces*, *Gymnomyces*, *Macowanites*, *Martellia* (Elliott and Trappe 2018), or *Lactarius*, containing *Zelleromyces* (Vidal et al. 2019). This suggests that these taxa should be regarded as neotenic, together with *Arcangeliella* (Zeller and Dodge 1936) and the sequestrate *Lactifluus* species (Lebel et al. 2021). However, progenetic origins must be ascribed to *Leucogaster* (Fogel 1990) and *Fevansia* (Trappe and Castellano 2000) due to the absence of true hymenia. Species of these two genera are putatively the result of the same sequestration event (Smith et al. 2013, Sanchez-García et al. 2020) confirming the phylogenetic pattern expected under progenesis. The case of *Hybogaster*, recently confirmed as a "sequestrate" form of *Bondarzewia guaitecasensis* (Palfner et al. 2020), might constitute a phenomenon other than sequestration since the contortion of the hymenophore is not associated with enclosure, but rather to an abnormal proliferation and folding, and it seems to be associated with latitude and temperature gradients. The position of many other taxa such as *Mycolevis* and *Leucophelps* (among others) is still doubtful due to their unresolved phylogenetic relationships and lack of information on the structure and ontogeny of their fertile tissues. The evolution of the truffle genera *Stephanospora* and *Mayamontana* is also unclear because they are among only a few truffle taxa where the closest known relatives are saprobic resupinate crust fungi (in this case, taxa in the genera *Cristinia* and *Lindtneria*). Although *Stephanospora* and *Mayamontana* species are characterized by a true hymenium and therefore likely of neotenic origin, the possibility of a progenetic origin cannot be ruled out (Martin et al. 2004; Castellano et al. 2007; Lebel et al. 2015).

Among the Phallomycetidae, the majority of described species of gasteroid genera lack a true hymenium and instead have a gleba that consists of a mass of spores that can either be gelatinous, slimy, or powdery (Hosaka et al. 2006). Similar and extensive morphological diversification is also evident in the Phallales and the Geastrales. This, together with the limited number of origins with numerous taxa (Sanchez-García et al. 2020; Davoodian et al. 2021), suggests a progenetic origin of almost all sequestrate groups in this clade. A notable exception is the genus *Gautieria*, which is characterized by having a true hymenium (Stewart 1974). *Gautieria* is inferred to be neotenic and derived from either a coralloid or gomphoid ancestor, but more phylogenetic data on Gomphales are needed to clearly infer the origin of this group (Xu et al. 2022a; Gianchi et al. 2010).

Within the order Agaricales the situation is more heterogeneous. The family Agaricaceae has an exceptionally large number of sequestration events, as evident from many works (e.g., Gube 2009; Lebel and Syme 2012). Puffballs in the genera *Abstoma, Arachnion, Bovista, Calvatia* (s.l.), *Disciseda*, *Lycoperdon* (including synonyms such as *Vascellum*), and *Mycenastrum* form a monophyletic clade (Gube 2009) and putatively represent a single progenetic evolutionary lineage. This conclusion is supported both by morphological and phylogenetic data. Members of this core puffball clade include taxa with a spongy, cottony, or powdery gleba and they exhibit diverse methods of peridial opening (apical pore, fragmentation, laceration, basal pore). The stalked genera, including species of *Battarrea, Chlamydopus, Dictyocephalos, Queletia,* and *Tulostoma* (incl. *Schyzostoma*), putatively form a monophyletic, highly diverse clade with numerous species (Gube 2009). Their powdery or cottony glebas, with no trace of true hymenial surface, seem to fit well with our concept of a progenetic group. A progenetic event must be inferred for the Nidulariaceae given the irregular organization of the hymenial palisade within the peridioles (Payer 1850), which would also explain the unique body plan and morphological diversification within this clade (Kraisitudomsook et al. 2021). However, the specialized morphology and the reduction of the number of propagules (the peridioles) suggest that this situation in Nidulariaceae might be more complex and needs to be further studied. Numerous secotioid and gasteroid taxa fitting the neotenic model occur independently in the genera *Agaricus* (Lebel and Syme 2012), *Chlorophyllum* (Loizides et al. 2020), and *Macrolepiota* (Lebel and Syme 2012). The anatomically unique genera *Asperosporus* (Karlsen-Ayala et al. 2021) and *Podaxis* (Morse 1933) seem to represent independent neotenic sequestration events. Genera in other families in the order Agaricales also show independent sequestration events that mostly fit the neotenic pattern, such as *Amanita* (Truong et al. 2017a), *Chlorophyllum* (Loizides et al. 2020), *Cortinarius* (Peintner et al. 2001), *Descolea* (Kuhar et al. 2017), *Entoloma* (Kinoshita et al. 2012), *Inocybe* (Caiafa et al. 2021a), *Laccaria* (Sheedy et al. 2013), and *Psathyrella* (Moreno et al. 2015). Members of the genus *Cribbea* are described as having a true hymenium and this is inferred as a neotenic innovation within the *Xerula-Oudemansiella* complex (Lebel and Catcheside 2009). Secotioid genera such as *Clavogaster* (Hennings 1896), *Galeropsis*, *Gastrocybe*, *Weraroa* (Watling and Martín 2003), and many others have been described and they all appear to be cases of neoteny. *Guyanagaster* (Henkel et al. 2010) seems to be a clear case of neoteny, although the development of the fertile tissues have not been described in depth, and the diversification pattern cannot be assessed from only the two known species (Koch et al. 2017). *Hymenogaster* is a well-known genus where the nature of the sequestration is unclear. Reijnders (1976, 2000) verified the presence of true hymenial surfaces within the basidiomes. There are many described species of *Hymenogaster* but nonetheless there is low morphological diversity and a single evolutionary origin as well as a tendency toward reduction in sterigmal number (Stielow et al. 2011). Overall, the data suggest that this is an exceptional case of a neotenic origin followed by a subsequent radiation. Of special interest within the order Agaricales is the preservation of the stipe in genera such as *Tulostoma* or *Battarrea*. We hypothesize that stipe loss is not a necessary consequence of progenesis. This heterogeneous distribution of heterochronic morphologies has been reported many times as "mosaic heterochrony"(e.g., David 1989). Environmental factors might play a role in selecting paedomorphic stipitate morphologies, as is suggested by the frequency of this stipitate morphology in arid environments.

Both types of sequestration can be also observed in the order Boletales. The genera *Astraeus*, *Calostoma*, *Diplocystis*, *Pisolithus*, *Scleroderma*, and *Tremellogaster* constitute the gasteroid representatives of the Sclerodermatineae. These likely arose from either one or two origins of the sequestrate morphology (Wilson et al. 2008). These taxa fit the progenetic pattern due to the absence of a true hymenium and they also exhibit a phylogenetic pattern of infrequent origins followed by successful evolutionary radiation along with morphological diversification. The complete lack of hymenium in the related genera *Alpova*, *Melanogaster*, *Neoalpova*, and *Paralpova* (Dodge 1931; Guzmán 1969; Xu et al. 2022a; Alvarado et al. 2021) suggests that they may have arisen from a progenetic origin. The work by Clémençon (2005) documenting the resemblance in development of the genus *Chamonixia* with non-sequestrate boletes suggests that a neotenic origin is likely. A layer resembling a true hymenium has been shown by Orihara et al. (2016) in *Turmalinea* and by Lebel et al. (2012) in *Rossbeevera*. These studies also showed the modest diversification of the two genera. The case of *Octaviania* is more confusing, since a rudimentary hymenium was mentioned by Orihara et al. (2012, 2021) but Reijnders (2000) thought that the fertile tissue consisted of bundles of basidia growing into cavities. A similar secondary hymenial proliferation was ascribed by Reijnders (2000) to *Rhizopogon*, a genus (discussed below) whose progenetic nature is also supported by the frequency of basidia with large numbers of sterigmata (Talbot 1973). The continuity of the centripetal tramal plates of the monotypic *Sedecula* with the center-base of the basidiome (Zeller 1941) strongly suggests that this genus is neotenic, although there are no known agaricoid intermediates between *Sedecula* and the corticioid *Coniophora* clade, from which the sequestrate genus has evolved (Trappe et al. 2015). Sequestrate species occur within the Serpulaceae and were classified in the genus *Gymnopaxillus* (e.g., *G. nudus*), together with morcheliform species (e.g., *G. morchelliformis*) (Skrede et al. 2011). We can assume that both forms result from a neotenic innovation, the former, retaining the enclosed morphology and the latter, arresting the development of the pileal surface so the hymenophore grows and completely covers the globose fruiting body to reach the morel-like shape described by Claridge et al. (2001).

A true hymenium seems to be present also in species of *Royoungia* (Castellano et al. 1992), *Mycoamaranthus* (Lumyong et al. 2003), *Solioccasus* (Trappe et al. 2013), and is evident in *Gymnogaster* (Gelardi et al. 2017), *Truncocolumella*, *Gastroboletus* (Smith and Singer 1959), *Jimtrappea*, *Castellanea*, and *Costatisporus* (Smith et al. 2015). These genera all seem to fit the neotenic evolution and diversification pattern. Other Boletaceae genera that fit the neotenic syndrome due to true hymenium development and low species diversity include *Carolinigaster* (Crous et al. 2018), *Kombocles* (Castellano et al. 2016), *Mackintoshia* (Pacioni and Sharp 2000), *Rhopalogaster* (Smith et al. 2018), and sequestrate taxa within *Xerocomellus* (Smith et al. 2018), e.g. *Heliogaster* (Frank et al. 2020). *Afrocastellanoa* shows no resemblance to the epigeous sister genus (*Porphyrellus*) and the hymenium is absent or disorganized (Orihara and Smith 2017). We thus hypothesize that *Afrocastellanoa* represents a progenetic innovation with little or no subsequent species-level diversification.

Finally, the rare sequestrate form of *Lentinus tigrinus* in the order Polyporales studied by Hibbett et al. (1994) is apparently neotenic innovation given the complete correspondence with the agaricoid body plan.

13.7 Discussion

Studies aiming to elucidate the driving forces behind basidiome sequestration, such as the one conducted by Sheedy et al. (2016), have revealed conflicting results regarding the age of different sequestrate lineages. Future studies discriminating between neotenic and progenetic events might resolve these apparent contradictions by studying the differences in diversification rates between both groups.

If our hypothesis is corroborated, different strategies should be chosen to study these processes in neotenic vs. progenetic taxa. The increased spore production of progenetic forms at the costs of the development of surrounding tissues seems evident. Increased spore production would be an argument in favor of r-selected strategies, which can be contrasted in the framework of early vs. late-successional conditions given the early colonization ability of r-strategic fungi (Boddy and Hiscox 2016). However, the overwhelming morphological diversification of these groups of progenetic taxa might be an obstacle to tracing the change in performance resulting from the ancestral sequestration event, since ancestral state reconstruction has proven difficult (Hosaka et al. 2006). The idea that neotenic sequestration events should theoretically be successful in highly competitive scenarios (e.g., stable habitats with high levels of competition) could potentially be tested because of their closeness to agaricoid forms. For example, we can speculate that the higher rate of sequestration in the genus *Cortinarius* in southern-Gondwana landmasses might be related to the extremely competitive reproductive niche that results from the dominance by *Cortinarius* among the mycorrhizal fungi in *Eucalyptus* (Horton et al. 2017) and Nothofagaceae (Nouhra et al. 2013; Truong et al. 2017b) forests. This neotenic evolutionary "escape route" could be related to advantages due to animal mycophagy, in contrast to the wind-dispersed ballistospores of the agaricoid species.

Further research is needed to corroborate these ideas, since a great number of sequestrate taxa are undescribed and many others still have not been phylogenetically placed based on molecular data. Isolated species do not provide a framework to verify whether they result from isolated sequestration events or not. For example, in genera such as *Cortinarius* or *Descolea,* sequestrate forms previously classified under *Protoglossum* or *Descomyces* have undergone a high degree of sequestration (e.g., completely enclosed basidiomes, compact gleba with very small galleries, and dramatic stipe reduction). They superficially resemble progenetic sequestrate fungi rather than neotenic secotioid taxa. However, they have true hymenial surfaces that are congruent with the secotioid (neotenic) nature of the other sequestrate representatives in these genera. Morphological studies are also needed to confirm the hymenial configuration and development of some groups. Reijnders (2000) explains that in some cases "the distinction between cavities surrounded by a hymenium and spaces filled with (partly) fructifying elements becomes vague." This makes it extremely difficult to determine if a true hymenium is present, and, consequently, whether the species is neotenic or progenetic. Reijnders (2000) provides evidence showing that the fertile surfaces found in the galleries of *Rhizopogon* constitute examples of a secondary invasion of the cavities by fertile hyphae, confirming the progenetic origin of these genera. Progeneticity is also supported by the large number of species described in these genera, the high spore production potential (Bruns et al. 2009), and the success of *Rhizopogon* species in r-strategic scenarios such as in the early stages of succession (Vlk et al. 2020). Reijnders (2000) discussed the situation of many ambiguous genera in the neotenic vs. progenetic scheme and resolved many of them. The diversification pattern confirms the progenetic character of these groups suggested by the absence of a hymenium in most cases.

The look for adaptive forces slowly driving the sequestration process resulted in a deep understanding of ecological and biological phenomena involving fungi, interacting plants and animals, and their environments (Sheedy et al. 2016; Danks et al. 2020; Caiafa et al. 2021b). In this framework, a series of logical and plausible adaptive hypotheses were produced in an attempt to explain the evolution of sequestrate basidiomes from epigeous, agaricoid ancestors. Factors such as aridification, animal mycophagy and spore dispersal, and enhanced protection of the fertile tissues have all been considered as potential factors (Savile 1955, 1968; Thiers 1984; Bruns et al. 1989; Varga et al. 2022).

A similar tendency in animal biology was challenged by Gould and Lewontin (1979) as a consequence of the "adaptationist program" dominating Western science in the mid-twentieth century. Gould and Lewontin (1979) "...fault the adaptationist program for its failure to distinguish current utility from reasons for origin." Gould and Lewontin (1979) also explain the flawed logic of this adaptationist program, by which, if an adaptive explanation fails, another (frequently suboptimal) one must be true, without ever questioning this "program" (in the sense of Lakatos 1970). They emphasize that this bias is much more evident in the work of Alfred Wallace (e.g., Wallace 1871) than in the work of Darwin (e.g., Darwin 1859), particularly since Darwin did not limit evolutionary explanations to only adaptive forces. Earlier criticisms of this adaptationist logic include the work of D'Arcy Thompson (Thompson 1942) against the "inappropriate atomization" of the studied characters, i.e. the delimitation of a character that might not be a unit of selection in nature. Another bias that is evident in the literature on sequestrate fungi is the pervasive concept of gradual change. This is frequently quoted in Latin as "Natura non facit saltum" (Nature does not jump) and is present in Darwin's work (Darwin 1859) as a fundamental idea. The opposing conceptions, such as macromutations leading to dramatic morphological changes as proposed by Goldschmidt (1933), found many obstacles until more recent times (Baura et al. 1992).

As possible alternatives to these oversimplifications, more interesting explanations for sequestration might be found by considering the constraints imposed by the body plan and the developmental process, the possibility of abrupt changes, and the importance of not reducing the origin to the "current utility" of traits.

The fact that neoteny and progenesis partition sequestrate taxa into two distinct developmental groups that were previously recognized by past mycologists (Malençon 1955; Pilát 1958; Heim 1971) is especially interesting, particularly since such distinctions provide a plausible explanation that was not easy to achieve before Gould's (1977) synthesis. These processes also provide a mechanistic understanding of the origin of morphological configurations that can be selected for in nature by different environmental or ecological forces. The paedogenetic origin of sequestrate fungi still has many implications that need to be studied in a more modern framework that includes developmental genomics. As Viragh et al. (2021) pointed out, the genetic mechanisms of sequestration are not well understood. No specific genes have been yet found that are linked to the process of gastromycetation. The study of genetic factors leading to paedomorphosis in salamanders suggest that the "single gene hypothesis" will certainly be abandoned in favor of a multigenic model (Voss 1995). We speculate that mutations changing mechanical resistance of the veils or triggering the differentiation of fertile tissues in immature stages are likely candidates. A complex multifactorial control of developmental mechanics explaining hymenophoral configurations in non-sequestrate mushrooms has been proposed (Kuhar et al. 2022) that would also explain rapid evolutionary change and convergence. A similar model could be extended to sequestrate fungi. However, a deeper knowledge of the genetic regulation of the developmental pathways of fungal basidiomes is necessary to test such models. The corroboration of this scheme would make the sequestration process a perfect model for the study of heterochronic evolution in the future.

Acknowledgments This work was supported by FONCyT (Grant PICT 2018 – 3781), a Fulbright Fellowship and a Friends of Farlow Fellowship (to FK) and also the US National Science Foundation grant DEB-1354802 (to M.E.S.) and NIFA-USDA award FLA-PLP-005289 (to MES).

References

Albee-Scott SR (2007) Does secotioid inertia drive the evolution of false-truffles? Mycol Res 111(9): 1030–1039

Alvarado P, Cabero J, Moreno-Mateos D, Vizzini A, Alonso J, Lebeuf R, Vidal JM et al (2021) Phylogenetic relationships among false truffle genera of *Alpova, Melanogaster, Neoalpova,* and *Paralpova,* gen. nov. Mycologia 113(4):828–841

Baura G, Szaro TM, Bruns TD (1992) *Gastrosuillus laricinus* is a recent derivative of *Suillus grevillei*: molecular evidence. Mycologia 84(4):592–597

Boddy L, Hiscox J (2016) Fungal ecology: principles and mechanisms of colonization and competition by saprotrophic fungi. In: Heitman J, Howlett JB, Crous PW, Stukenbrock EH, James TY, Gow NAR (eds) The Fungal Kingdom, Edition. American Society of Microbiology, Washington, DC, pp 293–308

Bougher NL, Castellano MA (1993) Delimitation of *Hymenogaster* sensu stricto and four new segregate genera. Mycologia 85(2):273–293

Braaten CC, Matheny PB, Viess DL, Wood MG, Williams JH, Bougher NL (2014) Two new species of *Inocybe* from Australia and North America that include novel secotioid forms. Botany 92(1):9–22

Britz R, Conway KW (2016) *Danionella dracula*, an escape from the cypriniform Bauplan via developmental truncation? J Morphol 277(2):147–166

Bruns TD, Fogel R, White TJ, Palmer JD (1989) Accelerated evolution of a false-truffle from a mushroom ancestor. Nature 339(6220):140–142

Bruns TD, Peay KG, Boynton PJ, Grubisha LC, Hynson NA, Nguyen NH, Rosenstock NP (2009) Inoculum potential of *Rhizopogon* spores increases with time over the first 4 yr of a 99-yr spore burial experiment. New Phytol 181(2):463–470

Caiafa MV, Sandoval-Leiva P, Matheny PB, Calle A, Smith ME (2021a) Four new species of sequestrate *Inocybe* from Chilean Nothofagaceae forests. Mycologia 113(3):629–642

Caiafa MV, Jusino MA, Wilkie AC, Díaz IA, Sieving KE, Smith ME (2021b) Discovering the role of Patagonian birds in the dispersal of truffles and other mycorrhizal fungi. Curr Biol 31(24):5558–5570

Castellano MA, Trappe JM, Malajczuk N (1992) Australasian truffle-like fungi. III. *Royoungia* gen. nov. and *Mycoamaranthus* gen. nov. (Basidiomycotina). Aus Syst Bot 5(5):613–616

Castellano MA, Trappe JM, Lodge DJ (2007) *Mayamontana coccolobae* (Basidiomycota), a new sequestrate taxon from Belize. Mycotaxon 100:289–294

Castellano MA, Elliott TF, Truong C, Séné O, Dentinger BT, Henkel TW (2016) *Kombocles bakaiana* gen. sp. nov. (Boletaceae), a new sequestrate fungus from Cameroon. IMA Fungus 7(2):239–245

Claridge AW, Trappe JM, Castellano MA (2001) Australasian truffle-like fungi. X. *Gymnopaxillus* (Basidiomycota, Austropaxillaceae). Aust Syst Bot 14 (2):273–281

Clémençon H (2005) New observations on the basidiome ontogeny of *Chamonixia caespitosa* (sequestrate Boletaceae). Persoonia 18(4):499–504

Coker WC, Couch JN (1974) The gasteromycetes of the eastern United States and Canada. University of North Carolina Press, Chapel Hill, North Carolina

Crous PW, Luangsa-Ard JJ, Wingfield MJ, Carnegie AJ, Hernández-Restrepo M, Lombard L, Roux J, Barreto RW, Baseia IG, Cano-Lira JF, Martín MP et al (2018) Fungal planet description sheets: 785–867. Persoonia 41:238

Danks MA, Simpson N, Elliott TF, Paine CT, Vernes K (2020) Modeling mycorrhizal fungi dispersal by the mycophagous swamp wallaby (*Wallabia bicolor*). Ecol Evol 10(23):12920–12928

Darwin C (1859) On the origin of species by means of natural selection, or, the preservation of favoured races in the struggle for life. J. Murray, London

David B (1989) Mosaic pattern of heterochronies: variation and diversity in Pourtalesiidae (deep-sea echinoids). Evol Biol 24:297–327

Davoodian N, Lebel T, Castellano MA, Hosaka K (2021) Hysterangiales revisited: expanded phylogeny reveals new genera and two new suborders. Fungal Systematics and Evolution 8(1):65–80. https://doi.org/10.3114/fuse.2021.08.06

De Bary A (1884) Vergleichende morphologie und biologie der Pilze. Mycetozoen und Bacterien, Wilhelm Engelmann, Leipzig, Germany

de Beer GR (1951) Embryos and ancestors, 3rd edn. Oxford University Press, Oxford

Dodge CW (1931) *Alpova*, a new genus of Rhizopogonaceae, with further notes on *Leucogaster* and *Arcangeliella*. Ann Mo Bot Gard 18(3):457–464

Dring DM (1973) Gasteromycetes. In: Ainsworth GC, Sparrow FK, Sussman AS (eds) The fungi, vol IVB. Academic Press, New York & London, pp 451–478

Elliott TF, Trappe JM (2018) A worldwide nomenclature revision of sequestrate *Russula* species. Fungal Systematics and Evolution 1:229

Fischer E (1933) Gasteromycetae. In *Engler and Prantle,* Die Naturlichen Pflanzen-familien

Fogel R (1990) *Leucogaster columellatus* is a *Sclerogaster*. Mycologia 82(5):655–657

Frank JL, Siegel N, Schwarz CF, Araki B, Vellinga EC (2020) *Xerocomellus* (Boletaceae) in western North America. Fungal Systematics and Evolution 6:265

Gelardi M, Fechner N, Halling RE, Costanzo F (2017) *Gymnogaster boletoides* JW Cribb (Boletaceae, Boletales), a striking Australian secotioid bolete. Austrobaileya 10(1):121–129

Gerber S, Hopkins MJ (2011) Mosaic heterochrony and evolutionary modularity: the trilobite genus *Zacanthopsis* as a case study. Evol Int J Org 65(11): 3241–3252

Giachini AJ, Hosaka K, Nouhra E, Spatafora J, Trappe JM (2010) Phylogenetic relationships of the Gomphales based on nuc-25S-rDNA, mit-12S-rDNA, and mit-atp6-DNA combined sequences. Fungal Biol 114(2–3):224–234

Goldschmidt R (1933) Some aspects of evolution. Science 78(2033):539–547

Gould SJ (1977) Ontogeny and phylogeny. Harvard University Press, Cambridge, MA

Gould SJ, Lewontin RC (1979) The spandrels of San Marco and the Panglossian paradigm: a critique of the adaptationist programme. Proc B Soc Lond B 205:581–698

Gube M (2009) Ontogeny and phylogeny of gasteroid members of Agaricaceae (Basidiomycetes). Dissertation, Friedrich-Schiller-Universität Jena

Gube M, Dörfelt H (2011) Gasteromycetation in Agaricaceae sl (Basidiomycota): morphological and ecological implementations. Feddes Repert 122(5–6): 367–390

Gull K (1981) Meiosis, basidiospore development and post-meiotic mitosis in the basidiomycete. *Archives of Microbiology* 128(4):403–406

Guzmán G (1969) *Veligaster*, a new genus of the Sclerodermataceae. Mycologia 61(6):1117–1123

Haeckel E (1866) Generelle Morphologie der Organismen: Bd. Allgemeine Entwicklungsgeschichte der Organismen. ed. G. Reimer

Harrower E (2017) Systematics and biogeography of the *Cortinarius violaceus* group and sequestrate evolution in *Cortinarius* (Agaricales). Dissertation,. University of Tennessee Knoxville

Harrower E, Smith ME, Mujic AB, Truong C, Henkel TW, Aime MC, Ovrebo C, Matheny PB (2016) Are sequestrate taxa evolutionary dead-ends? Assessing evolution and diversification of sequestrate *Cortinarius*. Mycological Society of America (MSA) Annual Meeting, Berkeley, California

Heim R (1971) Interrelationships between the Agaricales and Gasteromycetes. In: Petersen RH (ed) Evolution in the higher basidiomycetes. University of Tennessee, Knoxville, Tennessee, USA, pp 505–534

Henkel TW, Smith ME, Aime MC (2010) *Guyanagaster*, a new wood-decaying sequestrate fungal genus related to *Armillaria* (Physalacriaceae, Agaricales, Basidiomycota). Am J Bot 97(9):1474–1484

Hennings PC (1896) *Clavogaster* eine neue Gasteromyceten-gattung sowie mehrere neue Agaricineen aus Neu-Seeland. Hedwigia 35:303–305

Hibbett DS, Tsuneda A, Murakami S (1994) The secotioid form of *Lentinus tigrinus*: genetics and development of a fungal morphological innovation. Am J Bot 81(4): 466–478

Hornok L (2007) Sexual and vegetative compatibility/incompatibility in *Fusarium* species. Acta Phytopathol Entomol Hung 42(2):291–296

Horton BM, Glen M, Davidson NJ, Ratkowsky DA, Close DC, Wardlaw TJ, Mohammed C (2017) An assessment of ectomycorrhizal fungal communities in Tasmanian temperate high-altitude *Eucalyptus delegatensis* forest reveals a dominance of the Cortinariaceae. Mycorrhiza 27(1):67–74

Hosaka K, Bates ST, Beever RE, Castellano MA, Colgan W, Dominguez LS, Simpson NB et al (2006) Molecular phylogenetics of the gomphoid-phalloid fungi with an establishment of the new subclass Phallomycetidae and two new orders. Mycologia 98(6):949–959

Justo A, Morgenstern I, Hallen-Adams HE, Hibbett DS (2010) Convergent evolution of sequestrate forms in *Amanita* under Mediterranean climate conditions. Mycologia 102(3):675–688

Karlsen-Ayala E, Gazis R, Smith ME (2021) *Asperosporus subterraneus*, a new genus and species of sequestrate Agaricaceae found in Florida nursery production. Fungal Syst Evol 8(1):91–100

Kinoshita A, Sasaki H, Nara K (2012) Multiple origins of sequestrate basidiomes within *Entoloma* inferred from molecular phylogenetic analyses. Fungal BiolFungal Biol 116(12):1250–1262

Koch RA, Wilson AW, Séné O, Henkel TW, Aime MC (2017) Resolved phylogeny and biogeography of the root pathogen *Armillaria* and its gasteroid relative, *Guyanagaster*. BMC Evol Biol 17(1):1–16

Kraisitudomsook N, Healy RA, Smith ME (2021) Molecular systematics and taxonomic overview of the bird's nest fungi (Nidulariaceae). Fungal Biol 125(9): 693–703

Kretzer A, Bruns TD (1997) Molecular revisitation of the genus *Gastrosuillus*. Mycologia 89(4):586–589

Kuhar F, Smith ME, Mujic A, Truong C, Nouhra E (2017) A systematic overview of *Descolea* (Agaricales) in the Nothofagaceae forests of Patagonia. Fungal Biol 121(10):876–889

Kuhar F, Terzzoli L, Nohura E, Robledo G, Mercker M (2022) Pattern formation features might explain homoplasy: fertile surfaces in higher fungi as an example. Theory in Biosciences 2022:1–11

Lakatos I (1970) Falsification and the methodology of scientific research programmes. In: Lakatos M (ed) pp. 91–195

Larsson E, Jeppson M (2008) Phylogenetic relationships among species and genera of Lycoperdaceae based on ITS and LSU sequence data from north European taxa. Mycol Res 112(1):4–22

Lebel T, Castellano MA (2002) Type studies of sequestrate Russulales II. Australian and New Zealand species related to *Russula*. Mycologia 94(2):327–354

Lebel T, Catcheside PS (2009) The truffle genus *Cribbea* (Physalacriaceae, Agaricales) in Australia. Aust Syst Bot 22(1):39–55

Lebel T, Syme A (2012) Sequestrate species of *Agaricus* and *Macrolepiota* from Australia: new species and combinations and their position in a calibrated phylogeny. Mycologia 104(2):496–520

Lebel T, Orihara T, Maekawa N (2012) The sequestrate genus *Rosbeeva* T. Lebel & Orihara gen. nov. (Boletaceae) from Australasia and Japan: new species and new combinations. Fungal Divers 52:49–71

Lebel T, Castellano MA, Beever RV (2015) Cryptic diversity in the sequestrate genus *Stephanospora* (Stephanosporaceae: Agaricales) in Australasia. Fungal Biol 119(4):201–228

Lebel T, Cooper JA, Castellano MA, Nuytinck J (2021) Three independent evolutionary events of sequestrate *Lactifluus* species in Australasia. FUSE 8:9–25

Lohwag H (1926) Zur Entwicklungsgeschichte und morphologie der Gastromyceten. Beih Bot Centralbl 42(2):117–334

Loizides M, Alvarado P, Polemis E, Dimou DM, Zervakis GI, Thines M, Telle S, Konstantinou G, Gube M

(2020) Multiple evolutionary origins of sequestrate species in the agaricoid genus *Chlorophyllum*. Mycologia 112(2):400–422

Lumyong S, Sanmee R, Lumyong P, Yang ZL, Trappe JM (2003) *Mycoamaranthus cambodgensis* comb. nov., a widely distributed sequestrate basidiomycete from Australia and southeastern Asia. Mycol Prog 2(4): 323–325

Malençon G (1955) Le développement du *Torrendia pulchella* Bres. et son importance morphogénétique. Laboratoire de cryptogamie du Muséum national d'histoire naturelle

Mangold-Wirz K (1966) Cerebralisation und ontogenese modus bei eutherien. Cells Tissues Organs 63(4): 449–508

Martín MP, Raidl S, Telleria MT (2004) Molecular analyses confirm the relationship between *Stephanospora* caroticolor and Lindtneria trachyspora. Mycotaxon 90(1):133–140

McNamara KJ (1982) Heterochrony and phylogenetic trends. Paleobiology:130–142

Moreno G, Heykoop M, Esqueda M, Olariaga I (2015) Another lineage of secotioid fungi is discovered: *Psathyrella secotioides* sp. nov. from Mexico. Mycol Prog 14(6). https://doi.org/10.1007/s11557-015-1057-8

Morse EE (1933) A study of the genus Podaxis. Mycologia 25(1):1–33

Müller F (1864) Für Darwin. Wilhelm Engelmann

Nagy LG, Riley R, Tritt A, Adam C, Daum C, Floudas D, Sun H, Yadav JS, Pangilinan J, Larsson KH, Matsuura K (2016) Comparative genomics of early-diverging mushroom-forming fungi provides insights into the origins of lignocellulose decay capabilities. Mol Biology Evol 33(4):959–970

Neville P, Poumarat S, Ivaldi P (2004) Recoltes provencales de *Descolea tenuipes* (Setch.) Neville et Poumarat comb. nov. Etude sur la variabilité sporique. Bull Trimest Soc Mycol Fr 120(1–4):51–72

Nouhra ER, Dominguez LS, Becerra AG, Trappe JM (2005) Morphological, molecular and ecological aspects of the South American hypogeous fungus *Alpova austroalnicola* sp. nov. Mycologia 97(3): 598–604

Nouhra E, Urcelay C, Longo S, Tedersoo L (2013) Ectomycorrhizal fungal communities associated to *Nothofagus* species in Northern Patagonia. Mycorrhiza 23(6):487–496

Oizumi R, Kuniya T, Enatsu Y (2016) Reconsideration of r/K selection theory using stochastic control theory and nonlinear structured population models. PLoS One 11(6):e0157715

Orihara T, Smith ME (2017) Unique phylogenetic position of the African truffle-like fungus, *Octaviania ivoryana* (Boletaceae, Boletales), and the proposal of a new genus, *Afrocastellanoa*. Mycologia 109(2):323–332

Orihara T, Smith ME, Shimomura N, Iwase K, Maekawa N (2012) Diversity and systematics of the sequestrate genus *Octaviania* in Japan: two new subgenera and eleven new species. Persoonia - Molecular Phylogeny and Evolution of Fungi 28(1):85–112

Orihara T, Ohmae M, Yamamoto K (2016) First report of *Chamonixia caespitosa* (Boletaceae, Boletales) from Japan and its phylogeographic significance. Mycoscience 57(1):58–63

Orihara T, Healy R, Corrales A, Smith ME (2021) Multilocus phylogenies reveal three new truffle-like taxa and the traces of interspecific hybridization in *Octaviania* (Boletaceae, Boletales). IMA Fungus 12(1):1–22

Pacioni G, Sharp C (2000) *Mackintoshia*, a new sequestrate basidiomycete genus from Zimbabwe. Mycotaxon 75:225–228

Palfner G, Galleguillos F, Arnold N, Casanova-Katny A, Horak E (2020) Sequestrate syndrome in *Bondarzewia guaitecasensis* (Fungi, Basidiomycota)? The case of *Hybogaster giganteus* revisited. Phytotaxa 474(3): 272–282

Payer JB (1850) Botanique cryptogamique. F. Savy, Paris

Pegler DN, Young TWK (1979) The gasteroid Russulales. Trans. Brit. Mycol. Soc. 72(3):353–388

Peintner U, Bougher NL, Castellano MA, Moncalvo JM, Moser MM, Trappe JM, Vilgalys R (2001) Multiple origins of sequestrate fungi related to *Cortinarius* (Cortinariaceae). Amer J Bot 88(12):2168–2179

Pilát A (1958) Gasteromycetes Houby-Brichatky. Práce Ceskoslovenské Akademie Ved, Prague

Rauschert S (1964) *Montagnea arenaria* (DC. ex Fries) Zeller, ein für Deutschland neuer Steppenpilz. Westfälische Pilzbriefe 5(1):1–13

Rehsteiner H (1892) Beiträge zur entwicklungsgeschichte der Fruchtkörper einiger gastromyceten

Reijnders AFM (1976) Recherches sur le développement et l'histogénèse dans les Asterosporales. Persoonia 9(1):65–83

Reijnders AFM (1999) The formation of spores by metamorphosed basidia in *Mycocalia* and *Scleroderma*. Mycol Res 103(5):521–526

Reijnders AFM (2000) A morphogenetic analysis of the basic characters of the gasteromycetes and their relation to other basidiomycetes. Mycol Res 104(8): 900–910

Reznick D, Bryant MJ, Bashey F (2002) r-and K-selection revisited: the role of population regulation in life-history evolution. Ecology 83(6):1509–1520

Sáenz JA, Carranza J, Gómez VS (1983) Estudio comparativo al microscopio de luz y al microscopio electrónico de barrido de *Laternea triscapa, Laternea pusilla* y *Ligiella rodrigueziana*. Rev Biol Trop 31(2): 327–331

Saint-Hilaire IG (1837) Histoire générale et particulière des anomalies de l'organisation chez l'homme et les animaux: ouvrage comprenant des recherches sur les caractères, la classification, l'influence physiologique et pathologique, les rapports généraux, les lois et les causes des monstruosités, variétés et vices de conformation, ou traité de tératologie (Vol. 1). Établissement encyclographique

Sánchez-García M, Ryberg M, Khan FK, Varga T, Nagy LG, Hibbett DS (2020) Fruiting body form, not nutritional mode, is the major driver of diversification in mushroom-forming fungi. PNAS 117(51): 32528–32534

Sato H, Toju H (2019) Timing of evolutionary innovation: scenarios of evolutionary diversification in a species-rich fungal clade, Boletales. New Phytol 222(4): 1924–1935

Savile DBO (1955) A phylogeny of the Basidiomycetes. Can J Bot 33(1):60–104

Savile DBO (1968) Possible interrelationships between fungal groups. *The Fungi, An Advanced Treatise*, 649–675

Schröter J (1876) Über die Entwicklung und die systematische Stellung von *Tulostoma* Pers. Cohn's Beiträge zur Biologie d. Pflanzen 2:65–71

Shea BT (1989) Heterochrony in human evolution: the case for neoteny reconsidered. Am J Phys Anthropol 32(S10):69–101

Sheedy EM, Ryberg M, Lebel T, May TW, Bougher NL, Matheny PB (2016) Dating the emergence of truffle-like fungi in Australia, by using an augmented meta-analysis. Aust Syst Bot 29(5):284–302

Sheedy EM, Van de Wouw AP, Howlett BJ, May TW (2013) Multigene sequence data reveal morphologically cryptic phylogenetic species within the genus *Laccaria* in southern Australia. Mycologia 105(3): 547–563

Singer R (1958) The meaning of the affinity of the Secotiaceae with the Agaricales. Sydowia 12(1):43

Singer R, Smith AH (1960) Studies on Secotiaceous fungi IX the *Astrogastraceous* series. Mem Torrey Bot Club 21(3):1–112

Skrede I, Engh IB, Binder M, Carlsen T, Kauserud H, Bendiksby M (2011) Evolutionary history of Serpulaceae (Basidiomycota): molecular phylogeny, historical biogeography and evidence for a single transition of nutritional mode. BMC Evol Biol 11(1):1–13

Smith AH, Singer R (1959) Studies on Secotiaceous fungi-IV: *Gastroboletus, Truncocolumella* and *Chamonixia*. Brittonia:205–223

Smith ME, Schell KJ, Castellano MA, Trappe MJ, Trappe JM (2013) The enigmatic truffle *Fevansia aurantiaca* is an ectomycorrhizal member of the *Albatrellus* lineage. Mycorrhiza 23(8):663–668

Smith ME, Amses KR, Elliott TF, Obase K, Aime MC, Henkel TW (2015) New sequestrate fungi from Guyana: *Jimtrappea guyanensis* gen. sp. nov., *Castellanea pakaraimophila* gen. sp. nov., and *Costatisporus cyanescens* gen. sp. nov.(Boletaceae, Boletales). IMA fungus 6(2):297–317

Smith ME, Castellano MA, Frank JL (2018) *Hymenogaster macmurphyi* and *Splanchnomyces behrii* are sequestrate species of *Xerocomellus* from the western United States. Mycologia 110(3):605–617

Stewart EL (1974) The genus *Gautieria* Vitt. (hymenogastrales-basidiomycetes). Dissertation,. Oregon State University

Stewart EL, Trappe JM (1975) *Gautieria albida* and *Hymenogaster monosporus* sp. nov. Trans Brit Mycol Soc 65:330–332

Stielow B, Bratek Z, Orczán AKI, Rudnoy S, Hensel G, Hoffmann P, Göker M et al (2011) Species delimitation in taxonomically difficult fungi: the case of *Hymenogaster*. PLoS One 6(1):e15614

Sunhede SI (1993) Geastraceae (Basidiomycotina): morphology, ecology, and systematics with special emphasis on the North European species. Synopsis Fungorum 1

Takhtajan A (1969) Flowering plants: origin and dispersal. Smithsonian Institution Press, Washington, DC

Takhtajan A (1976) Neoteny and the origin of flowering plants. Origin and early evolution of angiosperms. CB Beck, ed.

Talbot PHB (1973) Basidial morphology and hymenophoral development in *Rhizopogon*. Persoonia 7(2):339–350

Theißen G (2009) Saltational evolution: hopeful monsters are here to stay. Theory Biosci 128(1):43–51

Thiers HD (1984) The secotioid syndrome. Mycologia 76(1):1–8

Thompson DW (1942) On growth and form. Cambridge University Press, Cambridge

Tilley SG (1973) Life histories and natural selection in populations of the salamander *Desmognathus ochrophaeus*. Ecology 54(1):3–17

Trappe JM, Castellano MA (2000) New sequestrate Ascomycota and Basidiomycota covered by the Northwest Forest Plan. Mycotaxon 75:153–179

Trappe JM, Castellano MA, Halling RE, Osmundson TW, Binder M, Fechner N, Malajczuk N (2013) Australasian sequestrate fungi 18: *Solioccasus polychromus* gen. & sp. nov., a richly colored, tropical to subtropical, hypogeous fungus. Mycologia 105(4): 888–895. https://doi.org/10.3852/12-046

Trappe MJ, Smith ME, Hobbie EA (2015) Exploring the phylogenetic affiliations and the trophic mode of *Sedecula pulvinata* (Sedeculaceae). Mycologia 107(4):688–696

Truong C, Sánchez-Ramírez S, Kuhar F, Kaplan Z, Smith ME (2017a) The Gondwanan connection–Southern temperate *Amanita* lineages and the description of the first sequestrate species from the Americas. Fungal Biol 121(8):638–651

Truong C, Mujic AB, Healy R, Kuhar F, Furci G, Torres D, Sandoval-Leiva P, Fernández N, Escobar JM, Moretto A, Palfner G, Pfister D, Nouhra E, Swenie R, Sánchez-García M, Matheny PB, Smith ME (2017b) How to know the fungi: combining field inventories and DNA-barcoding to document fungal diversity. New Phytol 214(3):913–919

Varga T, Földi C, Bense V, Nagy LG (2022) Radiation of mushroom-forming fungi correlates with novel modes of protecting sexual fruiting bodies. Fungal Biol 126(9):556–565

Vidal JM, Alvarado P, Loizides M, Konstantinidis G, Chachuła P, Mleczko P, Llistosella J (2019) A phylogenetic and taxonomic revision of sequestrate Russulaceae in Mediterranean and temperate Europe. Persoonia 42(1):127–185

Virágh M, Merényi Z, Csernetics Á, Földi C, Sahu N, Liu XB, Hibbet DS, Nagy LG (2021) Evolutionary morphogenesis of sexual fruiting bodies in Basidiomycota:

toward a new Evo-devo synthesis. Microbiol Mol Biol Rev 86(1):e00019–e00021

Vlk L, Tedersoo L, Antl T, Větrovský T, Abarenkov K, Pergl J, Albrechtová J, Vosátka M, Baldrian P, Pyšek P, Kohout P (2020) Early successional ectomycorrhizal fungi are more likely to naturalize outside their native range than other ectomycorrhizal fungi. New Phytol 227(5)

Voss SR (1995) Genetic basis of paedomorphosis in the axolotl, *Ambystoma mexicanum*: a test of the single-gene hypothesis. J Hered 86(6):441–447

Wake DB (1991) Homoplasy: the result of natural selection, or evidence of design limitations? Am Nat 138(3): 543–567

Wallace AR (1871) Contributions to the theory of natural selection. Macmillan and Company, London

Watling R, Martín MP (2003) A sequestrate *Psilocybe* from Scotland. Bot J Scotl 55(2):245–257

Wilson AW, Binder M, Hibbett DS (2008) Does gasteromycetation in the Agaricomycetes represent an evolutionary dead-end or a key innovation? A study with emphasis on the Sclerodermatineae. Annual Meeting of the Mycological Society of America, State College, PA

Wilson AW, Binder M, Hibbett DS (2011) Effects of gasteroid fruiting body morphology on diversification rates in three independent clades of fungi estimated using binary state speciation and extinction analysis. Evol Int J Org Evol 65(5):1305–1322

Wirz KM (1950) *Zur quantitativen Bestimmung der Rangordnung bei Säugetieren: Studien über Cerebralisation* Dissertation. Universität Basel

Xu YY, Jian SP, Mao N, Yang ZL, Fan L (2022a) *Gomphocantharellus*, a new genus of Gomphales. Mycologia 114(4):748–756

Xu YY, Yan XY, Li T, Zhao TY, Lv JC, Fan L (2022b) The taxonomic revision of *Melanogaster* (Paxillaceae, Boletales) in China based on molecular and morphological evidence. Mycol Prog 21:98

Zeller SM (1941) Further notes on fungi. Mycologia 33(2): 196–214

Zeller SM, Dodge CW (1936) *Elasmomyces, Arcangeliella*, and *Macowanites*. Ann Mo Bot Gard 23(4):599–638

Index

© The Author(s), under exclusive license to Springer Nature Switzerland AG 2023

S. Pöggeler, T. James (eds.), *Evolution of Fungi and Fungal-Like Organisms*, The Mycota 14,
https://doi.org/10.1007/978-3-031-29199-9

P

GPSR Compliance

The European Union's (EU) General Product Safety Regulation (GPSR) is a set of rules that requires consumer products to be safe and our obligations to ensure this.

If you have any concerns about our products, you can contact us on ProductSafety@springernature.com

In case Publisher is established outside the EU, the EU authorized representative is:

Springer Nature Customer Service Center GmbH
Europaplatz 3
69115 Heidelberg, Germany

Batch number: 10755744

Printed by Printforce, the Netherlands